Amorphous Materials

Organizing Committee for the Third International Conference on the Physics of Non-crystalline Solids, Sheffield, September, 1970.

Chairman	
Professor R. W. Douglas	University of Sheffield
Secretary	
Mr. B. Ellis	University of Sheffield
Professor G. Allen	University of Manchester
Dr I. G. Austin	University of Sheffield
Professor J. Lamb	University of Glasgow
Dr A. J. Leadbetter	University of Bristol
Professor N. March	University of Sheffield
Professor D. W. Saunders	College of Aeronautics, Cranfield

Amorphous Materials

Papers presented to the Third International Conference on the Physics of Non-crystalline Solids held at Sheffield University, September, 1970

Edited by

R. W. DOUGLAS

Professor of Glass Technology
University of Sheffield

and

BRYAN ELLIS

Senior Lecturer in High Polymers
Department of Glass Technology

WILEY—INTERSCIENCE

a division of John Wiley & Sons Ltd

LONDON · NEW YORK · SYDNEY · TORONTO

Library of Congress catalog card number 77-162326

ISBN 0 471 21988 6

Printed in Great Britain by Page Bros. (Norwich) Ltd., Norwich

Contributing Authors

F. Ackermann, Institut für Anorganische Chemie, Universität Stuttgart, Germany

G. Allen, Chemistry Department, University of Manchester, Great Britain

P. Andreatch, Bell Telephone Laboratories, Inc., Holmdel, New Jersey, U.S.A.

C. A. Angell, Department of Chemistry, Purdue University, Lafayette, Indiana, U.S.A.

A. J. Apling, School of Chemistry, University of Bristol, Great Britain

S. Arafa, Solid State and Materials Research Center, The American University in Cairo, Egypt, U.A.R.

P. F. Aubourg, Paris, France

R. C. Ayerst, Department of Chemistry, University of Manchester, Great Britain

L. W. Barr, Department of Physics, Paisley College of Technology, Renfrewshire, Great Britain

T. B. Bateman, Bell Telephone Laboratories, Inc., Holmdel, New Jersey, U.S.A.

J. Baugher, James Franck Institute, University of Chicago, U.S.A.

R. J. Bell, Division of Numerical and Applied Mathematics, National Physical Laboratory, Teddington, Great Britain

B. A. Bilby, Department of the Theory of Materials, University of Sheffield, Great Britain

A. Bishay, Solid State and Materials Research Center, The American University in Cairo, Egypt, U.A.R.

R. Bose, Vitreous State Laboratory, The Catholic University of America, Washington, D.C., U.S.A.

P. B. Bowden, Metallurgy Department, Cambridge University, Great Britain

R. C. Brown, Department of Physics, The University, Sheffield, Great Britain

H. R. Carleton, Department of Materials Science, State University of New York, U.S.A.

L. M. Castelliz, Atlantic Industrial Research Institute, Nova Scotia Technical College, Nova Scotia, Canada

A. R. Cooper, Case Western Reserve University Cleveland, Ohio, U.S.A.

D. E. Day, Ceramic Engineering Department, University of Missouri-Rolla, U.S.A.

P. Dean, Division of Quantum Metrology, National Physical Laboratory, Teddington, Great Britain

A. R. Dexter, National Institute of Agricultural Engineering, Silsoe, Bedford, Great Britain

R. W. Douglas, Department of Glass Technology, The University of Sheffield, Great Britain

M. P. van Duijkeren, Centraal Laboratorium TNO, Delft, The Netherlands

B. Eckstein, Technische Hochschule, Aachen, Germany

S. F. Edwards, Schuster Laboratory, University of Manchester, Great Britain

A. Eisenberg, Department of Chemistry, McGill University Canada

B. Ellis, Department of Glass Technology, University of Sheffield, Great Britain

J. Etchepare, Laboratory of Optics and Molecular Physics, Ecole Polytechnique, Paris, France

M. Fischer, Institut für Anorganische Chemie der Universität Stuttgart, Germany

V. D. Frechette, S.U.N.Y. College of Ceramics, Alfred University, New York, U.S.A.

G. H. Frischat, Technische Universität Clausthal, West Germany

G. Gee, Department of Chemistry, University of Manchester, Great Britain

M. Goldstein, Belfer Graduate School of Science, Yeshiva University, New York, U.S.A.

J. Götz, Atlantic Regional Laboratory, National Research Council of Canada, Nova Scotia, Canada

I. Gutzow, Institute of Physical Chemistry, Bulgarian Academy of Science, Sofia, Bulgaria

B. Hamilton, Department of Glass Technology, University of Sheffield, Great Britain

R. N. Haward, Shell Research Ltd., Carrington Plastics Laboratory, Manchester, Great Britain

J. HEIJBOER, Centraal Laboratorium TNO, Delft, The Netherlands

G. HIRSCH, Badische Anilin- und Soda-Fabrik A.G., Kunststoff-Labor, 67 Ludwig-shaften/Rh. Germany

M. L. HUGGINS, Arcadia Institute for Scientific Research, Northridge Lane, Woodside, California, U.S.A.

J. F. HUTTON, Shell Research Ltd., Thornton Research Centre, Chester, Great Britain

J. F. JANSSON, Department of Polymer Technology, Royal Institute of Technology, Stockholm, Sweden

G. M. JEFFS, I.C.I. Ltd., Petrochemicals and Polymers Laboratory, Runcorn, Cheshire, Great Britain

R. M. JOHN, Department of Chemistry, The University, Manchester, Great Britain

J. T. KRAUSE, Bell Telephone Laboratories, Inc., Murray Hill, New Jersey, U.S.A.

H. KREBS, Institut für Anorganische Chemie, Universität Stuttgart, Germany

C. R. KURKJIAN, Bell Telephone Laboratories, Inc., Murray Hill, New Jersey, U.S.A.

R. M. LATANISION, Research Institute for Advanced Studies, Martin Marietta Corporation, Baltimore, Maryland, U.S.A.

A. J. LEADBETTER, School of Chemistry, University of Bristol, Great Britain

T. A. LITOVITZ, Vitreous State Laboratory, The Catholic University of America, Washington D.C., U.S.A.

J. A. MCAINSH, I.C.I. Ltd., Petrochemicals and Polymers Laboratory, Runcorn, Cheshire, Great Britain

R. K. MACCRONE, Materials Division, Rensselaer Polytechnic Institute, Troy, New York, U.S.A.

P. B. MACEDO, Vitreous State Laboratory, The Catholic University of America, Washington, D.C., U.S.A.

H. J. MCSKIMIN, Bell Telephone Laboratories, Inc., Holmdel, New Jersey, U.S.A.

N. H. MARCH, Department of Physics, The University, Sheffield, Great Britain

C. R. MASSON, Atlantic Regional Laboratory, National Research Council of Canada, Nova Scotia, Canada

A. J. MATHESON, School of Chemistry, University of Essex, Great Britain

E. R. MICHALIK, P.P.G. Industries Inc., Pittsburgh, Pennsylvania, U.S.A.

S. MILLER, Department of Physics & Astronomy, Rensselaer Polytechnic Institute, New York, U.S.A.

J. N. MUNDY, Argonne National Laboratory, Argonne, Illinois, U.S.A.

S. M. OHLBERG, P.P.G. Industries Inc., Pittsburgh, Pennsylvania, U.S.A.

S. PARKE, Department of Glass Technology, University of Sheffield, Great Britain

G. H. PARR, JR., Research Institute for Advanced Studies, Martin Marietta Corporation, Baltimore, Maryland, U.S.A.

M. C. PHILLIPS, Shell Research Ltd., Thornton Research Centre, Chester, Great Britain

C. PRICE, Department of Chemistry, University of Manchester, Great Britain

M. PROD'HOMME, Laboratoire des Verres, C.N.R.S., Paris, France

V. PROVENZANO, Vitreous State Laboratory, The Catholic University of America, Washington, D.C., U.S.A.

S. RAHA, Metallurgy Department, Cambridge University, Great Britain

K. J. RAO, Department of Chemistry, Purdue University, Lafayette, Indiana, U.S.A.

H. RAWSON, Department of Glass Technology, University of Sheffield, Great Britain

G. REHAGE, Institute of Physical Chemistry, Technical University of Clausthal, Germany

S. REICH, The Weizmann Institute, Rehovot, Israel

G. ROSS, Department of Physics, Queen Elizabeth College, University of London, Great Britain

H. RÖTGER, Veb Jena Glaswerk Schott & Gen., Jena, D.D.R.

A. H. ROWE, Solid State Division, A.E.R.E. Harwell, Great Britain

G. L. SALINGER, Department of Physics and Astronomy, Rensselaer Polytechnic Institute, Troy, New York, U.S.A.

P. C. SCHULTZ, Research and Development Laboratories, Corning Glass Works, New York, U.S.A.

T. P. SEWARD, III, Research and Development Laboratories, Corning Glass Works, New York, U.S.A.

J. H. SIMMONS, Institute for Materials Research U.S. National Bureau of Standards, Washington, D.C., U.S.A.

H. T. SMYTH, Research and Development Laboratories, Corning Glass Works, New York, U.S.A.

J. M. STEVELS, Philips Research Laboratories, N.V. Philips Gloeilampenfabrieken, Eindhove, The Netherlands

R. A. STEWART, P.P.G. Industries Inc., Pittsburgh, Pennsylvania, U.S.A.

L. C. E. STRUIK, Centraal Laboratorium TNO, Delft, The Netherlands

M. TOMOZAWA, Materials Division, Rensselaer Polytechnic Institute, Troy, New York, U.S.A.

D. R. UHLMANN, Department of Metallurgy and Materials Science, Massachusetts Institute of Technology, U.S.A.

J. R. VARNER, S.U.N.Y. College of Ceramics, Alfred University, New York, U.S.A.

B. E. WARREN, Massachusetts Institute of Technology, Cambridge, Massachusetts, U.S.A.

H. A. WATERMAN, Centraal Laboratorium TNO, Delft, The Netherlands

A. R. C. WESTWOOD, Research Institute for Advanced Studies, Martin Marietta Corporation, Baltimore, Maryland, U.S.A.

R. C. WILLIAMS, Department of Chemistry, University of Manchester, Great Britain

T. C. WOO, Department of Mechanical Engineering, University of Pittsburgh, Pennsylvania, U.S.A.

A. C. WRIGHT, J. J. Thomson Physical Laboratory, Reading, Great Britain

Introduction

The term non-crystalline solids describes that class of materials in which, although the atoms are close together as in crystals and liquids, the arrangement of the atoms is not regular as in a crystal but possesses only the short range order of a liquid; these materials, however, do not flow as liquids but retain their shape like crystals. A familiar example of a non-crystalline solid is a typical inorganic glass, a material which has cooled below its freezing point, or liquidus temperature, without precipitating crystals and has subsequently cooled below a temperature at which the thermal motion of translation of the atoms has practically ceased.

The class of glassy materials includes low molecular weight organic compounds such as those studied about forty years ago by Tamman and G. S. Parks. The industrially important amorphous organic polymers would be included, many of which are not crystallizable because of permanent molecular stereo-irregularity. Also, there are many who would include the so-called amorphous films in the general class.

The temperature at which the material changes from a super-cooled liquid, or its equivalent, to a material in which thermal motion of translation of the atoms may be neglected is usually called the 'Glass Transition Temperature' although it is generally acknowledged that it requires operational definition in terms of particular experimental conditions. The glass transition temperature is often determined by studying some property such as the density at a steady rate of heating. When the translational motion of the atoms is fast enough the atomic configuration changes to approach the equilibrium arrangement for the liquid state, resulting in an increase in thermal expansion. The change in behaviour occurs over a small range of temperature and, for example, in a thermal expansion measurement at a constant heating rate the change is sufficiently sharp for the glass transition temperature to be defined. The glass transition, however, is found to vary with the time scale of the experimental probe; if this is much shorter than the time of relaxation of configurational changes the property will be observed to vary with time.

Models of the glass transition abound in which a thermodynamic interpretation is given based on a temperature being reached at which, for example, no more free volume is available, or a limiting configurational entropy is reached. There are discussions of the possible existence of such limiting situations but experimentally all the phenomena observed appear to be accountable for in terms of rates of approach to equilibrium. There is an almost 'philosophical' difficulty in trying to give meaning to situations which are accessible only after the lapse of very long periods of time. Strictly, such situations can only be described by the use of appropriate memory functions but we have not yet reached a universally accepted view of the situation.

The present state of the molecular theory of liquids only permits discussion of the very simplest liquids in which the interatomic potentials are spherically symmetrical and pair-wise additive. Glasses and other non-crystalline solids without exception have much more complicated interatomic potentials but while they are difficult materials to understand in this respect they are materials of tremendous technological importance. Much of the existing experimental data on these materials has been acquired in the search for the answer to some technological problem whether immediately or in the long term. This was certainly the situation when in the 1950's the industrial importance of materials and particularly the demand for new materials, resulted in encouragement of fundamental investigations. One aspect of this was the organization of the first conference on the Physics of Non-crystalline Solids organized by Professor V. D. Frechette at Alfred, New York, in 1958.

At the first conference scattering of radiation, relaxation phenomena, optical absorption, magnetic resonance, movements of ions in thin films and thermal conduction at low temperatures were considered. A second conference was arranged in Delft in 1964 under the chairmanship of Professor J. A. Prins, covering much the same field but with the deliberate inclusion of more papers on organic polymers. When the time came to organize the third conference in Sheffield in 1970 the volume of work in the field had grown considerably. There had been considerable advances in techniques of X-ray and neutron scattering and the theoretical treatment of liquids had also advanced in the analyses of vibrational processes, time correlation functions and with the use of large computers in molecular dynamics calculations. In technology the discovery of the semi-conducting glasses and the possibility of making electronic switches from them has resulted in a large volume of publications in this field. In commercial glass technology the development of the so-called glass ceramics encouraged much investigation of the fundamental glass-forming step, the step of 'not crystallizing'; the glass ceramic being a glass which by suitable heat treatment is converted to a crystalline material. It therefore became necessary to adopt some formula to limit the breadth of the subjects to be discussed at the Sheffield conference. Although nucleation and crystallization would by definition be excluded from a programme on non-crystalline solids it is fortunate that the Faraday Society organized a meeting to discuss electronic processes in glasses and nucleation and crystal growth which immediately followed the Sheffield conference. Moreover, in 1969 large conferences on the electronic properties of amorphous materials had been held in New York and Cambridge.

One of the main objectives of the organizers of all three conferences on the physics of non-crystalline materials has been to bring together people interested in a variety of non-crystalline materials from a number of different viewpoints. This objective is amply reflected in the papers that follow in this volume. The papers range from straightforward reports of experimental observations to up-to-date reviews of the state of theories of metallic liquids and those consisting of long chain molecules, from relaxation of glasses at room temperature to the viscosity-temperature relationship of super-cooled liquids. In spite of the deliberate limitation of the programme it is still a wide-ranging list of subjects which was covered. We have arranged the papers not quite in the order of presentation but so as to group together as far as possible the papers dealing with a particular part of the subject. In preparing the papers for publication it was decided that promptness of publication should take precedence over any attempt to obtain complete uniformity of style and presentation. The papers vary widely in content but they reflect faithfully the variety of the audience and also the purpose of these conferences which is to bring together a group of people of diverse backgrounds and motivation so that broader bases for their work may be encouraged.

Finally, our gratitude must be expressed to the authors all of whom responded courteously to our requests for abbreviations and changes. We also wish to record our thanks to our helpers in the preparations for the conference and particularly to Dr. Susan Frank who has worked with us from the beginning and has given great assistance in preparing the papers for publication.

R. W. DOUGLAS
B. ELLIS
Department of Glass Technology,
University of Sheffield

Contents

RELAXATION PROCESSES

THE TRANSFORMATION RANGE

VISCOSITY AND DIFFUSION PROCESSES

STRUCTURE

NUCLEAR MAGNETIC RESONANCE AND ELECTRON SPIN RESONANCE STUDIES

VIBRATIONAL PROPERTIES

FRACTURE

RELAXATION PROCESSES

1

The Rheology of Glassy Materials—A General Survey

R. W. DOUGLAS

Department of Glass Technology, The University of Sheffield, England

---

As the temperature of a supercooled liquid falls the atomic configuration changes which occur take longer and longer to approach completion. When the time required for these changes becomes longer than the experimental time, a measured property appears to change with time, and the temperature may then be said to lie within the transformation range. The transformation range occupies a small range of temperatures and if experiments are done at constant rate of heating or cooling a rather sudden change in the temperature coefficient occurs. This sudden change is frequently related to the 'glass transition temperature' which must therefore be defined operationally in terms of the particular experiment.

There have been several discussions of the thermodynamics of the transformation range in which the thermodynamic functions include parameters to describe the configuration, or more strictly, the progress of certain molecular processes. These arguments have recently been reviewed by A. J. Staverman (1968) who defined the limiting states as follows: the state in which the molecular process has proceeded for, say, less than 1% of the complete process can be considered as a 'frozen state' whereas the state in which the process has proceeded for, say, 99% of the complete process is the 'equilibrium state'. The equation

$$\frac{V}{T}\Delta K \Delta C_P - V^2(\Delta\alpha)^2 = \sum_i \sum_k \frac{S_i^2 S_k^2}{G_{ii} G_{kk}} \left(\frac{V_i}{S_i} - \frac{V_k}{S_k}\right)^2$$

is derived where:

ΔK $= K_e - K_f$, where e, f refer to the equilibrium and frozen states, K is the compressibility;

ΔC_P similarly refers to the difference in specific heats;
$\Delta\alpha$ is the difference in the coefficients of thermal expansion;
$G_{ii} = (\partial^2 G/\partial Z_i^2)$ where G_i, S_i, Z_i are the free energy, entropy and configuration parameters respectively, and the differentiation is taken with pressure, temperature and the remaining configuration parameters constant.
$S_i = \partial S/\partial Z_i$
$V_i = \partial V/\partial Z_i$

The right hand side can be interpreted as containing information about the distribution of V/S values of the frozen process, V/S being the ratio of the contribution to a volume change, to the entropy associated with that process. If there is only one configuration variable the right hand side is zero. It should be noted, however, that the equilibrium and frozen states are defined operationally; between these two states, that is in the transformation range, in order to describe properties completely, memory functions are required in an analogous manner to the creep and stress functions of linear viscoelasticity.

In the transformation range a glass, if heated long enough at atmospheric pressure and constant temperature, will eventually reach a state in which its volume appears no longer to vary with time. This is the equilibrium state or the stabilized condition. Even when a glass is in the stabilized condition, a change of the external force is followed by a strain which increases slowly with time to reach a new equilibrium. Such behaviour can, of course, be described in terms of creep functions or stress relaxation functions. Because these functions are not found experimentally to be simple exponentials, it is convenient to express them as the sum of a series of exponentials. However, it is found experimentally that, instead of the series of exponentials, over two or three decades of time the function can be written $\exp - (t/\tau)^b$ where b is a constant. We shall show that there is ample evidence to support the proposition that τ, a relaxation time, is a function of temperature and the configuration alone. Having demonstrated the convenience of this particular material constant τ, it can be used to discuss the accessibility of equilibrium states as the temperature is lowered.

EXPERIMENTS ON STABILIZED GLASSES

Kurkjian (1963) and de Bast and Gilard (1963) have described experiments on stabilized glasses. Commercial soda-lime-silica glasses were used but the differences in composition of these glasses and those used in other experiments referred to later make no appreciable contribution to the discussion. Kurkjian's establishment of the stress relaxation function will be considered first.

The experiments consisted of measuring the decay of torsional stress at constant strain, the strain being applied with an electromagnetic device which also enabled the stress at any given time to be measured. Particular care was taken over the definition of the stress at zero time, S_0, by measuring the rigidity modulus for very short times so that S_0 could be calculated from the initial strain. The results were then expressed in terms of the normalized stress $S_n = S/S_0$. A series of experiments was made at temperatures between 473° and 533 °C corresponding to equilibrium viscosities of approximately 10^{16} p and 10^{13} p.

By shifting the time scale of the various experiments according to the equation

$$\ln (t_1/t_2) = \frac{E}{R}\left(\frac{1}{T_1} - \frac{1}{T_2}\right)$$

all the observations could be brought to lie on a smooth curve, described approximately by the equation

$$S_n = \exp - \left(\frac{t}{\tau}\right)^b$$

where

$$\tau = \tau_0 \exp - \left(\frac{E}{RT}\right)$$

and b varies slowly with t/τ.

S_n could, of course, alternatively be expressed as a sum of exponential terms $S_n = \int_0^\infty F(\tau)e^{-t/\tau}\,d\tau$ where $F(\tau)$ is the contribution to the relaxation process of that portion of the system having relaxation time τ. This method of expressing the form of the relaxation curve is convenient for mathematical manipulation; it is however, no more than an expression of the fact that in order to account for the shape of the curve the strain must be expressed not only as a function of the stress alone but also as a function of the rate of change of stress and of higher time derivatives of the stress.

In order to emphasize the physical relationships in and below the transformation range it is convenient to use the approximate expression

$$S_n = \exp - \left(\frac{t}{\tau}\right)^b$$

Over two or three decades of t, b is constant within experimental error but it does appear probable that it varies continuously with t/τ. When t/τ is very large $b \to 1$ and the material behaves as a Newtonian liquid. At smaller values of t/τ, Kurkjian's data give the following values of b:

Table 1. b as function of t/τ, from Kurkjian's data

t/τ	10^2	10^1	10^0	10^{-1}	10^{-2}	10^{-3}
b	0·98	0·63	0·48	0·37	0·31	0·26

τ at 473 °C $= 10^{4\cdot 3}$ and $\tau = \tau_0 \exp - (E/RT)$, $E \approx 150$ kcal/mol

In experiments covering just over two decades of τ, de Bast and Gilard (1963) found reasonable experimental agreement using a constant $b = 0{\cdot}5435$. For example, they calculated the coefficient of viscosity from the stress relaxation function using Boltzmann's superposition principle

$$\eta = G \int_0^\infty S_n \mathrm{d}t \tag{1}$$

The temperature variation of viscosity obtained directly from observations of rate of strain was

$$\log \eta = 36{\cdot}10 \times (10^3/T) - 30{\cdot}58$$

and from the stress relaxation experiment, using equation (1)

$$\log \eta = 35{\cdot}73 \times (10^3/T) - 30{\cdot}38$$

Thus, to this accuracy, the material can be described as linearly viscoelastic and thermorheologically simple.

MEASUREMENT OF EXTREMELY HIGH VISCOSITY

By inserting $S_n = \exp - (t/\tau)^b$ into equation (1) and integrating we obtain

$$\eta = G\left(\frac{\tau}{b}\right)\Gamma\left(\frac{1}{b}\right)$$

If $b = 0{\cdot}5$ this becomes

$$\eta = 2G\tau = 2\eta_\infty \tag{2}$$

where η_∞ is the 'equilibrium viscosity' and the Maxwell equation $\eta = G\tau$ has been used.

Hopkins (1963) pointed out that if $b = \frac{1}{2}$ a determination of η from a stress relaxation experiment when $\eta_\infty = 10^{20}$ p would need to last about 300 years if the η estimated was to be within 1% of the equilibrium viscosity.

The stress relaxation function and the creep function are related through the superposition principle

$$\varepsilon_t = \int_{-\infty}^{t} \delta(t - \theta)\, \dot{\delta}(\theta)\, d\theta$$

and

$$S_t = \int_{-\infty}^{t} \sigma(t - \theta)\, \dot{\varepsilon}(\theta)\, d\theta$$

θ is a time variable, ε is the strain, S the stress, and $\delta(t)$, $\sigma(t)$ are the creep and stress functions defined respectively as the creep resulting from the application of a constant stress of magnitude unity at $t = 0$ and the stress required to maintain a constant strain of unity. As Zener (1948) demonstrates

$$\delta(t)\, \sigma(t) = 1 + \int_{0}^{t} \{\delta(t) - \delta(t - \theta)\}\dot{\sigma}\, d\theta$$

When t is small the product of the two functions is unity. Therefore, if t/τ is small, we may use

$$\delta(t) = \frac{1}{M} \exp\left(\frac{t}{\tau}\right)^{b} = \frac{1}{M}\left[1 + \left(\frac{t}{\tau}\right)^{b}\right]$$

or for the strain

$$\varepsilon = \frac{S}{M} + \frac{S}{M}\left(\frac{t}{\tau}\right)^{b}$$

where S/M is the instantaneous strain on applying the load at $t = 0$.

High viscosities are determined from the ratio of an applied stress to the resulting rate of strain. The viscosity measured at times when t/τ is small will be less than the true viscosity

$$\eta_{\text{apparent}} = \frac{S}{\dot{\varepsilon}} = \frac{M\tau}{b}\left(\frac{t}{\tau}\right)^{1-b}$$

But $\eta_{\infty} = M\tau$

$$\therefore \eta_{\text{apparent}} = \frac{\eta_{\infty}}{b}\left(\frac{t}{\tau}\right)^{1-b} = \frac{\eta_{\infty}^{b}}{b}(Mt)^{1-b} \tag{3}$$

G. O. Jones (1944) and Blaimont *et al.* (1951) have described measurements of η_{apparent} at temperatures where η_{∞} would be expected to be about 10^{20} p. The following typical values are calculated from their data.

Table 2. Comparison of estimated and experimental values of viscosity ($\eta_\infty = 10^{20}$ p)

	Time of experiment	t/τ	$\log \eta_{app}(b = 0{\cdot}3)$	$\log \eta_{expt}$
G. O. Jones	40 hr	10^{-3}	17·93	17·3
G. M. Blaimont	400 hr	10^{-2}	18·63	18·7

The experimental values were calculated for uniaxial stress, i.e.

$$3\eta = S/\dot{\varepsilon}$$

where $\dot{\varepsilon}$ is the rate of strain observed. η_{apparent} in Table 2 is therefore one third the value given by equation (3).

If the specimens were not fully stabilized there would be an additional contribution to the difference between the experimental viscosity and the estimated equilibrium viscosity, although it appears that the authors were of the opinion that their heat treatments were such that stabilization could make little contribution to their observations.

Goldstein (1968) has discussed some experiments of Zijlstra (1963) in which the 'initial' viscosity was measured, at a series of temperatures, of specimens which had been stabilized in the middle of the range of temperatures covered in the experiments. These observations were compared with similar measurements on quenched fibres. Typical results are shown in Figure 1. When the experiments were extended to lower temperatures, the temperature coefficient decreased to give an apparent activation energy of about 10 kcal compared with 100 kcal for the lines 2 and 3 in Figure 1. Goldstein remarks on the change in apparent activation energy with the amount by which the temperature of the experiment differs from the initial temperature of the specimen, i.e., before quenching or heat treating. Now the time taken by Zijlstra to measure the initial viscosity was apparently about ten minutes while the time taken to stabilize at 391 °C was apparently about 340 hours. The results thus demonstrated that for the 'quenched' fibres the relaxation time τ_0 in the expression

$$\tau = \tau_0 \exp - \left(\frac{E}{RT}\right)$$

is smaller than that for the stabilized fibres. The fall off of the activation energy at lower temperatures will be accounted for by the very high η_∞ at these temperatures and the arguments used in the previous paragraph in discussing Jones' and Blaimont's results.

We shall now consider experiments on stabilization and show, certainly within the accuracy of available experimental data, the effects of stabilization can be allowed for by treating τ as a function of configuration as well as temperature. Additionally, within the range of temperature considered, and provided the specimen has at the start of the experiment been in equilibrium configuration, τ can be taken as proportional to the difference between the actual and the equilibrium volume.

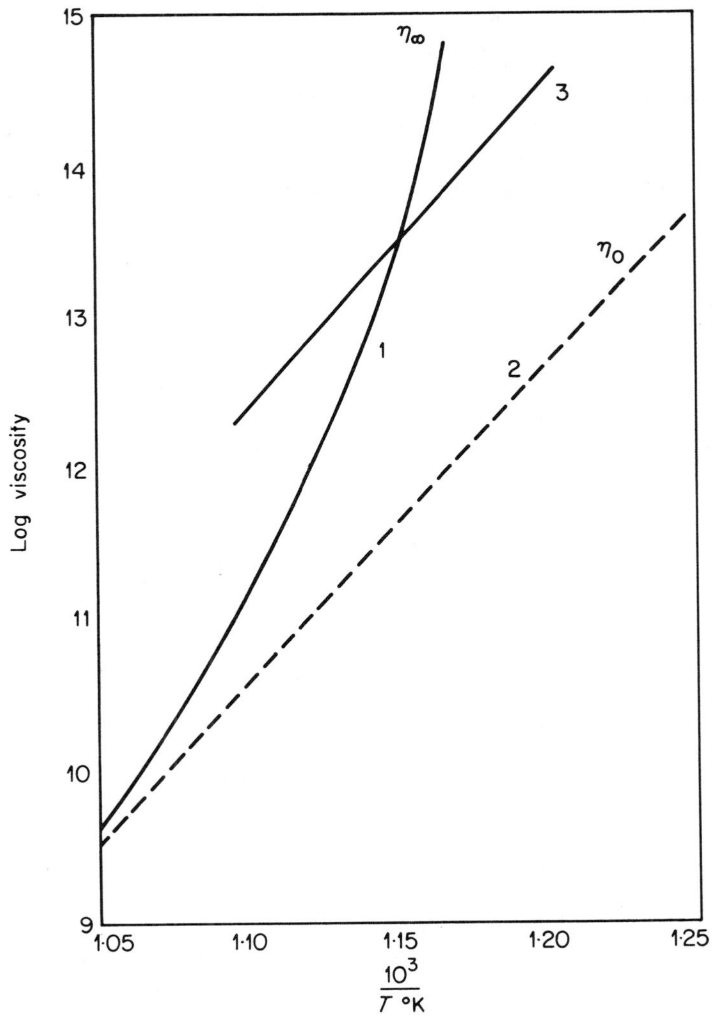

Figure 1. Viscosity–temperature dependence of borosilicate glass BSI
(1) equilibrium viscosity (η_∞),
(2) initial viscosity of quenched fibres (η_0),
(3) initial viscosity of fibres previously stabilized 500 hours at 601 °C
(after A. L. Zijlstra, *Physics Chem. Glasses*, 1963, **4** (4), 143, figure 12)

MEASUREMENTS DURING STABILIZATION

De Bast and Gilard (1963) devised an experiment in which the change in length of a specimen in uniaxial stress could be separated into the change due to applied stress and the change due to stabilization. Defining the viscosity as $S/\dot{\varepsilon}$ where $\dot{\varepsilon}$ is the instantaneous rate of strain they found that

$$\log \eta/\eta_{\infty} = a(v - v_{\infty})$$

They also made measurements of the relaxation of stress with stabilized and unstabilized specimens. In a stress relaxation experiment the variation in length which would result from the applied stress was compensated by varying the load. With non-stabilized glass an additional variation in length occurred owing to the volume change accompanying stabilization. Figure 2

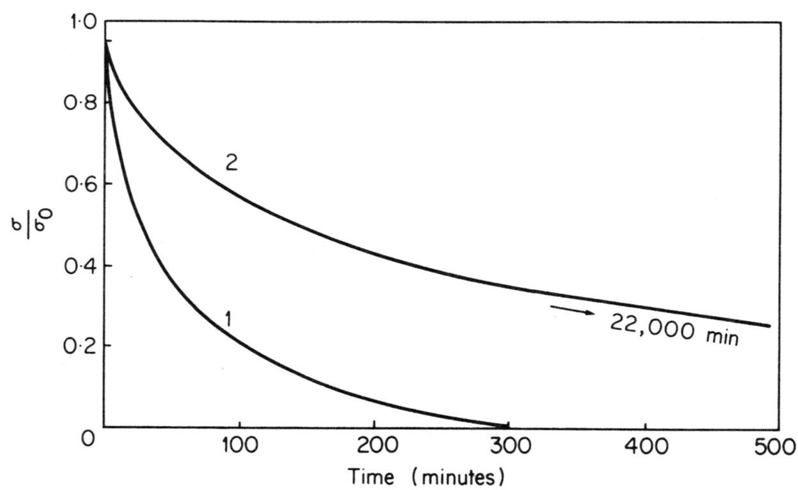

Figure 2. Comparison of relaxation curves for (2) stabilized and (1) non-stabilized glass at 503 °C (constant deformation)
(after J. de Bast and P. Gilard, *Physics Chem. Glasses*, 1963, **4** (4), 117, figure 16)

shows that when a specimen first stabilized at 531 °C is used at 503 °C the stress cancels out after 330 minutes despite the fact that stabilization would take about 20,000 minutes and that a stabilized glass takes over 22,000 minutes to relax at 510 °C. This apparent relaxation of the stress after 330 minutes results from the contraction due to stabilization; to keep the length of the initially unstabilized specimen constant after 330 minutes a tension would be needed. Using the same apparatus which enabled the viscous creep to be separated from the stabilization change in volume, it was possible to neutralize the stabilization volume change. However, during the stabilization the relaxation time τ was also changing. De Bast and Gilard treated this as a change of the effective viscosity with stabilization which they derived from a separate experiment. Thus, for the unstabilized specimen, the following equation gave satisfactory agreement with experiment

$$\frac{\sigma}{\sigma_0} = \exp - \left[\int_0^t \frac{Mt}{\eta(t)} \, dt \right]^b$$

Or, using

$$\text{og}\ \eta = \log \eta_\infty + a(v - v_\infty)$$

$$\frac{\sigma}{\sigma_0} = \exp - \left[\int_0^t \frac{Mt}{\eta_0 \exp (a\Delta v)} \, dt \right]^b$$

Equation (4) is similar in form to that derived by Kovacs (1961).

Kovacs' argument, modified to follow from the previous discussion, is as follows. When stabilization is studied by measuring the volume, the change in volume may be considered to be due to a pseudo-stress, P_p, proportional to $\Delta v = v_t - v_\infty$, i.e. $P_p = K\Delta v$ Thus $\Delta \dot{v}/v = K\Delta v/\eta_t$ where η_t is value of a volume viscosity at time t.

If
$$\eta_t = \eta_\infty \exp a\, \Delta v$$

$$\frac{\Delta \dot{v}}{v} = \frac{K\Delta v}{\eta_\infty} \exp - a\, \Delta v$$

Expressions of this kind fitted closely the observed stabilization curves. Kovacs' expression can be formulated from the superposition principle as follows:

$$\varepsilon_t = \int_\infty^t \delta(t - \theta)\, \dot{S}(\theta) \, d\theta$$

If ε_t is the 'configurational strain' $(v - v_\infty)/v_\infty$, the pseudo-stress $K(v - v_\infty)/v_\infty$, the creep function is simply $1/\eta_\theta$ or $1/\eta_v$

$$\therefore \frac{v - v_\infty}{v_\infty} = \int_{-\infty}^t \frac{1}{\eta_\infty} \frac{K(v - v_\infty)\, d\theta}{v_\infty \exp [a(v - v_\infty)]}$$

THE NEED FOR A MEMORY FUNCTION

Macedo and his colleagues (1970) have recently carried out some very interesting transformation range experiments. In a typical experiment on the relaxation of refractive index of B_2O_3 glass, two specimens were brought to equilibrium at 310·5 °C. They were then heat treated at various temperatures and the refractive index measured. The resulting changes in refractive index are shown in Figure 3.

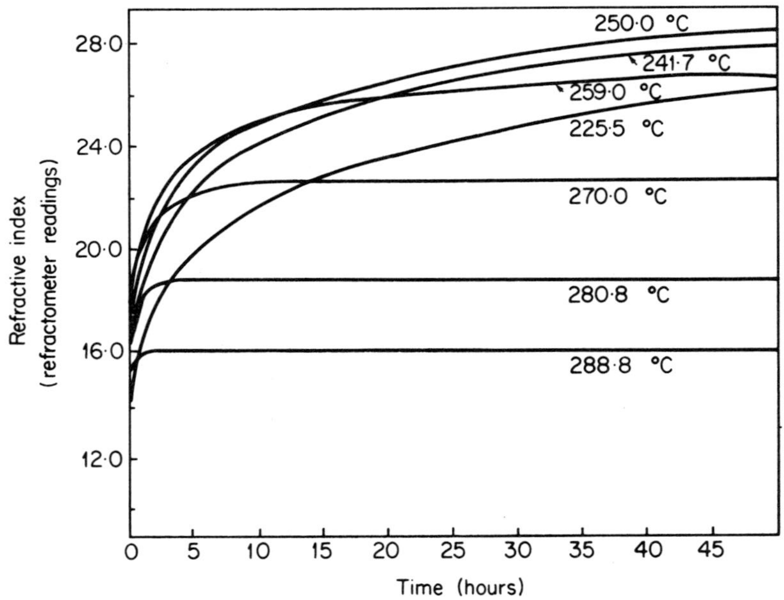

Figure 3. Approach curves for B_2O_3 glass starting from an equilibrium state at 310·5 °C (N_∞ = 9.50 scale reading) (after L. Boesch, A. Napolitano and P. B. Macedo, *J. Am. Ceram. Soc.*, 1970, **53** (3), 148, figure 2A)

Following these experiments, a specimen being heat treated at, say, 225 °C was removed from the furnace at the time when its refractive index reached the equilibrium value for, say, 259·7°C. The specimen was then reinserted in an oven at 259·7 °C. The subsequent change in the refractive index is shown in Figure 4.

Macedo accounts for this behaviour by postulating a fast and a slow relaxation process but this does not appear to be necessary. In the temperature range of the experiments the refractive index is nearly proportional to the density and we shall for convenience discuss the experiments in terms of volume. Let the experiment be defined as being conducted on a specimen initially in equilibrium at T_0, it is then heated at T_1 for time t_1, until its volume reaches the equilibrium value for T_2; $T_0 > T_2 > T_1$. The specimen is then transferred to T_2 and heating continued for time t_2. If the volume and configuration are uniquely related no further change of volume should occur at T_2, but the volume does change.

The course of events may be described as follows. During the heating at T_1 the pseudo-stress is initially

$$Kv\Delta\alpha \ (T_1 - T_0)/v$$

and subsequently

$$K(v - v_{\infty\, T_1})/v_{\infty}.$$

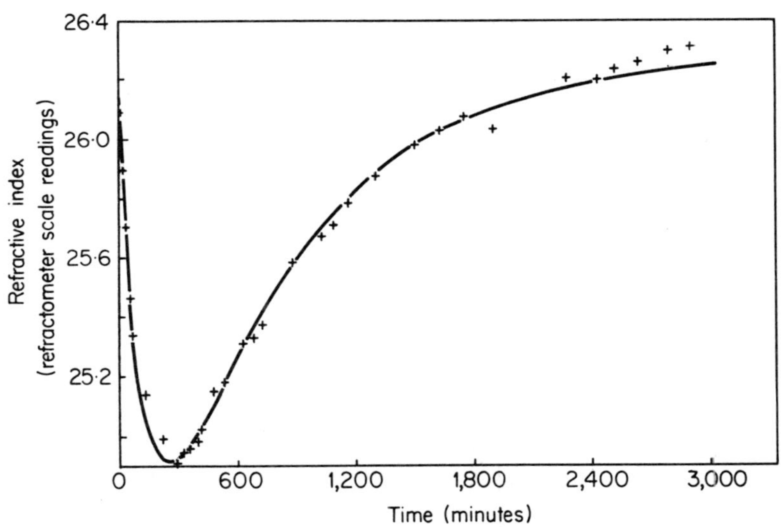

Figure 4 Crossover from 225·5 ° approach curve to 259·7 °C (after L. Boesch, A. Napolitano and P. B. Macedo, *J. Am. Ceram. Soc.*, 1970, **53** (3), 148, figure 6A)

At the crossover point there is a discontinuous change of $\eta_{\infty\, T_1}$ to $\eta_{\infty\, T_2}$ and the pseudo-stress is changed to $K(v - v_{\infty T_2})/v_{\infty}$ which is at first zero but then increases because, by the superposition principle, we cannot assume this stress will remain zero. Instead we have to consider that the original course of the change in volume continues but with the rate changed because of the change in η_{∞}. However, because the volume was $v_{\infty\, T_2}$ when the temperature was changed to T_2 we must take account of the effect of this change of v_{∞} on the pseudo-stress. First imagine that the original pseudo-stress is removed by applying an equal and opposite pseudo-stress related to T_1 at the time of the change in temperature. This immediately causes the volume to increase above $v_{\infty T_2}$, giving rise to a new pseudo-stress causing the volume again to approach $v_{\infty T_2}$. In Figure 5 the suggested explanation is illustrated in terms of refractive index where the changes are opposite in sign to the volume changes.

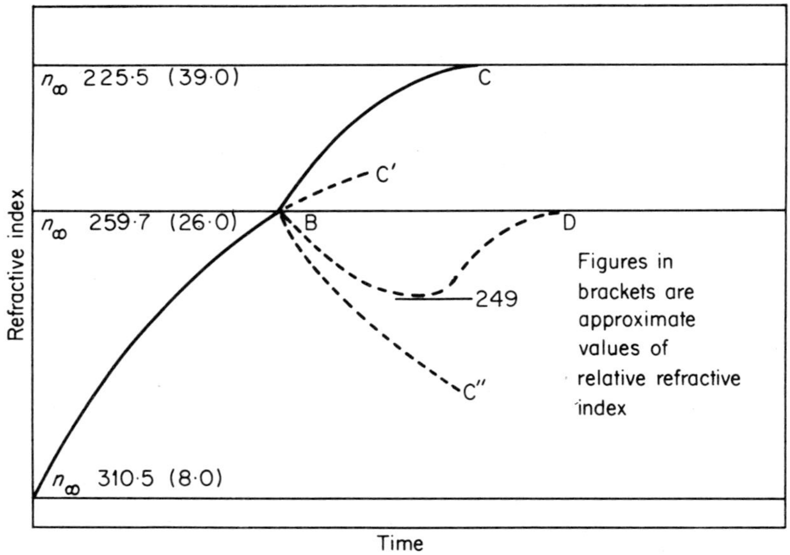

Figure 5. Suggested account of Macedo 'crossover' experiment
BC′—original path
BC —continuation of original path with $\eta_{\infty T_1} \rightarrow \eta_{\infty T_2}$
BC″—subtraction of BC due to 'removal of pseudo-stress'

EXPERIMENTS BELOW THE TRANSFORMATION RANGE

Using the approximate creep function $\delta(t) = (t/\tau)^b$ we may calculate for these soda-lime-silica glasses that at a temperature of about 350 °C, $\tau_{eq} \approx 10^{11}$ therefore experiments on a time scale of about one second can be said to take place in the frozen state. We shall now show again that there are strong indications that the same stress relaxation function can be used and that τ remains the controlling material constant. Experiments in which the decay of free torsional oscillations of a fibre are measured have time scales of the order of 1 second.

Zener (1948) shows that the logarithmic decrement is given, to a close approximation, by

$$\log \text{dec} = \left[-\frac{1}{2}\ \frac{\mathrm{d}\ln\sigma(t)}{\mathrm{d}\ln(t)}\right]_{t=\pi/4\omega} = \frac{b}{2^{(2b+1)}}\left(\frac{1}{\omega\tau}\right)^b$$

where ω is the angular frequency.

In a typical experiment of this kind on soda-lime glass the logarithmic decrement decreases rapidly as the temperature falls below 500 °C but the rate of decrease falls with the temperature until it becomes nearly constant at and below room temperature. Superimposed on this background are two peaks, one at about − 50 °C which is associated with the stress-induced diffusion of sodium ions, and one at about 250 °C which may be associated with oxygens ions. It is the background, however, which would be related to the stress relaxation function found at higher temperatures.

Mohyuddin and Douglas (1960) noticed that this background varied with frequency approximately as $1/\omega^{\frac{1}{3}}$, as predicted, i.e. $b = \frac{1}{3}$, a reasonable value. Douglas, Duke and Mazurin (1968) have checked this experiment carefully and obtained the same result. Thus again the behaviour indicates that the same stress function can be used provided τ is given its appropriate value depending upon the temperature and configuration.

The assumption will therefore be made that the stress relaxation function found for stabilized glass applies to glass that is not stabilized and an approximate form of the creep function for very small t/τ will be used to discuss the measurement of delayed elasticity by S. Pearson (1952) at temperatures near to room temperature.

CREEP NEAR ROOM TEMPERATURE, i.e. 500 °C BELOW THE TRANSFORMATION RANGE

Pearson (1952) applied a torque to a length of glass rod in the range 20–80 °C and measured the strain as a function of time. Like G. O. Jones (1944) he attempted to study the recoverable strain by removing the load after 200 hours and observing the decay rate. From the apparently linear portion of his curves he deduced viscosities of the order of 10^{20} poises at room temperature, with an activation energy of 4·5 kcal/mol. He drew attention to the difficulty of determining the recoverable strain with any great accuracy and as Hopkins (1963) has shown, the method adopted by Jones and Pearson to determine the recoverable strain is valid only if the creep can be described by a single exponential in time. Nevertheless, the ratio of the recoverable strain c, to the instantaneous strain a, found by Pearson, was very much smaller than the ratio found by G. O. Jones at temperatures between 400 and 500 °C. Both of them drew attention to the fact that the ratio c/a varied with temperature in the same way as the apparent viscosity.

There is a possibility that the processes which give rise to the second anelastic peak could contribute to the creep observed by Pearson, but this seems unlikely. Mohyuddin and Douglas (1960) give the following data:

Activation energy of 'oxygen' peak = 59·7 kcal/mol
Temperature of peak, $1/T$ °K = 19×10^{-4} for 0·93 sec/cycle

Thus for 100 hours per cycle the peak would shift to

$$\frac{1}{T} = 430, \text{ i.e. } 157\ ^\circ\text{C}$$

It seems unlikely therefore, that the relaxation associated with the oxygen peak would have affected Pearson's results, although the possibility cannot be ruled out.

Recalling the approximate creep function for small t/τ

$$\varepsilon_t = \text{Instantaneous strain} + \frac{1}{M}\left(\frac{t}{\tau}\right)^b$$

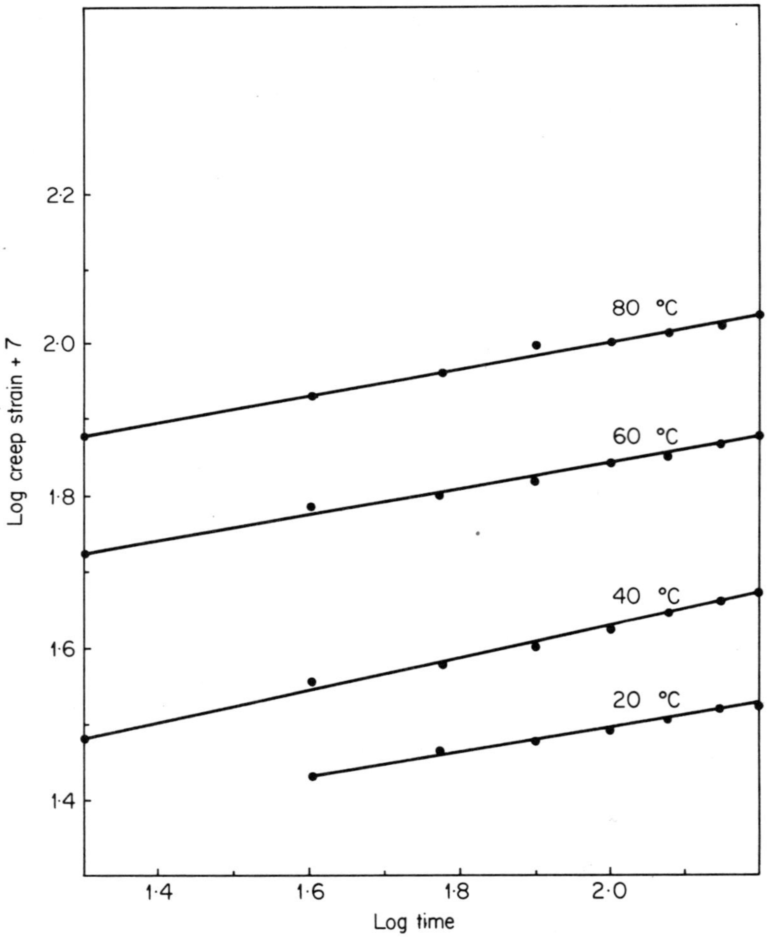

Figure 6. Log strain versus log t, using Pearson's results

The creep which Pearson observed should be proportional to t^b. The results are plotted as log strain versus log t in Figure 6. The mean slope of these lines turns out to be 0·2 and inserting Pearson's value of the stress and an approximate value of the rigidity modulus, values of τ were found as shown in Table 3.

Table 3. b and τ calculated from Pearson's data

T °C	20	40	60	80
b	0·19	0·21	0·18	0·18
τ	$10^{14\cdot7}$	$10^{13\cdot9}$	$10^{12\cdot7}$	$10^{11\cdot8}$

Assuming $M = 5 \times 10$ dyn/sq cm; $S = 4250$ lb/sq in

Thus Pearson's observations can be described adequately by the stress function and, indeed, the value b of about 0·2 at $t/\tau \approx 10^{-8}$ is perhaps a reasonable extrapolation of the data of Table 1.

ESTIMATION OF THE RATIO (DELAYED ELASTIC STRAIN/INSTANTANEOUS STRAIN)

Unfortunately it is not possible to make a direct comparison between the calculated values of this ratio and the values given by Pearson and Jones, both of whom considered that there was true viscous flow and therefore non-recoverable strain present. However, the calculation of c/a is of interest in itself.

If a load S is applied for time t and removed at t the strain at time $2t$ will be equal to the difference between S applied for time $2t$ and S applied for t

$$\varepsilon_{2t} = \frac{S}{M}\left[\left(\frac{2t}{\tau}\right)^b - \left(\frac{t}{\tau}\right)^b\right]$$

Then c, the recoverable strain is

$$c = \varepsilon_t - \varepsilon_{2t}$$

and the instantaneous strain $a = S/M$

$$\therefore \frac{c}{a} = \left(\frac{t}{\tau}\right)^b [2 - 2^b]$$

Comparison may be made with the experimental values of c/a as follows:

Table 4. Values of ratio c/a (delayed strain/elastic strain)

Assumed values t	τ	b	Temperature	c/a calc.	c/a expt.
200	10^4	0·18	40 °C	$6{\cdot}3 \times 10^{-2}$	$3{\cdot}7 \times 10^{-3}$
40	10^7	0·33	400 °C	0·18	0·20

Perhaps more important than these figures, in view of the experimental difficulties, is the prediction that temperature coefficients of c/a and $1/\eta$ must be the same as found by Pearson and Jones; for the apparent viscosity η_a is given by

$$\eta_a = \frac{S}{\dot{\varepsilon}} = \frac{M}{b}\tau^b t^{1-b}$$

and

$$\frac{c}{a} \propto \left(\frac{t}{\tau}\right)^b$$

Thus

$$\frac{c}{a} \text{ and } \frac{1}{\eta} \propto \tau^{-b}$$

RECAPITULATION

In this review I have attempted to show that the available experimental evidence on the rheological behaviour of glasses can be described systematically in terms of a stress function in which the time always appears divided by a material constant τ. $\sigma(t) = \exp - (t/\tau)^b$ and $1 > b > 0{\cdot}2$. When t/τ is large, $b \to 1$ and the behaviour is Maxwellian. When t/τ is very small, $b = 0{\cdot}2$. In stabilized glass $\tau = \tau_0 \exp - (E/RT)$, and for soda-lime glass $E \approx 150$ kcal/mol; this is the same activation energy as for viscosity in the range around 10^{11} poises; in the frozen, or glassy, state E is smaller, generally it is round about $E/2$.

Stabilized glasses are linear viscoelastic, thermorheologically simple materials. To a first approximation, in unstabilized glasses which have been brought to equilibrium in the transformation range and then have suffered a sudden change in temperature, τ can be assumed to vary linearly with Δv, the departure of the actual volume from the equilibrium volume. It is suggested that the behaviour during stabilization may be described by using the superposition principle and taking account of the variation of τ with configuration and temperature. Although the stress function could be expressed more formally as a sum of exponential terms the present form has the advantage of enabling the behaviour over two or three decades of time to be expressed in terms of two constants alone. It would of course be of great interest to know how b varies with t/τ for a number of glasses. The experimental results which have been discussed are relatively few and it cannot be claimed that the discussion does more than lay out a method which, if confirmed by subsequent experiments, brings all the rheological and transformation range behaviour into a coherent scheme.

VARIATION OF τ WITH TEMPERATURE

In stabilized glasses τ by definition has the same temperature coefficient as the coefficient of viscosity. When measurements of apparent viscosity,

$$\eta_{\text{apparent}} = \frac{\text{stress}}{\text{strain rate}},$$

are made at temperatures and times where t/τ is small,

$$\eta_{\text{apparent}} \propto \frac{1}{\tau^b}$$

and the apparent activation energy is bE. Thus some of the observed decrease in activation energy below 150 kcal/mol as the temperature falls below the transformation range is accounted for, but at room temperature, allowing for b, a value of 20 kcal/mol was obtained from Pearson's data. At these temperatures the glass is not stabilized and some of the difference between 150 and 20 kcal/mol can be attributed to the freezing out of the configuration contribution. De Bast and Gilard found experimentally that the configuration contribution is roughly half of the 150 kcal/mol. The remaining difference between 20 and 75 kcal/mol must be attributed to the fact that in interpreting $d \log \tau/d(1/T)$ as an activation energy the contribution of the change of volume is ignored. Indeed, the coefficient of expansion of the glass does decrease with temperature so that the contribution of volume change to the temperature coefficient would be reduced.

The data on the apparent activation energies is assembled in Table 5.

Finally, it should be mentioned that there is not yet available any satisfactory account of the variation of the viscosity of glass-forming melts with temperature. This topic has recently been reviewed by Macedo and Napolitano (1968) who concluded that none of the present theories of viscosity is adequate to fit their comprehensive data on B_2O_3 glass.

Table 5. Apparent activation energies

Calculated	Experimental	A.E. kcal/mol	Temp °C	Approx time (t/τ)
From creep function $(t/\tau)^b$	Pearson (torsional strain)	20	40	10^{-6}
From stress function $\exp-(t/\tau)^b$	Mohyuddin and Douglas (anelasticity)	33	400	10^{-4}
	Kurkjian (stress relaxation stabilized)	150	500	10^{+2}
Viscosity	de Best and Gilard			
	(thermal)	80		
	(configurational)	89	500	10^{+2}

VISCOSITY–TEMPERATURE RELATIONSHIPS

The important contribution of change in volume to the change in viscosity with temperature has long been recognized. Free volume theories have received much discussion but it is only in the last decade that accurate measurements of viscosity and specific volume have become available over a wide temperature range. However, it is interesting to note that over most of the temperature range (data of Macedo and Napolitano) there is simple empirical relationship between viscosity η and the specific volume V of B_2O_3 from 260–400 °C,

$$\log \eta = 13{\cdot}06–190\,(V - 0{\cdot}555) \tag{5}$$

and from 600–1400 °C,

$$\log \eta = 3{\cdot}08–45{\cdot}6\,(V - 0{\cdot}63) \tag{6}$$

It is reasonable to suppose that there is some change in configuration taking place in the region 400–600 °C and indeed Zarzycki (1956) has shown that in B_2O_3 there is X-ray diffraction evidence of a change in coordination between 20 ° and 1200 °C.

The behaviour of B_2O_3 is quite different from that of GeO_2 and SiO_2; for these oxides the viscosity varies exponentially with temperature. However, the volume changes in B_2O_3 are quite different from those of SiO_2 and GeO_2. The volume expansivities of SiO_2 and GeO_2, although not known accurately in the temperature ranges under discussion, are certainly less than 5×10^{-5} per °C. (Shartsis, Spinner and Capps (1953) and Riebling (1963)) while the expansivity of B_2O_3 is about 5×10^{-4}. These low expansivities may contribute to the linearity of $\eta = Ae^{B/T}$ plots for SiO_2 and GeO_2

but the high activation energies remain to be accounted for. In this connection it may be remarked that the relation between volume and viscosity must be expected to vary considerably from one substance to another, as, indeed, the B_2O_3 results suggest. However, the large temperature coefficients of η for SiO_2 and GeO_2 suggest that measuring the viscosity at constant volume would by no means result in removing most of the temperature variation observed at constant pressure. That this is so has long been known for certain organic liquids, alcohols and paraffins (Bridgman (1949), Jobling and Lawrence (1952)). Sperry and Mackenzie (1968) have recently shown that the viscosity of B_2O_3 has a strong temperature dependence at constant volume, almost as large as that at constant pressure.

In view of the remarkable relation between the volume and viscosity of B_2O_3 at constant pressure it might have been thought that at constant volume the temperature dependence should disappear. That it does not is but one example of the fact that in these materials the change of volume with temperature at constant pressure is accomplished in a different manner, on the atomic scale, from the same change in volume with pressure at constant temperature.

ACKNOWLEDGEMENT

I have enjoyed many interesting discussions of these topics with my colleague Mr. B. Ellis whom I also thank for drawing my attention to some papers which I had missed.

REFERENCES

De Bast, J., and P. Gilard (1963). *Physics Chem. Glasses*, **4** (4), 117.
Blaimont, G. M., J. R. G. de Bast, P. G. Migeotte, and H. P. C. Vandecapelle (1951). *J.S.G.T.*, **35**, 490.
Boesch, L., A. Napolitano, and P. B. Macedo (1970). *J. Am. Ceram. Soc.*, **53** (3), 148.
Bridgman, P. W. (1949). *Proc. Am. Acad. Arts Sci.*, **77**, 115, 129.
Douglas, R. W., P. J. Duke, and O. V. Mazurin (1968). *Physics Chem. Glasses*, **9** (6), 169.
Goldstein, M. (1968). *Trans. Soc. Rheol.*, **12** (1), 69.
Hopkins, I. L. (1963). *Physics Chem. Glasses*, **4** (4), (a) 137, (b) 139.
Jobling, A., and A. S. C. Lawrence (1952). *J. chem. Phys.*, **20** (8), 1296.
Jones, G. O. (1944). *J.S.G.T.*, **28**, 432.
Kovacs, A. J. (1961). *Trans. Soc. Rheol.*, **5**, 285.
Kurkjian, C. R. (1963). *Physics Chem. Glasses*, **4** (4), 128.
Macedo, P. B., and A. Napolitano (1968). *J. chem. Phys.*, **49** (4), 1887.
Mohyuddin, I. and R. W. Douglas (1960). *Physics Chem. Glasses*, **1** (3), 71.
Pearson, S. (1952). *J.S.G.T.*, **36**, 105.
Riebling, E. F. (1963). *J. chem. Phys.*, **39** (7), 19.
Shartsis, L., W. Capps, and S. Spinner (1953). *J. Am. Ceram. Soc.*, **36** (2), 35.
Sperry, L. L., and J. D. Mackenzie (1968). *Physics Chem. Glasses*, **9** (3), 91.
Staverman, A. J. (1966). *Rheol. Acta*, **5** (4), 283. (1968). *Rubb. Chem. Technol.*, **41** (3), 544.

Zarzycki, J. (1956). *Proc. IV Int. Congress on Glass*, Paris, pp. 323–30.
Zener, C. (1948). *Elasticity and Anelasticity of Metals.* University of Chicago Press.
Zijlstra, A. L. (1963). *Physics Chem. Glasses*, **4** (4), 143.

2

Secondary Relaxations and The Nature of the Glassy State

M. GOLDSTEIN

Belfer Graduate School of Science, Yeshiva University, New York, U.S.A.

I have recently (Goldstein, 1969) proposed a new approach to the problem of liquid viscosity and the glassy state, which involves a recognition of the near impossibility, for the foreseeable future, of solving the many-body problem necessary for an *ab initio* calculation of liquid viscosity. It is suggested therefore that the kind of naïve, intuitive theories we will of necessity have to use for some time will make precise quantitative prediction of transport properties of complex liquids impossible, and our theories had best be tested by the breadth of their range of semi-quantitative and qualitative predictive value.

I have given reasons for believing that flow in 'viscous' liquids (shear relaxation time $> 10^{-9}$ sec) is primarily a thermally activated process limited by high potential energy barriers, a view consistent with the Adam–Gibbs approach (Adam and Gibbs, 1965), but not with free volume theory (Ferry, 1961; Turnbull and Cohen, 1961). An analysis of the detailed mechanism by which flow in an amorphous substance by such a mechanism can occur leads to certain restrictions on the type of multidimensional potential energy surface (U plotted as a function of the atomic or molecular coordinates) possible; further restrictions arise from a wealth of experimental knowledge about volume relaxation and its relation to shear relaxation (Hertzfeld and Litovitz, 1959; Litovitz and Davis, 1965). It was found possible to construct potential energy surfaces that could represent a first approximation to the common features of glass-forming liquids, and yet be sufficiently tractable to allow calculation both of thermodynamic and rheological properties. Since the known thermodynamic and rheological properties of glass-forming liquids were used as input in constructing the surface, it is not surprising that many of the predicted properties are in qualitative accord with experimental behaviour. However one prediction could be made that was not input data, and was of a sufficiently surprising character that it provided an important test of the value of the approach.

This is the prediction that some molecular mobility would remain in the glassy state even though the glass transition has occurred, this mobility being of the type that shows up as secondary relaxations in dielectric, mechanical, or nuclear magnetic resonance experiments. Note the purely qualitative character of this prediction: no statement is made about the location of secondary relaxations in the temperature–frequency plane, or about their strength, only that they should occur. The origin of these relaxations may be described in terms of a not totally unfamiliar idea that in the potential energy surface there are barriers of different heights: what is new is the idea that viscous flow requires that there be sufficient thermal energy to overcome *all* of them at an appreciable rate, while the secondary relaxations arise from transitions over the lower ones. The important distinction between this statement and the usual generally accepted molecular interpretation of glassy relaxations (McCrum, Read and Williams, 1967; McCall, 1969) is that no *internal* molecular mode of motion is needed. the potential barriers resisting the molecular rearrangements can arise from intermolecular forces alone.

The obvious experimental test is of course to make glasses from rigid molecules, those that possess no internal hindering potentials, and see if glassy relaxations appear.

Dr. G. P. Johari in our laboratories has studied dielectric relaxation in some twenty glass-forming systems made up of simple molecules. Included were two fused salt glasses, three hydrocarbon liquids, four solutions of rigid polar molecules (e.g. chlorbenzene) in the rigid non-polar solvent decalin, four approximately eutectic mixtures of rigid polar molecules (e.g. 1-chloronaphthalene) with the rigid polar molecule pyridine, and seven alcohols. The alcohols are not rigid molecules; the reason for including them will become apparent shortly. Conventional dielectric equipment for the frequency range 50 Hz to 200 kHz was used. Measurements were made at temperatures near and below the glass transition temperatures T_g. Most liquids studied were subjected to purification procedures: no evidence of crystallization or phase separation was observed. Our results have been published elsewhere (Johari and Goldstein, 1970A); we describe them briefly here.

The results can be summarized in the statement that almost all (15 of 19 systems) showed evidence of a secondary (β) relaxation below the main (α) relaxation. What was surprising and unexpected was the marked degree of regularity of the behaviour of these systems of varied chemical make-up. The pattern is as follows:

(1) A secondary transition (peak in tan δ) appears some 10–30 °C below

T_g at a measuring frequency of 1 kHz (in some cases it appeared only as a clearly defined shoulder on the α relaxation).

(2) The 'activation energy' of the α transition is much greater than that of the β: $\sim$ 50 kcal *vs.* 5–10 kcal. This implies that either the two relaxations will cross in a temperature–frequency diagram, or else merge above T_g into a single relaxation region. We consider this latter event more probable.

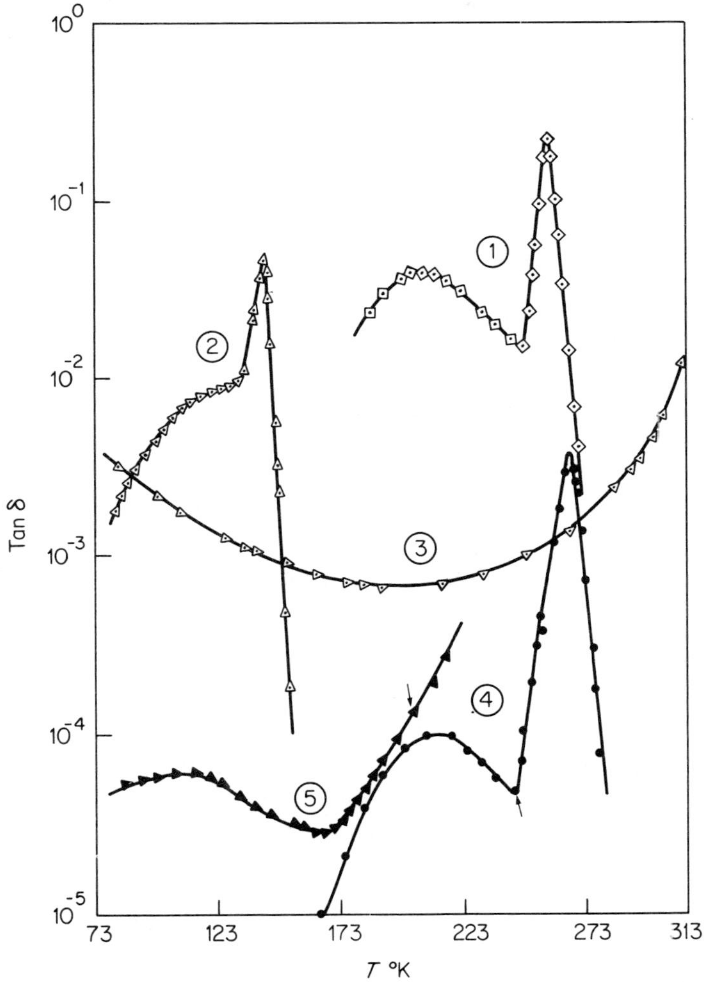

Figure 1. Dielectric loss factor of (1) 63·2 mole % pyridine in 1-chloronaphthalene (scale has been shifted to the right by 100°), (2) 16·5 mole % chlorobenzene in decalin, (3) 45 mole % Ca $(NO_3)_2$–KNO_3 mixture, (4) *o*-terphenyl and (5) $Ca(NO_3)_2.4H_2O$ (scale has been shifted downwards by two logarithmic cycles) at 1 kHz as a function of temperature. Arrows indicate the literature values for T_g when known.
(after G. P. Johari and M. Goldstein, *J. phys. Chem.*, (1970), **74**, 2034)

(3) Neither relaxation is described by a single relaxation time.

(4) The relaxation spectra are markedly temperature dependent, broadening as temperature is lowered.

(5) Tan δ at the maximum for the β process is about one order of magnitude smaller than for the α process.

Figure 1 shows tan δ at 1 kHz plotted on a logarithmic scale *vs.* temperature for five of the systems studied. The Ca $(NO_3)_2$–KNO_3 glass is atypical in that a secondary relaxation does not appear near T_g (60 °C) but instead tan δ shows a rise below 200 °K, indicative of a loss peak below liquid nitrogen temperatures. This fused salt becomes highly conductive near T_g, and it is possible that a β relaxation is buried in the main α loss region. An alcohol, 1, 2 propane-diol, and the ester propylene carbonate failed to show a β relaxation even as a shoulder on the main glassy relaxation: these two liquids also showed very large loss at the α maximum, which might swamp a β-relaxation if present.

If the experiments had been performed only on alcohols, the temptation would have been to explain the β-relaxation in terms of hindered rotation within the alcohol molecules. The close correspondence of the alcohol behaviour to that of the rigid molecules makes such an explanation implausible. Indeed the glassy polymers as a class behave in a strikingly similar way. One need only compare our measurements of ε' and ε'' (the real and imaginary parts of the dielectric permittivity), in the regions of the α and the β relaxation (Johari and Goldstein, 1970B) with corresponding measurements on relaxations in glassy polymers (McCrum, Read and Williams, 1967; McCall, 1969) to recognize the similarity.

We conclude that the β-relaxation, rather than being attributable to internal hindered rotational degrees of freedom within the molecule, is a characteristic relaxation of the glassy state.

I would like to conclude by making clear what is *not* claimed in this work.

First of all, it is not asserted that every single glass will necessarily show a relaxation of this type. There could always be unusual features in the molecular packing in specific glass-forming substances that might depress the β process, or cause it to disappear entirely. The belief that most glasses will show such a relaxation is supported by the observations that three quarters of the simple liquid systems we studied showed such a relaxation, perhaps the same fraction of glassy polymers show it, and a high proportion of silicate glasses do also (the so-called 'network' relaxation. (Hopkins and Kurkjian, 1965)).

Secondly, this is not offered as an explanation of *every* glassy relaxation. There are some relaxations in which wide-ranging evidence from a variety of experimental methods has convincingly identified a relaxation with the internal motion of a particular molecular group: the cyclohexyl relaxation in polymers is an outstanding example of this (McCrum, Read and Williams, 1967).

Thirdly even for the β-relaxation in polymers, I do not claim that explanations offered hitherto in terms of either side-group or 'crank-shaft' motions are totally in error. Certainly molecular mobility must involve either molecules or parts of them: there could always be details of the amorphous state packing that make the β relaxation in some one polymer primarily a side-chain motion, while in another it is mainly segments of the main chain that are moving. All that is claimed is that intermolecular forces, which of course create the glassy state, are clearly sufficient to provide the potential barriers for β-relaxations as well, and the possibility must be raised that a particular type of molecular motion that can be identified as taking place over an internal barrier in the isolated molecule is opposed in the glassy state by a barrier that is primarily intermolecular in origin.

Fourthly, while the model I have proposed is the source of the prediction of relaxations intrinsic to the glassy state, in fairness to the free volume theory I must concede that the experimental facts are also capable of an explanation in terms of a non-random distribution of local free volume, permitting islands of mobility to exist in an otherwise rigid matrix.

The relative merits of the various points of view on the glassy state can only be decided, as indicated earlier, by seeing which has the broader range of successful predictive value. We are currently working on the thermodynamic consequences of the model, in an attempt to establish a general origin for the cooperative character of the glass transition, which seems required by the entropy paradox of Kauzmann (1948), and which was explained by Gibbs and diMarzio as arising from packing problems with gradually stiffening chains (1960).

ACKNOWLEDGEMENT

We are grateful to the Owens–Illinois Foundation and to the Department of Defence (Grant No. N00014–69–A–0411) for the support of this research.

REFERENCES

Adam, G., and J. H. Gibbs (1965). *J. chem. Phys.*, **43**, 139.

Ferry, J. D. (1961). *Viscoelastic Properties of Polymers.* John Wiley, New York.

Gibbs, J. H. (1960). 'Nature of the glass transition and the vitreous state' in J. D. Mackenzie (Ed.), *Modern Aspects of The Vitreous State* Vol. I. Butterworth, London. pp. 152–187.

Goldstein, M. (1969). *J. chem. Phys.* **51**, 3728.

Herzfeld, K. F., and T. A. Litovitz (1959). *Absorption and Dispersion of Ultrasonic Waves.* Academic Press, New York.

Hopkins, I. L., and C. R. Kurkjian (1965). 'Relaxation spectra and relaxation processes in solid polymers and glasses' in W. Mason (Ed.), *Physical Acoustics*, Vol. 2B. Academic Press, New York. pp. 91–163.

Johari, G. P., and M. Goldstein (1970A). *J. phys. Chem.*, **74**, 2034.

Johari, G. P., and M. Goldstein (1970B). *J. chem. Phys.*, **53**, 2372.
Kauzmann, W. (1948). *Chem. Rev.*, **43**, 219.
Litovitz, T. A., and C. M. Davis (1965). In W. P. Mason (Ed.), *Physical Acoustics*, Vol. IIA. Academic Press, New York, pp. 282–350.
McCall, D. W. (1969). 'Relaxations in solid polymers' in R. S. Carter and J. J. Rush (Eds.), *Molecular Dynamics and Structure of Solids.* National Bureau of Standards, Washington, D.C. pp. 475–537.
McCrum, N. G., B. E. Read, and G. Williams (1967). *Anelastic and Dielectric Effects in Polymeric Solids.* John Wiley, New York.
Turnbull, D., and M. H. Cohen (1961). *J. chem. Phys.*, **34**, 120.

3

Study of the Molecular Motion of Polar Side-chains in Solid Polymers by Mechanical Measurements

H. A. WATERMAN, L. C. E. STRÚIK, J. HEIJBOER and M. P. VAN DUIJKEREN

Centraal Laboratorium TNO, Delft, The Netherlands

Methacrylate polymers in which the alkyl group is n-propyl, n-butyl, etc., show a low-temperature γ-transition (Hoff, Robinson and Willbourn, 1955, Heijboer, 1965, McCrum, Read and Williams, 1967). This transition is caused by the alkyl group, which relaxes independently of the oxycarbonyl group and the main chain.

This investigation is concerned with the effect of substitution of the CH_3-group of a n-alkyl side-chain by a polar group on the temperature location of the γ-transition and on its apparent activation energy. Table 1 gives the names of the side groups investigated, together with their formulae and the polymer code names.

Table 1

Name of side-group	Formula	Abbreviation of polymer name
n-Propyl	$-(CH_2)_2CH_3$	PnPrMA
β-Chloroethyl	$-(CH_2)_2Cl$	PβClEtMA
β-Bromoethyl	$-(CH_2)_2Br$	PβBrEtMA
β-Cyanoethyl	$-(CH_2)_2CN$	PβCNEtMA

The mechanical properties have been determined over a broad frequency range, viz. from 1 Hz up to 5×10^6 Hz. In order to cover this range four different experimental techniques had to be used.

EXPERIMENTAL

Preparation of monomers and polymers

n-Propyl methacrylate (nPrMA), β-chloroethyl methacrylate (βClEtMA) and β-bromoethyl methacrylate (βBrEtMA) were obtained by acid-catalysed transesterification of methyl methacrylate with the corresponding alcohols. β-Cyanoethyl methacrylate was obtained according to Yakubovich *et al.* (1958), from β-hydroxypropionitrile and methacrylyl chloride. Yields and properties of the monomers, purified by repeated distillation, are given in Table 2.

Table 2. Yields and physical constants of the monomers

Monomer	Yield %	Boiling range °C/mmHg	Refractive Index n_D^{20}	Density d_4^{20}
nPrMA	80	70–71/80	1·4186	0·9057
βClEtMA	70	61–62/10	1·4479	1·1003
βBrEtMA	65	42–42·5/0·6	1·4740	—
βCNEtMA	70	80/1·8	1·4465	1·0429

The polymerization was carried out between glass plates; initiator: 0·05% azobisisobutyronitrile. The polymerization began at 45 ° or 50 °C and was completed at 100 °C for about 16 hr. During the polymerization the liquid monomer changes to a solid glassy polymer. Specimens were machined from the polymers without prior purification.

MECHANICAL MEASURING TECHNIQUES

The measurements were performed by means of four different techniques:

(1) Torsional vibrations (0·3 to 3 Hz) (Nederveen and van der Wal, 1967).
(2) Flexural vibrations (300 to 2000 Hz) (P. Dekking, 1961).
(3) Longitudinal vibrations (10 kHz and 100 kHz) (Lanting and Waterman, 1964, and Waterman and Lanting, to be published).
(4) Propagating waves (5×10^6 Hz) (Waterman, 1963).

A more detailed description of these techniques is given by Heijboer *et al.* (1971). Young's storage modulus E' and loss modulus E'' were determined from flexural vibrations and longitudinal vibrations. From E' and E'' storage and loss moduli in shear G' and G'' were calculated from the approximation formulae:

$$E' = 2G'(1 + \mu')$$

$$E'' = 2G''(1 + \mu')$$

For the real part of Poisson's ratio μ', the value obtained from the propagating-wave measurements at 5×10^6 Hz was used. The real and imaginary parts of Poisson's ratio, μ' and μ'', were calculated from the formulae:

$$\mu' = \frac{1 - 2\left(\frac{C_t}{C_l}\right)^2}{2 - 2\left(\frac{C_t}{C_l}\right)^2}$$

and

$$\mu'' = \frac{\tan \delta_t - \tan \delta_l}{2\left(\frac{C_l}{C_t} - \frac{C_t}{C_l}\right)^2}$$

where C_l and C_t are the velocities of the longitudinal and transverse waves, respectively, and $\tan \delta_t$ and $\tan \delta_l$ are the loss tangents for shear waves and longitudinal waves, respectively.

RESULTS AND DISCUSSION

In Figure 1 the storage modulus in shear, G', the loss modulus in shear, G'', and loss factor, $\tan \delta$, are plotted *vs.* temperature for poly(β-chloroethyl methacrylate) (PβClEtMA) at five frequencies. To avoid overcrowding in the figure, not all the measuring points for the longitudinal vibrations (10^4 Hz and 10^5 Hz) are shown. The polymer clearly shows a γ-transition. As can be seen in the figure, neither the height of $\tan \delta_{max}$ nor that of G''_{max} vary much with frequency, the height of G''_{max} being less sensitive to frequency. Similar results are found for the other polymers, viz. poly(n-propyl methacrylate) (PnPrMA), poly(β-bromoethyl methacrylate) (PβBrEtMA) and poly(β-cyanoethyl methacrylate) (PβCNEtMA).

Figure 2 gives a comparison of the storage modulus in extension, E', and the loss modulus in extension, E'', at a frequency of about 10 kHz. From Figure 2 it follows that substitution of the CH_3-group by a polar group produces a shift of the γ-transition to higher temperatures. The effects of substitution by Cl and Br are very similar, the substitution by CN having a more pronounced effect. Substitution by CN not only causes the greatest temperature shift, but also a substantial broadening of the damping peak. The value of E' increases in the same order as the temperature of the γ-transition.

Figure 3 gives the real and imaginary parts, μ' and μ'', of Poisson's ratio as functions of temperature, at 5×10^6 Hz, for PβBrEtMA and PβCNEtMA. As can be seen the γ-dispersion is also observed for μ''. The difference in

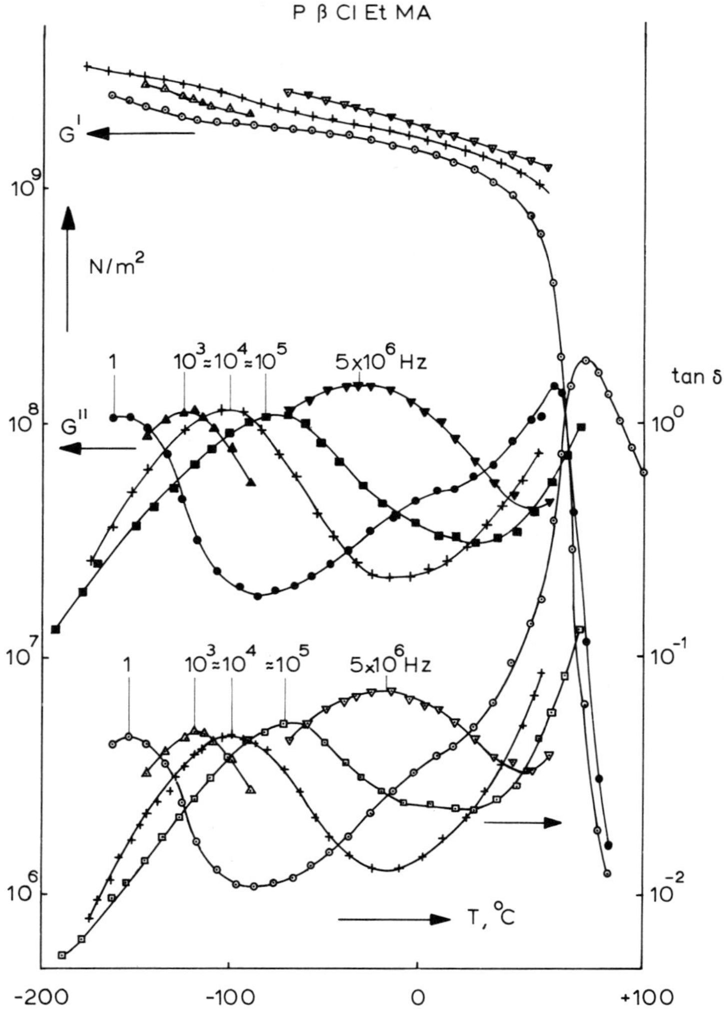

Figure 1. Storage modulus in shear, G', loss modulus in shear, G'', and loss factor, tan $\delta_{G'}$ *vs.* temperature for PβClEtMA at five frequencies: 1 Hz, $5{\cdot}10^2$ Hz, $\sim 10^4$ Hz, $\sim 10^5$ and $5{\cdot}10^6$ Hz

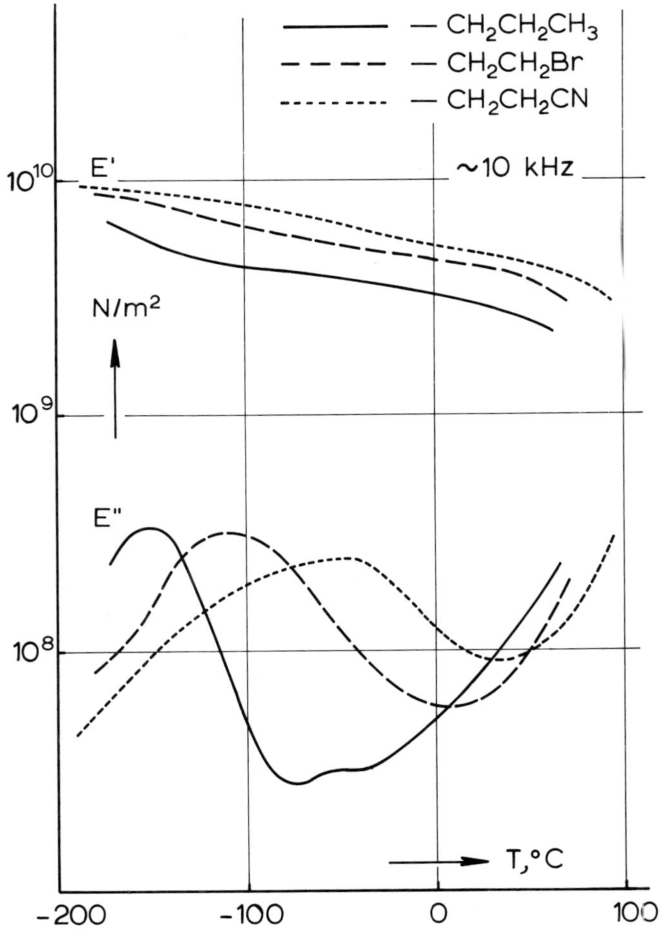

Figure 2. Storage modulus in extension, E', and loss modulus in extension, E'', *vs.* temperature for PnPrMA, PβBrEtMA and PβCNEtMA at a frequency of about 10 kHz

width of the γ-dispersions in the G'' *vs.* temperature curves (see Figure 2) for both polymers is clearly reflected in the μ'' *vs.* temperature curves. For both polymers μ''_{max} is located about 15 °C higher than the corresponding G''-maxima. On the high-temperature side, μ' begins to increase owing to the onset of the glass-rubber transition.

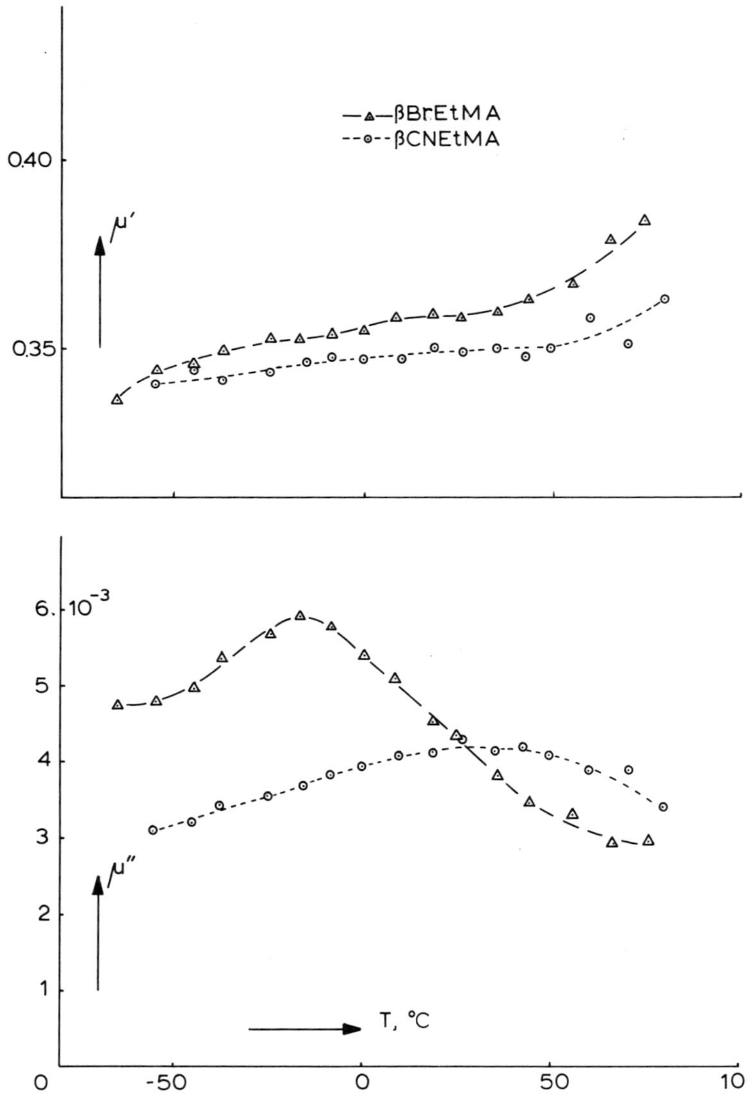

Figure 3. Real and imaginary parts of Poisson's ratio, μ' and μ'' *vs.* temperature, for PβCNEtMA and PβBrEtMA. Frequency 5×10^6 Hz

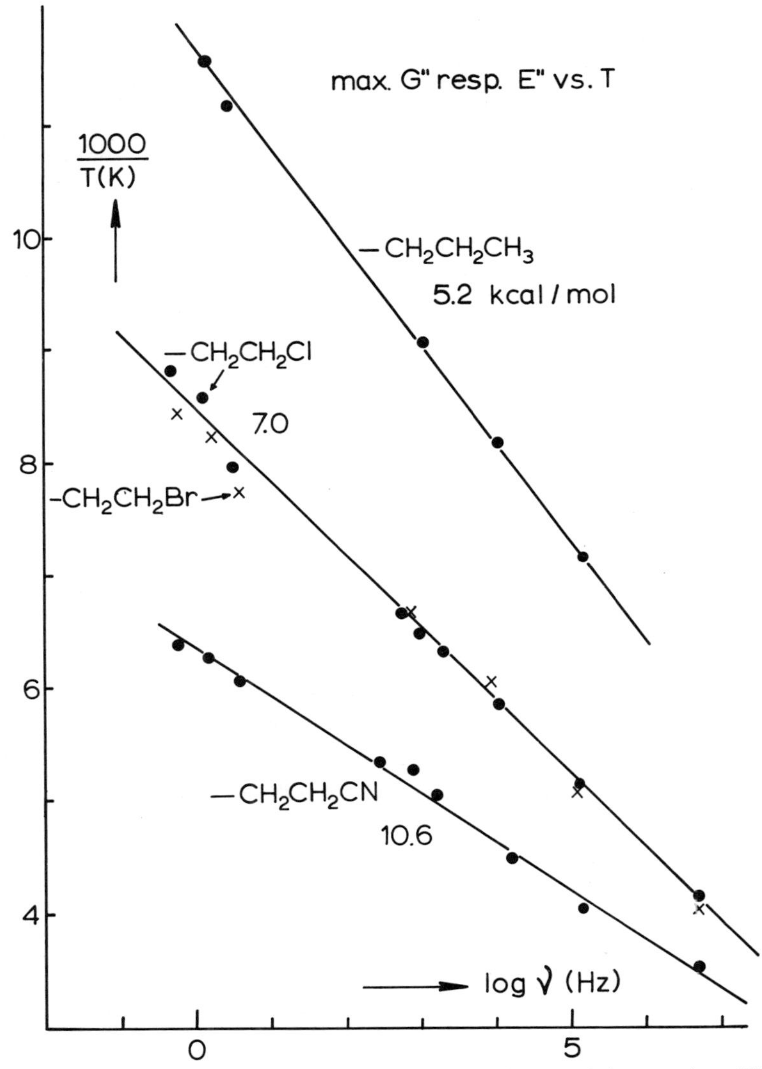

Figure 4. Arrhenius plot, $1/T$ against log ν, for loss modulus maxima. The temperatures of the maxima are obtained from a G'' or E'' *vs.* T curve

In Figure 4 Arrhenius plots are given for the values of $1/T$ at the maximum of G'' as a function of the logarithm of the frequency. (For PnPrMA the maximum of G'' at 5×10^6 Hz could not be determined for lack of a suitable, low-melting transmission liquid.) It is seen that the points fit rather well on straight lines. From these plots apparent activation energies E_a can be obtained. E_a can also be estimated from the relation $\nu = \nu_0 \exp(-E_a/RT)$, where ν_0 is of the order of the vibration frequency of a molecule (10^{12} Hz–10^{14} Hz) (see, e.g. Bueche, 1962). Both values are given in Table 3, together with data on the temperatures of the loss-maximum at various frequencies.

Table 3. Location of the loss modulus maxima and apparent activation energies

Polymer	Temperature (°C) of the loss modulus maximum at a frequency of					Apparent activation energy E_a	
	1 Hz	10^3 Hz	10^4 Hz	10^5 Hz	$5 \cdot 10^6$ Hz	kcal/mol	
						[a]	[b]
PnPrMA	−188	−164	−152	−136		5·2	5·1
PβClEtMA	−156	−119	−103	−81	−32	7·0	7·0
PβBrEtMA	−154	−121	−108	−77	−27	7·0	7·0
PβCNEtMA	−114	−76	−54	−28	+10	10·6	9·7

[a] from shift with frequency [b] from $\nu = 10^{13} \exp(-E_a/RT)$, for $\nu = 1$ Hz

Excepting PβCNEtMA, the agreement between the activation energies, obtained in different ways, is very good. This means that we are dealing with a molecular motion hindered by a constant activation energy E_a, whereas the Arrhenius plots extrapolate to $\log \nu = 13$ for $1/T \to 0$. From Table 3 it is seen that substitution of the CH_3-group by a polar group results in an increase in the apparent activation energy for the γ-transition. Substitution by the CN-group has the greater effect, the effects of substitution by Cl and Br being equal within the experimental error. Beside a change in polarity, the above-mentioned substitutions produce a change in the mass of the relaxing side group. In Table 4 the masses of the substituted side group and the dipole moments are compiled.

Table 4.

Polymer	Mass of moving side-group	Dipole moment[a] 10^{-18} e.s.u.	Activation energy for the γ-process in kcal/mol		
			Mechanical		Dielectr.
			This work	Lit.	
PnPrMA	43	0	5·2	4·0[b]	
PβClEtMA	63·5	2·0	7·0	6·0[b]	
PβBrEtMA	108	2·0	7·0		7·1[c]; 9·2[d]
PβCNEtMA	54	3·6	10·6		8·7[e]

[a] McClennan (1963).
[b] Hoff, Robinson and Willbourn (1955).
[c] Ong (1969).
[d] Mikhailov and Borisova (1964).
[e] Mikhailov, Artyukhov and Borisova (1968).

From Table 4 it follows that the polarity, and not the mass, of the side group is the dominant factor which determines the activation energy of the γ-process.

Kail (1965) found that an iodine atom at the end of the side-chain of a polyolefin-1 causes the γ-transition to shift to higher temperatures. In conflict with our findings Kail concludes that mass and not polarity, is a dominant factor in the γ-process. For theoretical reasons, we believe that mass is unlikely to have any significant effect on the temperature-location of a secondary transition.

Table 4 also shows literature data on the activation energies for the mechanical and the dielectric γ-process.

Our results are in qualitative agreement with those of Hoff *et al.* Mikhailov *et al.* (1968) conclude that the activation energy does not depend on the dipole moment, and that only the temperature location of the γ-process shifts to higher temperature for increasing dipole moment. This is in direct conflict with our own results. The value found by Ong supports our contention that the activation energy does increase for increasing dipole moment. The shift to higher temperatures of the relaxation process also constitutes evidence in favour of a higher activation energy.

It is interesting to note that Hara (1967) found a dielectric loss peak in cyanoethyl cellulose at -75 °C and 3 kHz, very near to the mechanical loss peak we found for PβCNEtMA. This is a further argument in favour of the idea that side groups move independently of the backbone chain. Mikhailov and Borisova (1967) already concluded this from their comparison of PβClEtMA and poly(β-chloroethyl chloroacrylate): they found the dielectric γ-maxima of both polymers to be identical.

REFERENCES

Bueche, F. (1962). *Physical Properties of Polymers.* Interscience, New York. p. 89.

Dekking, P. (1961). *Ph.D. Thesis*, University of Leiden. p. 32.

Hara, T. (1967). *Jap. J. appl. Phys.*, **6**. 147.

Heijboer, J. (1965). In J. A. Prins (Ed.), *Physics of Non-crystalline Solids.* North Holland Publishing Co., Amsterdam. p. 231.

Heijboer, J., L. C. E. Struik, H. A. Waterman, and M. P. van Duijkeren (1971). *J. macromolek. Sci. Phys. B*, **5**, 375.

Hoff, E. A. W., D. W. Robinson, and H. Willbourn (1955). *J. Polym. Sci.*, **18**, 161.

Kail, J. A. E. (1965). *Polymer* (London), **6**, 535.

Lanting, S. S. and H. A. Waterman (1964). *J. scient. Instrum.*, **41**, 605.

McClellan, A. L. (1963). *Tables of Experimental Dipole Moments.* Freeman and Co., San Francisco.

McCrum, N. G., B. E. Read, and G. Williams (1967). *Anelastic and Dielectric Effects in Polymeric Solids.* John Wiley and Sons, London. p. 271.

Mikhailov, G. P. and T. I. Borisova (1964). *Polym. Sci. U.S.S.R.*, **6**, 1971.

Mikhailov, G. P. and T. I. Borisova (1967). *Polym. Sci. U.S.S.R.*, **9**, 1310.

Mikhailov, G. P., A. T. Artyukhov, and T. I. Borisova (1968). *Polym. Sci. U.S.S.R.*, **8**, 1755.

Nederveen, C. J. and C. W. van der Wal (1967). *Rheol. Acta*, **6**, 316.

Ong, P. H. (1969). Centraal Laboratorium TNO. Personal communication.

Waterman, H. A. (1963). *Kolloidzeitschrift*, **192**, 1.

Waterman, H. A. and S. S. Lanting. To be published in *J. scient. Instrum.*

Yakubovich, A. Y., V. V. Razumovskii, and S. Rozenstein (1958). *Zh. obshch. Khim.*, **28**, 2292.

4

Review of Mechanical Relaxations in Mixed Alkali Silicate Glasses

D. E. DAY

Ceramic Engineering Department, University of Missouri–Rolla, U.S.A.

One of the interesting features of mixed alkali glasses is the way the properties change with the addition of a second alkali. The large deviation from linearity exhibited by many properties of mixed alkali glasses has been referred to as the 'mixed alkali', 'neutralization', or 'polyalkali' effect.

Several theories have been proposed for the mixed alkali effect, but they lack the general utility of being able to explain the many varied changes in properties. Properties involving ion movement show the most pronounced deviation from linearity with the addition of a second alkali. The electrical conductivity (Hakim and Uhlmann, 1967), viscosity (Poole, 1949), and alkali diffusion coefficients (Sendt, 1964; Evstrop'ev, 1960) for mixed alkali glasses are several orders of magnitude lower than for comparable single alkali glasses. Properties such as density and refractive index show relatively small deviations from linearity. Isard (1969) gives a comprehensive review of the mixed alkali effect in glass.

The mechanical energy absorption, or internal friction, also shows large changes with the addition of a second alkali, Figure 1. The internal friction curve for the $Na_2O . 3SiO_2$ glass shows the two internal friction peaks typically observed in a single alkali glass. The peak at -32 °C is due to the stress-induced movement of the alkali ions (alkali peak), whereas the smaller peak at 182 °C is attributed to the presence of non-bridging oxygen ions (NBO peak). Each internal friction peak is accompanied by a relaxation of the shear modulus, superimposed on a general decrease in modulus with increasing temperatures.

Rötger (1958) and Jadgt (1960) both noted the large increase in internal friction for mixed alkali silicate glasses, Na–K glass in Figure 1. Initially, the one large peak observed in mixed alkali glasses was thought to result from a rapid growth in the alkali peak. More recent work (Steinkamp, Shelby and Day, 1967) has established that it is a new relaxation (mixed alkali peak).

Depending upon composition, all three mechanical relaxations may be observed in mixed alkali silicate glasses. Usually only the mixed peak is observed, unless the concentration of second alkali is small, because of its

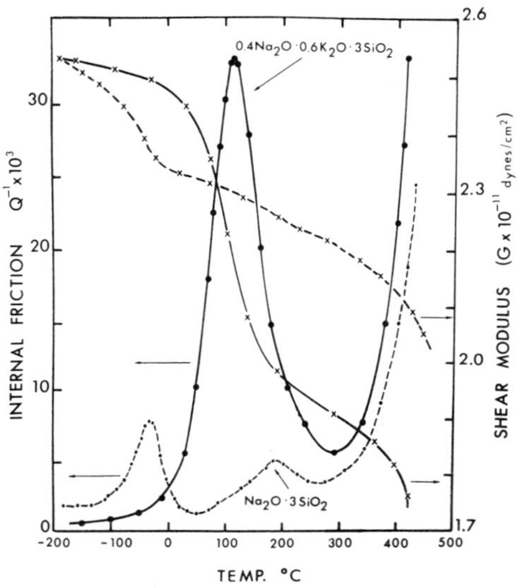

Figure 1. Comparison of internal friction for single and mixed alkali silicate glass. Frequency: 0·4 Hz

large magnitude. The relaxation of the shear modulus associated with the mixed alkali and single alkali peaks in Figure 1 is typical, being 22 % and 5·5 %, respectively.

ALKALI INTERNAL FRICTION PEAK

Single alkali glasses

The alkali internal friction peak is observed in single alkali glasses containing Li, Na, K, and Rb. In all single alkali silicate glasses the alkali peak becomes larger with increasing alkali content, as shown by the sodium silicate glasses in Figure 2. There is an apparent lack of agreement for the peak height between 25 % and 35 % Na_2O. Figure 2 also shows that the magnitude of the alkali peak changes in a manner generally corresponding to the change in the Na diffusion coefficient. This correlation between the magnitude of the alkali peak and the alkali diffusion coefficient has been reported for several different single alkali glasses (McVay and Day, 1970) and is considered important in understanding the behaviour of the alkali peak in mixed alkali glasses.

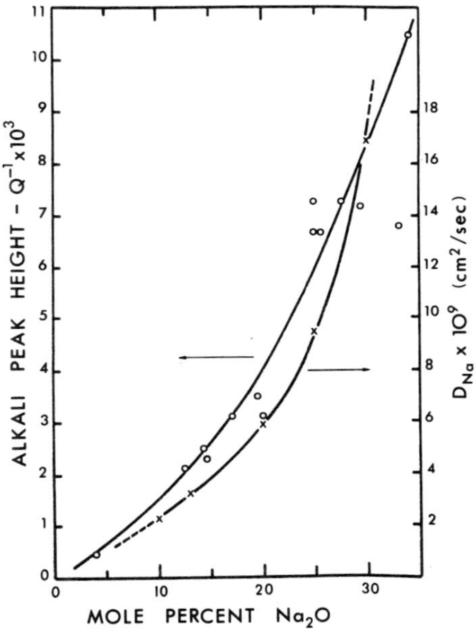

Figure 2. Height of alkali internal friction peak and sodium diffusion coefficient (415 °C) in sodium silicate glasses. Peak height data from various sources in the literature and diffusion data (from Evstrop'ev, 1960)

The activation energy for the alkali peak in single alkali glasses is usually close to that for electrical conductivity or alkali self-diffusion and shows the same general compositional dependence. In Li, Na and K silicate glasses the activation energy decreases with increasing alkali content and typically varies from about 12 to 20 kcal/mol.

Mixed alkali glasses

Adding a second alkali produces two changes in the alkali peak. The rapid reduction in magnitude, with small additions of the second alkali, is illustrated in Figure 3. The arrow on the left axis denotes the peak height in the $Na_2O \cdot 3SiO_2$ and $Li_2O \cdot 3SiO_2$ glasses. The lines having three or more data points were determined using a least squares fit. While additional data points are needed for some of the compositional systems, the peak height appears to be inversely proportional to the logarithm of the concentration of the second alkali. After a minimum concentration is exceeded, the value of which depends upon the second alkali, the peak height decreases at a rate essentially independent of the second alkali.

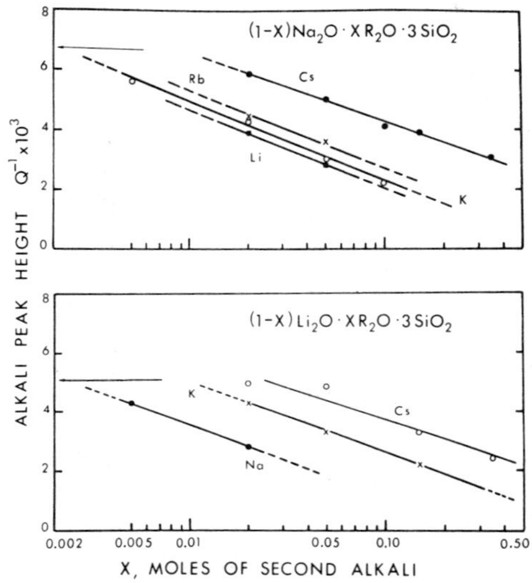

Figure 3. Change in height of alkali peak upon addition of second alkali. In sodium glass R = Li, K, Rb, and Cs. In lithium glass R = Na, K, and Cs.

The alkali peak is expected to become small with the substitution of a second alkali because of the decrease in concentration of the original alkali, Figure 2. If the change in concentration of the original alkali was the primary factor affecting the peak height, however, the peak height should be inversely proportional to the concentration rather than to the logarithm of the concentration of the second alkali. No explanation has been suggested for the semi-logarithmic relationship between concentration and peak height.

The second change in the alkali peak is an increase in the activation energy. In Li–Cs silicate glasses (Shelby and Day, 1970), the activation energy increases from 14 ± 1·5 to 23·4 ± 1·5 kcal/mol for a $Li_2O . 3SiO_2$ and 0·65 Li_2O · 0·35 $Cs_2O . 3SiO_2$ glass, respectively. The activation energy increases whenever a second alkali is introduced although the amount of increase varies with the different alkali ions.

There is a close correlation between the mechanical losses for the alkali peak in mixed alkali silicate glasses and the electrical loss characteristics. Several investigations (Hakim and Uhlmann, 1967; Mazurin, 1965; Ivanov,

1964; Myuller, 1959; Lengyel, 1940) have shown that the electrical dissipation factor (loss) decreases and the activation energy for d.c. or a.c. conduction increases with the addition of a second alkali. This correlation indicates that the mechanism for the mechanical loss alkali peak and that for electrical loss are basically similar.

MIXED ALKALI INTERNAL FRICTION PEAK

Magnitude

Silicate glasses containing dissimilar alkali ions exhibit an internal friction peak which is a distinct mechanical relaxation separate from the alkali peak (Shelby and Day, 1969). Generally, the maximum peak height is not at the composition containing equal concentrations of both alkalis. As the size or mass difference between the alkali ions increases, the skewness becomes more pronounced and the maximum moves closer to the end member glass containing the larger alkali ion. For a given substitution of a second alkali, a larger mixed peak is observed when the second alkali is smaller than the original alkali (Shelby and Day, 1970).

In Na–Rb silicate glasses, McVay (1970) found that the magnitude of the mixed peak was closely associated with the alkali diffusion coefficients. Figure 4 shows that the mixed peak is a maximum at that composition, within

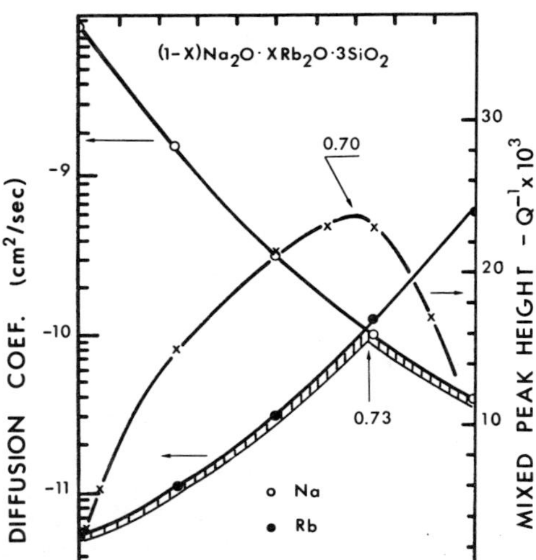

Figure 4. Alkali diffusion coefficients (350 °C) and height of mixed alkali internal friction peak in (1-X)$Na_2O \cdot X Rb_2O \cdot 3SiO_2$ glasses. Cross-hatched curve denotes slower moving alkali ion. (McVay, 1970)

experimental error, where the diffusion coefficients for Na and Rb are equal; i.e. Rb/(Rb + Na) = 0·73. The skewness of the peak height curve also seems to be related to the relative difference (or ratio) of the diffusion coefficients. The difference in the diffusion coefficients is much less for the Rb-rich glasses than for the Na-rich glasses. This may be the reason why a larger mixed peak is observed when Na (smaller ion) is substituted for Rb (larger ion) as opposed to the substitution of Rb for Na.

The internal friction and alkali diffusion coefficients have been measured for the same glasses in only the Na–Rb system. However, the Na and K diffusion coefficients in 80 mol % SiO_2 glasses (Evstrop'ev, 1960) also intersect like those for the Na–Rb glasses. Although the compositional difference prohibits an exact comparison, the skewness of the mixed peak height curve for Na–K glasses, 75 mol % SiO_2, (Shelby and Day, 1969) agrees reasonably well with the relative change in the Na and K diffusion coefficients.

Activation energy

The typical change in the activation energy for the mixed peak within a compositional system is shown in Figure 5. The alkali ratio where the activation energy is a minimum usually does not coincide with that where the

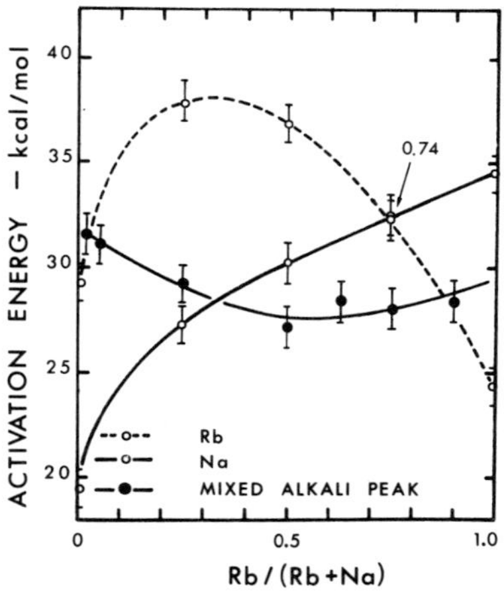

Figure 5. Activation energy for alkali diffusion, open circles, and mixed alkali internal friction peak in (1–*X*) $Na_2O \cdot XRb_2O \cdot 3SiO_2$ glasses. (McVay, 1970)

magnitude of the mixed peak is a maximum. In Na–Rb glasses the minimum in the activation energy is at a Rb/(Rb + Na) ratio of about 0·60, whereas the maximum peak height is at 0·70. In Li–Na glasses, 75 mol % SiO_2, the minimum activation energy and maximum peak height are at a Na/(Li + Na) ratio of 0·25 and 0·55, respectively (Shelby and Day, 1969). No correlation should be assumed between peak height and activation energy.

In the Na–Rb glasses there is no direct correlation between the activation energy for the mixed peak and the activation energy for alkali diffusion. The activation energies for Na or Rb diffusion, Figure 5, are significantly different from those for the mixed peak. The relaxation mechanism for the mixed peak is apparently more complicated than simply single alkali ion movement.

For glasses containing equal concentrations of dissimilar alkali ions, the activation energy varies directly with the radius ratio, Figure 6, whereas the

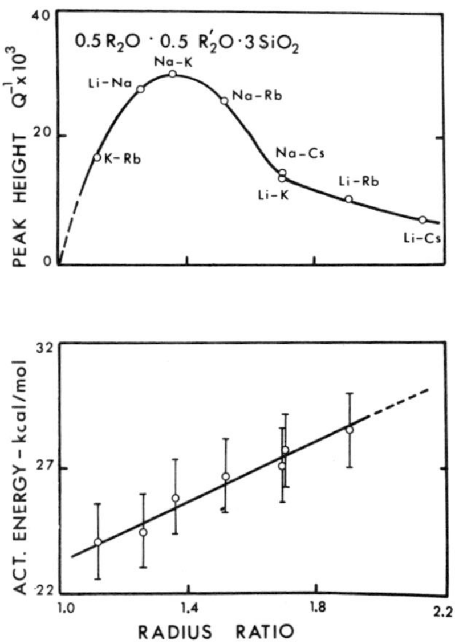

Figure 6. Height and activation energy of mixed alkali internal friction peak in mixed alkali glasses containing equal concentrations of two alkalis. (Shelby and Day, 1970)

peak height is a more complex function of the radius ratio. As noted previously for the variation in activation energy within a compositional system, there is no correlation between peak height and activation energy between compositional systems. For example, the Na-K glass has an intermediate activation energy but the largest peak whereas the K-Rb glass has a lower activation energy and a smaller peak. The mixed peak has also been compared to the alkali ion mass difference and mass ratio, but the radius ratio provides the most consistent trend in the experimental data.

Ion-exchanged glasses

Annealed ion-exchanged alkali silicate glasses also exhibit a mixed peak (Taylor and Rindone, 1968; de Waal, 1969; Shelby, 1970) very similar to that for conventionally melted glasses. There is apparently sufficient mixing of the dissimilar alkali ions in the thin ion-exchanged layer for this mechanical relaxation to occur. Removal of the ion-exchanged layer eventually results in the disappearance of the mixed peak (Taylor and Rindone, 1970).

Relaxation mechanism

A cooperative reorientation or coupled motion of the dissimilar alkali ions has been suggested as the mechanism for the mixed alkali peak. Shelby and Day (1970) propose that the dissimilar alkali ions interact and form electrically inactive elastic dipoles. No electrical loss mechanism analogous to the mixed alkali mechanical loss peak has been reported. De Waal (1969) suggests an interchange in position of the dissimilar alkali ions.

The general increase in activation energy with increasing radius ratio, Figure 6, is consistent with a coupled reorientation of the dissimilar alkali ions as this rearrangement should become more difficult with increasing size difference. The correlation between the maximum peak height and the diffusion coefficient of the slower moving ion in the Na–Rb glasses also suggests that the coupled reorientation is determined by the slower moving ion. In the Na–Rb glasses, the relaxation time for the mixed peak is a minimum when the diffusion coefficient of the slower moving ion is a maximum, i.e. when $Rb/(Rb + Na) = 0{\cdot}73$ (McVay, 1970). Both ions are involved in the relaxation process as indicated by the difference in the activation energy for the mixed peak and that for either Na or Rb diffusion.

REFERENCES

Evstrop'ev, K. K. (1960). *Structure of Glass*, Vol. 2. Consultants Bureau, New York, p. 237.
Hakim, R. M., and D. R. Uhlmann (1967). *Physics Chem. Glasses*, **8** (5), 174.
Isard, J. O. (1969). *J. non-cryst. Solids*, **1**, 235.
Ivanov, A. O. (1964). *Soviet Phys. solid St.*, **5** (9), 1933.
Jagdt, R. (1960). *Glastech. Ber.*, **33** (1), 10.
Lengyel, V. B. (1940). *Glastech. Ber.*, **18** (7), 177.
McVay, G. L., and D. E. Day (1970). *J. Am. Ceram. Soc.*, **53** (5), 284.
McVay, G. L. (1970). *Ph. D. Thesis*, University of Missouri–Rolla, Rolla, Missouri.
Mazurin, O. V. (1965). *Structure of Glass*, Vol. 4. Consultants Bureau, New York, p. 158.
Myuller, R. L. (1959). *Soviet Phys. tech. Phys.* **4** (11), 1219.
Poole, J. P. (1949). *J. Am. Ceram. Soc.*, **32** (7), 230.
Rötger, H. (1958). *Glastech. Ber.*, **31** (2), 54.
Sendt, A. (1964). *Glastech. Ber.*, **37** (2), 116.

Shelby, J. E., Jr., and D. E. Day (1969). *J. Am. Ceram. Soc.*, **52** (4), 169.
Shelby, J. E., Jr., and D. E. Day (1970). *J. Am. Ceram. Soc.*, **53** (4), 182.
Shelby, J. E., Jr. (1970). *J. Am. Ceram. Soc.*, **53** (4), 226.
Steinkamp, W. E., J. E. Shelby, Jr., and D. E. Day (1967). *J. Am. Ceram. Soc.*, **50** (5), 271.
Taylor, T. D., and G. E. Rindone (1968). *J. Am. Ceram. Soc.*, **51** (5), 289.
Taylor, T. D., and G. E. Rindone (1970). *J. Am. Ceram. Soc.*, **53** (4), 227.
de Waal, H. (1969). *Physics Chem. Glasses*, **10** (3), 108.

5

Relaxation of Torsional Stress of Silicate Glasses

T. C. WOO

Department of Mechanical Engineering, University of Pittsburgh, Pittsburgh, Pa., U.S.A.

S. M. OHLBERG, E. R. MICHALIK and R. A. STEWART

PPG Industries, Inc., Harmar Township, Pittsburgh, Pa., U.S.A.

Stress relaxation in the transformation range of glass has been reviewed comprehensively by Isard and Douglas (1955); Ritland (1955) and Kurkjian (1963). Apart from offering experimental results of torsional stress relaxation for different glasses, the purposes of this paper are, to predict the tempering capabilities for those glasses from the relaxation data, and to discuss interrelations among various viscoelastic functions. The former objective is motivated by the questions generally raised in a manufacturing industry as to the potentialities of thermal tempering of different glasses. The latter is intended to enhance the usefulness of those experimental data. The conversion from torsional relaxation to flexural relaxation, for instance, can be performed mathematically and yield results more directly applicable for thermal stress calculations. The basis throughout this work is linear viscoelasticity with temperature dependence taken into consideration.

In studying thermal stress generation it should be noticed that relaxation functions in the transformation range and their temperature dependence act quite independently and differently. The relaxation function describes the isothermal viscoelastic property and the temperature influence is assumed just to shift the relaxation curve along the logarithmic time scale from one temperature to another.

This shift property has been seen to be a good approximation for glass by Kurkjian (1963) and by Bagley and Woo (1965). The individual effects of the relaxation and shift characteristics on the thermal stress evolution have not yet been expounded fully. It seems conceivable that a glass with much slower stress relaxation rate may generate higher thermal stresses owing to the steeper slope of the shift function. With the actual data on hand for various glasses it would be appropriate to look into this kind of query.

APPARATUS

The experimental apparatus used for determining torsional relaxation consists of a tubular, split-wound, power positioning furnace with individual trimmers for each of the three heating zones. An auxiliary zone attached to the furnace bottom compensates for conduction heat losses associated with air bearing operation. Furnace control is by current proportioning through a silicon controlled rectifier (SCR) network capable of maintaining a set temperature to within $\pm$ 0·5 °F. Figure 1 illustrates the remaining apparatus.

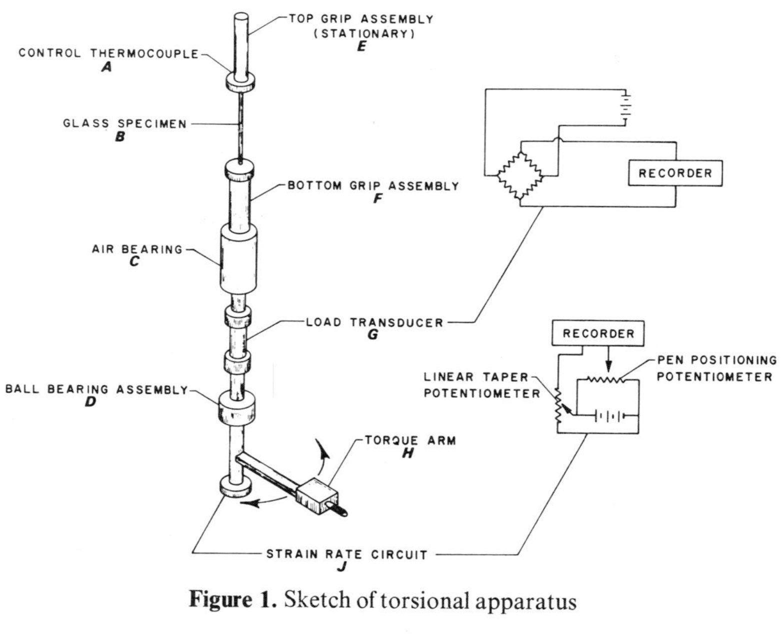

Figure 1. Sketch of torsional apparatus

Glass grip assemblies and tubular mounting shafts, E and F, are made from inconel metal because of its desirable expansion properties, high strength, and thermal stability at experimental temperatures. The load transducer G is fabricated from a low modulus, high sensitivity, high rigidity material and incorporates a full bridge measurement circuit as illustrated in the schematic. To assure frictionless load transmission from the specimen, B, to the load transducer, a radial journal type air bearing, C, is utilized. Angular displacement (of the torque shaft) versus time or strain-rate data are obtained by use of a linear taper potentiometer and associated apparatus illustrated at J. Torsional loads are applied in either direction by manually rotating the torque arm, H, against a stop.

EXPERIMENTAL PROCEDURES

Glass rod specimens 4·5 inches long and with diameters ranging from 0·115–0·200 in $\pm$ 0·0005 in are placed in the gripping arrangement. Specimen and grips are completely enveloped inside a stainless steel sleeve within the furnace. During this set-up, precautions are taken to eliminate the possibility of: (1) specimen deformation; (2) mechanical loading of specimen; (3) frictional drag on lower grip and shaft assembly.

Although temperature stability can be realized within one hour, in order to assure complete thermal stability, a three hour dwell time is used. This incidentally provides ample time for dead weight calibration of load transducer and associated instrumentation. During this same period the strain rate apparatus is also calibrated. Torsional loading is developed by manually rotating the torque arm through an angular displacement. The final position of the arm is limited by a fixed stop. At zero stress a cam on the torque arm automatically energizes the recorder sweep which then continuously provides plots of torque versus time. Data reproducibility for structurally stabilized glass is within $\pm$ 2–3 % mainly because of one inch ounce transducer sensitivity, and stress resolution of one inch ounce per 0·01 second.

Strain rate data are automatically recorded as the specimen is loaded.

RELAXATION FUNCTIONS AND THEIR CONVERSION

The torsional relaxation test has been conducted for five glasses; plate, sheet, float, flint, and pyrex glass; the results at various temperatures are given in Figures 2 to 6.

The torsional stresses have been normalized and the shape of the curves appears to be very similar. The shift functions for the glasses are obtained by the displacement of logarithmic time scale while coinciding the neighbouring relaxation curves. Figure 7 shows the shift functions for the pyrex and float glass as examples. A linear relationship seems to be satisfactory for the present data and the slopes are plate 0·0231/°F, sheet 0·0211/°F, float 0·0236/°F, flint 0·0232/°F, and pyrex 0·0207/°F.

The stress relaxation test is sometimes performed for pure bending (Van Zee and Noritake 1958) owing to the relative simplicity of the experimental set-up. Through linear viscoelasticity the torsional and flexural relaxation functions are functionally related just as the shear modulus and Young's modulus are related algebraically in elasticity. Denoting torsional and flexural

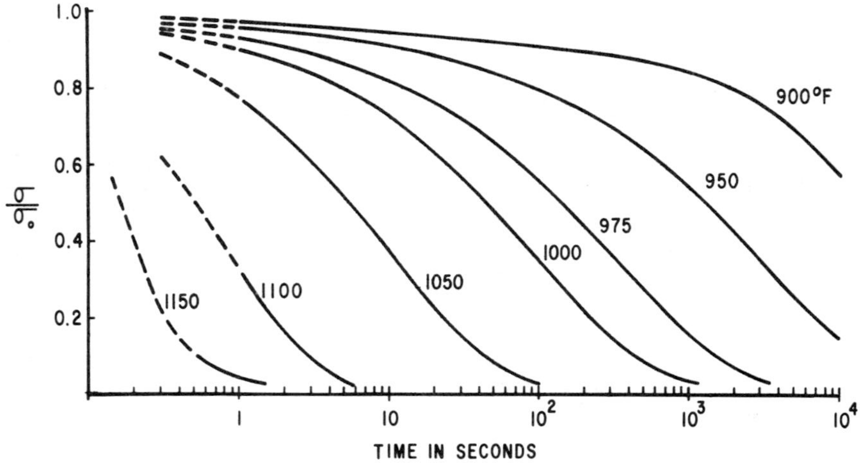

Figure 2. Relaxation curve for plate glass

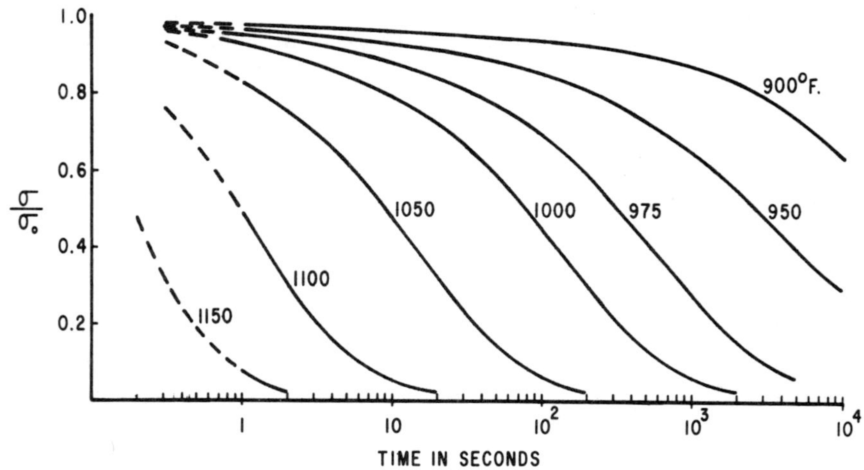

Figure 3. Relaxation curve for sheet glass

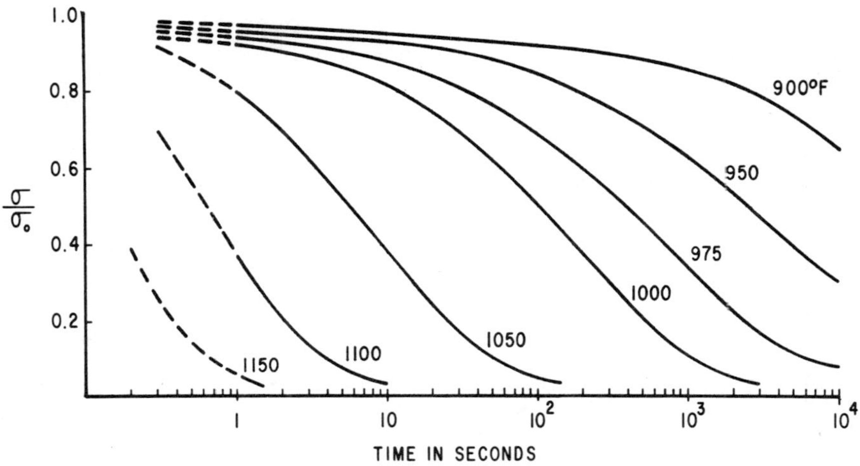

Figure 4. Relaxation curve for float glass

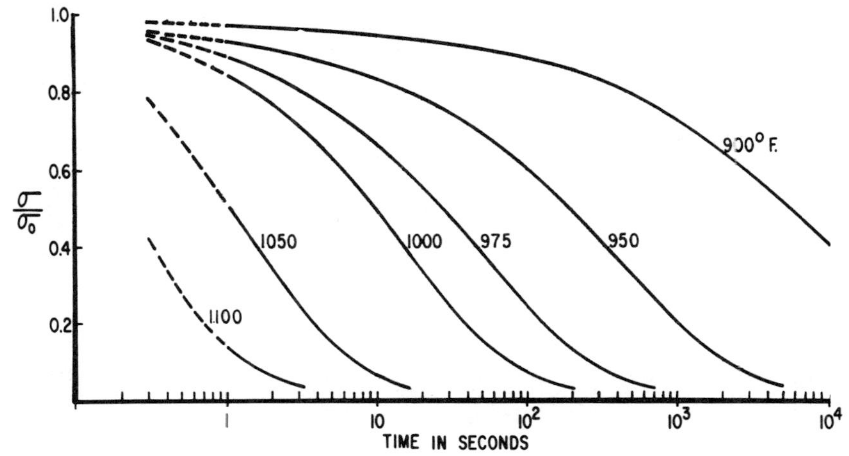

Figure 5. Relaxation curve for flint glass

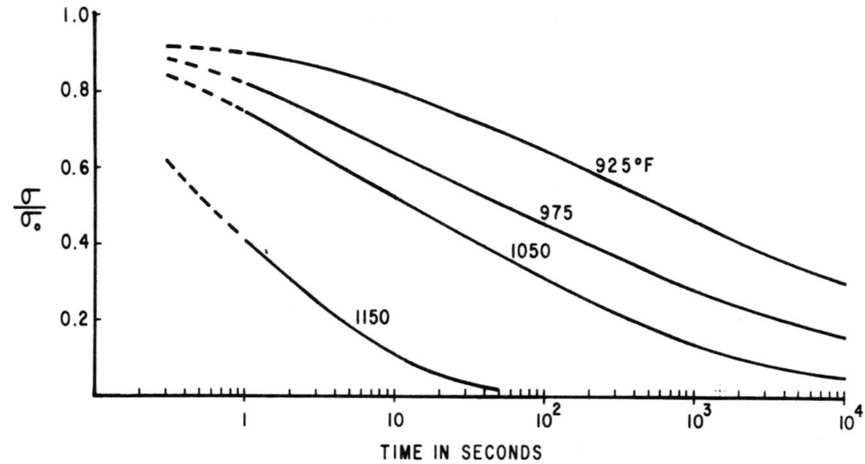

Figure 6. Relaxation curve for pyrex glass

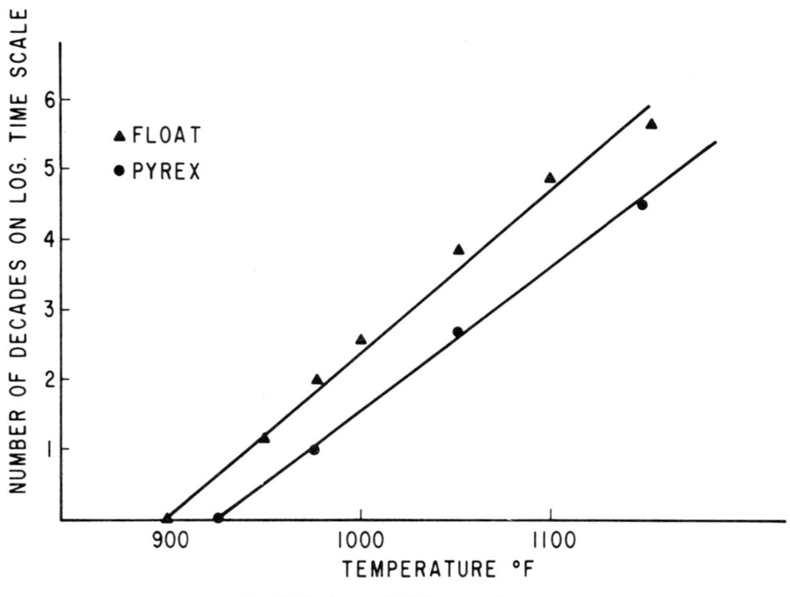

Figure 7. Shift data of different glasses

relaxation functions by $G(t)$ and $E(t)$ respectively the relationship between them is (Stuart, 1952)

$$E(t) = 2G(t)(1 + \nu_0) + 2\int_0^t \dot{\nu}(t - \xi)\, G(\xi)\, d\xi$$

where ν_0 means $\nu(0)$ and $\nu(t)$ is the viscoelastic function corresponding to Poisson's ratio in elasticity, with the dot denoting derivative with respect to its argument. If we assume elastic dilatational behaviour, which is roughly true for most glasses, we have

$$\nu(t) = (1/2)\left[1 - \frac{E(t)}{3K_0}\right]$$

where K_0 is bulk modulus. Therefore

$$E(t) = G(t)\left[3 - \frac{E_0}{3K_0}\right] - \frac{1}{3K_0}\int_0^t \dot{E}(t - \xi)\, G(\xi)\, d\xi$$

where E_0 denotes $E(0)$. The above is an integral equation of second kind and can be solved by iterative procedures. The thermal tempering stress build-up makes use of the flexural relaxation functions (Lee, Rogers and Woo, 1965) obtained from the integral equation, which therefore serve as inputs in the tempering stress studies in the sequel.

TEMPERING STUDIES

Basing our discussion on the flexural relaxation functions and the shift properties we will compare the merits of the thermal tempering capabilities of various glasses. The hypothetical case is chosen to be 0·125 inch thick glass plate of infinite expanse with isothermal initial temperature and heat transfer coefficient of 170 Btu ft^{-2} hr^{-1} $°F^{-1}$. The initial quenching temperature of 1200 °F is chosen for plate glass. Instead of using the same initial temperature for all the glasses we increase or decrease it according to the stress relaxation rate. The temperature at which 50% of the stress is relaxed in 10 seconds is taken as the initial temperature. From Figures 2 to 6 we obtain 1213, 1205, 1165 and 1221 °F as the appropriate starting temperatures for the sheet, float, flint and pyrex glass respectively. The residual centre tensions for the five glasses are given as:

Plate glass	10,084 psi
Sheet glass	9,913 psi
Float glass	9,998 psi
Flint glass	9,535 psi
Pyrex glass	2,252 psi

As previously pointed out the permanent stress depends on both the relaxation data and the shift function and it is not possible to correlate the tempering capability on either one of them alone. The low value for pyrex is due partly to its low shift property, but is mainly due to its low value of thermal coefficient of expansion $1{\cdot}8 \times 10^{-6}/^{\circ}F$ as compared to $5{\cdot}0 \times 10^{-6}/^{\circ}F$ for the other glasses.

REFERENCES

Bagley, F. L., and T. C. Woo (1965). *Proc. 4th Int. Congress on Rheol.*, Pt. 2, 643.
Isard, J. O., and R. W. Douglas (1955). *J.S.G.T.*, **39**, 61.
Kurkjian, C. R. (1963). *Physics Chem. Glasses*, **4**, 129.
Lee, E. H., T. G. Rogers, and T. C. Woo (1965). *J. Am. Ceram. Soc.*, **48**, 480.
Ritland, H. N. (1955). *J. S. G. T.*, **39**, 108.
Stuart, H. A. (1952). *Die Physik der Hochpolymeren*, vol. 1., Springer, p. 73.
Van Zee, A. F., and H. M. Noritake, (1958). *J. Am. Ceram. Soc.*, **41**, 164.

6

The High-Frequency Mechanical Response of Supercooled Liquids

A. R. DEXTER† and A. J. MATHESON

Department of Chemistry, University of Essex, Colchester, Great Britain

---

The object of this work was to compare the viscoelastic relaxation of the bulk modulus with the viscoelastic relaxation of the shear modulus in two simple glass-forming organic liquids.

In the limit of high frequencies of alternating strain a viscoelastic material behaves elastically both on compression and on shear. The resulting bulk and rigidity moduli are designated K_∞ and G_∞ respectively.

In the limit of low frequencies of alternating strain, the bulk and shear behaviours are not the same: a liquid cannot support a shear stress indefinitely ($G_0 = 0$), whilst it can support a compression indefinitely ($0 < K_0 < K_\infty$). At low frequencies, however, the atoms or molecules in the liquid are able to flow during the duration of the applied strain and viscosity effects are observed. Viscous stresses are proportional to the rate of strain, and since in an alternating experiment the time derivative $\mathrm{d}/\mathrm{d}t$ may be replaced by $\mathrm{i}\omega$ where $\mathrm{i}^2 = -1$ and $\omega = 2\pi f$ is the angular frequency, it can be shown that real viscosities are equivalent to imaginary moduli. For example, consider a liquid undergoing a slow change in volume. Then

$$\text{stress} = K_0 \times \text{strain} + \eta_B^* \times \frac{\mathrm{d}}{\mathrm{d}t}(\text{strain}) \qquad (1)$$

where η_B^* is the complex bulk or volume viscosity. Substituting for $\mathrm{d}/\mathrm{d}t$ gives

$$\text{stress} = (K_0 + \mathrm{i}\omega\eta_B^*) \times (\text{strain}) \qquad (2)$$

Hence

$$K^* - K_0 = K' + \mathrm{i}K'' = \mathrm{i}\omega\eta_B^* = \mathrm{i}\omega(\eta'_B - \mathrm{i}\eta''_B) \qquad (3)$$

† Present address: National Institute of Agricultural Engineering, Silsoe, Bedford.

and so the mechanical response of liquids may be discussed in terms of either a complex modulus or a complex viscosity. By analogy with equation (3),

$$G^* = G' + iG'' = i\omega\eta_s^* = i\omega(\eta_s' - i\eta_s'') \tag{4}$$

which gives the complex shear modulus in terms of the complex shear viscosity.

In the limit of high frequencies where the liquid is supposed to behave elastically, the dissipation components K'' and G'' tend towards zero, whilst the energy storage components $(K_0 + K')$ and G' tend towards K_∞ and G_∞ respectively. It is conventional to write the high frequency limiting value of K' as K_2 so that $K_0 + K_2 = K_\infty$.

At frequencies which are intermediate between the 'low' and the 'high' frequencies mentioned so far, the viscoelastic transition region is observed. In this region $0 < G' < G_\infty$ and $0 < K' < K_2$, and both G'' and K'' are finite. If the response to a step in strain in the liquid were an exponential decay of the resulting stress with time, then it can be shown that the frequency dependence of the modulus is given by (Herzfeld and Litovitz, 1959)

$$\frac{K^* - K_0}{K_2} = \frac{\omega^2\tau_B^2 + i\omega\tau_B}{(1 + \omega^2\tau_B^2)} \tag{5}$$

where τ_B is the bulk or volume relaxation time and is equal to the time constant of the stress decay mentioned above. A similar expression may be written for the normalized complex shear modulus G^*/G_∞ in terms of a shear relaxation time τ_s.

In most molecular liquids, however, an exponential stress decay does not provide a good description of the behaviour. In these cases, if the stress decay is still monotonic, a summation of exponentials having different time constants will always fit the data (Schwarzl and Struik, 1968). Equation (2) can then be separated and written as

$$\begin{aligned} K' &= K_2 \int_0^\infty \frac{g(\tau_B)\,\omega^2\tau_B^2}{1 + \omega^2\tau_B^2}\,d\tau_B \\ K'' &= K_2 \int_0^\infty \frac{g(\tau_B)\,\omega\tau_B}{1 + \omega^2\tau_B^2}\,d\tau_B \end{aligned} \tag{6}$$

where $g(\tau_B)$ is the density of relaxation times in the range τ_B to $(\tau_B + d\tau_B)$. Similar expressions can be written for G' and G''.

Since it is a very laborious process to calculate $g(\tau_B)$ for every set of experimental data (Schwarzl and Struik, 1968; Gross, 1953), resort is often made to empirical functions to describe the frequency dependence of the moduli. One of the more successful of these has been that of Barlow, Erginsav and

Lamb (1967, 1969). According to this model the complex shear compliance J^*, which is the reciprocal of the complex modulus

$$J^* = 1/G^* = J' - iJ'' \tag{7}$$

is given by

$$\begin{aligned} J &= \frac{1}{G_\infty} + \frac{k_s}{(\omega\eta_s G_\infty/2)^{\frac{1}{2}}} \\ J'' &= \frac{1}{\omega\eta_s} + \frac{k_s}{(\omega\eta_s G_\infty/2)^{\frac{1}{2}}} \end{aligned} \tag{8}$$

When $k_s = 0$, the frequency dependence is that of a single relaxation process and G^* varies with frequency as described by equation (5) with G^*, G_∞ and τ_s replacing $(K^* - K_0)$, K_2 and τ_B respectively. When $k_s > 0$ as is usually the case, the model predicts a distribution of relaxation times as in equation (6). The parameter k_s of equation (8) is therefore a measure of the width of the distribution of relaxation times which are required to describe the observed relaxational behaviour of the shear stress; the greater is k_s, the broader is the distribution.

In this paper we combine a previous determination of G^* for two supercooled liquids with values of the longitudinal modulus $M^*(= M' + iM'')$. The latter can be obtained from measurements of the absorption coefficient α and the phase velocity C of ultrasonic waves using the equations (Herzfeld and Litovitz, 1959)

$$\begin{aligned} M' &= \rho C^2 \left\{ \frac{1 - \left(\dfrac{\alpha C}{\omega}\right)^2}{\left[1 + \left(\dfrac{\alpha C}{\omega}\right)^2\right]^2} \right\} \\ M'' &= \frac{2\rho C^2 \left(\dfrac{\alpha C}{\omega}\right)}{\left[1 + \left(\dfrac{\alpha C}{\omega}\right)^2\right]^2} \end{aligned} \tag{9}$$

The bulk modulus K^* can then be extracted by application of the relation (Herzfeld and Litovitz, 1959)

$$K^* = M^* - (4/3)\, G^* \tag{10}$$

This procedure is necessary because it is not possible to determine K^* directly in liquids at high frequencies.

EXPERIMENTAL RESULTS

The amplitude absorption coefficient and the phase velocity of ultrasonic longitudinal waves were measured for *sec*-butylbenzene and 3-phenylpropyl chloride using a conventional variable-path apparatus (Andreae *et al.*, 1958; Dexter, 1967). The phase velocity was determined at temperatures above the viscoelastic transition region by an interferometer technique at a frequency of 5.000 MHz. The velocities were measured in this region to $\pm$ 0·1%. In the viscoelastic transition region the absorption coefficient was too high to enable this method to be used. In this case the phase velocity of a pulse of ultrasound was determined by electronically cancelling the received signal with a reference signal of the same frequency while the path length was varied. The velocities were obtained in this way to $\pm$0·5% at 5·25 and 15·55 MHz. At very low temperatures where the liquid viscosities were too high to enable the path length to be varied, the velocity was determined indirectly from impedance measurements. Here the reflection coefficient at the interface between a quartz delay line and the liquid was measured and this enabled the velocities to be obtained to an accuracy of $\pm$1·5%.

The absorption coefficient α was measured at 5·25, 15·55, 25·25, 35·3 and 55·1 MHz. by comparing the amplitudes of pulses of ultrasound which had traversed different path lengths in the liquid with the amplitude of a comparison pulse which could be accurately varied with a calibrated piston attenuator. The slope of a plot of attenuator setting against ultrasonic path length gave α to $\pm$2%. The experimental results have been presented graphically (Dexter, 1970; Dexter and Matheson, 1971b). Some of the absorption results are shown in Figure 3 below.

The data for G^* were taken to be those determined for these liquids by Davies (1968). He measured the real part R_s of the mechanical shear impedance by the normal incidence reflection technique (McSkimin, 1964). The impedance Z_s^* is related to G^* by

$$Z_s^* = R_s + iX_s = (\rho G^*)^{\frac{1}{2}} \tag{11}$$

The measurements were performed at frequencies of 10 and 30 MHz, and at viscosities from 10^{-2} to 10^{12} N s m^{-2}. Davies' measurements showed that the behaviour of R_s is well described by the Barlow, Erginsav and Lamb (1967, 1969) empirical model. We have therefore assumed that the predictions of G' and G'' from this model were sufficiently accurate for the purposes of the present analysis.

DISCUSSION

Experimental values for the complex bulk modulus

The values of α and C were smoothed, interpolated and tabulated as functions of temperature together with the corresponding values of ρ, η_s, C_0 and G_∞ (Dexter, 1970). The low frequency limiting velocities C_0 were obtained in the viscoelastic transition region by linear extrapolation with temperature of the 5·0 MHz velocities in accordance with usual practice (Barlow, Lamb and Taşköprülü, 1969). The values of G_∞ were extrapolated upwards in temperature through the viscoelastic transition region using the linear temperature dependence of $J_\infty = 1/G_\infty$ as determined by Davies (1968).

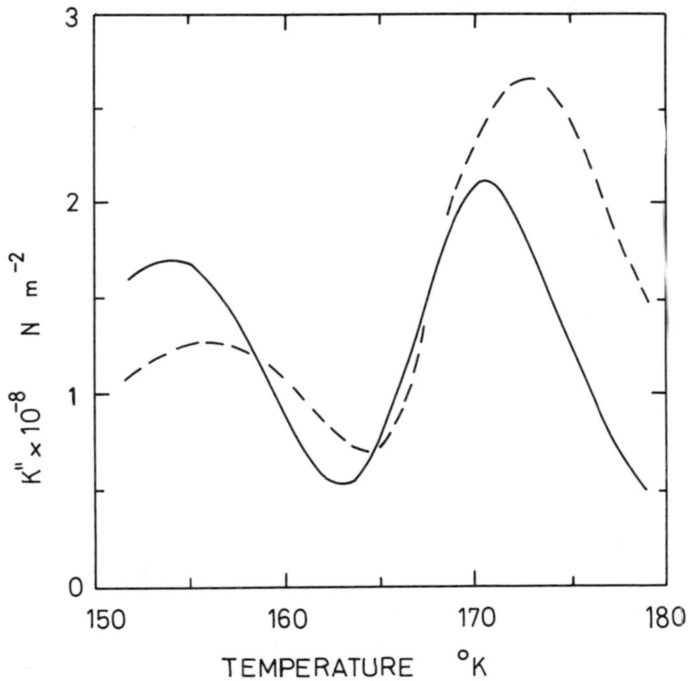

Figure 1. *Sec*-butylbenzene: temperature dependence of the imaginary part K'' of the bulk modulus in the viscoelastic dispersion region; ——, 5·25 MHz; – – –, 15·55 MHz

The smoothed tabulated data (Dexter, 1970) were inserted in equations (9) to give M' and M'' and into equations (7) and (8) to give G' and G''. The substitution of these values into equation (10) yielded the components of the bulk modulus ($K_0 + K'$) and K'' as functions of temperature. K'' for *sec*-butylbenzene is shown in Figure 1. The unusual feature of these results is the double-humped appearance of the graph of K'' against temperature. There is a corresponding dispersion in K' to match that in K'' (Dexter, 1970; Dexter and Matheson, 1971b). This dispersion continues up to at least $\eta_s =$

10^5 N s m^{-2} where all dispersion of G' is assumed to have ceased at these experimental frequencies. The results for 3-phenylpropyl chloride and *sec*-butylbenzene are qualitatively similar (Dexter, 1970).

The Barlow, Erginsav and Lamb viscoelastic model applied to the relaxing bulk modulus

A common argument in the literature of viscoelasticity is that if the same molecular motions are involved in the relaxation of both bulk and shear stresses through viscous flow processes, then the effective distribution of bulk relaxation times should be the same as the distribution of shear relaxation times. In this case, the Barlow, Erginsav and Lamb (1967, 1969) model which provides a good description of the relaxation of the shear modulus should also describe the relaxation of the bulk modulus. This model may be written out for the bulk modulus by replacing G^* by $(K^* - K_0)$, G_∞ by K_2, η_s by η_B, and k_s by a new width parameter k_B in equations (8). If the distributions of bulk and shear relaxation times are the same then

$$k_B = k_s$$

The model cannot be applied to the relaxation of $K^* - K_0$ unless K_2 and η_B are known. Here it is assumed that both K_2/G_∞ and η_B/η_s are constants and independent of temperature throughout the relaxation region. This allows a direct comparison to be made between the present results and those of Barlow, Lamb and Taşköprülü (1969) who made the same assumptions when analysing the ultrasonic results for a polyphenyl ether, bis-(*m*-(*m*-phenoxyphenoxy)phenyl) ether, in the viscoelastic transition region.

Both of these assumptions are probably sufficiently accurate over the small temperature range involved. The temperature dependence of the ratio η_B/η_s has been discussed fully elsewhere, (Dexter and Matheson, 1971b) and this ratio is likely to be sensibly independent of temperature throughout the viscoelastic region in these liquids. Recent theoretical studies (Dexter and Matheson, 1971a) have shown that K_∞/G_∞ is more likely to be independent of temperature than K_2/G_∞. However, since the temperature dependence of K_0/G_∞ is small over the temperature range involved, the choice of temperature independence of the ratio K_2/G_∞ makes an insignificant difference to the interpretation of the results.

If the temperature dependence of each of the relaxation times in a distribution is the same, then the normalized shear moduli when measured at different frequencies will all reduce onto a common curve when they are plotted against $\log(\omega\eta_s/G_\infty)$. The moduli resulting from equations (7) and (8) possess this property. If the assumptions that η_B/η_s and K_2/G_∞ are independent of temperature are valid, then

$$\log(\omega\eta_B/K_2) = \log(\omega\eta_s/G_\infty) + \text{constant}. \tag{12}$$

In this case the shape of the bulk modulus curve should be the same irrespective of whether it is plotted against log $(\omega\eta_B/K_2)$ or log $(\omega\eta_s/G_\infty)$, but with a slight shift along the frequency axis.

Since accurate K_∞ or K_2 data are not available because of the velocity dispersion still evident even at $\eta_s = 10^5$ N s m^{-2}, some estimated values (Dexter, 1970; Dexter and Matheson, 1971b) have been used to show the form of K'/K_2 and K''/K_2 when plotted against log $(\omega\eta_s/G_\infty)$. These curves are shown in Figure 2 together with G'/G_∞ and G''/G_∞ for comparison purposes. The higher temperature peak of Figure 1 is now on the left in Figure 2.

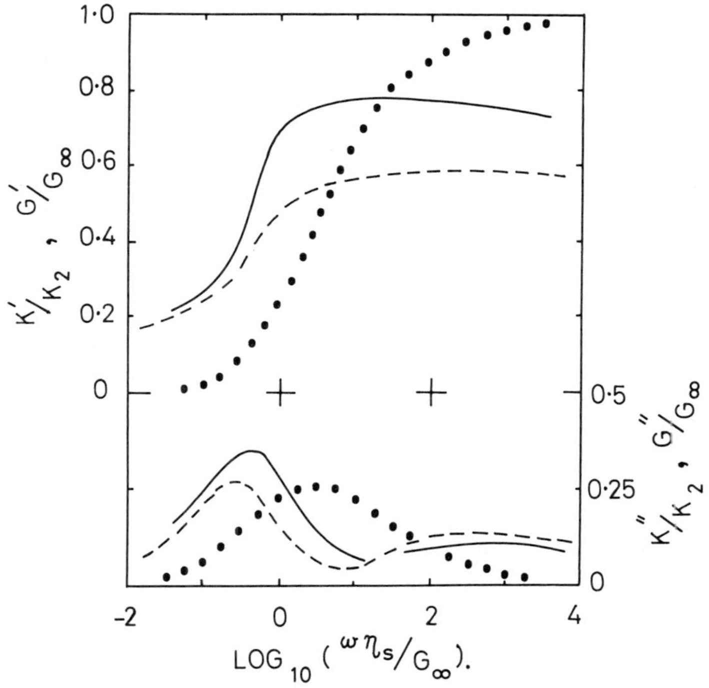

Figure 2. *Sec*-butylbenzene: comparison of the components of the reduced shear modulus (G'/G_∞, G''/G_∞) with the reduced bulk modulus (K'/K_2, K''/K_2); . . ., G'/G_∞ and G''/G_∞ calculated from the Barlow, Erginsav and Lamb model with $k_s = 0{\cdot}8$; - - -, 5·25 MHz and ——, 15·55 MHz values of K'/K_2 and K''/K_2

The Barlow, Erginsav and Lamb (1967) empirical model cannot fit results such as those for the bulk modulus shown in Figure 2, because the model produces only one peak in the imaginary part. We therefore conclude that there are two separate processes of bulk relaxation operative in these liquids and accordingly split the total relaxational modulus into two parts K_3 and K_4 such that

$$K_2 = K_3 + K_4$$

at a given temperature. The phrase 'at a given temperature' is important because the large variation in log $(\omega\eta_s/G_\infty)$ was obtained here by varying the temperature and thus the viscosity and not by varying the experimental frequency. We shall denote the lower frequency (higher temperature) part of K_2 by K_3, and the higher frequency (lower temperature) part by K_4.

Analysis of the relaxation of K_3

The Barlow, Erginsav and Lamb model was applied to the relaxation of K_3 because, judging by the shapes of the curves in Figure 2, K_3 had more in common with the relaxation of G_∞ than did K_4. The curve fitting for this process was performed by a least-squares technique (Dexter, 1970). This produced values for η_B/η_s, K_3/G_∞ and k_B for this process: these results are given in Table 1 together with the values of k_s determined by Davies (1968). The last column contains the ratio of bulk to shear relaxation times as determined from the ratio (Dexter, 1970; Dexter and Matheson, 1971b)

$$\frac{\tau_B}{\tau_s} = \frac{\eta_B\, G_\infty}{\eta_s\, K_3}. \tag{13}$$

Also included in the table for purposes of comparison are the results of Barlow, Lamb and Taşköprülü (1969) for bis-(*m*-(*m*-phenoxyphenoxy)-phenyl) ether.

3-phenyl propyl chloride is a very flexible molecule, *sec*-butyl benzene is slightly less so, whilst the polyphenyl ether is rigid. Davies (1968) pointed out that the activation energy for viscous flow in supercooled liquids is typically three times greater than that for molecular conformational changes. Flexible molecules can therefore change their shape in order to flow more easily, and Davies argued that this reduced the effective distribution of barrier heights to molecular flow and hence k_s. Table 1 clearly shows that the trend for k_B is the same as that for k_s which suggests that the same explanation holds for the relaxation of K_3 which occurs typically at

$$\tau_B = (4{\cdot}1 \pm 0{\cdot}8)\,\tau_s \tag{14}$$

The more flexible are the molecules in the liquid then the narrower is the distribution of relaxation times for this process, and the smaller are the fractions of K_2 and η_B which relax in this process. Corresponding increases occur in the amplitude of K_4.

To illustrate the excellent fit provided in this higher temperature region, Figure 3 compares the observed absorption coefficients with those calculated on the basis of G_∞ and K_3 relaxing with the parameters quoted in Table 1.

Table 1. Parameters of the Barlow, Erginsav and Lamb (1969) model for the relaxation of the higher temperature part K_3 of the bulk modulus

Liquid	k_s	k_B	η_B/η_s	K_3/G_∞	τ_B/τ_s
3-Phenylpropyl chloride	0·6	0·0	1·61	0·47	3·4
sec-Butylbenzene	0·8	0·4	3·86	0·79	4·9
A polyphenyl ether[b]	1·0	1·0	5·32	1·33[a]	4·1

[a] In this case $K_4 = 0$, and therefore $K_3 = K_2$. [b] Barlow, Lamb, and Tasköprülü (1969).

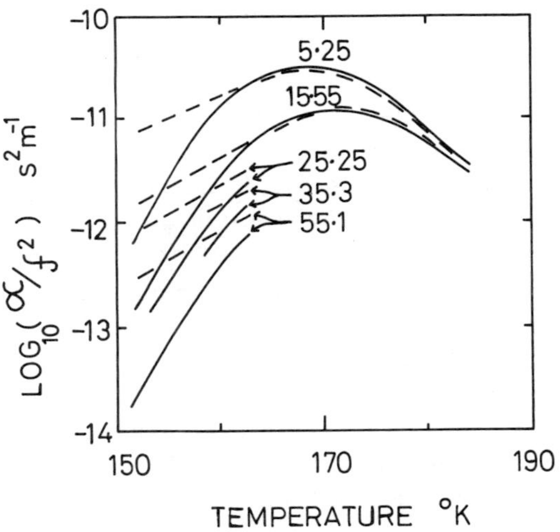

Figure 3. *Sec*-butylbenzene: observed (- - -) and calculated (——) ultrasonic absorption, the latter being obtained from the Barlow, Erginsav and Lamb model with $\eta_B = 3{\cdot}86\eta_s$ and $K_3 = 0{\cdot}79G_\infty$

The relaxation of K_4

In this low temperature region there are excess absorption coefficients remaining after those caused by the relaxation of K_3 have been subtracted off. This is illustrated in Figure 3. Corresponding dispersion of the velocity is observed (Dexter, 1970; Dexter and Matheson, 1971b). These excess absorptions are consistent within experimental error with a constant absorption per wavelength

$$\alpha\lambda = 2\pi(\alpha C/\omega) \tag{15}$$

which is equivalent to a constant absorption per cycle α/f if velocity dispersion is absent.

This type of behaviour has been observed at low temperatures in pentachlorobiphenyl, glycerol (Litovitz and Lyon, 1954), butanediol-1,3,2-methylpentanediol-2,4, hexanetriol-1,2,6 (Meister *et al.*, 1960), and several polyisobutylenes (Mason *et al.*, 1948). The frequency dependence of the ultrasonic absorption in most biological materials is also consistent with equation (15) (Goldman and Hueter, 1956).

Now the behaviour represented by equation (15) can be described in terms of a broad, flat distribution of relaxation times. It is important however to realize that such a distribution is not in itself a molecular mechanism, it is simply a method of describing the data. Two mechanisms have been proposed which would give rise to this kind of behaviour. These are hysteresis mechanisms (Dexter and Matheson, 1971b); Litovitz and Lyon 1954; Meister *et al.*, 1960; Mason *et al.*, 1948; Truell, Elbaum and Chick, 1969). and non-linear elasticity effects (Truell, Elbaum and Chick, 1969).

It is difficult to envisage a hysteresis mechanism occurring in liquids composed of such simple molecules as those considered here. On the other hand, theoretical studies (Dexter and Matheson, 1971a) have shown that non-linear effects are significant during compressive (but not shear) deformations of elastic liquids. This would lead to absorption of longitudinal ultrasonic waves but not of shear waves, as is observed. We therefore suggest that the region of dispersion caused by K_4 is caused by the relaxation of those intramolecular stresses which have not relaxed in the K_3 region and that non-linear effects also contribute to the ultrasonic absorption in this region. The internal molecular flexibility has a strong influence on K_4 as can be seen from Table 1. Further studies with differing ultrasonic amplitudes should show the relative importance of the molecular flexibility and non-linearity effects.

REFERENCES

Andreae, J. H., R. Bass, E. L. Heasell, and J. Lamb (1958). *Acustica*, **8**, 131.
Barlow, A. J., A. Erginsav, and J. Lamb (1967). *Proc. Roy. Soc.*, **A298**, 481.
Barlow, A. J., A. Erginsav, and J. Lamb (1969). *Proc. Roy. Soc.*, **A309**, 473.
Barlow, A. J., J. Lamb, and N. S. Taşköprülü (1969). *J. acoust. Soc. Am.*, **46**, 569.
Davies, D. B. (1968). *Ph.D. Thesis*, University of Essex, Colchester, England.
Dexter, A. R., (1967). *M.Sc. Dissertation*, University of Essex, Colchester, England.
Dexter, A. R. (1970). *Ph.D. Thesis*, University of Essex. Colchester, England.
Dexter, A. R., and A. J. Matheson (1971a). *J. chem. Phys.*, **54**, 203.
Dexter, A. R., and A. J. Matheson (1971b). *J. chem. Phys.*, **54**, 3463.
Goldman, D. E., and T. F. Hueter (1956). *J. acoust. Soc. Am.*, **28**, 35.
Gross, B. (1953). *Mathematical Structure of the Theories of Viscoelasticity*. Hermann et Cie., Paris.

Herzfeld, K. F., and T. A. Litovitz (1959). *Absorption and Dispersion of Ultrasonic Waves*. Academic Press, New York.

Litovitz, T. A., and T. Lyon (1954). *J. acoust. Soc. Am.*, **26**, 577.

Mason, W. P., W. O. Baker, H. J. McSkimin, and J. H. Heiss (1948). *Phys. Rev.*, **73**, 1074.

McSkimin, H. J. (1964). In W. P. Mason (Ed.), *Physical Acoustics*, Vol. 1A, Academic Press, New York.

Meister, R., C. J. Marhoeffer, R. Sciamanda, L. Cotter, and T. A. Litovitz (1960). *J. appl. Phys.*, **31**, 854.

Schwarzl, F. R., and L. E. C. Struik (1968). *Advances in molec. relax. Processes*, **1**, 201.

Truell, R., C. Elbaum, and B. B. Chick (1969). *Ultrasonic Methods in Solid State Physics*. Academic Press, New York.

7

Shear and Volume Structural Relaxation in Immiscible Oxide Melts

J. H. SIMMONS

Institute for Materials Research, U.S. National Bureau of Standards, Washington, D.C., U.S.A.

and P. B. MACEDO*

Vitreous State Laboratory, Catholic University of America, Washington, D.C., U.S.A.

Liquid–liquid phase separations occur in oxide glass systems and the phases can have very different compositions and thus widely different viscosities. Phase separation at the concentration which has the highest transition temperature, T_c (top of the immiscibility dome), occurs by means of a pseudo second-order phase transition which is characterized by the continuity of the free energy and its first derivatives across the transition boundary. As a consequence, the thermodynamically unstable fluctuations which lead to demixing below T_c, do not end abruptly when one raises the temperature to the critical point. These fluctuations, in fact, extend far into the supercritical ($T > T_c$) region, where they have well defined wavelengths and lifetimes. The transient domains resulting from these fluctuations have some structural relationship to the subcritical immiscible phases, and thus are characterized by large viscosity differences. The presence of these fluctuations in composition is expected to have large effects on the viscous flow processes of the materials.

Theoretical analyses, first derived to describe gas–liquid critical-point phenomena, were applied to liquid–liquid phase transitions by replacing the density fluctuations with fluctuations in composition (Fisher, 1964). Equations are thus available to describe the wavelengths, Λ, and lifetimes, τ, of the existent fluctuations as a function of temperature for $T > T_c$. The wavelength,

* Work done at Catholic University was supported by Office of Naval Research Contract #N000-14-68-A-0506-0002.

or range of fluctuations [equation (1)] varies from very small values far from the critical point, to macroscopic extent near T_c:

$$\Lambda = l/(T/T_c - 1)^{\frac{1}{2}} \tag{1}$$

The parameter l is a constant of the material and varies little among liquid–liquid non-polymeric systems (see Table 4 from Macedo *et al.*, 1968). Taking the average value, $l = 11$ Å, which is also the value measured for a $PbO \cdot B_2O_3 \cdot Al_2O_3$ glass (Zarzycki, 1968), we may plot the fluctuation wavelength, Λ, as a function of reduced temperature as will be shown later in Figure 2, and thus predict the typical size of the microstructure present in oxide systems above T_c. The lifetimes are found to be associated with the individual Fourier components $c(k, t)$ of the fluctuations, $c(r, t)$ which are space (r), and time (t) dependent:

$$c(r, t) = \int_{-\infty}^{\infty} c(k, t)\, e^{i\boldsymbol{k}\cdot\boldsymbol{r}}\, d\boldsymbol{k} \tag{2}$$

where k represents wave number. The lifetime, $\tau(k)$, of the k-th Fourier component is written as:

$$\tau(k) = \alpha\eta_s(0)/\{[a(T/T_c - 1) + bk^2]\, k^2\} \tag{3}$$

where $\eta_s(0)$ is the zero-frequency, or static shear viscosity and α is a proportionality constant. a and b are coefficients to the free energy expansion and are material constants. In order to extend this analysis to find a most probable or average fluctuation lifetime, a model must be proposed which gives a suitable distribution of Fourier components.

A number of theories have been proposed for the interaction of the supercritical $(T > T_c)$ fluctuations with the viscous flow processes. These models, sometimes called mode–mode coupling, formulated by Kawasaki (1966), and others, for the shear processes, and by Romanov and Solov'ev (1966), for the volume processes, propose that a relaxation region occurs in the modulus and the viscosity in addition to the structural relaxation. This additional relaxation results from direct energy exchange between the composition fluctuations and the flow processes. Figure 1 depicts the subsequent behaviour of the frequency-dependent viscosity, $\eta(\omega)$, in view of these two relaxation regions. Since the fluctuations are expected to have long lifetimes near T_c, we have located the anomalous fluctuation region, in Figure 1, at lower frequencies than the structural relaxation. It is evident, then, that, according to these models, the static viscosity value extrapolated from the ultrasonic data, η_u, corresponds to the expected non-critical material behaviour, η_0, and will be different from the measured value, $\eta_m(0)$, by an excess viscosity, $\Delta\eta(0)$, defined as:

$$\Delta\eta(0) \equiv \eta_m(0) - \eta_0 = \eta_m(0) - \eta_u \tag{4}$$

where $\eta_m(0)$ corresponds to the shear (η_s), or volume (η_v) viscosity. Both theories predict a divergence in $\Delta\eta(0)$ at T_c.

The average fluctuation lifetime $\overline{\tau_f}$ may be obtained by considering $\Delta\eta_v$, since $\Delta\eta_v \equiv \Delta K\overline{\tau_f}$ where ΔK is the difference between the instantaneous and static moduli.

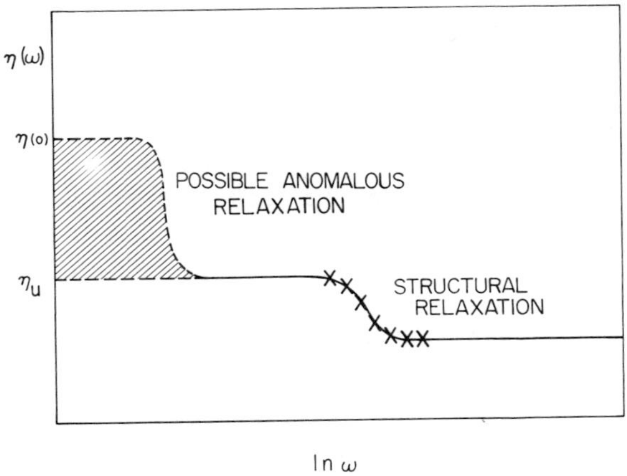

Figure 1. Frequency-dependent viscosity (shear or volume) at $T > T_c$, showing the structural relaxation region and the possible anomalous relaxation resulting from composition fluctuations. The crosses depict the range of ultrasonic frequencies and thus the information accessible to ultrasonic spectroscopy

The static, volume excess viscosity is given by Romanov and Solov'ev (1966) as:

$$\Delta\eta_v(0) = K_0 K_\infty \frac{1{\cdot}44\, k_B T}{8\pi^2 D l}\left[\frac{(\partial^2 v/\partial c^2)_{P,T}}{(\partial^2 G/\partial c^2)_{P,T}}\right]^2 \tag{5}$$

where K_0 and K_∞ are the static and instantaneous bulk moduli; $D = (T/T_c - 1)/\eta_s(0)$ is the interdiffusion coefficient, and G is the free energy which yields $(\partial^2 G/\partial c^2) = a(T/T_c - 1)$. v is the specific volume and $(\partial^2 v/\partial c^2)$ is a constant of the material. $\Delta\eta_v(0)$ thus diverges as $\eta_s(0)/(T/T_c - 1)^3$. The average composition fluctuation lifetime is $\overline{\tau_f} = \Delta\eta_v(0)/K_2$. The parameter K_2 is the strength of the relaxation $[\Delta K/K_0 K_\infty]$ and diverges as $(T/T_c - 1)^{-1}$. Therefore, the fluctuation lifetime is expected to diverge as:

$$\overline{\tau_f} = \alpha_0 \eta_s(0)/(T/T_c - 1)^2 \tag{6}$$

where α_0 is a constant of the material. [see also Garland (1970)]

Shear static viscosity measurements were recently conducted (Simmons, Napolitano *et al.*, 1970) on four doped sodium borosilicate mixtures with similar high temperature structures, but widely different critical temperatures.

The non-critical viscosity η_0 was represented by the data from one of the samples whose critical temperature was respectively 109, 187 and 98 °C lower than the three others. A large static excess shear viscosity, $\Delta\eta_s(0) \equiv \eta_s(0) - \eta_0$ was observed which appeared far above the critical temperature ($T/T_c \sim 1{\cdot}16$). The excess viscosity however approached a finite limit ($\Delta\eta_s \sim 2\eta_0$) near T_c, with no appearance of a divergence. The same behaviour was observed for all samples when the normalized excess shear viscosity $\Delta\eta_s(0)/\eta_s(0)$ was plotted as a function of reduced temperature, showing that the effect is indeed a critical point phenomenon. The lack of divergence at T_c indicates disagreement with the shear mode–mode coupling models.

In order to investigate the origin of this critical viscosity increase and the mechanisms of interaction between the fluctuations and the flow processes, structural relaxation measurements were undertaken in the same oxide mixtures. The results from transverse ultrasonic measurements conducted on the same four samples, and from longitudinal ultrasonic measurements conducted on one of these samples are reported here.

RELAXATION SPECTROSCOPY

General equations

The equations presented below are necessary to the analysis and are given for both the shear and volume relaxations. Their application to the proper mode can be effected by considering the appropriate normalization constants. The particular analysis of the modulus used below is chosen because it leads to the confinement of critical effects to one parameter, allows comparison of the results with other models, and leads to further physical interpretation of the parameters in a model proposed by the present authors (1971). Writing the frequency-dependent viscosity, $\eta(\omega)$ in terms of a distribution of relaxation times, $g(\ln \tau)$, we find (Herzfeld and Litovitz, 1959):

$$\eta(\omega) = \Delta m \left[\int_{-\infty}^{\infty} \tau/[1 + (\omega\tau)^2]\, g(\ln \tau)\, \mathrm{d} \ln \tau \right] \tag{7}$$

where τ is the volume or shear relaxation time, and Δm is the instantaneous modulus G_∞ for shear waves or the difference between the instantaneous and static moduli ΔK for volume relaxation. The zero-frequency limit of $\eta(\omega)$ is written as:

$$\eta(0) = \Delta m \int_{-\infty}^{\infty} \tau\, g(\ln \tau)\, \mathrm{d} \ln \tau \tag{8}$$

If we represent $g(\ln \tau)$ by the structural relaxation distribution only, then equation (8) yields the extrapolated value of η_u for the static viscosity, as shown in Figure 1. If a low frequency relaxation effect as predicted by the mode–mode coupling models does indeed occur, then an additional excess viscosity will be present.

Shear relaxation

Shear ultrasonic measurements were conducted on the four sodium-borosilicate mixtures. The discussion of the data and method of analysis is quite lengthy and is reported elsewhere (Simmons and Macedo, 1970). The measurements yielded values of G_∞, and relaxation time distributions, $g_s(\ln \tau)$ with their maxima at τ_s' for the four samples over a combined reduced temperature range of $1{\cdot}004 \leqslant T/T_c \leqslant 1{\cdot}81$. The shear distributions of relaxation times, $g_s(\ln \tau)$, were best represented by log-Gaussian functions of time for all temperatures down to $T/T_c = 1{\cdot}004$.

For a log-Gaussian distribution of relaxation times, the extrapolated shear viscosity, η_u, can be calculated from the ultrasonic parameters, G_∞, τ_s', and the variance of the distribution, θ_s, by the equation:

$$\eta_u = G_\infty \tau_s' \exp(\theta_s/2) \tag{9}$$

Comparison of the results, $\eta_u(T)$ (Simmons and Macedo, 1970) with the measured static viscosity values, only exhibited differences smaller than 10% which are within the experimental uncertainty of the ultrasonic data. Since η_u and $\eta_s(0)$ have the same values, the possibility of a low frequency relaxation in the shear viscosity, resulting from some direct energy exchange of the flow processes with the fluctuations in composition above T_c, is eliminated.

While we have discounted the mode–mode coupling model as a possible cause for the excess shear viscosity in these oxide mixtures, our static viscosity measurements have shown that an anomaly is indeed occurring in the oxide mixtures tested. Therefore the entire critical effect is contained in the structural relaxation and the parameters characterizing it. In order to investigate the interaction mechanism, the three parameters of equation (9) must be analysed to see which exhibits the associated critical-point effects.

The shear rigidity, G_∞, and the most probable relaxation time, τ_s' did not appear to be subject to critical point effects. The width of the distribution of shear relaxation times, however, demonstrated large changes as a function of reduced temperature (see Figure 2). The samples exhibited narrow distributions at high temperatures, followed by drastically broadening distributions as the critical point was approached. However, as the temperature was further lowered toward the critical temperature, the broadening saturated indicating a non-divergent width at T_c. The large increase in the width was associated with the viscosity anomaly, and represented some interaction between the composition fluctuations and the shear viscous relaxation process. Since the distributions were log-Gaussian, the critically-induced broadening resulted from the appearance of relaxation times shorter than the most probable, as well as longer ones. The appearance of these short relaxation times suggests that the interactions occurring with the fluctuations are vastly different from those proposed earlier. In fact we note that upon phase separation, the material decomposes into phases which are respectively more and less viscous than the homogeneous sample. The distributions then suggest a connection of the slow and fast relaxation times with the more and less viscous fluctuations. In other words, the distribution of relaxation times appears to originate from the existing critical-point-induced distribution

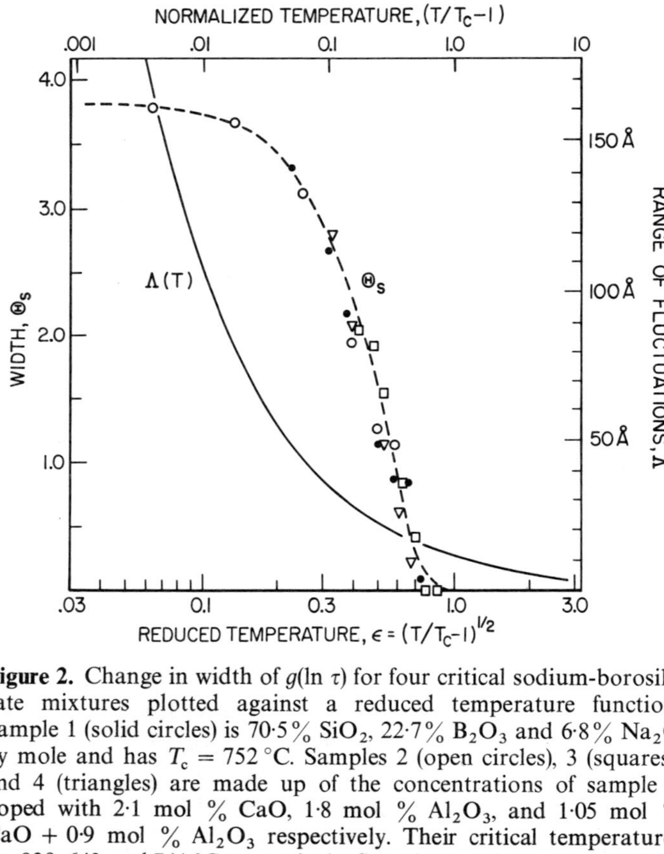

Figure 2. Change in width of $g(\ln \tau)$ for four critical sodium-borosilicate mixtures plotted against a reduced temperature function. Sample 1 (solid circles) is 70·5% SiO_2, 22·7% B_2O_3 and 6·8% Na_2O by mole and has $T_c = 752\ ^\circ C$. Samples 2 (open circles), 3 (squares), and 4 (triangles) are made up of the concentrations of sample 1 doped with 2·1 mol % CaO, 1·8 mol % Al_2O_3, and 1·05 mol % CaO + 0·9 mol % Al_2O_3 respectively. Their critical temperatures are 830, 643 and 741 °C respectively. Samples 2 and 3 demonstrated a high temperature residual width which was non-critical. This figure also shows the fluctuation size Λ as a function of reduced temperature

of microstructure. The saturation of the broadening very close to T_c provides a further clue to the interaction as it indicates that the latter weakens when the fluctuations grow to macroscopic sizes and Λ diverges. This result is consistent with the lack of divergence of the viscosity.

Volume relaxation

Longitudinal ultrasonic measurements were conducted on an oxide mixture (70·5 % SiO_2, 22·7 % B_2O_3 and 6·8 % Na_2O) with a critical temperature of 752 °C, (Haller *et al.*, 1970) from $T/T_c = 1{\cdot}1$ to 1·58, in order to investigate the interaction between the volume-sensitive fluctuations and the volume relaxation mechanisms, and to possibly calculate the average fluctuation lifetimes as a function of temperature. The parameters, ΔK, τ'_v, and θ_v, are found by subtracting 4/3 of the shear modulus from the longitudinal modulus, and then fitting the resulting complex volume modulus with a distribution of relaxation times. As in the case of the shear measurements, the distributions were log-Gaussian in time. The available frequency range,

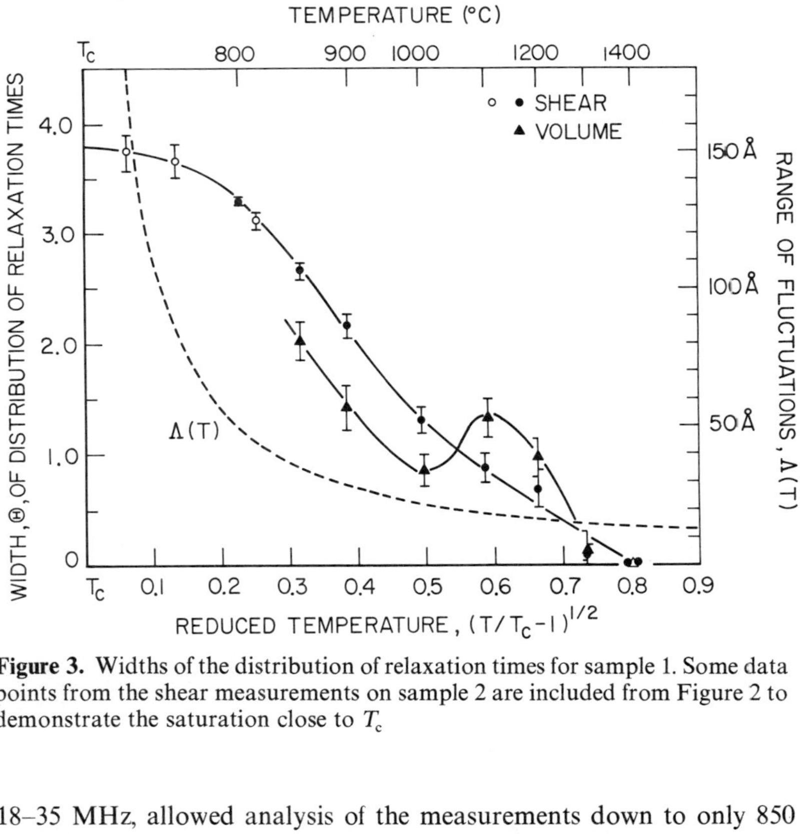

Figure 3. Widths of the distribution of relaxation times for sample 1. Some data points from the shear measurements on sample 2 are included from Figure 2 to demonstrate the saturation close to T_c

0·18–35 MHz, allowed analysis of the measurements down to only 850 °C ($T/T_c \sim 1{\cdot}1$). The width of the distribution of volume relaxation times is plotted in Figure 3 together with the shear data.

Two distinct critical-point effects appear in the behaviour of the width, θ_v. Below 1000 °C, the width follows closely the behaviour observed for the shear distribution of relaxation times. A sudden increase in width appears between 1000 and 1200 °C. This effect corresponds to the additional mode–mode coupling relaxation proposed by Romanov and Solov'ev (1966) and is a result of the overlap of the structural and the fluctuation relaxation regions (see Figure 1). As $\overline{\tau_f}$ approaches τ'_v, the ultrasonic experiment spans both relaxations and responds with an anomalous increase in width. Due to the frequency limitations of the apparatus, and the relatively low strength of the anomalous fluctuation relaxation, the two regions however could not be distinguished. Locating the mode–mode coupling relaxation near τ'_v between 1100 and 1200 °C, allows calculation of α_0 in equation (6) and yields the average fluctuation times, $\overline{\tau_f}$, as a function of temperature. Figure 4

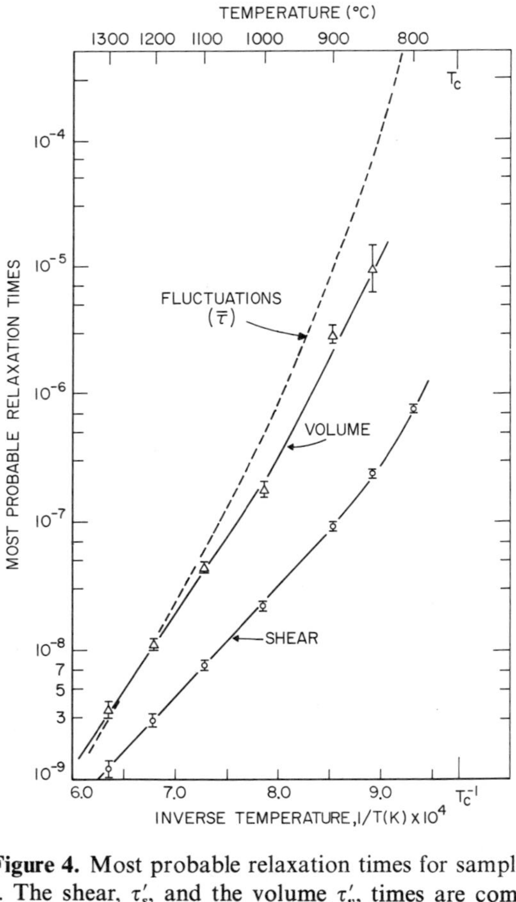

Figure 4. Most probable relaxation times for sample 1. The shear, τ'_s, and the volume τ'_v, times are compared to the average fluctuation time $\overline{\tau_f}$ as a function of inverse temperature

shows the three characteristic relaxation times τ'_s, τ'_v, and $\overline{\tau_f}$ as a function of inverse temperature. The average fluctuation lifetime is comparable to τ'_v in the temperature range 1100 to 1200 °C, but becomes longer for temperatures below 1000 °C. In the latter case, the fluctuation lifetimes have increased enough to locate the volume mode–mode coupling relaxation beyond the range of the ultrasonic equipment. The data, therefore, only reflects the volume relaxation, which now follows the behaviour of the shear process. Since $\overline{\tau_f} > \tau'_v$, the interaction occurring below 1000 °C is between the structural relaxation mechanism and a quasi-static distribution of local environments arising from fluctuations in composition between phases of widely different viscosities, as in the shear relaxation process at all temperatures ($\tau'_s < \overline{\tau_f}$). The distribution of volume structural relaxation times therefore appears to undergo the same broadening as the shear distribution.

A comparison between η_{uv} and $\eta_v(0)$ similar to the shear viscosity is not possible since the static volume viscosity is not known. It is expected, however, that the volume mode–mode coupling relaxation which appears in our analysis near 1100 and 1200 °C will occur at longer times for lower temperatures ($\overline{\tau_f} = 0{\cdot}59$ and 15 sec at 5 and 1 °C above T_c respectively). In fact, the equation of Romanov and Solov'ev predicts that $\overline{\tau_f}$ and $\Delta\eta_v(0)$ will diverge at T_c.

CONCLUSION

Analysis of the structural relaxation processes in immiscible glasses by ultrasonic spectroscopy yields information about the interaction of supercritical fluctuations in composition with the viscous flow mechanisms. A direct energy exchange resulting in an additional relaxation region in the frequency-dependent viscosities characterizing the material is found to occur only in the distribution of volume relaxations. This effect is detected in the analysis of the width of the distribution of volume structural relaxation times at high temperatures where the characteristic fluctuation time, $\overline{\tau_f}$, is of the order of the volume structural relaxation times, centred at τ'_v. At lower temperatures $\overline{\tau_f}$ is larger than τ'_v so that the fluctuations appear in a quasi-arrested state and lead only to temperature-dependent changes in the local environment of the flow species. At all temperatures investigated the shear structural relaxation times, centred at τ'_s, are shorter than $\overline{\tau_f}$ and τ'_v.

The results show that a large increase in the width, θ, of the distribution of relaxation times occurs in the shear and volume processes when the composition fluctuations appear quasi-arrested. Analysis of the behaviour of θ_s as a function of reduced temperature indicates that the broadening is a critical point effect, and suggests that it results from the microstructure in the melt caused by the fluctuations. This interaction does not involve energy exchange and should only depend upon the range of the fluctuations. The shear distribution is seen to be narrow at high temperatures and broadens drastically as T_c is approached. The broadening then saturates for temperatures very close to T_c. These three distinct regions in the behaviour of θ_s as a function of reduced temperature agree qualitatively with a model proposed by these authors (Macedo *et al.*, 1968). At temperatures below 1000 °C, where $\overline{\tau_f} > \tau_v'$, the distribution of structural volume relaxation times appears to follow the behaviour of the shear case.

REFERENCES

Fisher, M. E. (1964). *J. math. Phys.*, **5**, 944.

Fixman, M. (1962). *J. chem. Phys.*, **36**, 310.

Garland, C. W., D. Eden, and L. Mistura (1970). *Phys. Rev. Letters*, **25**, 1161.

Haller, W. K. D., D. H. Blackburn, F. E. Wagstaff, and R. J. Charles (1970). *J. Am. Ceram. Soc.*, **53**, 34.

Herzfeld, K. E., and T. A. Litovitz (1959). In *Absorption and Dispersion of Ultrasonic Waves*. Academic Press, New York and London, pp. 447–456.

Kawasaki, K. (1966). *Phys. Rev.*, **150**, 291.

Macedo, P. B., J. H. Simmons, and W. Haller (1968). *Physics Chem. Glasses*, **9**, 156.

Romanov, V. P., and V. A. Solov'ev (1966). *Collection: The Structure and Role of Water in the Living Organism* [in Russian], pp. 36–48.

Simmons, J. H., and P. B. Macedo (1970). *J. chem. Phys.*, **53**, 2914.

Simmons, J. H., and P. B. Macedo (1971). *J. chem. Phys.*, **54**, 1325.

Simmons, J. H., A. Napolitano, and P. B. Macedo (1970). *J. chem. Phys.*, **53**, 1165.

Zarzycki, J., and F. Naudin (1968). *C. r. hebd. Séanc. Acad. Sci. Paris*, **266**, 1005.

8

The Dynamic Viscoelastic Properties of some Non-Crystalline Metals

A. EISENBERG and S. REICH*

Department of Chemistry, McGill University, Montreal, Canada

It is known that rapid quenching of some alloys in the vicinity of the eutectic composition produces non-crystalline metals; (Duwez, 1967). Splat cooling techniques, developed by Duwez (1963) in which cooling rates of the order of 10^6 °C/sec are achieved, have been used to produce non-crystalline metallic phases in binary alloys, for example Au–Si (25 at. % Si); (Klement, Willens and Duwez, 1960; Chen and Turnbull, 1967), Pd–Si (20 at. % Si); (Chen and Turnbull 1969), and Pd–Ge (20 at. % Ge). Ternary glasses were also studied, for example: Pd Au Si; Pd Ag Si; Pd Cu Si. H. S. Chen and D. Turnbull (1967) showed the glassy nature of those materials, from ΔC_p versus temperature curves that exhibited glass transition behaviour (ΔC_p being the difference in specific heat in the amorphous and equilibrium crystalline states). This transition was also shown in specific conductivity measurements and qualitatively demonstrated by X-ray diffraction and transmission electron microscopy which indicated the onset of crystallization above T_g; (Chen and Turnbull, 1969). Viscosity measurements; (Chen and Turnbull, 1968) indicated $\eta = 10^{13}$ p at the glass transition.

The purpose of this study was to determine whether non-crystalline metals exhibited viscoelastic properties similar to those of non-metallic, non-crystalline materials such as polymers. Owing to limitations of sample size the vibrating reed technique was chosen. Most of the studies were made on $Pd_{80}\,Si_{20}$, although some work was done on $Pd_{80}\,Au_{11}\,Si_9$ and $Pd_{16}\,Au_{66}\,Si_{18}$.

* Present Address: The Weizmann Institute, Rehovot, Israel.

EXPERIMENTAL TECHNIQUES AND RESULTS

Sample preparation

Pure palladium (99·9995%) sponge, 99·99 + silicon and 99·9999 gold were used. The metals were mixed in a quartz tube, [total weight ~ 3 gm ±0·0005 gm, precision of composition ~1%]. The tube was evacuated, degassed and sealed under vacuum. The alloying was done at 1,100 °C. Samples were prepared by splatting loads of ~ 50 mg through a shock tube onto a copper block partly immersed in liquid nitrogen. The profile of the substrate block was stairlike so 5–8 thin reeds were produced from each splat. The reeds were 1·5 mm wide, 5–15 mm long and 20–120 microns thick. The spats were checked for crystallinity by X-ray diffraction; they gave diffuse rings below the glass transition temperature. Initiation of randomly oriented crystallites was observed in a characteristic asterism pattern upon storage above T_g.

A vibrating reed apparatus with a laser beam detection system was used to measure the mechanical loss as a function of temperature at the frequencies of the order of 100 cps. This apparatus is described in detail elsewhere; (Reich and Eisenberg, 1970).

Pd_{80} Si_{20}

Tan δ was measured as a function of temperature between 150 °K and 720 °K. Although great care was taken to prepare different splats in an identical manner, a wide range of tan δ *vs.* temperature profiles was obtained. The figures presented here are representative characteristic curves of a large number of runs.

In some samples a very narrow peak was observed (~ 10 °K in width) at the glass transition temperature as shown in Figure 1. The relaxation which occurs at 649 °K corresponds within 6 °K to the glass transition temperature for Pd_{80} Si_{20} as measured by Chen and Turnbull (1969) by D.S.C. A ten fold increase in the value of E'' is observed at T_g, while the real part of the modulus (E') is of the order of 8×10^{10} dyn/cm^2 up to T_g and also above it. The constancy in the value of E' across the transition region is probably due to the low intensity of the transition. This behaviour differs from the ordinary organic or inorganic glasses where a drop of 3–4 orders of magnitude is observed in the value of E' in going through the transition region. In addition to the glass transition peak a secondary relaxation at lower temperature is observed.

A sharply defined T_g peak is not, however, characteristic of all the samples which were studied. In some samples a broad peak is observed, as shown in Figure 2, while in still other samples the glass transition, though present, is very poorly defined. An example is shown in Figure 3. A multi-

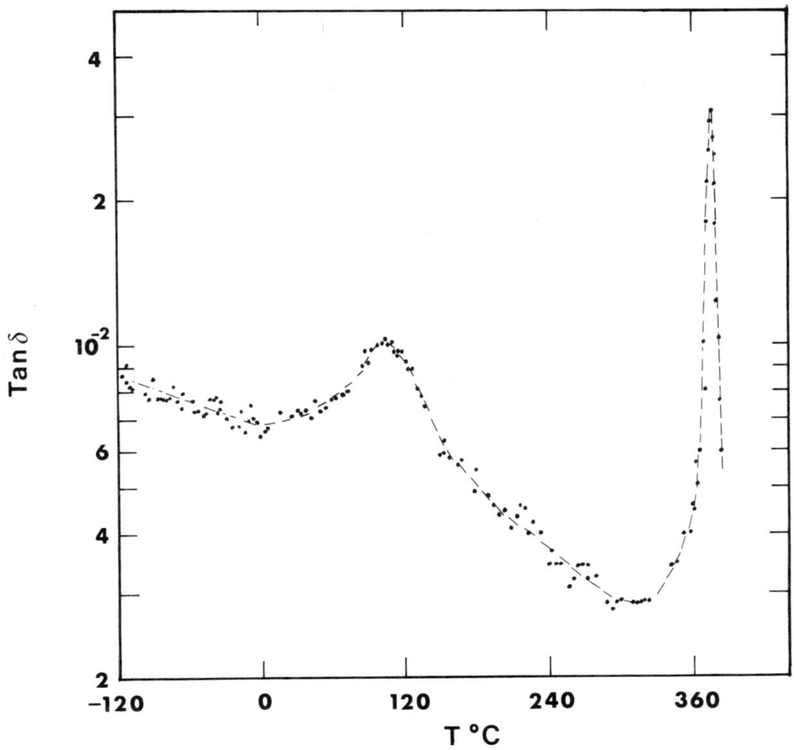

Figure 1. Tan δ *vs.* temperature for Pd_{80} Si_{20} glass measured at 88 cps

plicity of secondary peaks is observed in some samples, as shown in Figure 4.

The poor reproducibility of the mechanical loss curves is perhaps explained best in terms of phase separation occurring in the samples upon splatting, a phenomenon which would account for the different inelastic characteristics of different splats. The multiplicity of secondary relaxations is probably real and due to different compositions of the microphases. It may, however, be due to mechanical imperfections like microcracks in the samples; this could also affect the shift in the plateau value of tan δ in the range 10^{-2}–10^{-3} for different splats.

Pd_{80} Au_{11} Si_9 and Pd_{16} Au_{66} Si_{18}

The ternary alloys rich in palladium exhibit similar behaviour to the $Pd_{80}Si_{20}$ composition. A glass transition peak at 642°K is observed as shown in Figure 5. In these alloys, the reproducibility of mechanical properties is

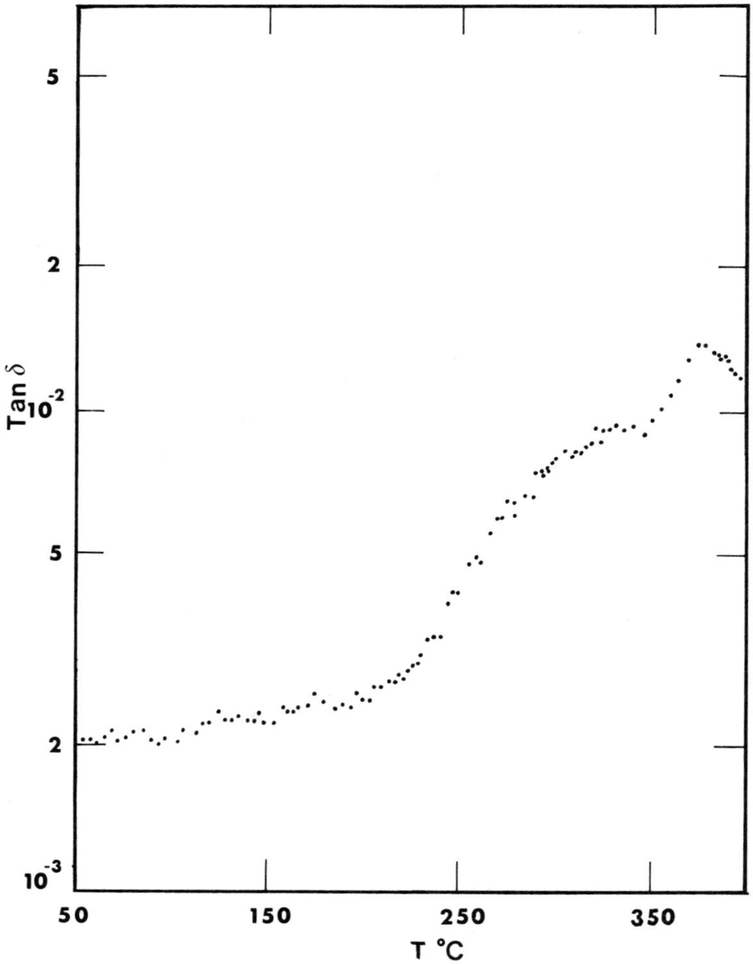

Figure 2. Tan δ *vs.* temperature for $Pd_{80}Si_{20}$ glass measured at 73 cps

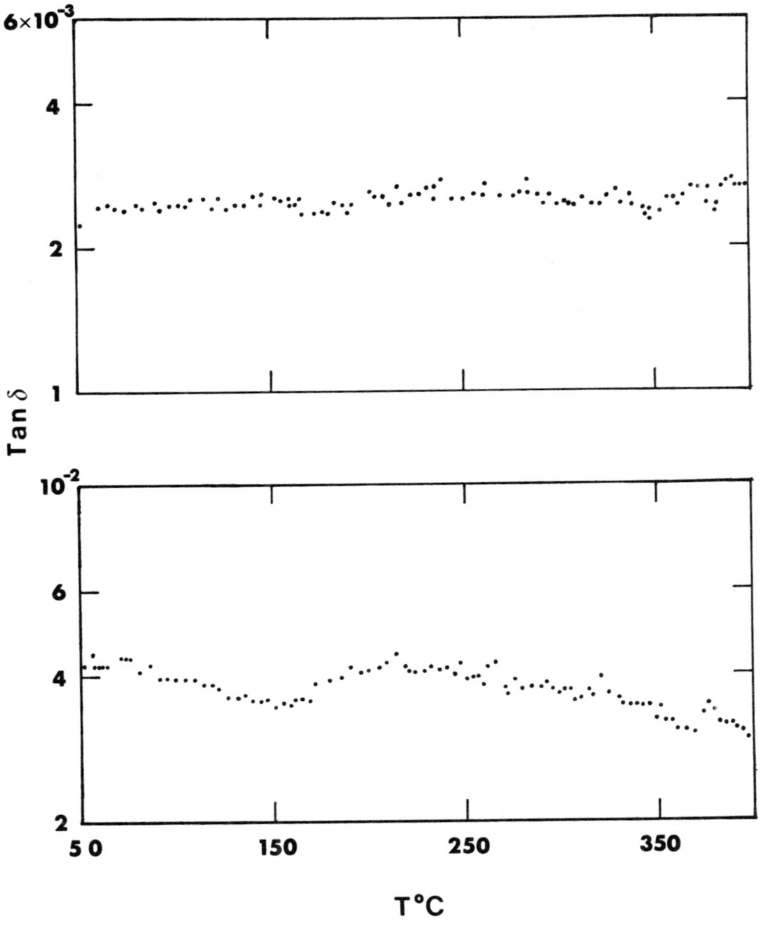

Figure 3. Tan δ *vs.* temperature for two different splats of $Pd_{80}Si_{20}$. Top curve measured at 165 cps. Bottom curve measured at 96 cps

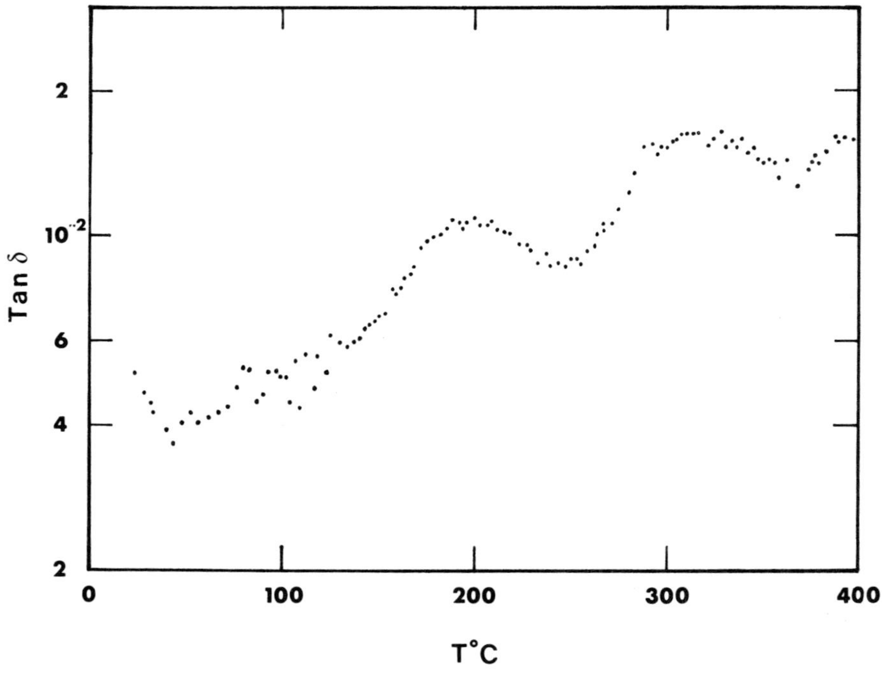

Figure 4. Tan δ *vs.* temperature for $Pd_{80}Si_{20}$ glass measured at 80 cps

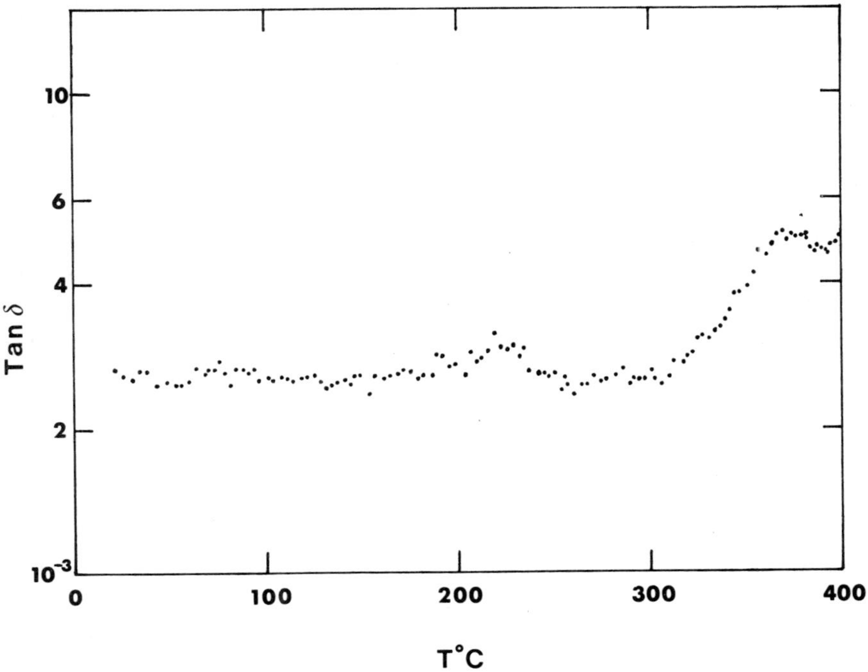

Figure 5. Tan δ *vs.* temperature for $Pd_{80}Au_{11}Si_9$ measured at 106 cps

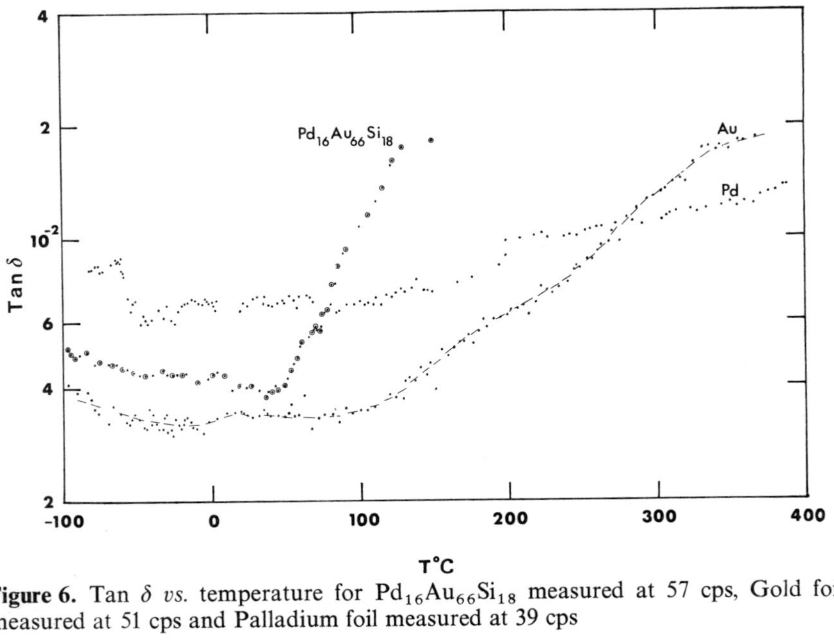

Figure 6. Tan δ *vs.* temperature for $Pd_{16}Au_{66}Si_{18}$ measured at 57 cps, Gold foil measured at 51 cps and Palladium foil measured at 39 cps

far better than in the Pd_{80} Si_{20} alloys, indicating their greater stability. Ternary alloys rich in gold, namely Pd_{16} Au_{66} Si_{18}, are unstable at room temperature and change colour from silver at liquid nitrogen temperature, to gold at room temperature. The kinetics of this process is, however, slow and takes at least 24 hours to be detected by eye. The dynamic curve for this splat is presented in Figure 6. In addition, the curves for pure gold and pure palladium metals are presented. The glass transition temperature for this alloy as detected by Chen and Turnbull (1969) using D.S.C. is 300 °K. We observe a steep increase in the mechanical loss around this temperature followed by a plateau which corresponds to the plateau observed in pure palladium and pure gold at high temperatures. This result suggests that crystallization occurs in these samples immediately after the glass transition. Such a behaviour could in fact explain the poor stability of the Pd_{16} Au_{66} Si_{18} glasses at room temperature.

CONCLUSION

Non-crystalline metals exhibit mechanical properties in the glass transition region which are very similar to those of non-metallic, non-crystalline materials except for the low intensity of the transition. Secondary peaks are probably also present, but their position and even magnitude may be obscured by mechanical imperfections of the samples. In general, reproducibility of mechanical properties from sample to sample is quite poor, possibly as a result of microphase separation upon splatting. It is noteworthy that in the Pd_{80} Si_{20} system, which exhibits a multiplicity of secondary transitions, reproducibility is quite poor, while in the Pd_{80} Au_{11} Si_9 system, in which the reproducibility is considerably better, the shape of the tan δ *vs.* temperature curve is quite flat. These two phenomena may be related. Thus, although the amorphous metals present a very tempting model for the study of the metastable glassy state, a good characterization of their inelastic properties can be accomplished only after a much better reproducibility in sample preparation techniques has been achieved.

ACKNOWLEDGEMENT

It is a pleasure to acknowledge the benefit of discussions with Professor D. Turnbull and Mr. D. Polk, who also kindly supplied several samples for a check of reproducibility. We also wish to thank Miss Polatayko and Mr. M. Kwan for their assistance with some of the experimental measurements. This work was supported, in part, by the National Research Council (Canada) and in part by the Office of Naval Research (USA).

REFERENCES

Chen, H. S., and D. Turnbull (1967). *Appl. Phys. Lett.*, **10**, 284.
Chen, H. S., and D. Turnbull (1967). *J. appl. Phys.*, **38**. 3646.
Chen, H. S., and D. Turnbull (1968). *J. chem. Phys.*, **48**, 2560.
Chen, H. S., and D. Turnbull (1969). *Acta metall.*, **17**, 1021.
Duwez, P. (1967). *Trans. ASM*, **60**, 606.
Duwez, P., and R. H. Willens (1963). *Trans. AIME*, **227**, 362.
Klement, W. Jr., R. H. Willens, and P. Duwez (1963). *Nature*, **187**, 869.
Reich, S., and A. Eisenberg (1970). *Rev. scient. Instrum.*, **41**, 1904.

9

Relaxations in Polymers at Low Temperatures

S. MILLER III

Department of Physics and Astronomy

M. TOMOZAWA and R. K. MacCRONE

Materials Division, Rensselaer Polytechnic Institute, Troy, New York, U.S.A.

The low-frequency vibrational mode spectrum of amorphous polymers has been shown to deviate strongly from the simple Debye spectrum. Specific heat measurements by Choy, Hunt, and Salinger (1970) on polymethyl methacrylate (PMMA) and polystyrene (PS), as well as by Reese (1965, 1966) and Noer (1959) show a value in excess of that expected from the Debye theory. This excess specific heat is assumed due to the presence of additional low frequency vibrational modes. It is to be expected that the presence of additional modes will affect other physical properties of the material. In this paper we present some low temperatures dielectric and anelastic measurements on representative polymers (PMMA and PS) which reveal the presence of several relaxation processes. We discuss the relation of these relaxations to the additional vibrational modes present in the material, and show theoretically, using models which may be easily generalized, how a vibrating polar group, mechanically coupled to the bulk through a Grüneisen constant, can give rise to relaxations of this kind.

Amorphous polymers are not the only amorphous materials which exhibit excess specific heat: excess specific heat has been observed in vitreous silica by Flubacher *et al.* (1959), in vitreous germania by Antoniou and Morrison (1965), and in glycerol glass by Craig *et al.* (1965). In all cases the excess specific heat is attributed to the presence of additional Einstein oscillators. In some cases, the presence of these low frequency vibrational modes is inferred from Raman spectroscopy (Flubacher *et al.*, 1959) and Krause and Kurkjian (1968), discussing amorphous oxides, point out that a broad mechanical loss peak ($\approx$ 80 °K) is also apparently characteristic of these materials. Fine *et al.* (1954) suggested that a characteristic mechanical loss peak observed in vitreous silica was due to some unusual vibrational mode present, and recently Scott and MacCrone (1970) have presented a detailed account of the relation between the low temperature dielectric and anelastic relaxations observed in this material.

Theoretical calculations on the vibrational mode spectrum of ideal simple amorphous materials are of interest. Dean (1960) has shown that the vibrational spectra of disordered solids are not smooth, and Rosenstock and McGill (1968) conclude that peaks in the vibrational spectra are to be expected. These 'excess' vibrational modes are supposed due to particular geometric groups of atoms which may vibrate almost independently of the surroundings. An earlier model by Rosenstock (1962) simply assumed the presence of a number of small masses attached, through weak springs, to an almost rigid lattice. This model is particularly appealing in the case of polymers, where the wagging or torsional motion of pendant groups or chain ends in various sized voids is a distinct possibility.

EXPERIMENTAL RESULTS

The dielectric measurements were carried out as described by Dominik *et al.* (1967). The anelastic measurements were obtained using a vibrating reed method previously described by Scott *et al.* (1968).

Figure 1 shows the capacitance and loss of PMMA as a function of frequency at 4·2 °K. The dispersive character is well marked. A distribution of relaxation times is involved, since the width of the tan δ *vs.* frequency curve for example is too broad to be described by a single relaxation time. The

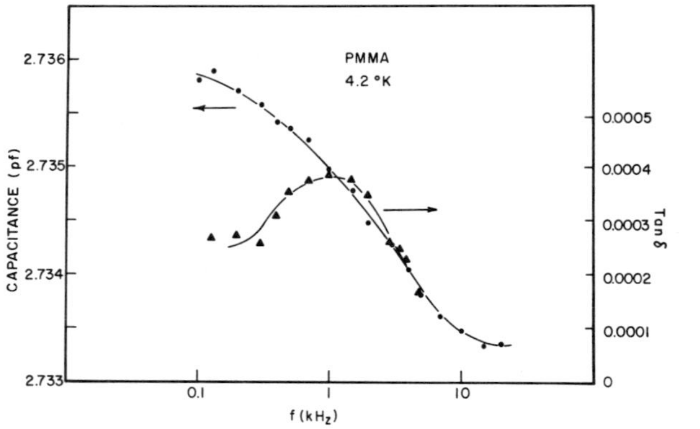

Figure 1. Capacitance and loss as a function of frequency at 4·2 °K for polymethyl methacrylate

relaxation time for this process is temperature dependent, as the variation of capacitance with temperature is also a function of frequency. The centre of the dispersion moves to higher temperatures with increasing frequency of measurement. At temperatures above this first low temperature dispersion, the capacitance again increases and indicates the possible presence of another relaxation at higher temperatures, see Figure 3 below.

The homologous anelastic quantities of PMMA are shown in Figure 2 where we have plotted $(f_T/f_{4\cdot 2})^2 = C_T/C_{4\cdot 2}$ for convenience, C_T is the elastic

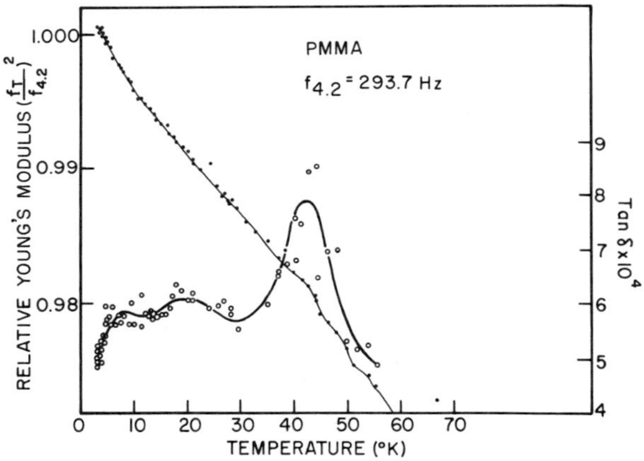

Figure 2. Mechanical loss and relative variation of elastic constant as a function of temperature for polymethyl methacrylate

constant at temperature *T*. Distinct relaxation processes at ~ 6 °K and at ~ 17 °K are apparent from the figure. Thus both dielectric and anelastic measurements on PMMA show relaxations in the same temperature and frequency range, which suggest that the relaxations are of the same structural origin.

The real part of the dielectric constant of PS is shown as a function of temperature in Figure 3, together with that of PMMA and polybutene, while the elastic modulus is shown in Figure 4. The anelastic measurements in Figure 4 clearly show the existence of several mechanical relaxations, the most pronounced, at low temperatures, being at ~ 7 °K, ~ 12 °K, and ~25 °K. The dielectric measurements also show what appears to be a dielectric relaxation, although in this case it is not clear what to take as the background behaviour. Thus again, anelastic measurements on PS, and possibly also dielectric behaviour, show relaxations in the same temperature and frequency range which suggests that the relaxations are of the same structural origin as the excess specific heat.

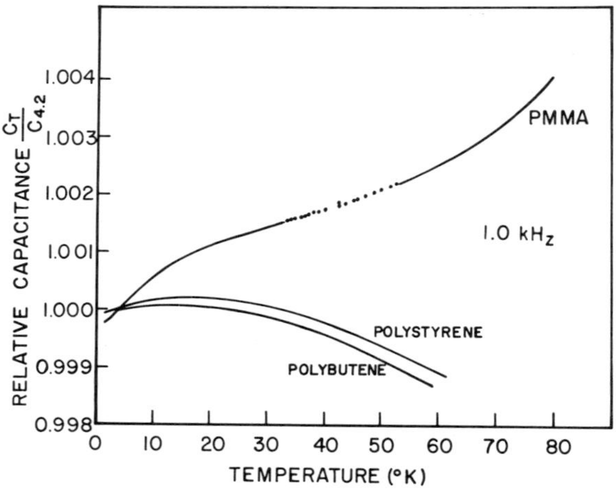

Figure 3. Capacitance as a function of temperature for polystyrene, polybutene, and polymethyl methacrylate

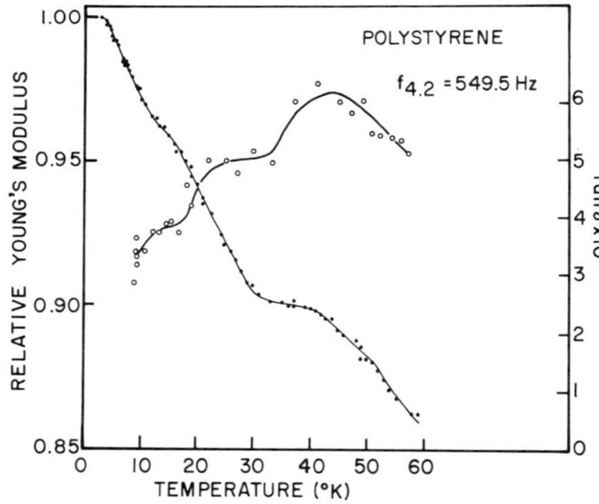

Figure 4. Mechanical loss and relative variation of elastic constant as a function of temperature for polystyrene

In this connection, polybutene is most interesting. It shows only a small excess specific heat. It has a dielectric constant which only decreases with increasing temperature (cf. Figure 3) while PMMA and PS show an increase with temperature. Finally, anelastic measurements by Crissman, Sauer, and Woodward (1964) on polybutene show no hint of a relaxation in this material in this temperature range, although their sensitivity was sufficient to indicate the relaxation in PMMA.

These experiments, we believe, establish a correlation between excess specific heat, and dielectric and anelastic relaxations which occur close to the temperature ($kT_\omega \approx \hbar\omega_0$) corresponding to the Einstein oscillators whose existence is inferred from the excess specific heat.

THE MODEL

We suppose that there exists in these polymers a significant number of structural units which undergo localized oscillations with a characteristic vibrational frequency ω_0. In the case of polymers, it seems reasonable to suppose that these structural units are some of the side groups in particular environments, such as for example the phenyl side group in PS. These side groups will oscillate, more and more independently of the chain as the ratios m/M and $u/U \to 0$, where m and u are the mass and spring constant of the side group and M and U the corresponding quantities of the chain section to which the side group is attached. As these conditions are approached, the side group oscillations will behave more and more as independent Einstein oscillators. Indeed Krumhansl (1968) has pointed out that many localized modes behave as if they were simple Einstein oscillators, and discusses theoretically how this occurs.

We therefore analyse this problem, as a first approximation, under the assumption m/M and $u/U \to 0$. The effect of an adiabatically applied mechanical stress or adiabatically applied electric field is shown to alter the resonant frequency ω_0 of the side group and so produces temporary non-equilibrium between the oscillator and the bulk phonons. This, therefore, results in a difference between the relaxed and unrelaxed elastic constants, C_0 and C_∞ and a difference between the relaxed and unrelaxed dielectric constants, K_0 and K_∞. The approach to equilibrium after an adiabatically applied stress or electric field is an entropy generating process, and thus gives rise to an associated mechanical or dielectric loss. The approach to equilibrium, however, implies $u/U \neq 0$, i.e. some interaction between the side group and the chain section.

DIELECTRIC RELAXATION

We suppose that the side groups in question undergo torsional or pendant (wagging) oscillations about their equilibrium positions. For both pendant and torsional oscillations, the Hamiltonian is of the form

$$H = \tfrac{1}{2}I\,(d\dot{\theta})^2 + \tfrac{1}{2}\,u_a\,(d\theta)^2$$

where the coordinate $d\theta$ is the angular displacement. Here u_a is the restoring force per unit angular displacement and I the appropriate moment of inertia of the side group. The solutions are

$$E_n = (n + \tfrac{1}{2})\,\hbar\omega_0 \text{ where } \omega_0 = \sqrt{(u_a/I)}$$

In the presence of an applied electric field, the Hamiltonian includes the perturbation term

$$H' = \sum_i \bar{F}\, r_i e_i \{\cos(\theta_i + \mathrm{d}\theta) - \cos\theta_i\}$$

For pendant oscillation, $\bar{F}$ is the electric field component in the plane of oscillation, r_i the component in the oscillation plane of the vector position of charge e_i with respect to the fulcrum and θ, the angle between $\bar{F}$ and r_i. For torsional oscillation, $\bar{F}$ is the electric field component perpendicular to the torsional axis, r_i the perpendicular-to-axis vector position of charge e_i, and θ_i the angle between $\bar{F}$ and r_i. The summation is over all charges i of the side group. On expansion to second order, the perturbation Hamiltonian is

$$H' = -\sum_i \bar{F}\, e_i r_i \left(\sin\theta_i\,\mathrm{d}\theta + \cos\theta_i \frac{(\mathrm{d}\theta)^2}{2}\right)$$

The first term on the right is linear in the coordinate $\mathrm{d}\theta$ and gives rise to a second order correction to the energy

$$E_n^{(2)} = \sum_m{}' \frac{|H'\,nm|^2}{E_n^0 - E_m^0} = -\frac{(\sum_i \bar{F}\, e_i r_i \sin\theta_i)^2}{2I\omega_0^2}$$

This energy correction does not depend on the quantum number n and is the same for all eigenstates. The effect of this term is to simply shift the energy zero by this amount. The second term on the right shifts each energy state by the amount $E_n^{(1)} = (n + \frac{1}{2})\,\hbar\Delta\omega$. Thus the new energy level is given by:

$$E_n = -\frac{(\Sigma \bar{F}\, e_i r_i \sin\theta_i)^2}{2I\omega_0^2} + (n + \tfrac{1}{2})\hbar(\omega_0 + \Delta\omega)$$

where

$$\omega_0 + \Delta\omega = \sqrt{\frac{u_a}{I}\left(1 - \frac{\sum_i \bar{F}\, e_i r_i \cos\theta_i}{2}\right)}$$

and hence

$$\Delta\omega \approx -\frac{\omega_0}{2}\,\frac{\sum_i \bar{F}\, e_i r_i \cos\theta_i}{u_a}$$

The Gibbs free-energy per oscillator is thus given by

$$G = -\frac{(\Sigma \bar{F} e_i r_i \sin\theta_i)^2}{2I\omega_0} + \frac{\hbar}{2}(\omega_0 + \Delta\omega) + kT\ln\left[1 - \mathrm{e}^{-\hbar(\omega_0+\Delta\omega)/kT}\right]$$

using the relation

$$e^{-G/kT} = \sum_n e^{-E_n/kT}$$

Hence the susceptibility

$$\Delta\chi = -\tfrac{1}{2}\frac{\partial^2 \Delta G}{\partial \bar{F}^2} = -\tfrac{1}{2}\frac{\partial^2 \Delta G}{\partial \Delta\omega^2}\left(\frac{\partial \Delta\omega}{\partial \bar{F}}\right)^2$$

is found to be

$$\Delta\chi = \frac{\hbar^2}{8kT}\frac{1}{\left[\sinh\left(\dfrac{\hbar\omega_0}{2kT}\right)\right]^2}\left(\frac{\sum_i e_i r_i \cos\theta_i}{2I\omega_0}\right)^2$$

$$= \frac{\hbar}{4\omega_0}\left(\frac{\sum_i e_i r_i \cos\theta_i}{2I\,\omega_0}\right)^2 \chi'\left(\frac{Z}{2}\right)$$

where

$$\chi'(Z) = \frac{Z}{[\sinh(Z)]^2}$$

and

$$Z = \frac{\hbar\omega}{kT}$$

(For convenience we have neglected the temperature independent instantaneous contribution due to the first term in G). Thus, a temperature dependent dielectric response is expected from a material containing N oscillators of this sort. The difference between the relaxed and unrelaxed susceptibility is given by:

$$\chi_0 - \chi_\infty = \frac{N\hbar}{4\omega_0}\left(\frac{\sum_i e_i r_i \cos\theta_i}{2I\,\omega_0}\right)^2 \chi'\left(\frac{Z}{2}\right) \tag{1}$$

The function $\chi'(Z/2)$ is shown plotted as a function of $Z^{-1} = kT/\hbar\omega_0$ in Figure 5.

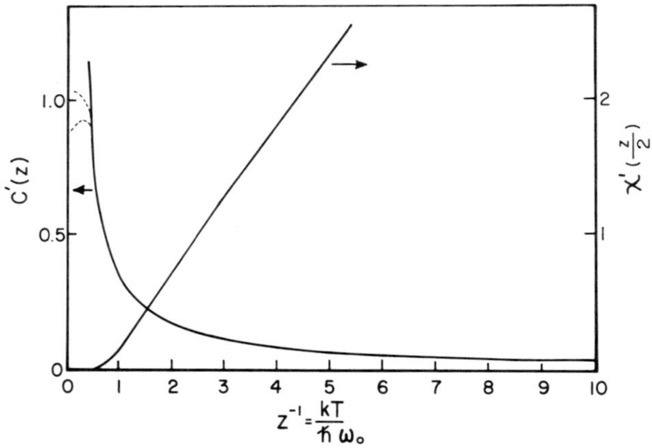

Figure 5. The function of $\chi'(Z/2)$ and $C'(Z)$ *vs.* Z^{-1}. See text for explanation of symbols

ANELASTIC RELAXATION

The most straightforward way of relating the strain ε to the Einstein oscillators of unperturbed frequency ω_0 is to use the conventional Grüneisen approach, i.e. $\omega = \omega_0(1 + \gamma\varepsilon)$, γ being the Grüneisen constant, and to assume that each oscillator per unit volume contributes a term α_d to the coefficient of thermal expansion. The quantity $N\alpha_d$ then, represents the contribution to the coefficient of expansion due to the presence of N of these oscillators per unit volume. For convenience we shall suppose that the coefficient of thermal expansion of the specimen containing no oscillator is zero. This arbitrarily sets the background thermoelastic losses, in which we are not interested here, at zero.

Thus, following Zener (1948), we write $\varepsilon = C_0\sigma + N\alpha_d T'$ for the strain relative to some standard state $\sigma = 0$, $T' = T_0 - T = 0$. Here C_0 is the relaxed isothermal elastic constant. A temperature difference between the Einstein oscillators and the reservoir may be induced by an adiabatically applied strain.

$$\frac{\partial T'}{\partial t} = \left(\frac{\mathrm{d}T'}{\mathrm{d}\varepsilon}\right)_{\text{adiabatic}} \frac{\mathrm{d}\varepsilon}{\mathrm{d}t} = -\beta\frac{\mathrm{d}\varepsilon}{\mathrm{d}t}.$$

The temperature difference is usually assumed to approach equilibrium through a relaxation time expression, i.e.

$$\frac{\partial T'}{\partial t} = -\frac{T'}{\tau}$$

Combining these three equations, the well known relaxation expression

$$\varepsilon/C_0 + \tau\dot{\varepsilon}/C_\infty = \sigma + \tau\dot{\sigma}$$

is obtained with the unrelaxed elastic constant

$$C_\infty = C_0/(1 + N\alpha_d\beta)$$

Now

$$\beta = -\left(\frac{\partial T}{\partial \varepsilon}\right) = \left(\frac{\partial T}{\partial U}\right)_\epsilon \left(\frac{\partial U}{\partial \omega}\right)_T \left(\frac{\partial \omega}{\partial \varepsilon}\right)$$

$$= \omega_0\gamma \left(\frac{\partial U}{\partial \omega}\right)_T \bigg/ C_V$$

$$= \frac{\hbar\omega_0\gamma}{2k}\,(e^Z - 2Z - e^{-Z})/Z^2$$

where U is the energy of the mode, C_ϵ is the specific heat at constant ϵ, and where the last expression applies specifically to an Einstein oscillator and where we have set $C_V = C_\epsilon$. Thus the difference between the relaxed and unrelaxed elastic constants is

$$C_0 - C_\infty = C_\infty N\alpha_d\beta$$

$$\therefore \frac{C_0 - C_\infty}{C_\infty} = \frac{N\alpha_d\hbar\omega_0\gamma}{2k} C'(Z) \tag{2}$$

where $C'(Z) = (e^Z - 2Z - e^{-Z})/Z^2$. The function $C'(Z)$ is shown plotted as a function of $Z^{-1} = kT/(\hbar\omega_0)$ in Figure 5.

This expression is probably a fairly good expression at high temperatures as $Z \to 0$. However, at low temperatures as $Z \to \infty$, the expression diverges since $C_V \to 0$ and we have not taken into account the temperature dependence of $N\alpha_d$, the coefficient of thermal expansion. Depending upon the temperature dependence of $N\alpha_d$, two possible variations, as $Z \to \infty$, of $C'(Z)$ are sketched in Figure 5.

In a practical experiment where sinusoidal stresses and electric fields of frequency ω_m are used, the compliances become complex. The real part contribution is given by the relaxation strength, equations (1) and (2), multiplied by the Debye factor $(1 + \omega_m^2\tau^2)^{-1}$ and the imaginary part is given by the appropriate relaxation strength multiplied by the corresponding Debye factor $\omega_m\tau/(1 + \omega_m^2\tau^2)$. The quantitative discussion of the relaxation time τ is given elsewhere (Scott *et al.*, 1970).

DISCUSSION

The analysis above has shown that a wagging or torsionally vibrating polar group can be expected to give rise to anelastic and dielectric relaxations. We suppose these to be the origin of the observed relaxations described in this paper.

In PMMA, there are two well defined relaxation processes, one at ~4–6 °K (Figures 1 and 2), and another at ~17 °K (Figure 2). In PS, mechanical relaxation measurements show three relaxations centred at ~7 °K, ~12 °K and ~25 °K (Figure 4), while the dielectric measurements show possibly a single broad relaxation, (Figure 3). The relaxations, we believe, occur mainly through the variations of the Debye factors as $\omega_m \tau$ decreases from values greater than unity to values less than unity as the temperature increases.

The relaxation strengths are given by:

$$\frac{C_0 - C_\infty}{C_\infty} \approx \frac{N\alpha_d \gamma \hbar \omega_0}{2k} C'(Z)$$

$$\chi_0 - \chi_\infty \approx \frac{N\hbar}{4\omega_0}\left(\frac{\Sigma' e_i r_i \cos\theta_i}{2I\omega_0}\right)^2 \chi'(Z/2)$$

so that comparison between theory and experiment is possible. In Table 1 we show the estimates of the theoretical values and the experimental values for comparison. The theoretical estimation is made using the values N and ω_0 determined from specific heat measurements, (Choy *et al.*, 1970; Salinger, 1970), and setting arbitrarily $\chi'(Z/2) = C'(Z) = 1$. For the strength of the mechanical relaxation we used the values 3 for γ, and the coefficient of thermal expansion $N\alpha_d \approx 1 \times 10^{-5}$. Agreement between theory and experiment is satisfactory. For the dielectric relaxation, it is necessary to estimate $\Sigma_i e_i r_i \cos\theta_i$ and I; the values used are listed in Table 1 and are appropriate to the ester and phenyl groups (Minkin *et al.*, 1970). Again, agreement is considered good, particularly since the parameters used cannot be widely varied without departing from structurally realistic groups.

Of interest is the striking difference between the relaxation structure as revealed by anelastic measurements, (Figures 2 and 4), compared to that revealed by dielectric measurements, (Figure 3). In contrast to the several distinct relaxations over the temperature range as revealed by anelastic measurements, the dielectric measurements reveal only rather broad smooth relaxation. We believe that this is due to the isotropic Grüneissen constant γ but anisotropic $\Delta\omega$, see the perturbation Hamiltonian.

A detailed discussion of the relaxation time τ, describing the approach to equilibrium is not possible in view of the uncertainties in the anharmonic coupling parameters. However, using fairly general arguments, Scott and

Table 1. Parameters and theoretical and experimental relaxation strengths

	$N(\text{gm}^{-1})$	$(\sum_i e_i r_i \cos\theta_i)^2$ (erg cm^3)	I (gm cm^2)	$(C_0 - C_\infty)/C_\infty$		$\chi_0 - \chi_\infty$	
				Theoretical	Experimental	Theoretical	Experimental
PMMA	$5{\cdot}6 \times 10^{18}$	$6{\cdot}6 \times 10^{-35}$	$\sim 10^{-38}$	$7{\cdot}3 \times 10^{-5}$	$5{\cdot}8 \times 10^{-4}$	2×10^{-3}	$1{\cdot}5 \times 10^{-4}$
	$2{\cdot}3 \times 10^{20}$			$2{\cdot}6 \times 10^{-4}$	6×10^{-4}	2×10^{-3}	2×10^{-4}
PS	$2{\cdot}1 \times 10^{18}$	$3{\cdot}2 \times 10^{-36}$	$\sim 10^{-38}$	$4{\cdot}5 \times 10^{-5}$	—	$1{\cdot}5 \times 10^{-4}$	—
	$7{\cdot}4 \times 10^{19}$			$1{\cdot}5 \times 10^{-4}$	$3{\cdot}5 \times 10^{-3}$	$1{\cdot}5 \times 10^{-4}$	$\sim 10^{-4}$
	$5{\cdot}5 \times 10^{20}$			$3{\cdot}0 \times 10^{-4}$	5×10^{-3}	$1{\cdot}3 \times 10^{-4}$	—

MacCrone (1970) have shown that a two chain-phonon–one oscillator-phonon interaction dominates the equilibrium. Thus we write

$$\tau^{-1} \propto [N(\omega_1)\, N(\omega_2)\, N(\omega_0)]_{\text{average}}$$
$$\approx \left[\frac{1}{e^{\hbar\bar{\omega}/kT} - 1}\right]^2 \frac{1}{e^{\hbar\omega_0/kT} - 1}$$

where we have used an average chain phonon frequency $\bar{\omega}$ instead of ω_1 and ω_2.

The temperature dependence of the relative relaxation time is shown in Figure 6 for three values of the parameter $\bar{\omega}/\omega_0$. As energy is conserved in the interaction, ω_1 and/or $\omega_2 > \omega_0$, and we believe a reasonable estimate is $\omega_1 \approx \omega_2 = \bar{\omega} = 10\,\omega_0$, see Scott and MacCrone (1970). Thus the centre of

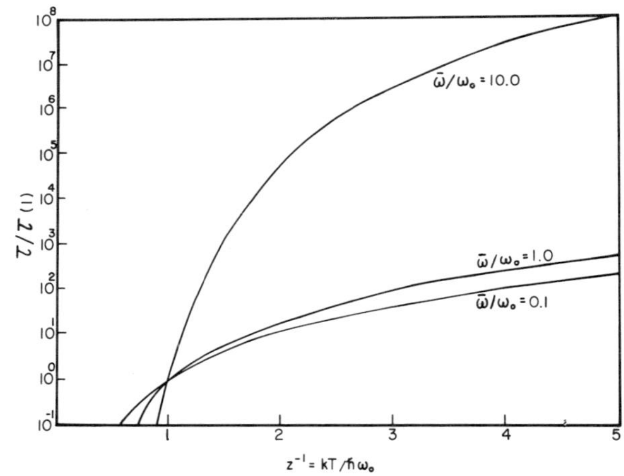

Figure 6. The temperature variation of the relative relaxation time for three values of $\bar{\omega}/\omega_0$

the relaxation is most likely to occur when $kT \approx \hbar\omega_0$, where the relaxation strengths are appreciable and when τ is changing rapidly. (In view of the two curves in Figure 5, however, the mechanical and dielectric relaxations are not necessarily expected to occur at exactly the same temperatures). On this assumption, the frequencies of the Einstein modes may be very roughly determined and are listed in Table 2, together with the values used in describing the excess specific heat. It should, however, be pointed out that these Einstein frequencies (temperatures) probably represent some average or dominant frequency in some fairly narrow distribution of excess modes. The discrete frequency hypothesis should not be taken too literally.

Table 2. Einstein frequencies from specific heat measurements and relaxation measurements

ω_0 (1/sec)

	From specific heat measurement (Choy *et al.*, 1970; Salinger, 1970)		Present work	
PMMA	4·9 °K→	6·44 × 10^{11}	6 °K→	7·89 × 10^{11}
	17·5 °K→	23·0 × 10^{11}	17 °K→	22·3 × 10^{11}
PS	3 °K→	3·94 × 10^{11}	7 °K→	9·17 × 10^{11}
	10 °K→	13·1 × 10^{11}	12 °K→	15·8 × 10^{11}
	20 °K→	26·3 × 10^{11}	25 °K→	32·9 × 10^{11}

ACKNOWLEDGEMENTS

Many helpful discussions with Dr. G. L. Salinger are gratefully acknowledged. For financial support we gratefully acknowledge an N.S.F. U.R.P. programme (S.C.M.), O.N.R. contract NR 032–519 (M.T.) and also N.S.F. contract GK 4120 for partial support (R.K.M.). The experimental facility was provided by a NASA Interdisciplinary Laboratory Grant.

REFERENCES

Antoniou, A. A., and J. A. Morrison (1965). *J. appl. Phys.*, **36**, 1873.
Choy, C. L., R. G. Hunt, and G. L. Salinger (1970). *J. chem. Phys.*, **52**, 3629.
Craig, R. S., C. W. Messena, and R. M. Mallya (1965). *J. appl. Phys.*, **36**, 108.
Crissman, J. E., J. A. Sauer, and A. E. Woodward (1964). *J. Polym. Sci.*, **A2**, 5075.
Dean, P. (1960). *Proc. R. Soc.*, **A254**, 507.
Dominik, L. A. K., and R. K. MacCrone (1967). *Phys. Rev.*, **156**, 910.
Fine, M. E., H. van Duyne, and N. T. Kennedy (1954). *J. appl. Phys.*, **25**, 402.
Flubacher, P., A. J. Leadbetter, J. A. Morrison, and B. P. Stoicheff (1959). *J. Physics Chem. Solids*, **12**, 53.
Krause, J. T., and C. R. Kurkjian (1968). *J. Am. Ceram. Soc.*, **51**, 226.
Krumhansl, J. A. (1968). In R. F. Wallis (Ed.). *Localized Excitations In Solids.* Plenum Press, New York, p. 17.
Minkin, V. I., O. A. Osipov, and Y. A. Zhdanov (1970). Translated by B. J. Hazzard. *Dipole Moments in Organic Chemistry.* Plenum Press, New York. p. 91.
Noer, R. J., C. W. Dempseym, and J. E. Gordon (1959). *Bull. Am. phys. Soc.*, **4**, 108.
Reese, W. (1966). *J. chem. Phys.*, **52**, 3629.
Reese, W., and J. E. Tucker (1965). *J. chem. Phys.*, **43**, 105.
Rosenstock, H. B. (1962). *J. Physics Chem. Solids*, **23**, 659.
Rosenstock, H. B., and R. E. McGill (1968). *Phys. Rev.*, **176**, 1004.
Salinger, G. L. (1970). See this volume, Chapter 46. Also, private communication.

Scott, W. W., and R. K. MacCrone (1968). *Rev. scient. Instrum.*, **39**, 821.
Scott, W. W., and R. K. MacCrone (1970). *Phys. Rev.*, **B1**, 3515.
Zener, C. M. (1948). *Elasticity and Anelasticity in Metals.* The University of Chicago Press, Chicago. p. 69.

10

The Magnitude and Relaxation of the Strain-Optic Constants in Amorphous Materials in Relation to Structure

H. R. CARLETON

Department of Materials Science, State University of New York, Stony Brook, New York, U.S.A.

The physical properties of non-crystalline solids are susceptible to study by optical techniques such as infrared and Raman spectroscopy, reflectance and absorption measurements, and light scattering. In addition, highly refined measurements of the dependence of refractive index on pressure, density, and other parameters provide a sensitive indicator of changes in structure. Current developments in optical instrumentation and coherent optics when combined with the inherent high resolution of optical measurements, can be expected to produce improved methods of materials characterization.

In particular, Brillouin and Debye–Sears light scattering from hypersonic and ultrasonic waves can readily be applied to the study of both solid, liquid, and transition range phases of the amorphous state. The high sensitivity of the scattering process to any optical anisotropy that may be present in the substance under study offers the potential of directly observing the relaxation processes associated with longitudinal and transverse wave propagation and viscous flow.

The evolution of the elastic moduli of liquids and the formation of a shear modulus upon approach to the glass transition provide an important link in the study of glass formation. Ultrasonic attenuation measurements in liquids (Litovitz and Davis, 1965) provide a direct measure of the dispersion of compressional waves and the associated structural relaxation but are limited to the study of complex liquids with long relaxation times since the upper frequency range of these techniques in liquids is of the order of 100 MHz. Brillouin scattering from thermal excitations extends the frequency range to

10^4 MHz, where the onset of faster relaxation processes can readily be observed. In addition, it can be expected that transverse acoustic modes can be supported by a fluid at these frequencies, a feature which appears to be verified by the recent observations of Stegeman and Stoicheff (1968). Additional features of depolarized scattering from ultrasonic waves in viscous media have been demonstrated by Riley and Klein (1969) and by Mangla (1970) who have used Debye–Sears scattering to determine the relaxation time of viscous fluids. In this case the depolarization is attributed to flow birefringence.

Quantitative interpretation of data obtained in light scattering experiments or in refractive index measurements must ultimately rest on a theory of the optical constants which includes the magnitude and dispersion of the strain-optic coefficients. Although such a theory has been developed for cubic crystals (Maradudin and Burstein, 1967), an adequate theory is not available for amorphous systems. In the following paragraphs, after a brief review of present theories of optical constants, it will be shown that a general form of the strain-optic constants can be derived from an adaption of the Kirkwood–Yvon theory of the dielectric constant. In this writing, the calculation is limited to include only the major features of a simple amorphous structure which contribute to the optical constants. In this context, a simple structure is taken to be an ensemble of isotropic, spherically symmetric, non-polar molecules characterized by a fixed polarizability α. Although such an assumption is a radical oversimplification of interesting non-crystalline systems, it will prove possible to infer several features which may bear on the structural evolution of more complex systems.

THEORIES OF THE OPTICAL CONSTANTS

The Lorentz–Lorenz theory of the refractive index n, given in terms of the molecular density ρ by

$$\frac{n^2 - 1}{n^2 + 2} = \frac{\alpha\rho}{3\epsilon_0} \tag{1}$$

where ϵ_0 is the permittivity of the vacuum, has been extensively used to relate optical data to changes in the state of a material. As the only structural factor in this expression is due to the modified local field of the Lorentz cavity, which accounts for the factor $n^2 + 2$ in the denominator of the left hand side, it cannot be expected to illuminate subtle structural changes, particularly those occurring at constant density. Several authors have extended the theory of the refractive index to include the correlations of neighbouring molecules using both phenomenological (Bottcher, 1952) and statistical mechanical arguments (Mazur and Jansen, 1955).

The theory of the strain-optic constants have been developed to a lesser degree. The density dependence of the refractive index can be approximated from (1) to be

$$\rho \frac{\partial n}{\partial \rho} = \frac{(n^2 - 1)(n^2 + 2)}{6n} \tag{2}$$

This expression suffers the same limitations as that for the refractive index. Furthermore, there is no possibility of developing the tensor properties of the strain-optic constants from an expression for an isotropic refractive index.

Using arguments similar to those of the Lorentz local field correction, Mueller (1935) was able to derive expressions for the anisotropy of the refractive index induced by a uniaxial strain S. From his results,

$$(n_{\parallel}^2 - 1)^{-1} = \frac{\epsilon_0}{\rho\alpha} - \frac{1}{3} + \frac{4}{15} S \tag{3}$$

$$(n_{\perp}^2 - 1)^{-1} = \frac{\epsilon_0}{\rho\alpha} - \frac{1}{3} - \frac{2}{15} S \tag{4}$$

the birefringence for light polarized parallel and perpendicular to the strain can be derived. This derivation also provided the first valid theoretical estimate of the strain-optic constants of isotropic solids. The second and third terms on the right-hand side of these equations represent the isotropic and anisotropic local field correction.

In order to test the general validity of (3) and (4) and to extend these theories to a variety of systems which may exhibit unique structural characteristics, more general expressions for the dependence of the refractive index on local mechanical disturbances must be developed. Progress in this direction can be readily made if it is assumed that the mechanical disturbances are infinitesimal strains which are instantaneous responses to applied uniform forces. This corresponds to the response of a system in the high-frequency limit, where the elastic moduli display solid-like behaviour.

CALCULATION OF OPTICAL CONSTANTS

The refractive index and strain-optic constants of a simple amorphous system may be calculated in the high frequency limit by using the dielectric formalism of Kirkwood and Yvon as extended to optical frequencies by Fixman (1955). Assuming that the location of each polarizable centre in a system of microscopic, isotropic, non-polar molecules is known and given by a hierarchy of multiparticle correlation functions, Fixman has shown that the optical dielectric constant is given by the expression

$$\overline{\overline{(\epsilon - \epsilon_0)}}^{-1} = \left[\frac{1}{\alpha\rho} - \frac{1}{3\epsilon_0}\right]\overline{\overline{U}} - \alpha \int_v (g_{12} - 1)\,\overline{\overline{T}}_{12}\cdot\overline{\overline{T}}_{21}\,\mathrm{d}^3R_2$$
$$+ \alpha\rho \int (g_{123} - g_{12}g_{23})\,\overline{\overline{T}}_{12}\cdot\overline{\overline{T}}_{23}\,\mathrm{d}^3R_2\,\mathrm{d}^3R_3 + \ldots \tag{5}$$

where α is the assumed free-particle polarizability, ρ is the particle density, g_{12}, g_{123}, etc., are the two and three particle correlation functions, $\overline{\overline{U}}$ is the unit matrix, and $\overline{\overline{T}}_{ij}$ is the dipole interaction matrix given by

$$\overline{\overline{T}}_{ij} = \frac{1}{4\pi\epsilon_0}(\overline{\overline{\nabla}}_i\overline{\overline{\nabla}}_i + k_0^2\overline{\overline{U}})\frac{e^{ik_0\tilde{R}_{ij}}}{\tilde{R}_{ij}} \tag{6}$$

where $\tilde{R}_{ij} = \tilde{R}_i - \tilde{R}_j$, and k_0 is the wavenumber of light in the vacuum. Using this formalism to calculate fluctuations in the refractive index, Fixman was able to relate the molecular theory of light scattering to the phenomenological Einstein–Smoluchowski theory.

Since the expansion parameter in equation (5) is the inverse density, convergence can be expected to be slow in the range of liquid and solid densities. Nevertheless, we may gain insight into the role that molecular correlations play in structural studies of amorphous systems by considering only terms in the expansion including the first non-vanishing correlation term and ignoring, for the moment, higher order terms. Although this may seem a drastic assumption, it is consistent with numerous studies of the role of density on refractive index, and will be adopted for the balance of this discussion.

Consider now the dependence of the dielectric constant on mechanical disturbances of the system. In particular, let us assume that the medium undergoes a sudden uniaxial strain S in the z direction. Under such circumstances it can be expected that the density and the correlation functions of a suitably truncated form of equation (5) will undergo change to correspond to this perturbation. In the absence of relaxation it follows that the new correlation function $g'_{12}(\tilde{R})$ can be written in terms of the initial g_{12} by

$$g'_{12}(x, y, z) = g_{12}(x, y, z(1 - S)) \tag{7}$$

Following this description, and assuming that S is infinitesimal, g_{12} can be expanded in terms of its derivative so that the strain dependence of expression (5) can be evaluated. After several changes of variable, integrating over the angular variables of the volume integral, and integrating by parts over the radius vector, the expression

$$\frac{\partial}{\partial S_z}\overline{\overline{\epsilon - \epsilon_0}}^{-1} = \frac{1}{\alpha\rho}\overline{\overline{U}} - \frac{2}{15\epsilon_0}\overline{\overline{B}} + \frac{2}{3\epsilon_0}\Gamma[\overline{\overline{U}} - \tfrac{1}{5}\overline{\overline{B}}] \tag{8}$$

is obtained. Here B is the diagonal matrix with trace zero and $B_{11} = B_{22} = 1$, $B_{33} = -2$ and Γ is the integral,

$$\Gamma = \frac{3\alpha}{4\pi\epsilon_0}\int_0^\infty g_{12}(r)\frac{dr}{r^4}, \tag{9}$$

which appears in other discussions of the refractive index.

THE OPTICAL CONSTANTS

The optical constants of the isotropic system may now be written in terms of the correlation integral. Since there are only two independent strain-optic constants in an isotropic system, it is necessary to consider only one uniaxial strain condition to obtain both. In addition, the density dependence of the refractive index is also of interest since it is this quantity which is readily obtained from pressure measurements and in Debye–Sears experiments.

REFRACTIVE INDEX

The refractive index obtained from direct evaluation of (5) is given most conveniently in the form

$$\frac{n^2 + 2}{n^2 - 1} = \frac{3\epsilon_0}{\alpha\rho} - 2\Gamma \tag{10}$$

which can be compared directly to the Lorentz–Lorenz form of equation (1).

STRAIN-OPTIC CONSTANTS

To the extent that out approximations are valid, equation (8) can be used to predict both of the high frequency strain-optic constants of simple molecular systems. This can be done by comparing (8) to Pockel's expression for p_{11}, p_{12} (Nye, 1957) with the result that

$$p_{11} = \frac{[n^2 - 1]^2}{n^4}\left[\frac{\epsilon_0}{\alpha\rho} + \frac{4}{15} + \frac{14}{15}\Gamma\right], \tag{11}$$

and

$$p_{12} = \frac{[n^2 - 1]^2}{n^4}\left[\frac{\epsilon_0}{\alpha\rho} - \frac{2}{15} + \frac{8}{15}\Gamma\right]. \tag{12}$$

In these expressions, the fourth-rank tensor p_{ijkl} has been written in the reduced notation and the refractive index n has been substituted for the dielectric constant through the Maxwell relation $n^2 = \epsilon/\epsilon_0$.

The density dependence of the refractive index can be written in terms of (11) and (12) or derived directly from the refractive index (10) with the result

$$\rho\frac{\partial n}{\partial \rho} = \frac{(n^2 - 1)(n^2 + 2)}{6n} + \frac{2(n^2 - 1)^2}{3n}\Gamma. \tag{13}$$

DISCUSSION OF THE HIGH FREQUENCY OPTICAL CONSTANTS

The physical mechanisms which have a dominant effect on the optical constants of an isotropic molecular system are evident in expressions (11) and (12). The first term in the brackets of each expression accounts for the change of refractive index due to density changes accompanying a strain. The second term is the anisotropic correction to the refractive index which is attributable to the elliptical distortion of the Lorentz cavity and is the same term derived by Mueller from phenomenological arguments.

The correlation term reflects the additional polarizing effect of neighbouring molecules on each other which is due to the difference between the polarization produced by a dipole spread over the volume outside of the Lorentz cavity and the polarization produced by a dipole distributed over a spherical surface centred with the Lorentz cavity.

It may be inferred that as $p_{11} \neq p_{12}$ in the high frequency limit, local optical anisotropy may be induced by either longitudinal or transverse hypersonic disturbances in the medium. Hence, depolarized optical scattering may be observed which is due to purely structural anisotropies which are unrelated to any anisotropies of molecular structure. This property is due to both the anisotropy of the fluctuating Lorentz cavity and to the fluctuations in the correlation term.

THE ROLE OF Γ IN THE RELAXATION OF THE OPTICAL CONSTANTS

We may now enquire what role the correlation term Γ may play in the refractive index and strain-optic constants of a substance. We will first show, from the evaluation of a class of correlation functions $g(r)$, that Γ will increase the Lorentz cavity contribution to the refractive index by 25 to 50% and that it may hence be expected to have a measurable effect on the observed strain-optic constants. We will also argue that the Γ contribution is the lowest order term that may be expected to exhibit relaxation and reflect structural changes in isotropic, central-force systems.

EVALUATION OF Γ

Mazur and Jansen have considered the correlation contributions to the refractive index for Lennard–Jones potentials. Since the behaviour of Γ under changes of the internal state of a substance are of primary interest, we have chosen to evaluate Γ by choosing a number of correlation functions which typify a range of system states. Bali (1969) evaluated Γ for a number of systems for which $g(r)$ could be determined. These include three systems exhibiting a continuous $g(r)$, i.e. an uncorrelated system, and the Kirkwood–Yvon system of hard spheres in both the Kirkwood and Born–Green parametrization. For the sake of comparison, three systems with discrete distributions were also evaluated. These include the hexagonal and cubic close-packed lattices in which averages over angular variables were carried out in order to make $g(r)$ isotropic. The third discrete system evaluated was the random close-packed rigid sphere model of Bernal as evaluated by Scott (1962).

In Figure 1, the computed values of Γ_0 have been plotted as a function of density in units of a_0^3, where a_0 is the hard sphere diameter. Since the polarizability α is unknown, Γ_0 has been defined in terms of the classical polarizability

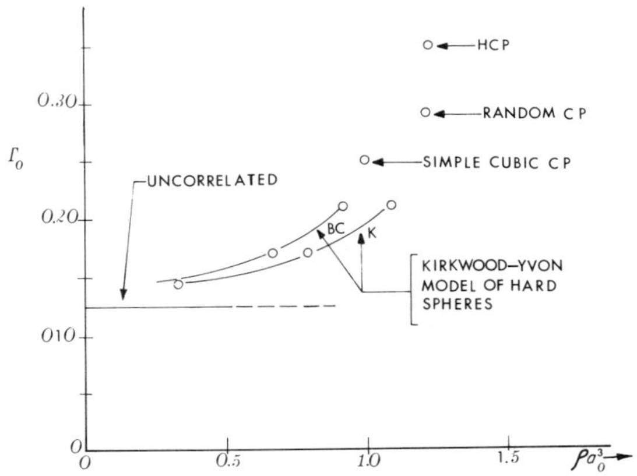

Figure 1. The dependence of the integral Γ_0 on the reduced density ρa_0^3. (From Bali, 1969)

$\alpha_0 = r_0^3$, where r_0 is the classical molecular radius, so that $\alpha_0 \Gamma = \alpha \Gamma_0$. It may be expected that the results of Figure 1 typify a wide range of physical systems. The three discrete models correspond to the limit $T \to 0$, while the uncorrelated model should correspond to all systems in the limit of high temperatures. The Kirkwood–Yvon model displays the behaviour of a thermodynamic system over a trajectory of temperatures (and hence density) at finite pressure. These results show that Γ_0 will lie between 0·125 and 0·375, indicating a correction to the Lorentz field term of between 25 and 50%.

It will be remembered that the strain-optic properties as here derived were valid only in the high frequency limit—i.e. under the assumption that no relaxation of the molecular correlation would occur in response to a strain. Hence a knowledge of Γ at the initial density is sufficient to evaluate the strain-optic constants in the high frequency limit. On the other hand, if the predictions of Figure 1 can be taken seriously, a qualitative view of the isothermal or adiabatic changes in the optical constants which occur with mechanical stresses of a substance can be inferred.

In the simplest case, the value of Γ_0 is seen to be independent of density in an uncorrelated system. This is true in both the isothermal and high frequency limits and implies that no relaxation may be expected to occur in the strain-optic properties of the substance, regardless of the rapidity with which mechanical disturbances may occur. In the other extreme, the density dependence of Γ_0 for well defined structures such as hcp, ccp, and the random close-packed model at $T = 0^0$ is given by

$$\Gamma_0(\rho) = \rho\Gamma_0(\rho_0)/\rho_0. \tag{14}$$

This condition makes it possible to derive the coefficient $\rho(\partial n/\partial \rho)$ directly from the index expression (13). Since the above assumption is equivalent to solid-like behaviour, it is not surprising that relaxation is not present in this case either.

It should be noted that an estimate of structural adjustments at finite temperatures can be made by evaluating Γ_0 for harmonic motion of the lattice sites. A calculation including thermal expansion would lead directly to a quantitative evaluation of the dispersion of Γ. At $T = 0^0$ it is clear that equation (14) is valid for these systems, while the contributions at finite temperatures will be additive, increasing the value of Γ_0 at given density. It can be expected that relaxation of this additive term will accompany processes of thermal conduction.

The thermodynamic case is represented by the Kirkwood–Yvon system which illustrates the range of behaviour that may be expected in the density dependence of the refractive index. Derivation of the strain-optic properties in the high frequency limit is again consistent with equation (14). This implies that the slope of the Γ_0 *vs.* ρ curve in Figure 1 is given by the line passing from the origin to the value of Γ_0 at the assumed initial density. From the figure it is clear that the dependence of Γ_0 on density in thermodynamic equilibrium is lower than that predicted in the high frequency limit. Hence, under rapid compression from a given initial density, the system will follow the trajectory corresponding to solid-like behaviour and subsequently relax (at constant density) to the equilibrium value of Γ_0. This process is associated with the process of structural relaxation in the substance. From this point of view, the dispersion in the correlation contribution to the strain-optic constants of a substance, which we have argued is the only relaxing term in simple systems, increases at lower densities, so that its contribution to the optical properties will be considerably weaker at frequencies below the structural relaxation time. For example, in the case of the Born–Green, $(\partial\Gamma/\partial\rho)_{\omega=0} \simeq 0{\cdot}5\,(\partial\Gamma/\partial\rho)_{\omega=\infty}$. This corresponds to the approximate relative density of carbon tetrachloride at standard temperature and pressure.

MOLECULAR COORDINATION

Owing to the rapid decrease of the factor r^{-4} in the integrand of Γ, this integral is most sensitive to the density of nearest neighbours. This will be true even for continuous distribution functions, since an integral equivalent of the coordination number can be defined. Hence, although the wavelength of light is almost three orders of magnitude greater than the interparticle spacing, precise determination of the optical constants can be expected to reflect the coordination number. This suggests that the refractive index of a simple, glass-forming system may indicate the development of increased coordination upon approaching the glass transition. A similar interpretation should be applicable to clustering of molecules in the liquid phase.

REFERENCES

Bali, S. G. (1969). *The Dependence of Strain-Optic Constants on Structural Correlation.* Master's Thesis, College of Engineering, Stony Brook, New York.

Bottcher, C. J. F. (1952). *Theory of Electric Polarization.* Elsevier, Amsterdam.

Fixman, M. (1955). *J. chem. Phys.*, **23**, 2074.

Litovitz, T. A., and C. M. Davis (1965). In W. P. Mason, *Physical Acoustics*, Vol. IIA. Academic Press, New York. pp. 282–350.

Mangla, B. S. (1970). *A Study of Birefringence in Liquids.* Master's Thesis, College of Engineering, Stony Brook, New York.

Maradudin, A. A., and E. Burstein (1967). *Phys. Rev.*, **164**, 1081.

Mazur, P., and L. Jansen (1955). *Physica,'s Grav.*, **21**, 208.

Mueller, H. (1935). *Physics*, **6**, 179.

Nye, J. F. (1957). *Physical Properties of Crystals.* Oxford Univ. Press, London. pp. 243–254.

Riley, W. A., and W. R. Klein (1969). *J. acoust. Soc. Am.*, **42**, 1258.

Scott, G. D. (1962). *Nature*, **194**, 956.

Stegeman, G. I. A., and B. P. Stoicheff (1968). *Phys. Rev. Lett.*, **21**, 202.

THE TRANSFORMATION RANGE

11

The Nature of the Glass Transition in Organic Polymers

G. GEE

Department of Chemistry, University of Manchester, Great Britain

The term 'glass temperature' (T_g) is widely quoted as an important characteristic of a polymer. It is important to realize that T_g can only be unambiguously defined in terms of the experiment in which it is measured: it is not a satisfactory material constant. More accurately one should speak of a transition region, above which the polymer is fluid or rubbery, below which it is a glass. The properties of the glass are dependent on its conditions of formation, in a way which is too complex to be specified by any single parameter.

The transition region is characterized by changes (idealized to discontinuities) in derivative functions such as specific heat and coefficient of expansion. Any molecular theory of these properties therefore leads to a molecular interpretation of the glass transition. Thus for example ΔC_p has been shown to be generally close to 11·2 JK^{-1} per 'bead', where the bead is identified as the smallest unit in a polymer chain capable of free station.

The most widely accepted of these molecular theories associates the glass transition with a given fractional free volume, generally taken to be about 0·025. It has been pointed out that exactly equivalent theories can be formulated, in which the role of volume is played instead by enthalpy or entropy, and that some facts contrary to the free volume theory then become explicable.

Attempts to relate T_g to polymer structure have focused on one or both of two properties: molecular cohesion, and chain stiffness. The latter has been particularly stressed by Gibbs and his coworkers, who have associated a theoretical glass transition (T_2) with the loss of configurational entropy. T_2 is however some 60 °K below T_g, and is more nearly equal to the temperature at which the extrapolated viscosity of the liquid tends to infinity.

Further evidence on the molecular significance of the transition region comes from the study of dynamic or dielectric relaxation. It is frequently found that a polymer shows a number of relaxations, of which one can usually be readily identified with glass transition, while others occur at lower temperatures. It is clear therefore that molecular motions can occur in glassy polymers, though it is by no means obvious in many cases what they are. The glass transition is certainly associated with large scale movement, but the description of this in molecular terms is difficult, and perhaps not susceptible to generalization in any but the broadest terms.

REFERENCES

Gee, G. (1966). *Polymer*, **7**, 177.
Gee, G. (1970). *Contemp. Phys.*, **11**, 313.

12

Physical Properties of Polystyrene Glasses Formed from the Fluid State at High Pressures

C. PRICE, R. C. WILLIAMS and R. C. AYERST

Department of Chemistry, University of Manchester, Great Britain

Renewed attention has been given in recent years to the effect of pressure on the temperature of glass formation in polymers (Bueche, 1962; Goldstein, 1964; Breuer and Rehage, 1967; Gee, 1966). Much of the work appears to have been directed towards testing the validity of the equations,

$$\frac{dT_g}{dP} = \frac{\Delta\kappa}{\Delta\alpha} \tag{1}$$

$$\frac{dT_g}{dP} = \frac{TV\Delta\alpha}{\Delta C_p} \tag{2}$$

where $\Delta\kappa$, $\Delta\alpha$ and ΔC_p are the discontinuities in the compressibility, the thermal expansion coefficient and the heat capacity at the transition, and dT_g/dP is the dependence of glass transition temperature on pressure. It is well established that if a single ordering parameter (Z) was sufficient to characterize the excess thermodynamic properties of the fluid state over those of the glass, then equations (1) and (2) and hence equation (3) below would hold (Prigogine and Defay, 1961; Davies and Jones, 1953)

$$\frac{\Delta\kappa}{\Delta\alpha} = \frac{TV\Delta\alpha}{\Delta C_p} \tag{3}$$

If more than one parameter is required, it has been shown (Davies and Jones, 1953) that equation (3) must be weakened to an inequality $\Delta\kappa/\Delta\alpha \geqslant TV\Delta\alpha/\Delta C_p$; where the sense of the inequality is governed by the conditions of thermodynamic stability. Goldstein (1963) has pointed out that regardless of the validity of the one parameter argument if the transition occurs at a constant value of the excess volume V_e equation (1) would hold, whilst if the transition occurs at a constant value of the excess entropy, or the excess enthalpy, equation (2) would hold. It follows of course that if a one parameter description was sufficient then constant V_e would necessarily imply constant S_e, H_e etc and we would not then be in a position to distinguish between the critical variables.

Unfortunately even for the more extensively studied organic polymers there is still much confusion with respect to both the reliability and the interpretation of experimental data. Some fairly recent assessments of the available data (Gee, 1966; Goldstein, 1963) seem to suggest that $\mathrm{d}T_g/\mathrm{d}P < \Delta\kappa/\Delta\alpha$ whilst $\mathrm{d}T_g/\mathrm{d}P \simeq TV\Delta\alpha/\Delta C_p$, which if substantiated would indicate that a one parameter description is not possible, and that it is S_e, or H_e, rather than V_e which determines T_g. Gee (1966) has pointed out, however, that in order to obtain unequivocal answers to the questions we are asking it would be far better to compare directly the properties of a series of glasses formed by cooling a given substance from the fluid state at different pressures. For example by comparing the densities (ρ) of such a series at atmospheric pressure it would be possible to test the assumptions underlying equation (1); for a limited range of pressure, so that $\Delta\kappa$ can be assumed to remain constant, Beuche (1962) showed that $\mathrm{d}\ln\rho/\mathrm{d}P = \Delta\kappa - \Delta\alpha(\mathrm{d}T_g/\mathrm{d}P)$. Following the original work of Tammann (1933) many workers (Breuer and Rehage, 1967; Shishkin, 1960; O'Reilly, 1962) have investigated the dependence of $\rho(1\ \mathrm{atm}, T < T_g)$ on the cooling pressure. There seems to be general agreement that $\mathrm{d}\ln\rho/\mathrm{d}P$ is positive. It follows therefore that $\Delta\kappa/\Delta\alpha$ must be greater than $\mathrm{d}T_g/\mathrm{d}P$.

There is no reason why a similar method to the above should not also be used to test the basis of equation (2). Thus, in the present investigation we undertook to compare the relative enthalpies of glasses formed at different pressures. All the measurements we report were carried out on an atactic sample of polystyrene having a narrow molecular weight distribution ($M_w = 390{,}000$, $M_w/M_n < 1{\cdot}1$) The sample had been synthesized anionically using n-butyl lithium as initiator. We determined the relative enthalpies of the polystyrene glasses by measuring their heats of solution in toluene at 1 atm pressure and 30 °C. Since polymers dissolve rather slowly it was necessary to use a calorimeter in which it was possible to measure small quantities of heat being given out over relatively long periods of time. The Tian–Calvet microcalorimeter, which under suitable conditions can measure heat fluxes down to 0·001 cal hr^{-1} proved ideal for this purpose.

DENSIFICATION PROCEDURE

The polymer was first heated at atmospheric pressure to a predetermined temperature T_x above the ordinary glass transition temperature ($\sim$ 97 °C). The pressure was then built up over a period of 15 minutes to a value ΔP_x, which was insufficient to bring about glass formation at T_x. To ensure the latter it was arranged that $(dT_g/dP)\,\Delta P_x = T_x - 120\,°C$ where dT_g/dP was taken equal to 0·03 °C bar^{-1}. Maintaining the pressure constant at ΔP_x the system was then slowly cooled, through the elevated glass transition region, down to 20 °C. At this stage the pressure was released and the sample recovered. The densities of the samples were measured by a flotation technique using sodium sulphate solutions at 30 °C.

Initially, the samples were taken through the above cycle using equipment in which pressures were applied directly to the sample by a piston arrangement. With such a system it was possible to affect marked changes in density, but some difficulty was encountered in obtaining complete reproducibility for fixed values of T_x and ΔP_x. It was therefore decided to apply all pressures hydrostatically. A pressurized reservoir fed oil into the low pressure end of a high compression pump which was driven by compressed air. The pump taken in conjunction with the intensifier could generate pressures up to 8000 bar. The pressure was transmitted to the pressure vessel containing the sample via a high pressure line using a hydrocarbon oil as the hydrostatic fluid. The vessel could be thermostated to $\pm$ 0·1 °C and the pressure maintained to better than $\pm$ 15 bar for the range up to 3000 bar. The polymer was isolated from the hydrostatic fluid either by submerging it in mercury, or by

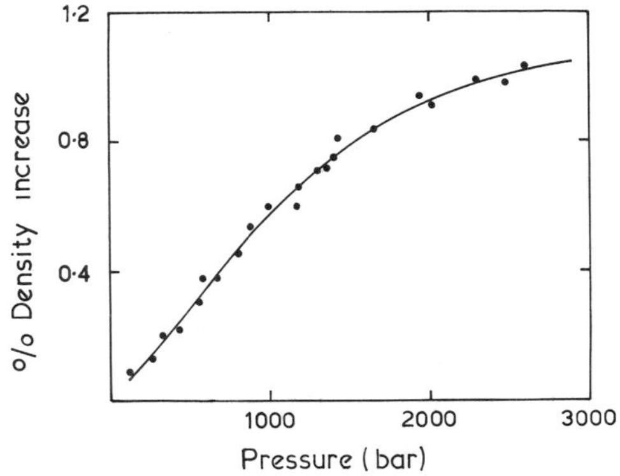

Figure 1. Percentage increase in density of polystyrene glasses (measured at 30 °C and 1 atm pressure) as a function of the pressure applied during cooling from the fluid state

sealing it in gold foil. The dimensions of the internal cylindrical cavity of the pressure vessel were 2 cm diameter × 15 cm length.

In Figure 1 the percentage density increase at 1 atm pressure is plotted against the pressure at which the sample was cooled from the fluid state. It is seen that as the pressure increases $d \ln \rho/dP$ decreases. Such an effect has been observed by other workers (Breuer and Rehage, 1967; Shishkin, 1960) and it has been asserted that there is a limiting pressure at which $d \ln \rho/dP$ becomes zero. However to date we have not carried out measurements at sufficiently high pressures to observe this limiting behaviour.

MICROCALORIMETRY

The Tian–Calvet microcalorimeter used in the investigation was a commercial instrument and its construction and technical operation are described in detail elsewhere (Calvet and Prat, 1963). Briefly the calorimeter consists of two identical elements (a laboratory element and a reference element) thermostated to the same temperature. Each element has two thermoelectric piles, one serving to detect the heat flux between the internal and external boundaries and the other providing Peltier cooling. To obtain optimum accuracy approximately 95% of the heat output from the laboratory element was accurately compensated by Peltier cooling, whilst the remaining thermal flux was measured by a Sefram verispot galvanometer and photodyne spot follower. The instrument was calibrated electrically.

In carrying out the experiment a glass mixing-cell containing toluene was positioned inside the laboratory element. Mixing was initiated by screwing down a microsyringe until a push rod forced a teflon stopper out of the roof of the cell, which in turn allowed the polystyrene sample to fall into the solution. Screwing down the push rod inevitably generated a small quantity of heat. However, by setting up a dummy system in the reference cell we were able to compensate for the effect. In the dummy system a piece of pyrex glass having the same volume as the polystyrene sample in the laboratory cell was released into an equal volume of toluene. The complete mixing process usually took about 10 hours.

Heats of solution ΔH_s are given in column 4 of Table 1 for a range of polystyrene glasses; the final concentration of each solution was $15{\cdot}6\ g\ l^{-1}$. Each ΔH_s recorded is a mean of at least three independent determinations agreeing to $\pm 0{\cdot}03\ cal\ g^{-1}$. Taking the state of a glass cooled at atmospheric pressure as reference, the enthalpy difference ΔH_d given in column 5 was calculated. In Figure 2, ΔH_d is plotted against the pressure at which the glass was formed from the fluid state. Over the lower part of the pressure range (up to 1200 bar) ΔH_d does not deviate significantly from zero, which would (if this behaviour were general) be in keeping with the assertion that the transition occurs at a constant value of the excess entropy or excess enthalpy;

Table 1. Calorimetric data

Sample	Pressure (bar) during cooling	Product at 30 °C and atmospheric pressure Density g cm^{-3}	$-\Delta H_s$ cal g^{-1}	ΔH_d cal g^{-1}
1 (original)	—	$1{\cdot}049_6$	6·39	(0)
2	250	$1{\cdot}051_0$	6·37	−0·02
3	430	$1{\cdot}051_9$	6·40	0·01
4	660	$1{\cdot}052_6$	6·34	−0·05
5	800	$1{\cdot}054_4$	6·40	0·01
6	1300	$1{\cdot}057_1$	6·52	0·13
7	1440	$1{\cdot}057_5$	6·53	0·14
8	1940	$1{\cdot}059_5$	6·66	0·27
9	2300	$1{\cdot}060_0$	6·85	0·46
10	2580	$1{\cdot}060_4$	7·00	0·61
11	2600	$1{\cdot}060_4$	6·96	0·57

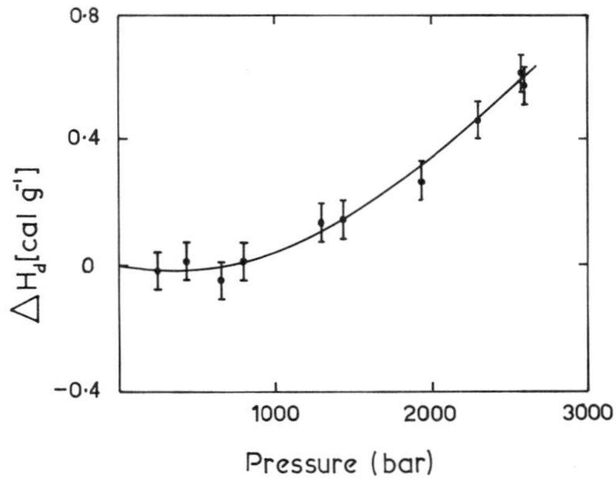

Figure 2. The dependence of the enthalpy of polystyrene glasses on the pressure applied during cooling from the fluid state

such an effect has already been reported by two of us (Allen *et al.*, 1968). If we take in account the complete range covered by the present investigation we must come to a different conclusion. For the higher pressures the deviation of ΔH_d from zero is well outside experimental error and hence the enthalpy as well as the density of a sample seems to depend on the pressure at which it is formed from the fluid state.

On first sight it might appear odd that on packing the molecules of a glass closer together the energy does not decrease. The solution must clearly lie in changes we have brought about in the internal structure of the glass, an obvious explanation for part of this compensating factor must be associated with the fact that we have frozen in the rotational isomerism at different levels.

RELAXATION BEHAVIOUR

Let us now consider the question of the relative thermodynamic stability of the densified glasses. It has been argued that by taking a sample through the densification cycle one is able to reach a state much further down the fluid line than is normally possible (Shishkin 1960). If this was the case densification would imply a lowering of the free energy and hence a state approaching closer to thermodynamic equilibrium. Volume relaxation

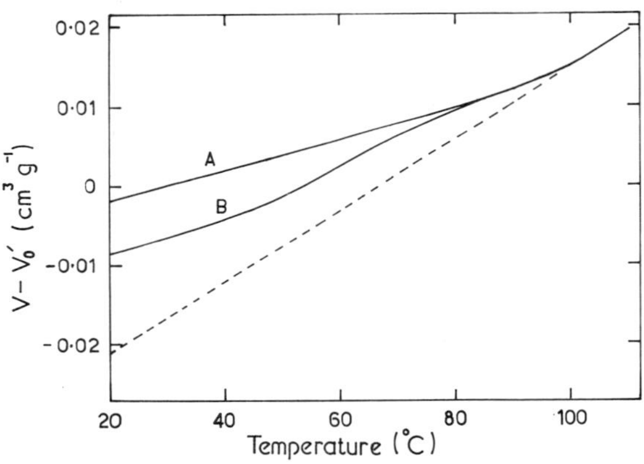

Figure 3. Effect of temperature on the volume of a densified (B) and undensified (A) glass as a function of temperature (— — — linear extrapolation of fluid line: V'_0 volume at 30 °C of the undensified glass)

experiments carried out by us do not support this argument. Below approximately 43 °C dilatometric studies of sample volume as a function of temperature (see Figure 3) were found to be quite reversible (the heating and cooling rates used in these experiments were 0·1 ° min^{-1}). However above this temperature volume relaxation started to occur, and Figure 4 illustrates the rate of volume relaxation at 55 °C. Since on relaxation of the densified glass (which implies decrease in free energy) the volume is found to increase, we must accept that volume cannot be taken as a meaningful measure of the thermodynamic stability. A similar conclusion can also be reached on the basis of the calorimetric data.

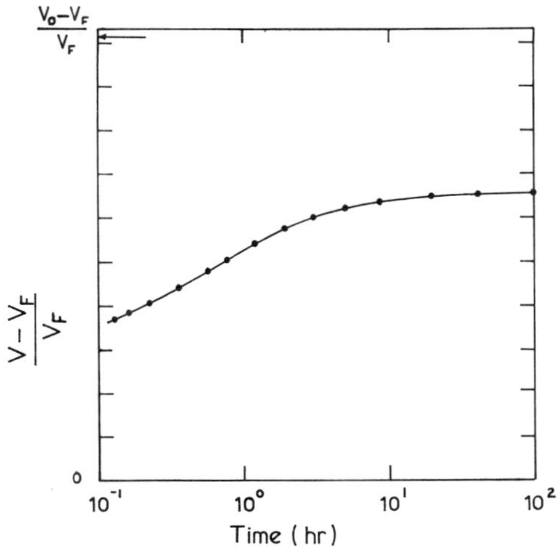

Figure 4. Isothermal volume relaxation of a densified sample at 55 °C. (V_0 and V_F are the volumes of the undensified glass, and the hypothetical fluid respectively at 55 °C)

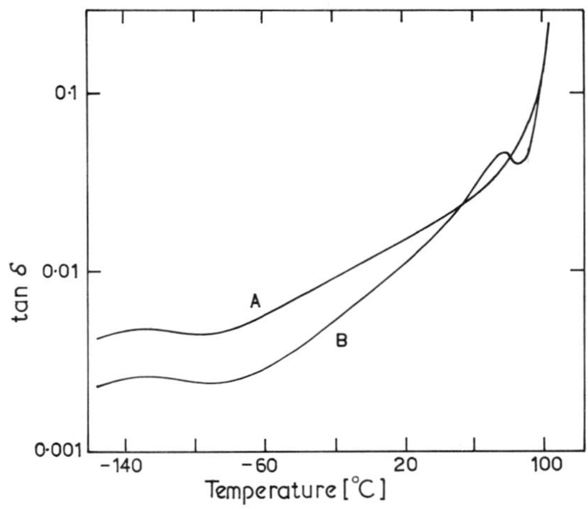

Figure 5. Plots of tan δ *vs.* temperature for a densified (B) and undensified (A) polystyrene glass

In the investigation we have also carried out some dynamic mechanical measurements using a torsion pendulum ($\sim$1 cps). Results of tan δ for both an ordinary and a densified polystyrene glass are given in Figure 5. The densified sample has a reduced loss factor over the whole of the lower temperature range. Apart from this the glasses show similar features, both having a weak γ peak (Illers and Jenckel, 1958) centred at approximately $-129\ °C$. The peak at 79 °C for the densified sample is a common feature of such plots. However it is merely a manifestation of the relaxation phenomenon observed dilatometrically and its intensity is found to vary with time of annealing.

REFERENCES

Allen, G., R. C. Ayerst, J. R. Cleveland, G. Gee, and C. Price (1968). *J. Polym. Sci. C*, **23**, 127.

Breuer, H., and G. Rehage (1967). *Kolloidzeitschrift*, **216**, 159.

Bueche, F. (1962). *J. chem. Phys.*, **36**, 2940.

Calvet, E., and H. Prat (1963). *Recent Progress in Microcalorimetry*, H. A. Skinner (Trans.). Pergamon Press, New York.

Davies, R. O., and G. O. Jones (1953). *Proc. R. Soc.* (London), **26**, A217.

Gee, G. (1966). *Polymer*, **7**, 177.

Goldstein, M. (1963). *J. chem. Phys.*, **39**, 3369.

Goldstein, M. (1964). In J. D. Mackenzie (Ed.), *Modern Aspects of the Vitreous State*. Vol. III. Butterworths, London. p. 90.

Illers, K. H., and E. Jenckel (1958). *Rheol. Acta*, **1**, 322.

O'Reilly, J. M. (1962). *J. Polym. Sci.*, **57**, 429.

Prigogine, I., and R. Defay (1961). *Chemical Thermodynamics*, D. H. Everett (Trans.). John Wiley, New York.

Shishkin, N. (1960). *Soviet Phys. solid St.*, **2**, 322.

Tammann, G. (1933). *Der Glaszustand*, Leipzig.

13

Glass Expansion Behaviour and 'Freezing' Temperature

H. RÖTGER

VEB Jenaer Glaswerk Schott & Gen., Jena, DDR

The basic differences between the transition of glass from the fluid to the solid state and the reverse process are discussed in this paper. A practical way to determine the 'freezing' temperature which marks this transition is described. The curves obtained from expansion measurements extending far beyond the transformation range confirm the theoretical findings.

Figure 1 (Salmang, 1957) shows the usual presentation of the linear expansion behaviour of a glass rod on increasing the temperature. The upper curve marked A illustrates the change in length of an annealed borosilicate glass with increasing temperature, with a steeper rise of expansion starting at about 650 °C. The behaviour of a chilled glass rod (curve B) shows a slight, temporary *reduction* in length above 600 °C. In the case of a chilled glass fibre of 6 μ thickness, a length reduction occurs even at relatively low temperatures, as shown by curve C. This graph, whilst showing the dependence of the expansion curve on previous sample treatment, does not provide any clue for understanding the nature of such pronounced differences.

Salmang neglects the fact that, owing to varying thermal treatments, the initial lengths are not the same and may not be immediately equated. Exact scientific studies can be based only on the supercooled liquid state above the 'freezing' temperature, as that state is thermodynamically defined, i.e. the glass volume and, consequently, the length of a glass rod are functions of temperature only, at constant pressure. The reference point, therefore, should be chosen somewhere on this equilibrium curve.

In the following, the rates of change of temperature are always assumed to be constant. In Figure 2, the change in length of a BK7 glass sample is shown as a function of temperature at a constant cooling rate. At high temperatures, this change in length takes place along the equilibrium curve with

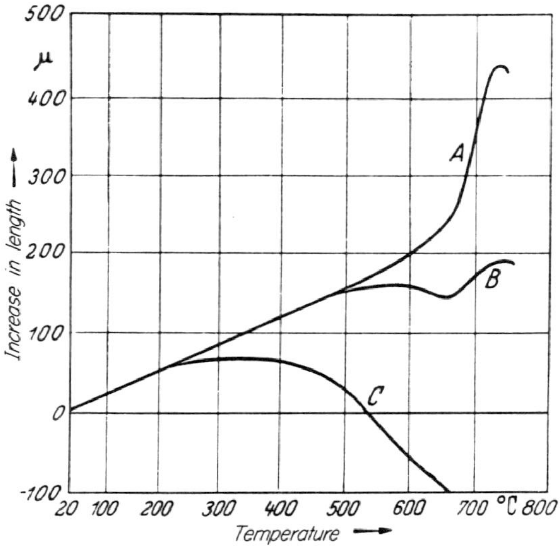

Figure 1. Length increase of a borosilicate glass sample after different thermal treatments: A annealed, B chilled glass rod, C chilled glass fibre, 0·006 mm thick (after Salmang, 1957)

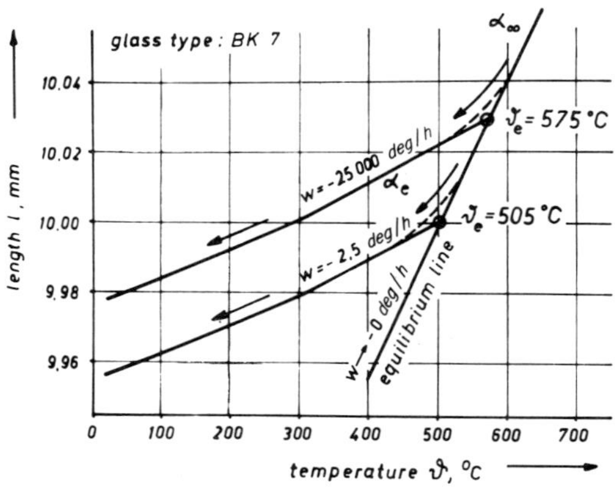

Figure 2. Length of a glass rod reducing with temperature at different cooling rates (schematic) (after Rötger, 1969)

an expansion coefficient of α_∞; at a certain (extrapolated) temperature the expansion curve bends to follow a curve having the markedly lower expansion coefficient of α_e. With $w = -25{,}000$ deg/h (corresponding to the upper curve), this bending takes place at a temperature $\vartheta_e = 575$ °C. As the sample is cooled down further towards the ambient temperature, the expansion coefficient is practically independent of the cooling rate.

At a cooling rate $w = -2{\cdot}5$ deg/h (corresponding to the lower curve), the expansion curve bends at a lower temperature, viz. at $\vartheta_e = 505$ °C. At this point, the length change assumes the behaviour known in the solid state, with practically the same expansion coefficient of α_e, and takes a course parallel to the upper curve down to ambient temperature. At the ambient temperature a difference in length is observed, which is characteristic of the difference in cooling rates. The glassy state is characterized by the temperature of the curve break (the extrapolated transition point between α_∞ and α_e). Let us designate this 'freezing' temperature correlated to the cooling rate w^- as 'rate-dependent "freezing" temperature $\vartheta_{e;\,w^-}$'. There exists a relationship that is linear, within wide limits, between $\vartheta_{e;\,w^-}$ and the length measured at ambient temperature on the one hand and the cooling rate w^- on the other.

In all cases of constant-rate annealing the state prevailing at room temperature is completely defined by a *single* parameter (w^-).

Figure 3 illustrates what happens if the same two BK7 glass rods are heated up at identical heating rates of 300 deg/h. The first part of the curve shows the same behaviour as that of Figure 1.

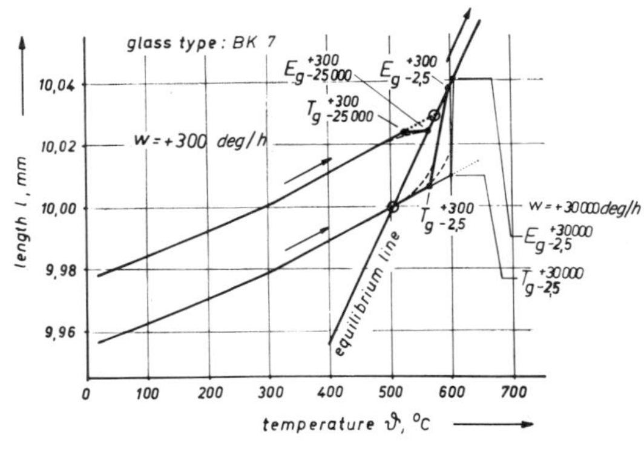

Figure 3. Length change of glass rods heated up at constant rates of 300 deg/h and 30,000 deg/h after pre-annealing as in Figure 2 (schematic) (after Rötger, 1969)

The expansion curve of the annealed rod (lower curve) deviates from the solid glass heating curve only at high temperatures. At the so-called 'transformation temperature' or 'thawing temperature' T_g the curve takes a steep upward course and then again follows a less steep, straight equilibrium line. Figure 1 does not show this bending, indicating that the information obtainable from conventional measurements is insufficient.

The expansion curve of the chilled rod (upper curve) deviates from that of the solid state at relatively low temperatures. The break in the curve formally corresponds to the 'transformation' or 'thawing' temperature, although in this case the deviation is a bending *downward*. Thus the curve reaches the equilibrium line which it then follows upward.

In either case, the difference in behaviour is characterized by the tendency to reach the equilibrium state (Rötger, 1959). The 'thawing' of the annealed rod takes place only after the equilibrium line has been crossed—i.e. above the 'freezing' temperature ϑ_e—and owing to this overheating a pronounced break in the curve occurs, which for some decades has induced scientists to attach physicochemical significance to that temperature. Now it is generally known that, in addition to the previous thermal history of the measured rod, this point also depends on the heating rate, although this dependence is a complicated one. Therefore, that break point must be characterized by means of two rate indices $T_{g\,w-}^{\,w+}$, with the subscript indicating the rate of the previous cooling of the glass and the superscript the heating rate employed during the measurement.

In the following, the temperature change rate w is always given in terms of deg/h. In our example, $T_{g-2\cdot5}^{+300}$ lies at 564 °C, whereas $T_{g-25,000}^{+300}$ lies at 521 °C. Thus, the thawing temperature T_g of the chilled specimen is lower than the corresponding 'freezing' temperature $\vartheta_{e;\,-25,000} = 575$ °C. After passing the thawing temperature $T_{g\,w-}^{\,w+}$, the heating curves join the line of equilibrium at the 'junction temperatures' $E_{g\,w-}^{\,w+}$, which are also qualified by two rate indices. Analogous, but converse differences can also be observed with identical cooling rates (subscript) and different heating rates during the test (Bartenev, 1966) (superscript). For example, when the heating rate of the lower curve is multiplied by 100, $T_{g-2\cdot5}^{+30,000} = 599$ °C.

From the above it is evident that all heating curves, irrespective of the difference in their initial stages change over into one common straight line of equilibrium. The point of intersection of the heating curve or the extrapolated heating curve, respectively with the equilibrium line will always indicate the 'freezing' temperature $\vartheta_{e;\,w-}$ introduced above, which is characteristic of the rate w^- of the previous cooling process. If the ratio of the amounts of w^+ and w^- is greater than about one, the thawing temperature T_g is always higher than ϑ_e. In our present example (BK 7 glass), with $w^- = -2\cdot5$ deg/h and $w^+ = +300$ deg/h, the difference is 564 °C − 505 °C = 59 deg, which is typical of normal transformation temperature tests. Thus, the common transformation temperature should by no means be equated with the 'freezing' temperature of the glass type used.

This knowledge can be utilized to give a more informative presentation of the Salmang heating curves shown in Figure 1. For this purpose the equilibrium line ∞ is drawn in Figure 4 more or less arbitrarily. Of the curves A, B and C, curve A as a reference curve remains in its old place. Curve B is shifted upwards equidistantly until the bending point, to be interpreted as 'junction temperature', lies on the line of equilibrium (dashed line B′). In the case of very rapidly cooled glass fibre sample C, the curve had to be shifted upwards through a considerably greater distance in order to make the weakly marked turning point (interpreted as E_{gC}) lie on the equilibrium line

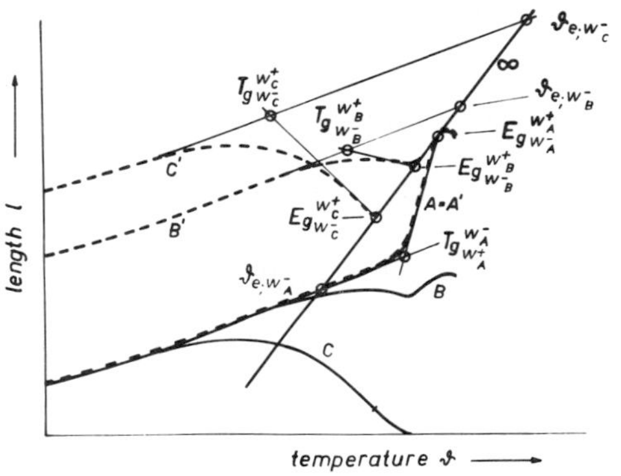

Figure 4. Salmang curves (Figure 1) displaced in parallel, with arbitrary equilibrium line (∞), for determination of freezing temperature ϑ_e, thawing temperature T_g, junction temperature E_g and length changes at ambient temperature at three cooling rates w^- (schematic)

(C′). The cooling and heating cycle for the three cooling rates $|w^-_A| < |w^-_B| < |w^-_C|$ may now be described as follows: at slow cooling (w^-_A) the sample length would change along the equilibrium line up to $\vartheta_{e;w^-A}$ and then (in an *equalized* form not plotted in the figure) pass over into the expansion curve of the solidified phase. Because of its very slow cooling its viscosity on re-entering the transformation range is very high and its expansion corresponds to that of the solidified glass in a range extending beyond the equilibrium line, and thawing takes place as 'late' as at $T_{gw^-A}^{w^-A}$. With the high superheating rate applied subsequently, the length change curve joins the equilibrium line at $E_{gw^-A}^{w^+A}$.

The length of the glass fibre cooled down at the very high rate of w^-_C would depart from the equilibrium line already at the temperature $\vartheta_{e;w^-C}$, so that a relatively low viscosity would be frozen in in that case. On reheating, this structure therefore will thaw at the very low thawing temperature $T_{gw^-C}^{w^+C}$ and, while contracting, will approach the equilibrium line, which is reached at $E_{gw^-C}^{w^+C}$ and along which subsequent expansion takes place. The glass rod cooled at the medium rate w^-_B, the $|w^+/w^-|$ ratio of which is considerably smaller than 1, also shows the phenomenon described for glass rod C but to a lesser degree.

Thus Figure 4 illustrates how the largely differing expansion behaviours may be united in one common scheme by means of the three quantities ϑ_e, T_g and E_g, which is likely to be important for our understanding of the relations involved. For practical applications, e.g. for ascertaining the expansion behaviour for w^+, w^- being given, it is necessary to know the functions

$$\vartheta_e = f(w^-),\ T_g = f(w^-;\ w^+) \text{ and } E_g = f(w^-;\ w^+).$$

If the three quantities are known at rated conditions (e.g. $w^- = -2{\cdot}5$, $w^+ = +300$), it is relatively simple to determine the *displacements of temperatures* by means of $\lg|w^-|$ or $\lg w^+$, respectively, according to experience described in the literature or gained from our own previous measurements. For example, the quantity $d\vartheta_e/d \lg|w^-| \equiv \Delta_{10}\vartheta$ is practically constant within a wide range of cooling rates. Further it is apparent that $\partial T_g/\partial \lg w^+ = \Delta_{10}\vartheta(T_g > \vartheta_e)$ and $\partial T_g/\partial \lg|w^-| = -\Delta_{10}\vartheta(T_g < \vartheta_e)$. The basic quantity ϑ_e can be found directly from a cooling curve. It is possible, however, to determine the three quantities named above, as well as the equilibrium line (α_∞) and the expansion coefficient α_e below ϑ_e from a single test graph, such as is represented by Figure 5. The heating curve at $w^+ = +300$ deg/h of a 16^{III} glass sample previously cooled at 2·5 deg/h has been recorded by means of an interference dilatometer. This instrument (Pulfrich, 1898) permits measurements free from deformation far beyond junction temperature. The thawing temperature found is $T_{g-2\cdot5}^{+300} = 572$ °C, which in this case corresponds to the normal transformation temperature. Further, $E_{g-2\cdot5}^{+300} = 592$ °C, and the intersection point of $\alpha_e = 11{\cdot}4 \times 10^{-6}\ \text{deg}^{-1}$ and $\alpha_\infty = 33{\cdot}2 \times 10^{-6}\ \text{deg}^{-1}$ yields the 'freezing' temperature $\vartheta_{e;-2\cdot5} = 507$ °C.

It is also possible to obtain the 'freezing' temperature by determining the fictive temperature $\vartheta_{f;w^-}$ (which is equivalent to the cooling rate w^-) from measurements of refractive index on samples cooled at constant rates (Besen, 1967; Hagy and Ritland, 1957). For the present 16^{III} glass sample we obtained $\vartheta_{f;-2\cdot5} = 507$ °C (Rötger, 1969), i.e. the same value as that obtained from the expansion test illustrated in Figure 5. This comparison of $\vartheta_{f;-2\cdot5}$ and $\vartheta_{e;-2\cdot5}$ has been made on 56 optical glass types and coincidence within 10 deg has been observed. In a few more glasses, major deviations have been found (Rötger, 1969).

In coupling cooling specifications for glasses it has been customary to proceed essentially from transformation temperatures. According to the considerations reported in the present paper, the 'freezing' temperature for a specified, adequately high cooling rate is more suitable. The scattering occurring in different glasses of the viscosities corresponding to the 'fixed points' is a measure of that suitability. A cooling rate of 250 deg/h has been chosen, and for the equilibrium viscosities η_∞ corresponding to $\vartheta_{e;-250} = \vartheta_{e;-2\cdot5} + 2 \times \Delta_{10}\vartheta$ it was found that $\lg \eta_{\vartheta_{e;-250}} = 13{\cdot}2 \pm 0{\cdot}3$ in 21 glass types. The same glasses were found to have a transformation temperature

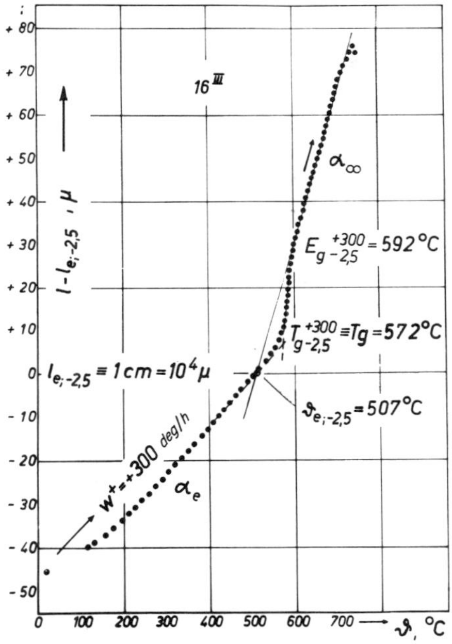

Figure 5. Interference dilatometer test curve of a 16^{III} glass sample heated at 300 deg/h (previously cooled at 2·5 deg/h), for determination of the expansion coefficient of solid glass in the freezing range (α_e), the equilibrium line of the supercooled liquid phase (α_∞), the thawing temperature $T_{g-2\cdot5}^{+300}$, the junction temperature $E_{g-2\cdot5}^{+300}$ and the 'rate-dependent freezing temperature" $\vartheta_{e;-2\cdot5}$ (after Rötger, (1969)

with $\lg \eta \ {}^{+300}_{-0\cdot5} = 12{\cdot}3 \pm 1{\cdot}1$. At the 'freezing' temperature corresponding to the cooling rate of 250 deg/h, all glasses tested showed practically the same viscosity within the measuring accuracy, whereas there were differences of more than two decimal places in the viscosities prevailing at the related transformation temperatures. This underlines the superiority of the 'freezing' temperature $\vartheta_{e;-250}$ over the 'transformation temperature', which in accordance with its nature as a thawing temperature lacks the quality of a fixed point.

REFERENCES

Bartenev, G. M. (1966). *Structure and mechanical properties of inorganic glasses.* Moscow. p. 25.

Besen, H. (1967). *Über die Abhängigkeit der Brechzahl optischer Gläser von der Temperatur und Wärmebehandlung.* Thesis, Friedrich-Schiller-Universität Jena. (Abstract). (1969). *Silikattechnik*, **20**, 334.

Hagy, H. E., and H. K. Ritland (1957). *J. Am. Ceram. Soc.*, **40**, 436.

Pulfrich, C. (1898). *Z. f. Instrumentenkunde*, **18**, 261. See also Lehmann, H. and Schulze, M. Th. (1936). *Zeiss-Nachr.*, **10**, 6.

Rötger, H. (1959). *Silikattechnik*, **10**, 524, and *Corrigenda* (1960). *Silikattechnik*, **11**, 146.

Rötger, H. (1969). *Silikattechnik*, **20**, 404.

Salmang, H. (1957). *Die physikalischen und chemischen Grundlagen der Glasfabrikation.* Springer Verlag, Berlin–Göttingen–Heidelberg. p. 50.

14

Repeatability Number, Deborah Number and Critical Cooling Rates as Characteristic Parameters of the Vitreous State

J. M. STEVELS

Philips Research Laboratories, N.V. Philips' Gloeilampenfabrieken, Eindhoven, the Netherlands

A glass is a solid formed from a liquid by cooling without showing first order phase transition. When a melt is cooled down slowly through its thermodynamic melting point or its liquidus temperature, T_1, nucleation followed by crystal growth may occur. However, if the crystallization does not occur on cooling, the structure of the melt becomes more and more rigid, the viscosity increasing approximately exponentially with decreasing temperature. When the viscosity reaches 10^{13}–10^{14} p at the 'transformation' temperature, T_g, the material has become 'rigid' from a practical point of view and is now called a glass.

Zachariasen (1932) in his classical paper stressed that crystals and the corresponding glasses are built up from the same structural units, the difference being that in crystals the structure is characterized by long- and short-range periodicity, whereas in glasses only short range order is present and long range order is absent.

Prins (1965) defined a glass as a solid showing absence of long-range order and absence of short time fluidity. Or perhaps, in a more positive way, glass is a solid showing both short-range order and long time relaxation phenomena only.

An old concept is to say that glass is an undercooled or supercooled liquid, but from the foregoing it may be concluded that there is a distinction between the two. Supercooling is a very important way of producing glasses but it is only one of many other methods. Secrist and Mackenzie (1965) showed that non-crystalline forms of SiO_2 can be made by many methods other than the conventional one, viz. the cooling of a melt. Roy (1970) also mentioned a number of non-crystalline forms of SiO_2 of which some coincide with the forms given by Secrist and Mackenzie. Combining the two lists it may be said that at least eleven different methods of producing non-crystalline silica are mentioned in the literature.

They are:

(1) vapour-phase hydrolysis;
(2) desiccating gels:
(3) deposition from organic solutions;
(4) conventional melting;
(5) neutron bombardment;
(6) reactive sputtering;
(7) thermal decomposition;
(8) glow-discharge decomposition;
(9) shock-wave transformation;
(10) formation of shear-amorphized layers;
(11) formation of reaction-amorphized solids.

Of these methods, number 1 starts from vapour, numbers 2–4 from liquids and numbers 5–11 from solids (number 6 is effected *via* the vapour phase).

It is not very likely that all these forms have the same microstructures. Roy states very clearly that as it is fully accepted that polymorphy exists in the different crystalline arrangements of SiO_2, we must expect analogous 'polymorphs' with different non-crystalline structures.

According to Roy some of these structures, for instance that of a shear-amorphized solid-phase and a vapour deposited phase of the same composition, will differ radically from one another. There will be minor differences in structure, even within the range of samples of fused silica obtained from a conventional melt cooled over a period of many weeks; or by quenching in 2 to 3 seconds over the same temperature range, or by quenching from, for instance, a 5 kb pressure.

These examples are well-chosen and everybody who has worked with these different methods of producing fused SiO_2 will affirm that differences *do* indeed exist, but it seems difficult to express them in a quantitative way. Yet it appears that there are a few parameters which must be important in this context. These are the Deborah Number (D.N.) and the Repeatability Number (R.N.), two concepts not very well-known so far in the glass literature, though the former is in the literature on rheology.

REPEATABILITY NUMBER

The concept of the distribution function is well-known in the literature on glass. It gives the number of atoms $g(r)$ round a central atom at a distance r.

Figure 1 gives five different distribution functions in one dimension for a perfect crystal, a colloidal system, a liquid system and a liquid system near the critical point showing fluctuation. They have been discussed earlier by Prins (1965).

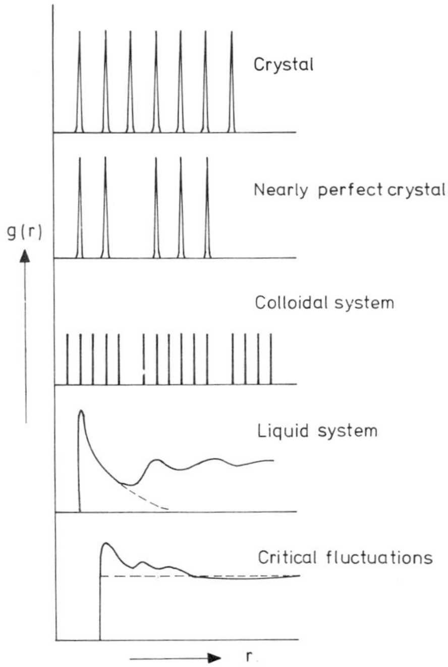

Figure 1. Distribution functions $g(r)$ for five different arrangements in one dimension as a function of the radius vector r. Four of them have been published earlier by Prins (1965)

The nearly perfect crystal is also shown in Figure 1 and is characterized by the fact that Schottky defects are present (or Frenkel defects, not represented here to avoid complicated drawings).

It is difficult to make similar models for three-dimensional distribution functions on theoretical considerations. However, it is not so difficult to visualize them in an experimental way, as they are the Fourier inversions of the X-ray diffraction patterns after subtracting their asymptotic values. The distribution functions obtained in this way are in fact electron density distribution curves. In a number of simple cases pictures may be given of such electron density curves for vitreous systems, especially so in the case of a uniform radial distribution, so that the electron density can be plotted as a function of the distance r from an 'averaged' central atom.

It would be useful to dispose of preferably *one* quantity that expresses the degree of order in such a system. It is proposed that this quantity is the repeatability number denoted by $(\mathrm{R.N.})_{r_0}$.

$(\mathrm{R.N.})_{r_0}$ gives the degree of order in a limited area, its size being characterized by a radius r_0. It is derived via the concept of CRS, the coefficient of repetition in space. In fact CRS is a quantity which has a certain relation to the distribution function $g(r)$. It may be called the averaged and normalized distribution function. CRS is still a function of r as shown in Figure 2. A more extensive treatment of the derivation of $(\mathrm{R.N.})_{r_0}$ will be given elsewhere (Stevels, 1971).

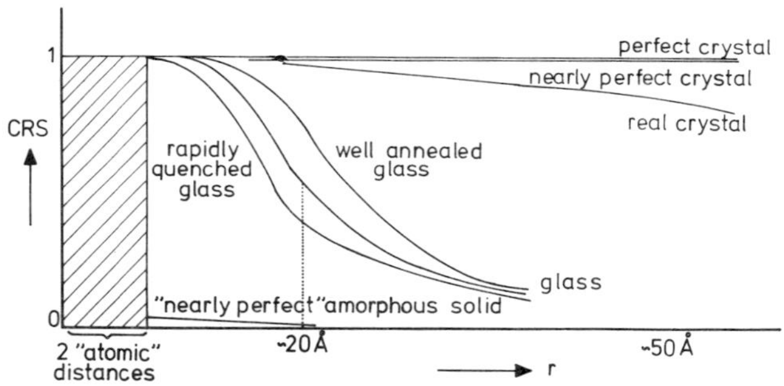

Figure 2. Coefficient of repetition in space (CRS) as a function of the distance (r) to an 'averaged' central atom in a number of different situations

If the degree of order is the same everywhere, one can define a value R.N., the repeatability number, which is a measure of the overall degree of order in that material.

In many cases, and especially in glasses, the degree of order will not be the same everywhere. Therefore, it is useful to speak of 'local' R.N. values. An overall R.N. value may be defined as the weighted average of all 'local' R.N. values. The knowledge of the variations in local R.N. values will be of great importance to the study of such questions as supercritical fluctuations, nucleation, and phase separation. For the perfect crystal the overall R.N. value will be 1, for the nearly perfect crystal it is slightly smaller than 1, for vitreous systems it may be between 1 and 0, with a great range of variations.

It seems appropriate to restrict the name of amorphous materials to those for which the value of R.N. is very low indeed over the whole area considered.

One can distinguish then, 'perfect' amorphous solids, for which R.N. = 0 (a situation that will hardly ever occur) and nearly perfect amorphous solids, where R.N. has a very small value. Vitreous or non-crystalline solids normally are characterized by a finite value of R.N., *therefore they should not be called amorphous.*

Finally, Figure 3 gives a survey of a number of 'situations' shown in the sequence of increasing R.N. values. Not much imagination is required to see that the eleven different methods mentioned earlier are, generally speaking,

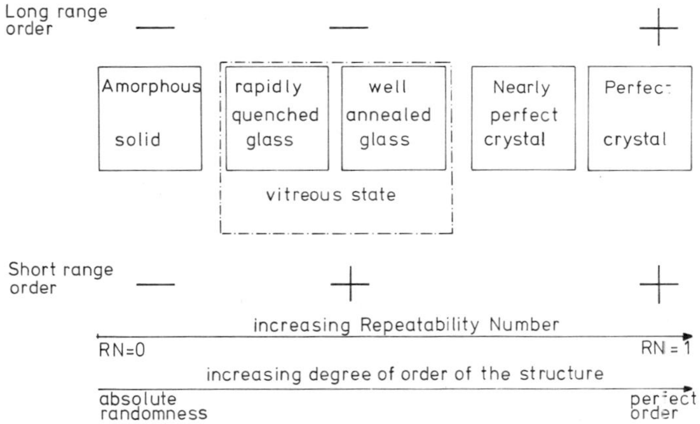

Figure 3. A number of situations shown in the sequence of increasing R.N. values

characterized by an increasing R.N. in the order 1–11 (though there may be an overlap) and that the difference between the three samples of fused silica obtained by conventional melting, but annealed in a different way, as mentioned by Roy and quoted above in the order as given, are characterized by a decreasing R.N.

DEBORAH NUMBER

The concept of the Deborah Number was introduced by Reiner (1964) during the 4th International Congress in Rheology, from which a few paragraphs are quoted with minor alterations.

The name goes back to a story in the Bible. In Judges 5, verse 5, the Prophetess Deborah states in the original language: 'The mountains *flowed* before the Lord'. When the Bible was translated into English, the translators made an error and translated the passage as 'The mountains *melted* before the Lord', and so it is written now in the authorized English version.

Anyhow, Deborah knew—and so do we nowadays—two things. First of all that the mountains *do* flow, as everything flows. But secondly, that they flow before the Lord, and not before man, for the simple reason that man in his short lifetime cannot observe this flow, while the time of observation of God is infinitely long.

We may therefore well define a dimensionless number, the Deborah Number as

$$\text{D.N.} = \frac{\text{time of relaxation}}{\text{time of observation}}$$

The difference between solids and fluids is then defined by the magnitude of the D.N.

If the time of observation is large, or conversely, if the time of relaxation of the material under observation is small, in other words if the D.N. is small, one sees the material flowing. On the other hand, if the D.N. is much larger, the material behaves as a solid.

It therefore appears that the D.N. is destined to become a fundamental number, bringing solids and fluids (and also glasses) under a common concept. Suppose a liquid is cooled from a high temperature, with the cooling rate of 2 °C min^{-1} which is normal practice for the formation of the usual

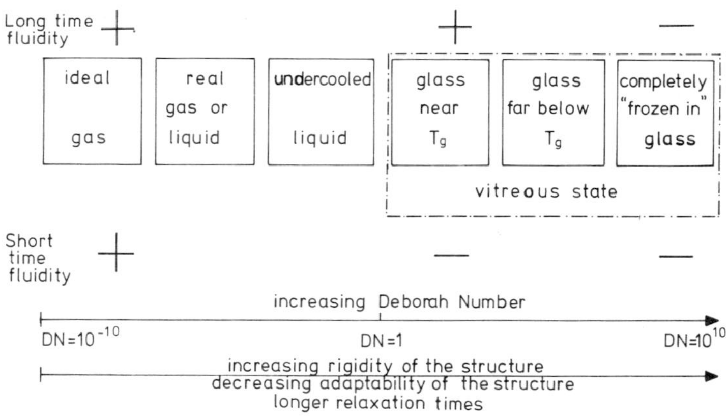

Figure 4. A number of situations shown in the sequence of increasing D.N. values

types of silicate glasses. In the liquid range the relaxation times may increase from 10^{-13} to 10^{-8} sec, to a few minutes at the transformation 'temperature' T_g. Thus, the transformation temperature is defined as the temperature where the D.N. = 1. In the vitreous state, (that is *below* T_g) the relaxation time of the material increases rapidly with the temperature and therefore the structure of the material cannot adapt itself to the equilibrium position belonging to the corresponding temperature. Therefore, in the region below T_g the D.N. increases rapidly with decreasing temperature to values of about 10^9 at room temperatures.

Little can be said at the moment about the real values of the D.N., but some rough estimates can be made. However, it is possible to give a survey diagram of a number of situations shown in the sequence of increasing D.N. values (Figure 4).

RELATION WITH CRITICAL COOLING RATES

In the last few years it has been found that more and more groups of materials can be obtained in a vitreous form, provided the cooling rate is sufficiently high. The lowest cooling rate for which a system does not show crystallization, either partial or complete, is called the *critical cooling rate* (CCR). Many CCR values have been measured and an interesting correlation exists between the CCR and the tendency to glass formation (Gelsing *et al.*, 1966).

Generally speaking the CCR for silicates is of the order of 0·03 °C sec^{-1}, for tungstates of the order of 10^2 °C sec^{-1}, for molybdates of the order of 10^4 sec^{-1} and by modern methods such a splat cooling, groups of materials have become known where the CCR is of the order of 10^6 °C sec^{-1} (Sarjeant and Roy, 1968). Generally speaking the CCR will be smaller, the less the liquid is ordered already, i.e. the lower the R.N. value of the liquids is. In other words, CCR and R.N. will have the tendency to show the same trend.

The D.N. and the CCR will tend to show an opposite trend. The higher the D.N. of the liquid the lower CCR, and conversely.

PRACTICAL RESULTS TO DATE

Finally the question can be put what are the practical results to date from these rather vague considerations?

Figures 3 and 4 clearly depict glass as a system characterized by the absence of short time relaxation phenomena and having structures characterized by the absence of long range order. It also shows that 'glass' does not always mean the same thing. A large variety of situations exists for systems of the same composition which all are called glass, their differences being characterized by different D.N. and R.N. values. A few examples have been given in this paper.

A number of non-conventional methods of glass formation can be better understood in terms of these new concepts. Secrist and Mackenzie (1964) have summed up a number of these non-conventional methods. Some of them we have already mentioned, namely shock-wave and neutron-bombardment transformation which give rise to the formation of glass from a crystalline solid. The order of the material is disturbed by these processes and the R.N. is decreased.

Glasses formed from direct vaporization or by indirect vapour techniques are obtained by a process where the R.N. of the system is increased. However, glasses obtained in such a way have a considerably lower R.N. than those of the category first mentioned.

Liquids with relatively high D.N. values will tend to form glasses on cooling. Liquids with low D.N. values will tend to crystallize, but with the recent technical possibilities of realizing high d (D.N.)/dt values, it is possible to pilot these liquids through the 'dangerous' zone and bring them into the vitreous state.

REFERENCES

Gelsing, R. J. H., H. N. Stein, and J. M. Stevels (1966). *Physics Chem. Glasses*, **7**, 185; van der Wielen, J. C. Th. G. M., H. N. Stein, and J. M. Stevels (1968). *J. non-cryst. Solids*, **1**, 18.

Primak, W. (1958). *Phys. Rev.*, **110**, 1240.

Prins, J. A. (1965). *Proc. Internl. Conf. Physics of Non-crystalline Solids*. North-Holland Publishing Company, Amsterdam. p. 1.

Reiner, M. (1964). *Physics to-day*, **17**, 62.

Roy, R. (1970). *J. non-cryst. Solids*, **3**, 33.

Sarjeant, P. T., and R. Roy (1968). *Mater. Res. Bull.*. **3**, 265; (1967). *J. Am. Ceram. Soc.*, **50**, 500.

Secrist, D. R., and J. D. Mackenzie (1964). *Glass Ind.*, **45**, 408.

Secrist, D. R., and J. D. Mackenzie (1965). *J. Am. Ceram. Soc.*, **48**, 487.

Stevels, J. M. (1971). To be published.

Weyl, W. A., and E. C. Marboe (1962). *The Constitution of Glasses*. John Wiley and Sons, New York and London. pp. 132–138.

Zachariasen, W. H. (1932). *J. Am. chem. Soc.*, **54**, 3841.

15

A New Viscoelastic Function for Studies of Transition Phenomena in Polymers

J. F. JANSSON

Department of Polymer Technology, Royal Institute of Technology, Stockholm, Sweden

---

In studying viscoelastic transition phenomena in polymers 'mechanical spectroscopy', i.e. analysis of periodic or transient viscoelastic functions, is often used.

In the following a new viscoelastic function is described with makes possible frequency(time)–temperature transformations, calculation of creep compliance from complex compliance, and separation of viscoelastic mechanisms.

A linear viscoelastic system gives compliances $J_1(t)$ and $J_2(t)$ in response to the periodic functions:

$$\sigma_1(t) = \begin{cases} \sigma_0; & 2n\tau < t < (2n+1)\tau \\ -\sigma_0; & (2n+1)\tau < t < (2n+2)\tau \end{cases} \qquad \text{square wave}$$

$$\sigma_2(t) = \begin{cases} (2n+1)\sigma_0 - \dfrac{\sigma_0}{\tau}t; & 2n\tau < t < (2n+1)\tau \\ -(2n+3)\sigma_0 + \dfrac{\sigma_0}{\tau}t; & (2n+1)\tau < t < (2n+2)\tau \end{cases} \qquad \text{saw tooth}$$

where $n = 0, 1, 2, 3\ldots$

If the period of these functions is 2τ then the time $\tau/2$ is of special interest because the sum of the compliances at this point agrees approximately with the creep compliance for the same time. This has been shown by Jansson (1967) and by Fourier transformation of the stress functions it is possible to calculate the creep compliance from the complex compliance

$$J(\tau/2) = J_1(\tau/2) + J_2(\tau/2) \equiv$$

$$\equiv \frac{4}{\pi} \sum_{n=0} (-1)^n \frac{1}{2n+1} J'_{2n+1} + \frac{8}{\pi^2} \sum_{n=0} (-1)^n \frac{1}{(2n+1)^2} J''_{2n+1}$$

where J'_{2n+1} and J''_{2n+1} respectively are the real and imaginary parts of the complex compliance at the frequencies $f = 1/(2n+1)2\tau$ Hz and $n = 0, 1, 2, 3 \ldots.$

The accuracy of this interrelation is of the same order as the corresponding equations by Ninomiya and Ferry (1959) and Struik and Schwarzl (1969).

Owing to the fact that J_1 and J_2 are the solutions to an odd and even periodic function, $J_1(\tau/2)$ and $J_2(\tau/2)$ are measures of the 'storage' and 'loss' mechanisms respectively. Thus the function $\log J_2 = g\{\log J_1\}$ reflects those special processes which cause the linear viscoelastic behaviour of the system.

The compliance function $\log J_2 = g\{\log J_1\}$ has been studied for the series of polymers listed with references in Table 1.

Table 1. Studied polymers

Polyvinylacetate PVAc,	(Williams, 1954)
PVAc plasticized	(Williams, 1955)
Polymethylacrylate PMA	(Williams, 1955)
Polyethylmethacrylate PEMA	(Ferry, 1957)
Poly-n-butylmethacrylate P-n-BMA	(Child, 1957)
Poly-n-octylmethacrylate P-n-OMA	(Dannhauser, 1958)
Poly-n-dodecylmethacrylate P-n-DMA	(Karuth, 1959)
Polyvinylchloride PVC plasticized	(Fitzgerald, 1953)
PVC gel	(Fitzgerald, 1953)
Polymethylpenten PMP	
Natural rubber NR	(Ferry, 1969)
Neoprene CR	(Yin, 1963)
Hypalon	(Kurath, 1959)
Polyisobutylene PIB	(Fitzgerald, 1953)

In order to be able to make a reasonably rapid study of the function for a large number of polymers direct measurements of J_1 and J_2 have not been made, but the compliances have been calculated from dynamic data in accordance with equation (1).

To avoid the uncertainties of the 'method of reduced variables', especially when superposed transition mechanisms occur, the calculations have been based on unshifted data when available. Thus, J_1 and J_2 are, in general, studied as functions of temperature at constant frequency (time).

Figure 1 shows log J_1 and log J_2 as functions of temperature for some polymers with monosubstituted carbon atoms in the main chains. The curves cover mainly the polymers' glass transitions. It is of interest to notice that log J_2 varies linearly with temperature in a region around the glass point, and that the slopes of the linear parts correspond approximately.

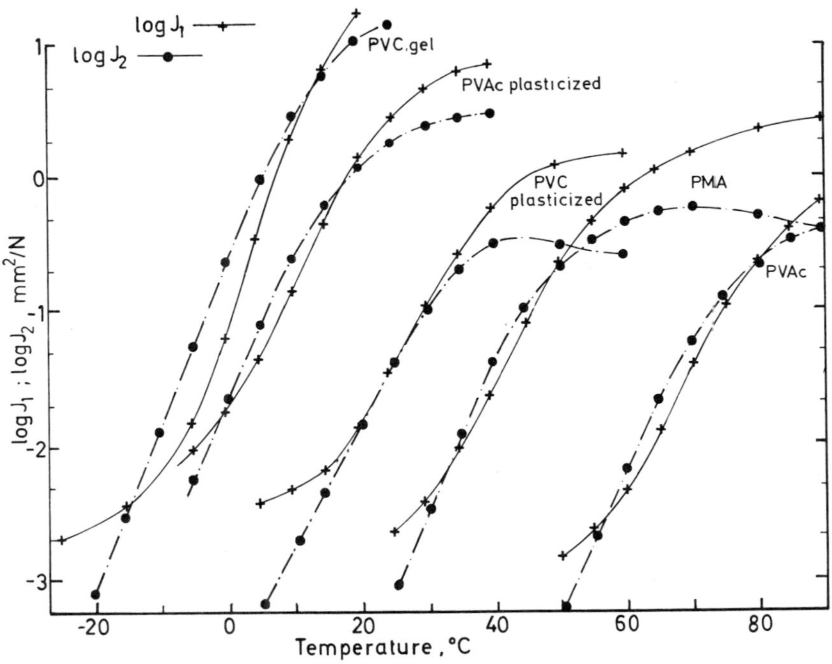

Figure 1. Log J_1 and log J_2 as functions of temperature at $t = 2{\cdot}5 \times 10^{-3}$ s

In Figure 2 the function log $J_2 = g\{\log J_1\}$ has been plotted for the polymers in Figure 1 and for some additional polymers with monosubstituted carbon atoms. In polymers with monosubstituted carbon atoms usually only one transition mechanism occurs in the main transition region. Thus in this respect the conditions for the method of reduced variables to be valid are fulfilled. PMP and CR are exceptions and both show secondary mechanisms in the plateau zone. For these polymers only master curves have been available and the authors consider that the method of reduced variables is valid.

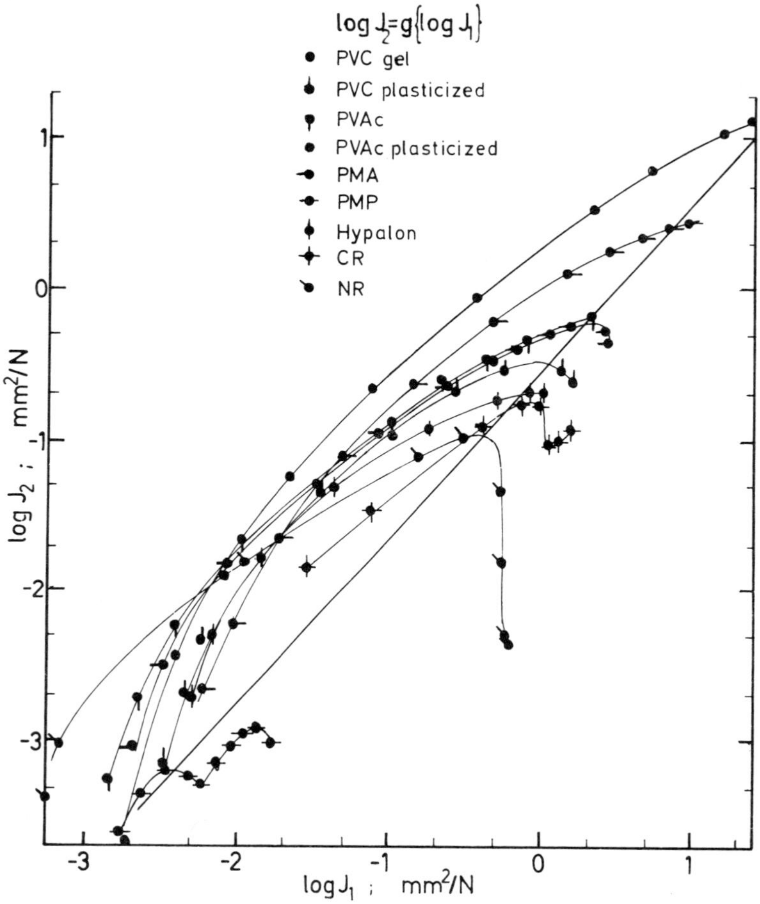

Figure 2. Log J_2 *vs.* log J_1 for various polymers; log $J_2 = g\{\log J_1\}$

It is of interest to notice that the maxima of all the curves in Figure 2 approximately fall along the straight line:

$$\log J_2 = a \log J_1 + b \tag{2}$$

where $a \approx 1{\cdot}10$ and $b \approx -0{\cdot}52$.

In Figure 3 the curves in Figure 2 have been shifted along the straight line until the maxima coincide or, in some cases, until the regions of the curves near the maxima coincide. From these curves a general function log $J_2 = g_0\{\log J_1\}$ can be derived from the relation

$$\log J_2 = g_0\,\{\log\,(J_1 - J_{10})\} \tag{3}$$

where J_{10} is the value of J_1 in the region where the main transition mechanism is 'frozen in' and J_2 goes to zero. The function $\log J_2 = g_0\{\log J_1\}$ is drawn in the figure.

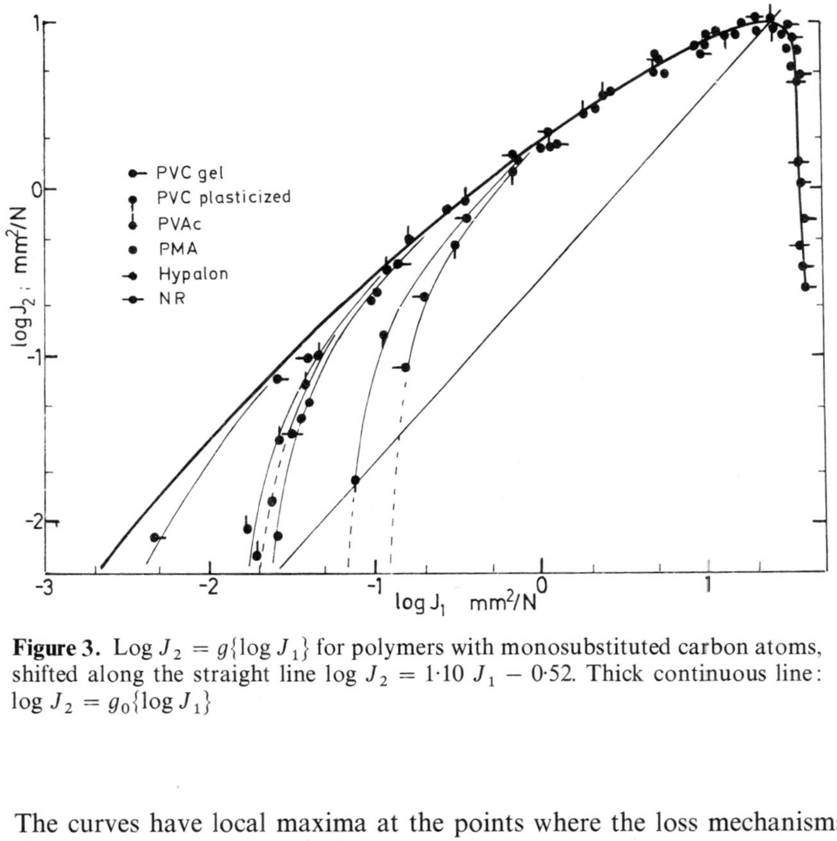

Figure 3. Log $J_2 = g\{\log J_1\}$ for polymers with monosubstituted carbon atoms, shifted along the straight line $\log J_2 = 1{\cdot}10\ J_1 - 0{\cdot}52$. Thick continuous line: $\log J_2 = g_0\{\log J_1\}$

The curves have local maxima at the points where the loss mechanisms attain their maxima. Above this point J_1 approaches a constant value and J_2 goes to zero in the plateau zone. Viscous flow appears in J_2 as an increase linear with time and can be separated in this way.

The curves of $\log J_2 = g\{\log J_1\}$ coincide whether the function is drawn at constant frequency and varying temperature or vice versa. The function can therefore be used for frequency (time)–temperature transformations and can be made the basis of calculations of shift functions and activation energies.

In a paper from 1955 Hoff *et al.* indicate the occurrence of loss mechanisms near the glass point in a series of polymethacrylates. Hoff ascribes these loss mechanisms to coupled movements between the main chain and side groups. Furthermore, the phenomena have been studied theoretically by Koppelmann (1955, 1967), experimentally by Ferry *et al.* in a series of papers from

1957–1959, and by Heijboer (1954, 1956. 1960, 1965). Ferry has also made an attempt at separating the different dispersion processes for PEMA, P-n-BMA and P-n-HMA. In the case of P-n-OMA and P-n-DMA he has merely demonstrated their existence. Information referring to the phenomena has been summarized by McCrum (1957).

It is of interest to consider the basic course of this transition mechanism in a series of polymers with side groups of gradually increasing volume: PP, PMA, PVAc and PIB, PNMA, PEMA, P-n-BMA, P-n-HMA, P-n-OMA, P-n-DMA. In this way an evaluation can be made of the conditions for a disubstituted methyl group to be required for the coupled transition mechanism to appear.

In Figure 4, log J_2 and log J_1 are drawn as functions of temperature for PIB and the series of polymethacrylates. For all the polymers the log J_2

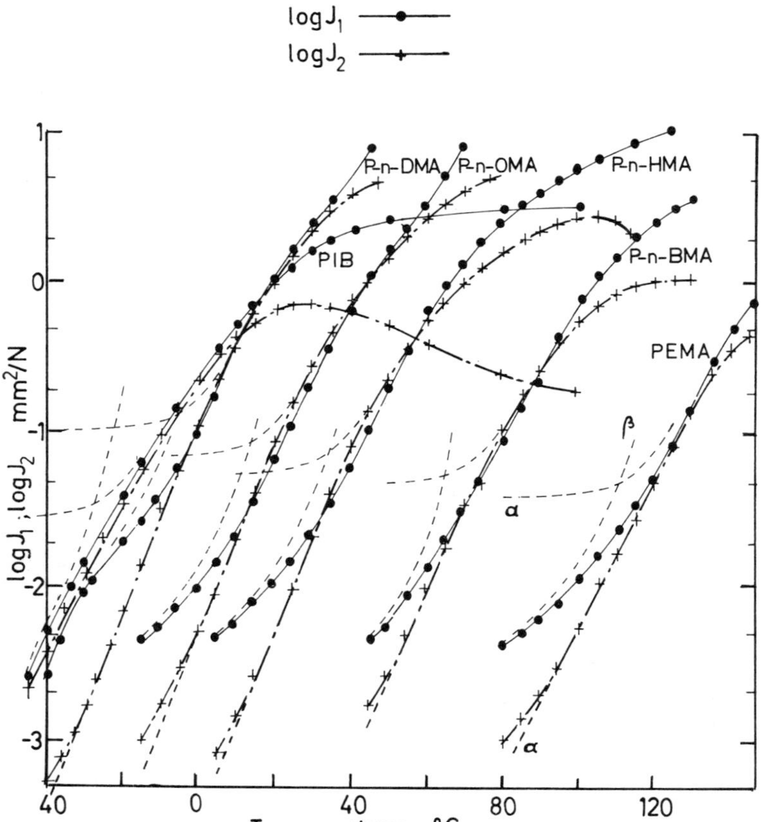

Figure 4. Log J_1 and log J_2 as functions of temperature at $t = 2{\cdot}5 \times 10^{-3}$ s

curves deviate from the linear relation which has been noticed in Figure 1. However, an intermediate linear region exists and the slope of the curves is here approximately the same for all the polymers with disubstituted carbon atoms and is about half the value of the corresponding slope for the monosubstituted polymers in Figure 1. In Figure 5 $\log J_2 = g\{\log J_1\}$ is shown for

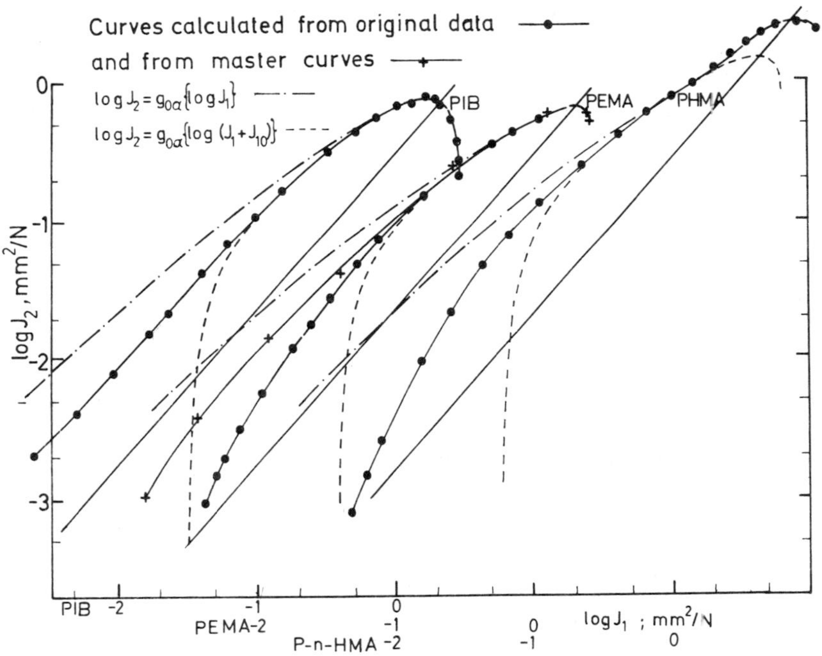

Figure 5. Log $J_2 = g\{\log J_1\}$ for PIB, PEMA and P-n-HMA

some of the disubstituted polymers. All the curves are calculated from the original data of the complex compliance. For PEMA the curve has also been calculated from Ferry's master curves. Considerable differences occur owing to the fact that the method of reduced variables is not applicable because of overlapping transition mechanisms with different shift factors. At high temperature the two mechanisms overlap but at high temperatures in combination with low frequencies J'_α and J''_α will dominate. There is, however, no way of extrapolating J'_α and J''_α to higher frequencies. In P-n-BMA and higher polymethacrylates a third mechanism is noticed in the plateau zone.

The weakness of Ferry's separation method is mainly the assumptions that the frequency dependence of J'_α and J''_α for the polymethacrylates can be approximated by the corresponding curves for PIB, and that the separation after this can be carried out based on the anomalies of the master curves. If PIB (this will be later demonstrated) has a secondary transition at T_g an inaccuracy will occur. The presence of overlapping mechanisms in PIB is possibly indicated by the asymmetry of tan δ near its maximum.

From the properties of the compliance function $\log J_2 = g_0\{\log J_1\}$ an alternative approximation method has been developed for the separation of the α and β mechanisms.

From experimental considerations it is assumed:

$$\log J_2 \equiv \log\{J_{2\alpha} + J_{2\beta}\} = g\{\log J_1\}$$

where

$$\frac{1}{J_1} = \frac{1}{J_{1\alpha}} + \frac{1}{J_{1\beta}} \tag{4}$$

i.e.

$$\log J_2 = \log\{J_{2\alpha} + J_{2\beta}\} = g\left\{\log \frac{J_{1\alpha} \cdot J_{1\beta}}{J_{1\alpha} + J_{1\beta}}\right\} = g\{\log J_1\}$$

where $\log J_{2\alpha} = g\{\log J_{1\alpha}\}$ corresponds to the main-chain mechanism and is supposed to be identical to the equivalent function for polymers with monosubstituted carbon atoms. From this the resultant of the superposed α and β mechanisms can be considered as the coupling of two general functions:

$$\log J_2 = g_\alpha\{\log J_1\} \quad \text{and} \quad \log J_2 = g_\beta\{\log J_1\}$$

Because the functions $\log J_2 = g_0\{\log J_1\}$ appear to be uniform for all polymers studied it is probable as well that the functions $\log J_2 = g_\beta\{\log J_1\}$ correspond for polymers with similar β transitions.

To extrapolate $J_{2\alpha}$ to lower temperatures it is assumed that the linear relation between $\log J_2$ and temperature is valid over a sufficient range. (It is well known, however, from theoretical investigations, that this linearity does not exist for very low temperatures.)

The slopes are then taken from the linear intermediate portions of the $\log J_2$ curves where the influence of the β mechanism can be supposed to be negligible and $J_{2\beta}$ is calculated from $J_{2\beta} = J_2 - J_{2\alpha}\big|_{T=\text{const.}}$

From the general shape of the function $\log J_2 = g_0\{\log J_1\}$ the α functions are determined after shifting the curves along the line $\log J_2 = a \log J_1 + b$.

The β functions so calculated are drawn in Figure 6. For PIB no $J_{2\beta}$ function can be derived because of the observed linear or almost linear relation between $\log J_{2\beta}$ and temperature and can be supposed to be $\ll J_{2\alpha}$. As can be expected, identical curves result for the β functions of all polymethacrylates except P-n-DMA. The deviations of this curve correspond to crystallization of the side chains. The α and β functions have been drawn in Figure 5.

It seems likely that the linear viscoelastic behaviour of any polymer can be described as a combination of a series of compliance functions, log $J_2 = g_{tr}\{\log J_1\}$, each of which has a form characteristic of the particular transition mechanism and which are combined in the way shown in equation (4).

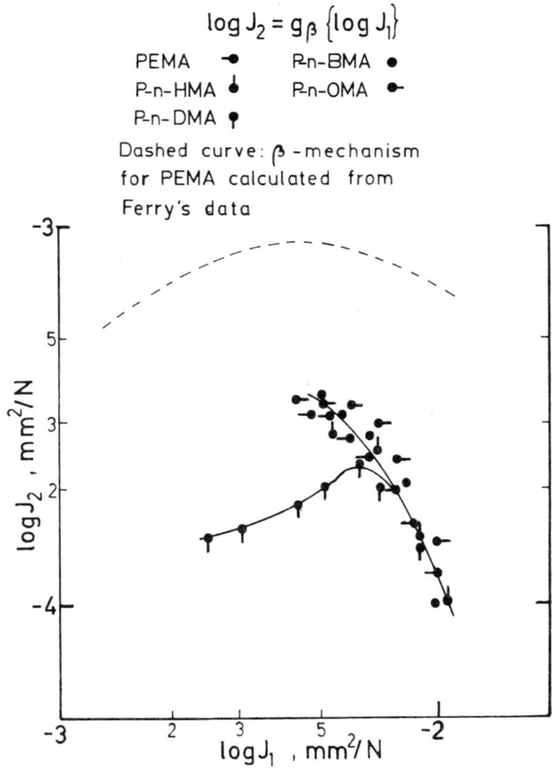

Figure 6. Log J_2 *vs.* log J_1 for various polymers: log $J_2 = g_\beta\{\log J_1\}$

The separation method shows β mechanisms of the coupled type in PIB and all polymethacrylates, but none can be seen in PP, PMA or PVAc. Thus the addition of a disubstituted methyl group seems to give enough steric hindrance for the β mechanism to appear even in PIB.

REFERENCES

Child, Jr., W., and J. D. Ferry (1957). *J. Colloid Sci.*, **12**, 327.
Child, Jr., W. C., and J. D. Ferry (1957). *J. Colloid Sci.*, **12**, 389.
Dannhauser, W., W. C. Child, Jr., and J. D. Ferry (1958). *J. Colloid Sci.*, **13**, 103.
Ferry, J. D., W. C. Child, Jr., R. Zand, D. M. Stern, M. L. Williams, and R. F. Landel (1957). *J. Colloid Sci.*, **17**, 53.
Fitzgerald, E. R., and J. D. Ferry (1953). *J. Colloid Sci.*, **8**, 1.
Fitzgerald, E. R., L. D. Grandine, Jr., and J. D. Ferry (1953). *J. appl. Phys.*, **24**, (5), 650.
Heijboer, J. (1956). *Kolloidzeitschrift*, **134**, 149; *Kolloidzeitschrift*, **148**, 36.
Heijboer, J. (1960). *Makromolek. Chem.*, **35A**, 86.
Heijboer, J. (1965). In *Physics of Non-Crystalline Solids.* North Holland, Amsterdam. p. 231.
Heijboer, J., P. Dekking, and A. J. Staverman (1954). In V. G. W. Harrison (Ed.), *Proc. 2nd Int. Congr. Rheol.* Academic Press, New York. p. 123.
Hoff, A. E. W., D. W. Robinson, and A. H. Willbourn (1955). *J. Polym. Sci.*, **18**, 161.
Jansson, J. F. (1967). In Wetton, Whorlow (Eds.), *Polymer Systems, Deformation and Flow.* Macmillan and Co Ltd, London. p. 85.
Koppelmann, J. (1955). *Kolloidzeitschrift*, **144**, 12.
Koppelmann, J. (1967). *Kolloidzeitschrift, Z. Polym.* Band 216–117, 6.
Kurath, S. F., E. Passaglia and R. Pariser (1959). *J. appl. Polym. Sci.*, **1**, 2, 150.
Kurath, S. F., T. P. Yin, J. W. Berge, and J. D. Ferry (1959). *J. Colloid Sci.*, **14**, 147.
McCrum, N. G., B. N. Read, and G. Williams. (1967). *Anelastic and Dielectric Effects in Polymeric Solids.* John Wiley, New York.
Ninomiya, K., and J. D. Ferry (1959). *J. Colloid Sci.*, **14**, 36.
Struik, L. C. E., and F. R. Schwarzl (1969). *Rheol. Acta.* Band 8, Heft 2, 134.
Williams, M. L., and J. D. Ferry (1954). *J. Colloid Sci.*, **9**, 479.
Williams, M. L., and J. D. Ferry (1955). *J. Colloid Sci.*, **10**, 1; *J. Colloid Sci.*, **10**, 474.
Yin, T. P., and R. Pariser (1963). *J. appl. Polym. Sci.*, **7**, 667.

16

On the Temperature Dependence of the Thermal Conductivity of Amorphous Polymers

G. REHAGE and G. HIRSCH*

Institute of Physical Chemistry, Technical University of Clausthal, Germany

In order to obtain some information on energy transport in polymers, the thermal conductivity of several amorphous polymers above and below their glass transition temperatures was determined. The measurements were made under steady-state conditions (Hirsch, 1966; Hirsch and Rehage, to be published). The samples were hollow cylinders, heated electrically from the inside and cooled from the outside by thermostated air. The thermal conductivity was then calculated from the temperature difference, the geometrical dimensions of the sample and the electrical power input. By varying the temperature of the cooling air the measuring temperature was varied also. In order to avoid heat losses, the ends of the cylindrical sample were in contact with protective heaters, the temperatures of which were adjusted to those of the sample. Observations were checked by measurements on known samples and were found to be very accurate and reproducible.

Investigations on amorphous polymers, differing in polarity, showed that with increasing temperature the thermal conductivity rises at an increasing rate below the glass transition temperature and linearly above it. The glass transition temperature itself is characterized by the beginning of this linear part of the curve. The change in slope between this linear part and the curved part decreases in the order polystyrene, polyvinylcarbazole and poly-*p*-chlorostyrene to polymethylmethacrylate and thus, with increasing polarity of the polymer (Figure 1).

* Present address: Badische Anilin- und Soda-Fabrik AG., Kunststoff-Labor, 67 Ludwigshafen/Rh.

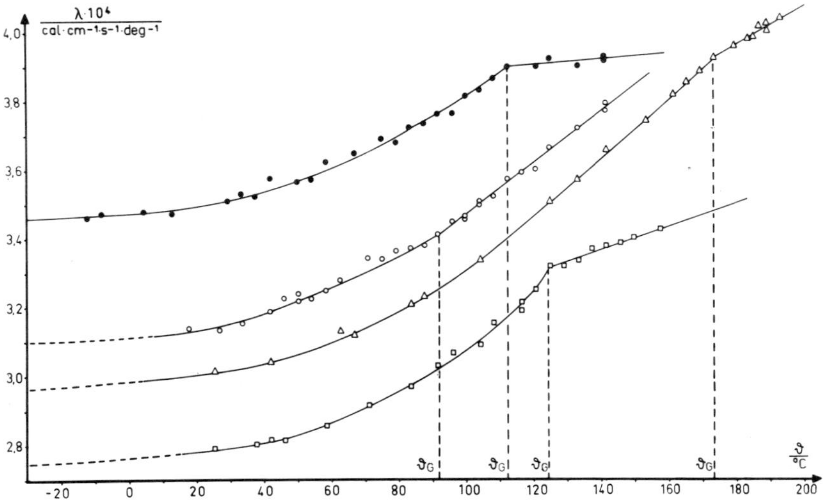

Figure 1. Temperature dependence of the heat conductivity λ of amorphous polymers.

ϑ = temperature in °C; ϑ_G = glass temperature.

● Polymethylmethacrylate
○ Polystyrene
△ Polyvinylcarbazole
□ Poly-*p*-chlorostyrene

The thermal conductivity, λ, of some corresponding metal-containing and crosslinked polymers have a very similar temperature dependence. The values for polystyrene are a little higher than those for crosslinked polystyrene but a little lower than those for a polystyrene with 12% aluminium (Figure 2). The particles of aluminium presumably provide easy paths for the heat flow.

The thermal conductivities of polymethylmethacrylate are all higher than those of the polymethylmethacrylates containing a metal organic compound, namely the tri-n-propyl-tin-methacrylate (Figure 3); here it is probable that the large tri-n-propyl-tin-group is not able to take part in the energy transfer as strongly as the methyl-group for reasons of steric hindrance.

It is known that the mobility of the polymer chains increases at the glass transition temperature and this point is clearly shown on the curves of thermal conductivity. The increase in the mobility of the side groups which occur at lower temperatures can also be recognized in the curves, but not so sharply. The ranges of temperature, in which the side groups become mobile are known from torsional vibration measurements (Illers, 1961; Meyer, 1962; Rehage and Schäfer, 1968; Schmieder and Wolf, 1955). These ranges are the same as the regions of temperature in which the thermal conductivities

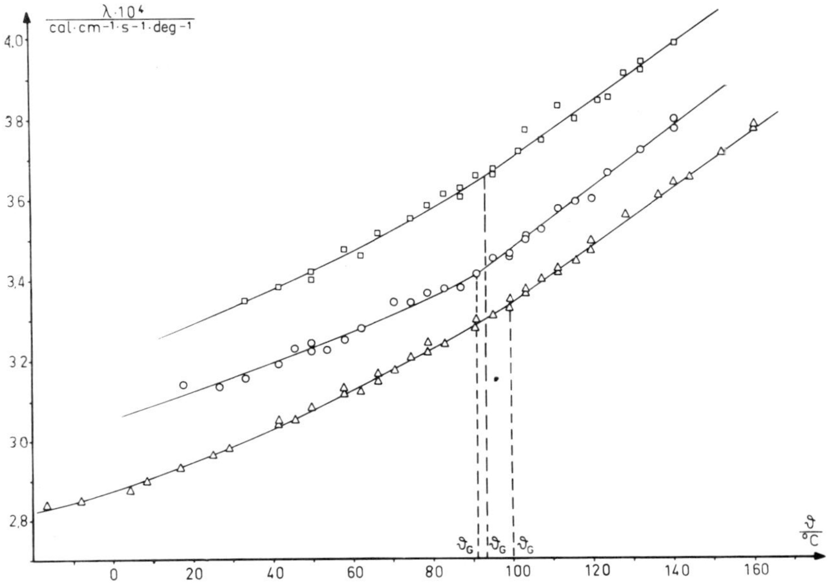

Figure 2. Temperature dependence of the heat conductivity λ of pure polystyrene (PST), PST with filler and cross-linked PST.
○ Polystyrene
△ Polystyrene crosslinked with ethyleneglycol-dimethacrylate (10%)
□ Polystyrene with aluminium containing filler (12·3%)

of the polymers show the first pronounced increase, starting from low temperatures. The increase can be best seen if $d\lambda/d\vartheta$ is plotted versus the temperature ϑ(°C) (see Figure 4). Heat transport must therefore be increased also by the increase in mobility of the side groups.

With all these facts one can try to construct a simple model of heat conduction for amorphous high polymers. The polymer chains are regions of good conduction but the passages between these chains are characterized by a low heat flux. These low flux paths therefore will determine the velocity of heat transport for the whole polymer network (Eiermann, 1964; Holzmüller and Altenburg, 1961). Starting from low temperatures there are always a certain number of side groups in energetic interaction which transfer a certain quantity of energy, the amount of which may increase slowly with increasing temperature. This explains the slow increase of the curves of thermal conductivity at lower temperatures. With increasing temperature it becomes possible for the side groups to change their position and existing interactions may be enlarged and new interactions, perhaps only for a short time, may be formed. The polymer network will be penetrated by a fluctuating system of additional or more effective energetic interactions, which

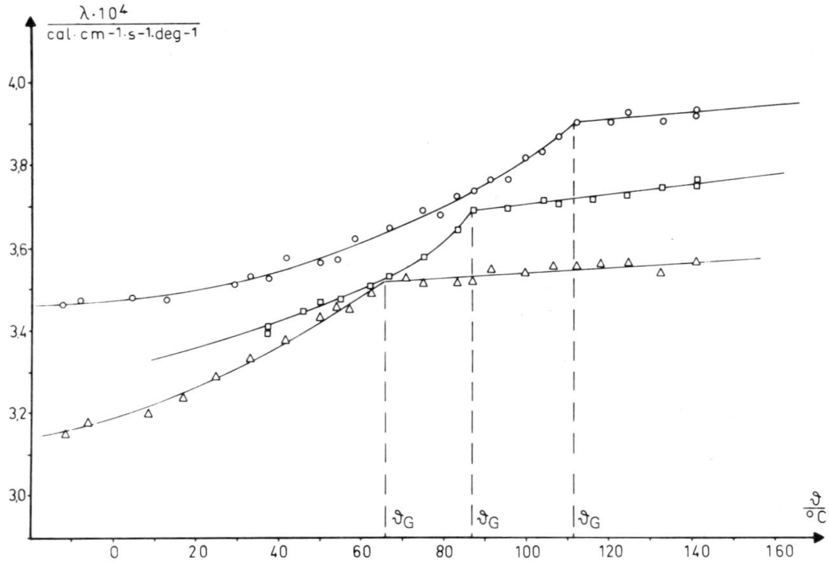

Figure 3. Temperature dependence of the heat conductivity of polymethylmethacrylate and copolymers with different tin content.

○ Polymethylmethacrylate

□ Copolymer of methylmethacrylate and 9·9 mol % of tri-n-propyl-tin-methacrylate

△ Copolymer of methylmethacrylate and 24·6 mol % of the tin compound

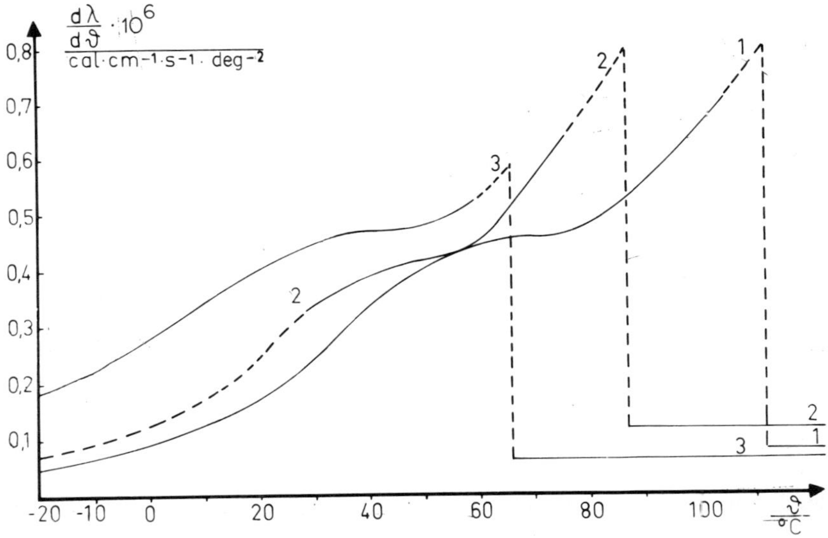

Figure 4. Slopes of the heat conductivity curves in dependence of the temperature.
(1) Polymethylmethacrylate
(2) Copolymer with 9·9 mol % of the tin compound
(3) Copolymer with 24·6 mol % of the tin compound

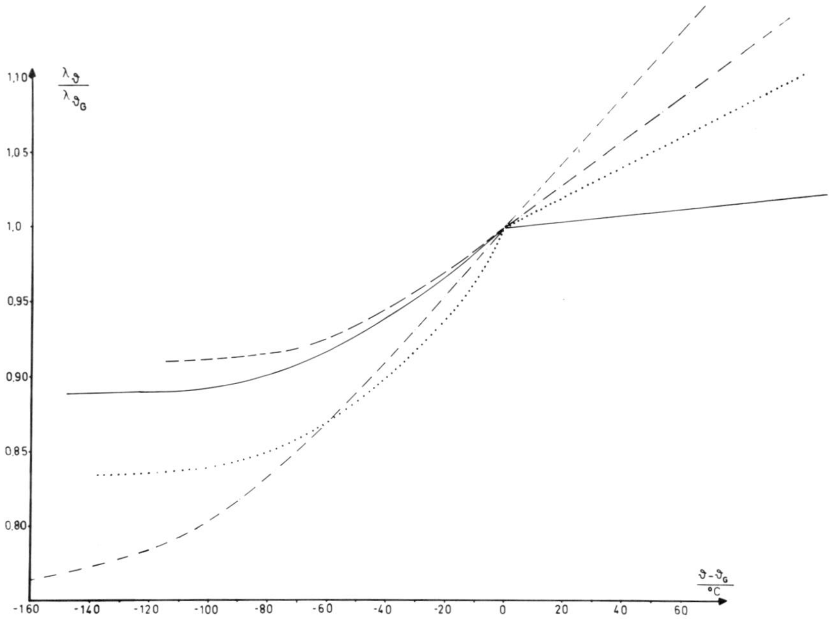

Figure 5. Normalized heat conductivity $\lambda_\vartheta/(\lambda_{\vartheta_G})$ of pure amorphous polymers versus $(\vartheta - \vartheta_G)$.
----- Polystyrene
-·-·- Polyvinylcarbazole
······· Poly-*p*-chlorostyrene
——— Polymethylmethacrylate

Table 1. Increase of thermal conductivity, polarity and thermal expansion of amorphous polymers above the glass transition temperature.

Polymer	$(d\lambda/d\vartheta) \times 10^6$; $\vartheta > \vartheta_G$ [cal cm^{-1} s^{-1} deg^{-2}]	μ [Debye]	$\alpha \times 10^4$ [deg^{-1}]
Polystyrene	0·750	0·4	1·30
Polyvinylcarbazole	0·550	0·94	1·45
Poly-*p*-chlorostyrene	0·338	1·54	1·80
Polymethylmethacrylate	0·075	1·97	2·75

increase the number of possibilities for intermolecular energy change. Thus thermal conductivity increases at an increasing rate below the glass transition temperature. At the glass transition temperature the polymer chains begin to become mobile. But this does not lead to an increase of the slope of the heat transport, as the competing effect of thermal expansion, which weakens the interactions and hence also the effects in change of position, becomes more pronounced. If the thermal expansion of the polymer is greater, this weakening of the interactions becomes more effective. It may be that this is a reason for the different increases of the thermal conductivity of the amorphous polymers above the glass transition temperature (Figure 1 and Figure 5). In the same sequence the thermal conductivity and the polarity of these polymers differ also (Table 1).

REFERENCES

Eiermann, K. (1964). *Kolloidzeitschrift*, **198**, 5.
Hirsch, G. (1966). *Thesis*, Aachen.
Hirsch, G., and G. Rehage. *Kolloidzeitschrift*, *Z. Polym.* (to be published).
Holzmüller, W., and K. Altenburg (1961). In *Physik der Kunststoffe*. Akademie–Verlag, Berlin.
Illers, K.-H. (1961). *Z. Elektrochem.*, **65**, 679.
Meyer, H. H. (1962). *Thesis*, Aachen.
Rehage, G., and E. E. Schäfer (1968). *Z. an. Chem.*, **235**, 137.
Schäfer, E. E. (unpublished measurements).
Schmieder, K., and K. Wolf (1953). *Kolloidzeitschrift*, **134**, 141.

17

The Thermodynamic Functions of Supercooled Glassforming Liquids and the Temperature Dependence of their Viscosity

I. GUTZOW

Institute of Physical Chemistry, Bulgarian Academy of Science, Sofia, Bulgaria

---

The present paper shows that it is possible to describe simply and in a generalized form both the thermodynamic functions and the temperature dependence of the viscosity of undercooled, glassforming melts. This is carried out on the basis of a simple statistical model, the so called 'hole theory of liquids', formulated by Eyring (1941) and Frenkel (1955).

Several attempts have been made to determine thermodynamic parameters of glassforming melts and glasses from the hole theory. Application has been made of the elementary theory of simple liquids (Hirai and Eyring, 1958, 1959; Wunderlich, 1960) and of more sophisticated lattice models (Gutzow, 1962, 1966), taking into account the configurational statistics of polymer chains. The theory of Gibbs and DiMarzio (1958, 1960; Adam and Gibbs 1965) describing the vitrification of liquids is also based on a similar model. Recently, the model proposed in 1962 has been applied by the author (1969, 1970) to the thermodynamics of crystallization of undercooled melts, and Bartenev *et al.* (1969) used the hole theory to describe molecular relaxation in glassforming liquids.

Even the use of simple statistical models leads to very complicated expressions for the thermodynamic functions of undercooled liquids and therefore considerable simplifications will be introduced here. It is then possible to derive simple mathematical expressions for the thermodynamic functions, and to give a detailed discussion of the basic premises of the statistical models used and of the limits of their application.

THERMODYNAMIC FUNCTIONS OF UNDERCOOLED LIQUIDS

According to modern lattice theories, each thermodynamic function of the liquid can be separated into two terms—vibrational and configurational. As the vibrational component is equal to the value of the corresponding thermodynamic function of the crystal at the same temperature T, the calculation of the difference between the thermodynamic functions of the undercooled liquid and the crystal is equivalent to the determination of the configurational term of the thermodynamic functions of the liquid.

Following the hole theory, the configurational entropy of a simple liquid can be determined by the expression

$$\Delta S_{(T)}{}^{\text{config}} = -R\left[\frac{\mathscr{V}_{(T)}}{1-\mathscr{V}_{(T)}}\ln \mathscr{V}_{(T)} + \ln(1-\mathscr{V}_{(T)})\right] \tag{1}$$

and as stated $\Delta S_{(T)}{}^{\text{config}}$ will be directly equal to the difference between the entropy of the liquid and the crystal $\Delta S_{(T)}$. Here $\mathscr{V}_{(T)}$ is the relative free volume of the liquid which, according to the equation

$$\mathscr{V}_{(T)} = \frac{n_0}{n_1 + n_0} \tag{2}$$

is determined by the number of vacant sites n_0 in a quasi-crystalline lattice having $n_1 + \mathrm{n}_0$ sites. As is known, in this case the temperature dependence of $\mathscr{V}_{(T)}$ is given by

$$\mathscr{V}_{(T)} = \exp\left(-\frac{\mathscr{U}_0}{kT}\right) \tag{3}$$

where $\mathscr{U}_0$ is the energy of hole formation.

When $\mathscr{V}_{(T)} = 0{\cdot}35$ equation (1) yields $\Delta S_{(T)} = R$, which is the entropy of melting of simple liquids, i.e. $\mathscr{V}_{(T)} = 0{\cdot}35$ is the free volume of the liquid at the temperature of melting T_m. Equation (1) can be simplified by expanding the logarithmic functions in the power series

$$\ln x = 2\left[\frac{x-1}{x+1} + \frac{1}{3}\cdot\left(\frac{x-1}{x+1}\right)^3 + \frac{1}{5}\left(\frac{x-1}{x+1}\right)^5 + \dots\right]$$

Confining ourselves to the first term, it follows that

$$\Delta S_{(T)} \cong \frac{6R\mathscr{V}_{(T)}}{2 + \mathscr{V}_{(T)} + \mathscr{V}^2_{(T)}} \cong 3R\mathscr{V}_{(T)} \tag{4}$$

It can be seen from Figure 1a that the latter expression is a good approximation for $\Delta S_{(T)}$ in the whole interval from $\mathscr{V}_{(T)} = 0$ to $\mathscr{V}_{(T)} = 0{\cdot}35$, i.e. from the absolute zero to T_m.

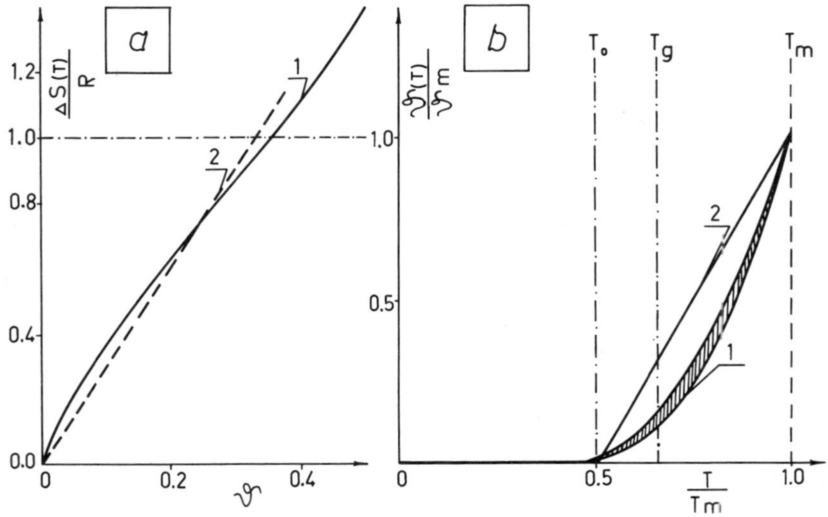

Figure 1. Illustration of the approximations for $\Delta S_{(T)}$ and $\mathscr{V}_{(T)}$
(a) Dependence of $\Delta S_{(T)}$ on $\mathscr{V}$. Curve 1—according to the precise equation (1) Dashed curve 2—according to the approximation (4).
(b) Dependence of $\mathscr{V}$ on T/T_m. Curve 1—according to the precise equation (7) with $\varepsilon = 3{\cdot}3$ and $\varepsilon = 4{\cdot}0$; the straight line (2)—according to the approximation (8) with $a_0 = 2$.

In the hole theory of simple liquids the value of $\mathscr{U}_0$ is assumed to be constant and determined by

$$\mathscr{U}_0 \cong (\tfrac{1}{5} \text{ to } \tfrac{1}{6}) \, . \, \lambda' \tag{5}$$

where λ' is the molar heat of evaporation of the melt. With Trouton's rule $\lambda'/RT_m \cong 20$ so that

$$\frac{\mathscr{U}_0}{RT_m} = \varepsilon = 3{\cdot}3 \text{ to } 4{\cdot}0 \tag{6}$$

where $\mathscr{U}_0$ now refers to a gram mole.
According to equations (3) and (6) the temperature dependence of the ratio $\mathscr{V}_{(T)}/\mathscr{V}_m$ where $\mathscr{V}_m$ is the free volume at T_m will be expressed by the following steep exponential function

$$\frac{\mathscr{V}_{(T)}}{\mathscr{V}_m} = \exp\left(-\varepsilon \, . \frac{\Delta T}{T}\right) \tag{7}$$

where $\Delta T = T_m - T$ denotes the undercooling. This relation is shown in Figure 1 for $\varepsilon = 3{\cdot}3$ and $\varepsilon = 4{\cdot}0$. When the values of T/T_m are less than $\frac{1}{2{\cdot}3} \div \frac{1}{1{\cdot}8}$, it follows from equation (7) that $\mathscr{V}_{(T)} \simeq 0$. For larger values of

T/T_m the exponent in equation (7) may be approximated by a linear dependence. Thus the expression

$$\frac{\mathscr{V}_{(T)}}{\mathscr{V}_m} \simeq \begin{cases} \dfrac{a_0 T}{T_m} + 1 - a_0, & \text{for } T_m \geqslant T \geqslant T_0 \\ 0 & \text{, for } T_0 > T \geqslant 0 \end{cases} \tag{8}$$

is obtained where $T_0 = (1 - 1/a_0)\, T_m$ is the temperature, below which it can be assumed with sufficient accuracy (e.g. 5%) that $\mathscr{V}_{(T)} = 0$.

With these two approximations (equations (4) and (8)) equation (1) yields

$$\frac{\Delta S_{(T)}}{\Delta S_m} \simeq \begin{cases} \dfrac{a_0 T}{T_m} + 1 - a_0, & \text{for } T_m \geqslant T \geqslant T_0 \\ 0 & \text{, for } T_0 > T \geqslant 0 \end{cases} \tag{9}$$

Using the thermodynamic relations

$$\Delta C_{p(T)} = \frac{\partial[\Delta S_{(T)}]}{\partial T}; \quad \Delta\Phi_{(T)} = \int_T^{T_m} \Delta S_{(T)}\, dT$$

$$\Delta H_{(T)} = H_m + \int_T^{T_m} \Delta C_{p(T)}\, dT \tag{10}$$

simple expressions can be derived also for the differences between the specific molar heats (ΔC_p), the enthalpies (ΔH), and the thermodynamic potentials ($\Delta\Phi$) of the supercooled liquid and the crystal.

Strictly speaking the deductions made so far refer to simple liquids. composed of non-polymeric and non-associated molecules. In practice, however, glassforming melts are either built of polymer molecules or undergo complex association and aggregation when undercooled. In such cases, as indicated in previous papers (Gutzow, 1962, 1966, 1969, 1970) other terms should be added to the basic entropy term, defined by equation (1) which accounts for the free volume $\mathscr{V}$, so that

$$\widetilde{\Delta S_{(T)}} = \Delta S^{\mathscr{V}}_{(T)} + \Delta S^{M}_{(T)} + \Delta S^{C}_{(T)} + \Delta S^{S}_{(T)} \tag{11}$$

Here ΔS^M, ΔS^C, ΔS^S, account respectively for the entropy of mixing, the possible spatial configurations and the various orientations of chain molecules with a different degree of polymerization (association). The statistical model, described by the author in 1962, makes possible the evaluation of each of these terms. It has also been shown (Gutzow 1969, 1970) that the entropy $\Delta S^{\mathscr{V}}$, expressed by equation (1) will also determine to a great extent the temperature course of $\widetilde{\Delta S_{(T)}}$ in these more complicated cases. This follows from the principle, formulated by Gibbs and DiMarzio (1958, 1960), according to which when $\mathscr{V} = 0$, i.e. when $\Delta S^{\mathscr{V}}_{(T)} \to 0$ all additive terms in equation (11) will also tend to zero. Hence it can be demonstrated that in principle approximation (9) should also be valid for more complicated models of polymer and associative melts, which are closer to reality. All these additional factors will be taken into consideration by defining approximation (9) in accordance with experimental data on the temperature course of $\Delta S_{(T)}$ for real glassforming melts in the temperature range from T_m to the temperature of vitrification T_g. This means that a_0 from equation (9) will be calculated according to the experiment.

SPECIFICATION OF THE VALUE OF a_0. COMPARISON WITH EXPERIMENT

According to the empirical rule established by Kauzman and Beaman (Kauzman 1948, Bartenev 1966), the following simple relation exists between the absolute temperature of melting and T_g:

$$\frac{T_g}{T_m} \simeq \tfrac{2}{3} \simeq 0{\cdot}667 \tag{12}$$

As can be seen in Figure 2, this rule holds good for all substances shown in the figure, for which T_m varies within the limits between 100 and 2000 °K. Since T_g/T_m has a constant value, it follows from equation (9) that the ratio $\Delta S_g/\Delta S_m$ should also be constant for all values. According to Figure 2, which is based on data given by Kauzman (1948), Bestul and Chang (1964) and Gutzow *et al.* (1969), the latter condition is actually met and

$$\frac{\Delta S_g}{\Delta S_m} \simeq \tfrac{1}{3} = 0{\cdot}33 \tag{13}$$

According to equation (9), a_0 can be determined directly by the expression

$$\frac{(\Delta S_m - \Delta S_g)}{\Delta S_m} \frac{T_m}{(T_m - T_g)} = a_0 \tag{14}$$

The values of this ratio, alternating within the limits between 1·5 and 2·5 (Figure 2) are well grouped around the average value $a_0 = 2$.

With this value of a_0 it follows that $T_0 = \frac{1}{2}T_m$ and from equation (8) and (9) one obtains

$$\frac{\mathscr{V}_{(T)}}{\mathscr{V}_m} = \frac{\Delta S_{(T)}}{\Delta S_m} \begin{cases} \dfrac{2T}{T_m} - 1, & \text{for } T_m \geqslant T \geqslant T_0 = \frac{1}{2}T_m \\ 0 \quad , & \text{for } T_0 > T \geqslant 0 \end{cases} \tag{15}$$

With the above relation and equation (10) the other thermodynamic functions can be written in the form

$$\Delta C_{p(T)} = \begin{cases} \dfrac{2\Delta S_m}{T_m} \cdot T \\ 0 \end{cases}; \qquad \Delta \Phi_{(T)} = \begin{cases} \Delta S_m \cdot \dfrac{T}{T_m} \cdot \Delta T \\ \frac{1}{4}\Delta S_m T_m \end{cases};$$

$$\Delta H_{(T)} = \begin{cases} \Delta S_m \cdot T_m \cdot \left(\dfrac{T}{T_m}\right)^2 \\ \frac{1}{4}\Delta S_m T_m \end{cases} \tag{16}$$

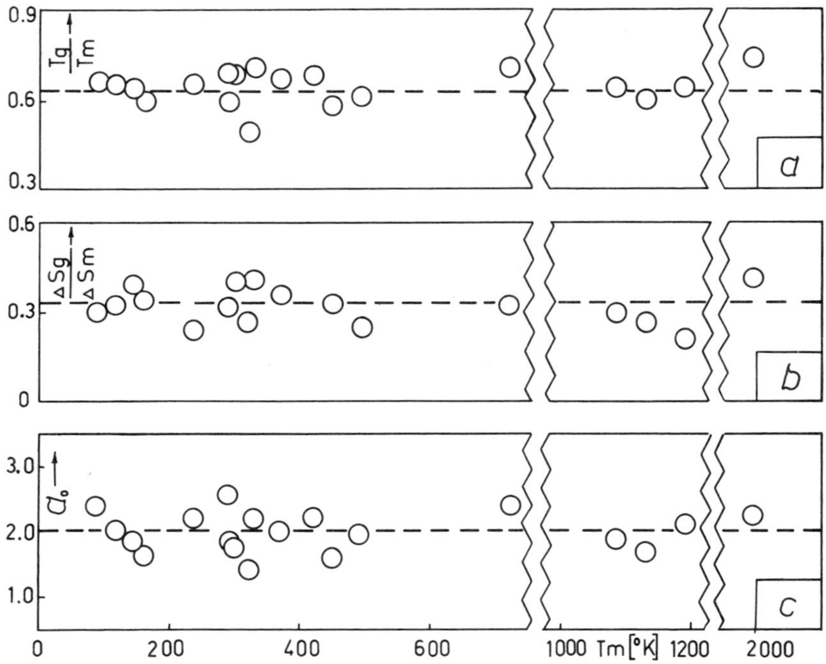

Figure 2. Experimental data on the thermodynamic properties of glassforming melts.

(a) The ratio T_g/T_m. The dashed line corresponds to 0·67.

(b) The ratio $\Delta S_g/\Delta S_m$. The dashed line corresponds to 0·33.

(c) Values of a_0. The dashed line is $a_0 = 2{\cdot}0$.

The substances are (in the order of increasing T_m): butene-1, 2-methylpentane, n-propanol, ethanol, $H_2SO_4 \cdot 3H_2O$, *d-l* lactic acid, glycerol, rubber, benzophenone, terphenyl, betol, glucose, polypropylene, Se, B_2O_3, $Na_2O{\cdot}4B_2O_3$, $K_2O{\cdot}4B_2O_3$, $Li_2O{\cdot}2B_2O_3$, SiO_2

within the limits indicated above. These equations are represented graphically in Figure 3. From equation (16) it follows that the ratio $\Delta H_g/\Delta H_m \simeq \frac{4}{9} = 0{\cdot}445$ and that $\Delta C_{p(T_g)}/\Delta S_m = \frac{4}{3} = 1{\cdot}33$, which, as shown on Table 1 are in a

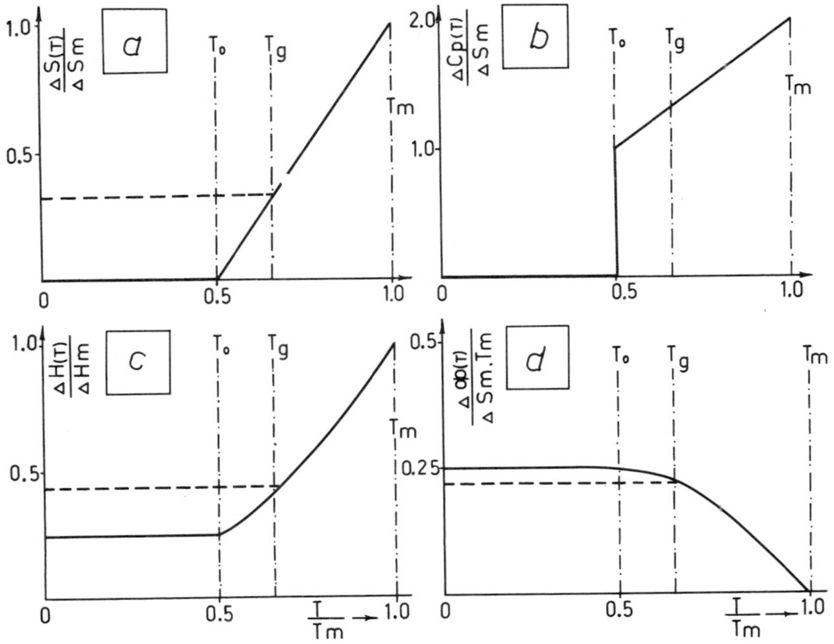

Figure 3. Temperature dependence of the thermodynamic functions of undercooled melts in internal equilibrium
(a) $\Delta S_{(T)}/\Delta S_m$ (b) $\Delta C_{p(T)}/\Delta S_m$ (c) $\Delta H_{(T)}/\Delta H_m$ (d) $\Delta \Phi_{(T)}/\Delta S_m T_m$. The dashed line shows the course of the respective function after vitrification

satisfactory agreement with the existing experimental data, compiled by Tammann (1933), Kauzman (1948), Wunderlich (1960) and Gutzow (1970).

Table 1

Substance	T_m, °K	$\frac{\Delta C_p(T_g)}{\Delta S_m}$	Substance	T_m, °K	$\frac{\Delta H_g}{\Delta H_m}$
Butene-1	88	1·508	n-Propanol	146	0·506
Pentane	119	1·258	Ethanol	156	0·549
n-Propanol	146	1·028	Glycerol	292	0·476
Ethanol	156	1·030	Benzophenone	321	0·479
Glycerol	292	1·301	Betol (β-naphthyl salicylate)	368	0·372
Selenium	491	1·336	$Na_2O\cdot4B_2O_3$	1073	0·457
$H_2SO_4\cdot3H_2O$	237	1·714	$Li_2O\cdot2B_2O_3$	1182	0·385

TEMPERATURE DEPENDENCE OF THE VISCOSITY OF UNDERCOOLED LIQUIDS

Cohen and Turnbull (1959) showed that the time for molecular relaxation τ of a liquid depends exponentially on the relative free volume $\mathscr{V}$ so that

$$\tau = \tau_0 \exp\left[\frac{nv_0}{\mathscr{V}_{(T)}}\right] \tag{17}$$

Here v_0 is the volume of a building unit in the liquid and n is the number of building units taking part in the process of molecular translation. As

$$\tau = \frac{V_M}{RT}\cdot\eta \tag{18}$$

where V_M is the molar volume of the liquid, equation (18) is usually considered as the theoretical background of the well known empirical equation given by Doolittle (1951) for the temperature dependence of the viscosity η.

According to the approximation [equation (15)], deduced in the previous paragraph, it follows from equation (17) that

$$\tau \simeq \begin{cases} \tau_0 \exp\left[\dfrac{nv_0}{\mathscr{V}_m(2T/T_m - 1)}\right], & \text{for } T_m \geqslant T \geqslant T_0 \\ \infty & , \text{ for } T_0 > T \geqslant 0 \end{cases} \tag{19}$$

In practice, before reaching $\tau = \infty$ the melt vitrifies at $T = T_g$ where at normal cooling rates $\tau \simeq 10^2$ sec. With the same approximation and equations (17) and (18) it can be shown that from T_m to T_g

$$\log\eta = \log\left(\frac{RT}{V_M}\cdot\tau_0\right) + 0{\cdot}43\frac{nv_0}{2\mathscr{V}_m}\cdot\frac{T_m}{(T - \frac{1}{2}T_m)} \tag{20}$$

must be valid. As $\log (RT/V_M)\,\tau_0 \cong$ const. this relation is analogous to the empirical equation of Vogel, Fulcher and Tammann, which is usually written in the form

$$\log\eta = A + \frac{B}{T - T_\infty} \tag{21}$$

From the interpretation given here it is obvious that the constants in equation (21) should have the following meanings:

$$A = \log\left(\frac{RT}{V_M}\cdot\tau_0\right); \qquad B = 0{\cdot}43\cdot\frac{nv_0}{2\mathscr{V}_m}\cdot T_m; \qquad T_\infty = T_0 = \tfrac{1}{2}T_m.$$

Equation (19) written first for the temperature $T > T_g$ and then for the temperature $T_g > T_0$ yields

$$\log\left[\frac{\tau(T)}{\tau(T_g)}\right] = -0{\cdot}43\cdot\frac{nv_0}{2\mathscr{V}_m}\cdot\frac{T_m}{(T_g - \frac{1}{2}T_m)}\cdot\frac{(T - T_g)}{(T - \frac{1}{2}T_m)} \tag{22}$$

As, according to equation (12), $T_m/(T_g - \frac{1}{2}T_m) = 6$, equation (22) is in fact equivalent to the empirical relation given by Williams, Landel and Ferry (1955)

$$\log\left[\frac{\tau(T)}{\tau(T_g)}\right] = -\varphi\cdot\frac{(T - T_g)}{(T - T_g + \beta)} \tag{23}$$

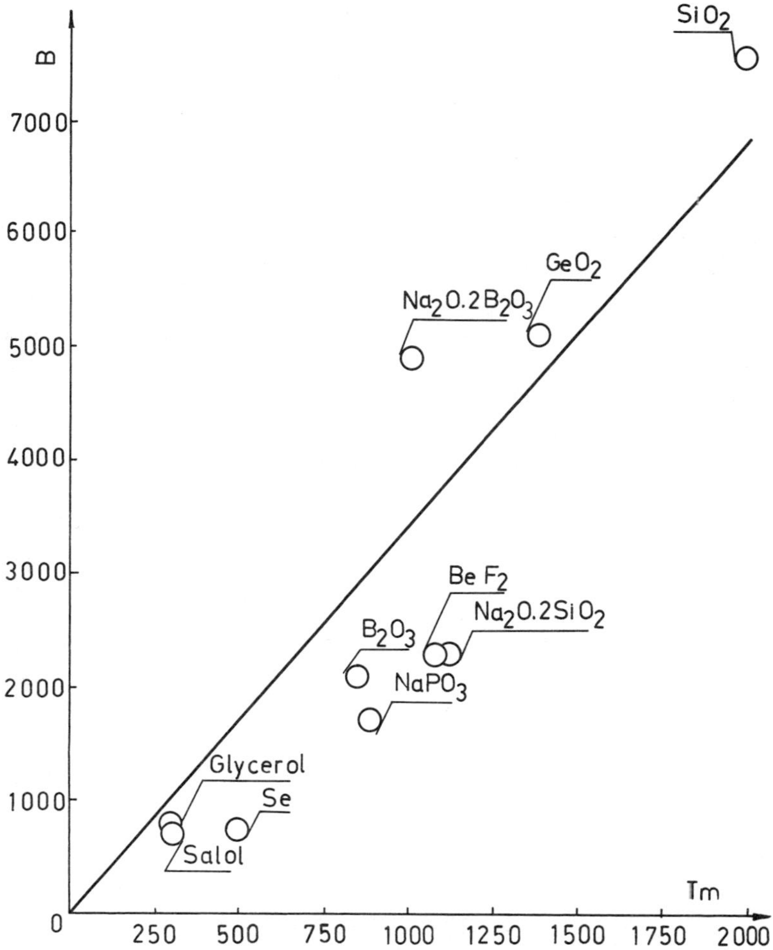

Figure 4. Dependence of B on T_m according to experimental data for the viscosity of various substances

Now, from equations (22), (23) and (12) it follows that the constants of the WLF-equation are

$$\varphi = 0{\cdot}43 \cdot \frac{3nv_0}{\mathscr{V}_m}; \qquad \beta = T_m\left(\frac{T_g}{T_m} - \frac{1}{2}\right) \cong \frac{1}{6}T_m$$

It is known that the temperature T_g corresponds to a viscosity of $10^{13{\cdot}5}$ poise. Then according to equation (20) one obtains for this temperature

$$\frac{T_g}{T_m} = \frac{1}{2} + 0{\cdot}43 \cdot \frac{nv_0}{2\mathscr{V}_m} \cdot \left[13{\cdot}5 - \log\left(\frac{RT_g}{V_M} \cdot \tau_0\right)\right]^{-1} \tag{24}$$

As all values on the right-hand side of equation (24) are approximately constant, and as simple estimations show that the second term in this equation should vary from 0·1 to 0·2, the empirical rule of Kauzman and Beaman is easily understood.

COMPARISON WITH EXPERIMENT

As seen in Figure 4 a simple correlation exists between the constant B and T_m, as predicted by equation (20). Each point in Figure 4 has been obtained by inserting viscosity data for the corresponding substances into coordinates $\log \eta \div 1/(T - \frac{1}{2}T_m)$, and determining B from the slope of the resulting straight line. The necessary η-values were taken from Jenkel (1939), Bestul (1959), Jackson *et al.* (1967) and Bartenev *et al.* (1969). So, following the present procedure it is possible to determine the constant B and T_∞ in equation (21) in advance, and A remains the only individual constant in equation (21) subject to experimental determination.

The detailed data, collected by Williams, Landel and Ferry (1955), as well as by Bestul (1959) give the value $\varphi = 17$ for one of the constants in equation (23), whence, according to equation (22) and $\mathscr{V}_m = 0{\cdot}37$ one obtains $nv_0 = 4{\cdot}6$ which is a very agreeable result. In contrast to the assumption made by Williams, Landel and Ferry and some other authors, it follows from the present derivation that β from equation (23) also depends on T_m. Only for substances with melting temperature about 300 °K, as is the case with most organic glassforming systems, it would follow that $\beta = 50$, as has been asserted by Williams, Landel and Ferry. The treatment given here is in accordance with the β-values, calculated by Bestul (1959), for substances with a considerably higher temperature of melting T_m, respectively T_g.

DISCUSSION

Thus, simple expressions for all thermodynamic functions of undercooled glassforming liquids in internal equilibrium can be derived on the basis of a simple or more elaborate hole theory, if appropriate approximations are made. The results so obtained are in satisfactory agreement both with the results of the experimental measurements (from T_m to T_g) and with more general theoretical considerations (e.g. the Third Law of Thermodynamics) for T_0 to 0 °K. The basic empirical relations, reflecting the viscosity and the relaxational behaviour of glassforming melts can also be expressed on the basis of the same theoretical premises. Moreover, the viscosity data on undercooled melts show definitely that at temperatures below $T = T_0$ the relative free volume of the fictive melt in internal equilibrium would in fact be $\mathscr{V}_{(T)} = 0$. Thus the temperature dependence of the viscosity corroborates the thermodynamic deductions.

It must be noted, however, that the considerations put forward here have disclosed also a considerable disadvantage of the model used, which is typical also for the more elaborate lattice models cited in the introduction. According to some very general conclusions (see, for example, Frenkel–1955) it might be expected that the thermodynamic functions in particular $\Delta S_{(T)}$, should have an S-shaped van-der-Waals temperature dependence in the proximity of T_m. Only such a dependence of the entropy difference would define a first order phase transformation at T_m and would make it possible not only to undercool the liquid, but also to overheat the crystal. Instead of this the temperature T_0 appears in the lattice models, formally corresponding to a second order phase transformation. It can be shown that this is not a result of the approximations introduced here, but merely a property of the lattice models themselves. But as a more precise approach to these problems does not exist at present, the results discussed here can be taken as a first step towards a more elaborate theory.

REFERENCES

Adam, G., and J. H. Gibbs (1965). *J. chem. Phys.*, **43**, 193.

Bartenev, G. M. (1966). *Stroenie i mekhanitcheskie svoistva neorganitcheskikh stekol. Izd. literatury po stroitelstvu*, Moskva. pp. 8–43.

Bartenev, G. M., D. S. Sanditov, I. V. Razumovskaya, and I. A. Lukyanov (1969). *Ukr. fiz. Zh.*, **14**, 1529.

Bestul, A. B. (1959). *Glasstech. Ber.* (Sonderband) **32 K**, VI 59.

Bestul, A. B., and S. S. Chang (1964). *J. chem. Phys.*, **40**, 3731.

Cohen, M. H., and D. Turnbull (1959). *J. chem. Phys.*, **31**, 1164.

Doolitle, A. K. (1951). *J. appl. Phys.*, **22**, 1471.

Eyring, H. (1941). In S. Glasstone, K. J. Laidler, and H. Eyring, *Theory of Rate Processes.* McGraw–Hill Book Co, Inc. New York, London, pp. 477–552.

Frenkel, J. (1955). *Kinetic Theory of Liquids.* Dover Publications, Inc., New York. pp. 93–182.

Gibbs, J. H. (1960). In J. D. Mackenzie (Ed.), *Modern Aspects of the Vitreous State.* Butterworths, London. p. 158.

Gibbs, J. H., and E. A. DiMarzio (1958). *J. chem. Phys.*, **28**, 373.

Gutzow, I. (1962). *Z. phys. Chem.*, **A221**, 153.
Gutzow, I. (1966). In E. A. Porai-Koshits (Ed.), *The Structure of Glass*, Vol. 6. Consultants Bureau, New York. p. 55.
Gutzow, I. (1970). Unpublished paper.
Gutzow, I., I. Konstantinov, and R. Kaischew (1969). *Comm. Deptmt. Chem.* (Bulg. Acad. Sci.) **2**, No. 2, 249.
Gutzow, I., I. Konstantinov, and R. Kaischew (1969). In *Proc. Fifth All-Union Conf. Glassy State*. Leningrad. 26–30. V. 1969. In preparation.
Hirai, N., and H. Eyring (1958). *J. appl. Phys.*, **29**, 810.
Jackson, K. A., D. R. Uhlmann, and J. D. Hunt (1967). *J. cryst. Growth*, **1**, 1.
Jenkel, E. (1939). *Z. phys. Chem.*, **A184**, 309.
Kauzmann, W., (1948). *Chem. Rev.*, **43**, 219.
Tammann, G. (1933). *Der Glaszustand.* Leopold Voss Verlag, Leipzig. pp. 82–86.
Williams, M. L., R. F. Landel, and J. D. Ferry (1955). *J. Am. chem. Soc.*, **77**, 3701.
Wunderlich, B. J. (1960). *J. phys. Chem.*, **64**, 1052.

18

Thermodynamic and Relaxational Aspects of the Glass Transition from a 'Bond Lattice' Model

K. J. RAO and C. A. ANGELL

Department of Chemistry, Purdue University, Lafayette, Indiana, U.S.A.

It has become clear, since Kauzmann (1948) pointed out that a number of glassforming liquids he examined were rapidly approaching the same state of total entropy as the crystal when the glass transition occurred, that a proper understanding of the thermodynamics of supercooled liquids would play an important part in understanding both the glass transition itself, and the relaxational characteristics of viscous liquids in general.

Progress in this direction has been made by Gibbs and co-workers (Gibbs and Dimarzio, 1958; Adam and Gibbs, 1965), who showed by a lattice theory that chain polymeric liquids would encounter a configurational ground state at temperatures in the vicinity of the experimental glass transition, and by Cohen and Turnbull (1959) and Bueche (1959) whose theories have emphasized the role of 'free volume'. Free volume plays a minor role in the Gibbs–Dimarzio theory. Free volume theories encounter difficulties with experimental findings on pressure effects (Goldstein, 1963) while the entropy theory of Gibbs could not be extended to small molecule, or network liquids.

The present paper will present a simple heuristic model for the glass transition in network liquids which with a little imagination may be considered to apply to simple molecular and ionic liquids as well. Here also excess or 'free' volume will be found to play only an incidental role, although chain stiffness *per se* does not enter the model at all, An important feature of the theory is the manner in which the glass transition is identified as a variant of the more common order–disorder (A transition) phenomena, being distinguished by the importance of lattice vibrational entropy changes which accompany the disordering, and by the loss of cooperative behaviour before the disordering is complete.

Before discussing details of the theory, (some zeroth order aspects of which have been published elsewhere [Angell and Wong; 1970; Angell, 1971]), it is desirable to summarize briefly what features of experience the theory should describe.

In the first place it should show how a substantial difference in heat capacity between supercooled liquids and glasses can arise, yet be able to account for certain cases such as GeO_2, BeF_2, where very little if any difference is detectable. Secondly, it should require that the heat capacity of the internally equilibrated supercooled liquid become progressively larger than that of the corresponding crystalline phase with decreasing temperatures and then, at temperatures not far below the experimental glass temperature, drop suddenly, or at least rapidly, to a value close to that of the crystal in order to resolve Kauzmann's equilibrium problem. The rate of decrease at this low temperature limit should be such as to explain the measured residual entropies which amount to some 25–50% of the fusion entropies. Thirdly the theory should provide an explanation of why the T_0 parameter of the Fulcher equation for viscosity of moderately viscous liquids should so frequently correspond with the temperature at which calculations based on extrapolated equilibrium liquid heat capacities, (Angell and Moynihan, 1969; Rao and Angell, 1971) indicate the excess entropy of the liquid would fall to zero.

BOND LATTICE MODEL FOR THE THERMODYNAMIC ASPECTS OF THE GLASS TRANSITION

From a spectroscopic study of $ZnCl_2$ glass into which CN^- ions were introduced as probes of the chloride ion bonding situation, Angell and Wong (1970), concluded that the glass contains mainly fully bridging chloride ions, but also smaller fractions of both singly *and doubly broken* bridges. The spectra of these three states are of similar half-width to the crystal bands, suggesting the bonded states are well defined in the glass, i.e. there is no partial bonding.

These observations have suggested, as a 'picture' of the $ZnCl_2$ glass, a disordered lattice-like arrangement of particles restrained from expansion under thermal motion by a tensed 3-dimensional network of well defined bonds. In this array, breaking of a bond appears as the key energetic step in an elementary lattice excitation which consists of the bond rupture and a concommittent long range readjustment of particle positions which reduces local strains to a minimum. It has been shown elsewhere (Angell and Wong, 1970; Angell, 1971), by invoking a lattice representation of the glass which focuses attention on the bonds between particles, rather than on the particles themselves, that the possibility of such excitations in a lattice of this type requires a temperature dependence of the thermodynamic properties which much resembles the thermodynamic manifestation of the glass transition phenomenon.

The treatment of this lattice, termed the 'bond lattice', in which each 'element' is taken as the region of interaction between network centre atoms in quasi-lattices in which a single identifiable bond couples the atoms or molecules, (e.g. Ge atoms in amorphous Ge, H_2O molecules in vitreous ice) leads to a distribution law for broken bonds ('off' lattice elements) of the form

$$n_x = 2N(1 + \exp{(h_x - Ts_x)/kT})^{-1} \qquad (1)$$

where n_x = is the number of 'off' elements at temperature T, h_x is the enthalpy change involved (mainly determined by the short range contribution to the bond energy), and s_x is the change in *vibrational* entropy due to changes in quasi-lattice vibration frequencies on bond rupture (Bell, Bird and Dean, 1968).

With this distribution is associated a heat capacity,

$$C_{\mathrm{p(conf)}} = 2R\left(\frac{\Delta H}{RT}\right)^2 \frac{A\exp{(\Delta H/RT)}}{[1 + A\exp{\Delta H/RT}]^2} \qquad (2)$$

where the constant $A = \exp{(-\Delta S/R)}$, which is a modified (by the constant A) version of the well known Schottky anomaly heat capacity (Schottky, 1921, 1922). The inclusion of the vibrational entropy term, however, is important since it allows the heat capacity to exceed the maximum value of 0·9 cal/mole-deg reached at $T = 0{\cdot}42\ \Delta\mathrm{H}/R$ in the Schottky case. Some representative plots of this heat capacity function for different values of the parameters are shown in Figure 1. The maximum in C_{p} depends only on ΔS, while the temperature at which the maximum is reached depends on ΔH for a given ΔS.

It is noted in practice that $C_{\mathrm{p(conf)}}$, taken as the amount, ΔC_{p}, by which the heat capacity of the supercooled liquid exceeds that of the crystal, consistently rises (under non-equilibrium conditions) to a maximum value at the glass transition temperature and thereafter decreases steadily with increasing temperature. From the proximity of the 'ideal' glass temperature in most cases, it may be judged that the fall in liquid heat capacity in a hypothetical equilibrium (slow) measurement must be quite rapid if not precipitous as often drawn. Thus the form of the configurational heat capacity given by the bond lattice model in zeroth order approximation is qualitatively that required to describe the thermodynamic manifestation of the glass transition. Energy and entropy parameters giving approximate descriptions of a particular experimental case may be assessed from plots of $C_{\mathrm{p(max)}}$ *vs.* ΔS, and $T(C_{\mathrm{p(max)}})$ *vs.* ΔH, (Rao and Angell 1971) *bearing in mind that the excess heat capacity to be fitted should be one half the molar excess heat capacity for a four coordinated matter-point quasi-lattice* (since there are $2N$ bond elements per N matter points).

Note from Figure 1 that where ΔS is very small, the total ΔC_p developed may be as small as 0·9 cal deg^{-1} mole^{-1}, and this may develop over a range of hundreds of degrees when ΔH is large. The non-appearance of a ΔC_p in

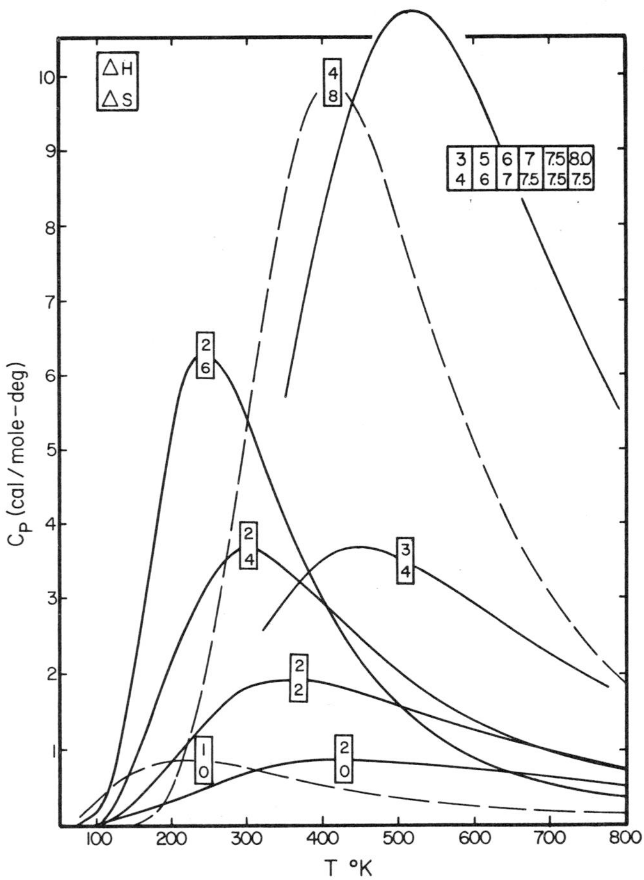

Figure 1. Heat capacity temperature relations for single excitation bond lattice model in zeroth order approximation

such glasses as BeF_2 (Taylor and Gardner, 1965), GeO_2 and Ge (de Neufville and Turnbull, 1970), may be understood in these terms. Although these are all tetrahedral lattices, the large configurational heat capacity of water suggests that tetrahedral lattices do not necessarily have unique behaviour in this respect.

Whether or not it is reasonable to consider non-network liquids in the simple terms of this model will be considered below. First, however, we return to $ZnCl_2$ to consider the consequences of bond lattice elements of the bridge-atom type (—Cl—). It would appear that elements of a bond lattice representing a bridge-bonded structure could exist in any of three states, viz., fully bridging ('on'), one side of bridge broken ('$\frac{1}{2}$ off'), and both sides broken ('off'). The statistical treatment of this case leads to 'three-state' thermodynamic equations according to which

$$N_{x(1)} = N_{gr} \exp - ([\Delta H_1 - T\Delta S_1]/RT)$$
$$N_{x(2)} = N_{gr} \exp -([\Delta H_2 - T\Delta S_2]/RT) \qquad (3)$$

$$N_{gr} = 1 - (N_{x(1)} + N_{x(2)})$$
$$= [1 + \exp(- \Delta G_{(1)}/RT) + \exp(- \Delta G_{(2)}/RT]^{-1} \qquad (4)$$

where N_{gr} is the fraction of ground state, fully bridging, bond elements, $N_{x(1)}$ the fraction of singly excited ('$\frac{1}{2}$ off') elements, and $N_{x(2)}$ the fraction of doubly excited ('off') elements. From (3) and (4)

$$C_{p(conf)} = R\left(\frac{\Delta H_1}{RT}\right)^2 N_{x(1)}(1 - N_{x(1)}) + R\left(\frac{\Delta H_2}{RT}\right)^2 N_{x(2)}(1 - N_{x(2)}) + \frac{2\Delta H_1\,\Delta H_2}{RT^2} N_{x(1)} N_{x(2)} \qquad (5)$$

where $C_{p(conf)}$ is the excess heat capacity per mole *of elements*. Similar equations have been given earlier by Leidecker *et al.* (1969) from consideration of a 3-species mixture model, $C_{p(conf)}$ in their case having been treated as the excess heat capacity per mole of compound. Figure 2 shows the effect the additional possible states can have according to the thermodynamic parameters characterizing them. One notes that the presence of additional higher enthalpy states can produce marked changes in the temperature dependence of the heat capacity at high temperatures of the sort needed to understand the temperature invariant liquid heat capacities often observed in practice (including cases where the vibrational heat capacity is almost constant). The increase at low temperatures is not affected, however, and it is in the rate of increase at low temperatures and the breadth near the maximum that the weakness of the present model in its zeroth order form becomes apparent.

Cooperative effects

To demonstrate the weakness referred to above, we examine curve 1 on Figure 3 in which a two state heat capacity plot, parameters $\Delta H = 3000$ cal mole^{-1}, $\Delta S = 5{\cdot}0$ e.u., is plotted on a logarithmic temperature scale. We

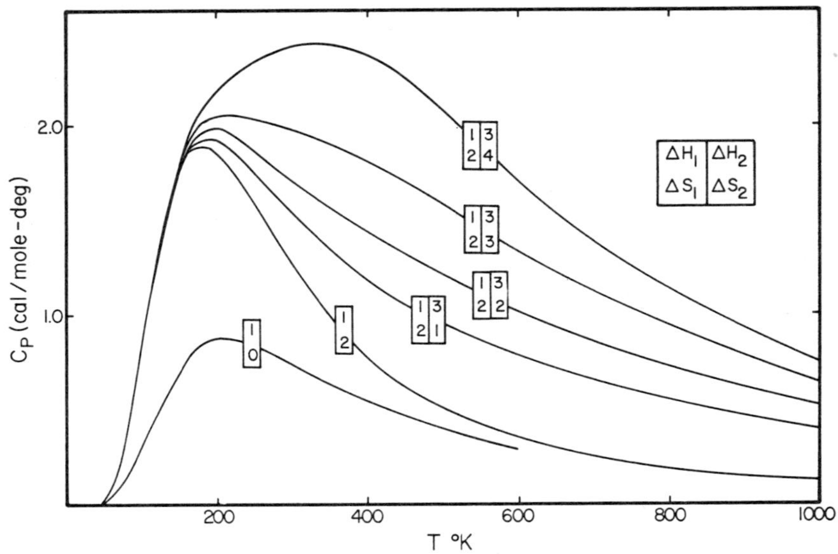

Figure 2. Heat capacity temperature relations for double excitation bond lattice model, zeroth order approximation

suppose, to be consistent with the experimental observation that ΔC_p is usually steadily increasing with decreasing temperature when T_g intervenes, that T_g for the substance must be placed at a temperature somewhat above that of the C_{conf} maximum. Also, to accord with a typical glassforming pure component, we take the melting point to be 1·7 T_g. The fraction of the entropy of fusion which is residual in the hypothetical glass at 0 °K can then be assessed from the comparison of the areas under the C_{conf} curve below T_g and below T_m respectively. For curve 1, the residual entropy would be ~60% of the entropy of fusion, whereas in practice values of 20–50% are more common. Thus a reason why the equilibrium C_{conf} should increase more rapidly at low temperatures must be found if the model is to provide a satisfactory interpretation of the actual glass transition.

We therefore invoke a cooperative effect, similar to those responsible for the lambda transition in many crystalline transformations and other critical phenomena now frequently described by Ising model statistics. In order to avoid the spike in heat capacity common to such transitions we must, however, require that the cooperative effect 'wash out' before disordering is complete. We have achieved this result in a somewhat artificial manner, suitable for demonstrating the improvement in theoretical predictions, by requiring that ΔH be made up of two contributions, $\Delta H_0 + \Delta H_1$, one of which vanishes with increasing temperature according to

$$\Delta H_{1(T)} = \frac{\Delta H_1}{1 + \exp (T - T_r)/D} \qquad (6)$$

Using one-dimensional representations of the particle interactions and their changes on excitation, we have elsewhere (Angell, 1971; Rao and Angell,

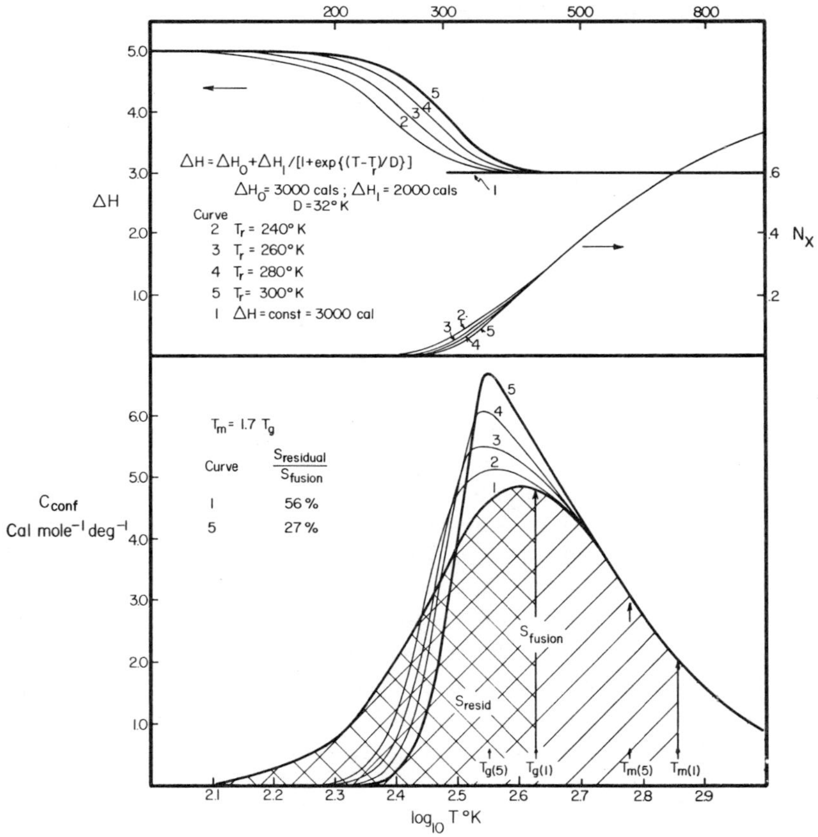

Figure 3. Heat capacity–temperature relations for bond lattice model with cooperative effects included

1971) rationalized this form in terms of characteristic differences between Ising and glass-bond lattices. We introduce therewith two additional adjustable parameters which is a disadvantage, but it appears that their values must be closely correlated with the values of ΔH_0 and ΔS, and improvements in the theory may lead to their reduction.* The theoretical heat capacity which now satisfies the requirements noted earlier is now given by

$$C_{\text{conf}} = \frac{d}{dT}[N_x . \Delta H] = N_x \frac{\partial \Delta H}{\partial T} + \Delta H \frac{\partial}{\partial T}(N_x) \tag{7}$$

where

$$N_x = \left[1 + \exp\left(\frac{\Delta H_0 + \Delta H_{1(T)} - T\Delta S}{RT}\right)\right]^{-1} \tag{8}$$

Differentiation and collection of terms now leads to

$$C_{\text{conf}} = \frac{\Delta H^2}{RT} N_x(1 - N_x) - \left[\frac{\Delta H N_x(1 - N_x)}{RT} - N_x\right] \frac{\partial \Delta H}{\partial T} \tag{9}$$

and

$$\frac{\partial \Delta H}{\partial T} = - \frac{\Delta H_1 \exp (T - T_r)/D}{D[1 + \exp (T - T_r)/D]^2} \tag{10}$$

Note that the first term in equation (9) is the normal zeroth order heat capacity the second term being a correction which increases in importance as T_g is approached from above.

The new C_{conf} *vs.* T relation has been computed for several values of the parameters T_r and D and is plotted in Figure 3. Also shown are the temperature dependences of ΔH_T and of N_x for these parameters. The form of curve 5 in particular is very similar to that of the excess heat capacity of 3-methyl pentane, and to the plots of the experimental 'structural heat capacities' *vs.* temperature assessed by Dexter and Matheson (1968) for a variety of super-cooled hydrocarbon liquids. The residual entropies are also now in harmony with experimental findings. One notes that the temperature range over which the cooperative effect has to operate is quite small.

Extension to the general case of the glassforming liquids

The above comparisons imply the possibility that the expressions derived for a simple network glass may apply to the general case of glassforming liquids. We seek briefly to justify this notion. Although λ transitions in crystalline phases are much less common than first order phase transitions, their existence, e.g. in sodium nitrate, is sufficient to show that ionic, or molecular groupings capable of rearranging in stepwise fashion can exist

* Space does not permit discussion of an alternative explanation (Rao, 1970) of the temperature dependence of ΔH_1 in which T_r and D are related to a critical amplitude in the vibrational motions of particles involved in the rearranging groups or breaking bonds.

in condensed phases. Such crystals always contain asymmetric ions or molecules, and the forces opposing the rearrangements are short range repulsions. The passage from one configuration to the other is thus energetically equivalent to breaking a bond. There seems no reason to deny the availability of such energetically distinct short range orderings to the liquid state, it being expected that their importance should be most marked at low temperatures where the packing density approaches that of the crystal. Thus it is probably reasonable to apply the bond lattice concept to glass formation in such liquids but perhaps to expect a larger number of states, or a greater energy distribution to be available to the lattice elements. Equation (5) may be generalized to arbitrarily large numbers of states yielding

$$C_{\text{conf}} = \sum_{i=1}^{n} \frac{\Delta H_i^2}{RT^2} N_{x(i)} (1 - N_{x(i)}) - \frac{2}{RT} \sum_{\substack{i,j \\ i>j}} \Delta H_i \Delta H_j N_{x_i} N_{x_j} \tag{11}$$

We have computed heat capacity curves for up to eight state systems in zeroth order approximation, using ascending values of both energy and entropy for the states and find the shape of the heat capacity curve to be little different in form from a two-state system. An example for a six state system is given in Figure 1.

In the case of such a distribution, states favoured by a change in temperature will not necessarily be those favoured by a change in pressure, and the experimentally observed deviations of the Prigogine–Defay ratio from the value unity is to be expected.

RELAXATIONAL ASPECTS

It has been shown elsewhere (Angell, 1971) how the bond lattice model implies non-Arrhenius relaxational behaviour if molecular rearrangements are assumed to occur in regions of the structure containing a critical number, or concentration of 'broken bonds' ('off' elements). A simple relation between relaxation time and mole fraction of 'off' elements, N_x was suggested by the zeroth order theory,

$$\tau = \tau_0 \exp 1/N_x \tag{12}$$

where $N_x(=n_x/2N)$ is given by equation (1).

In the temperature regions in which viscosity, or conductance, measurements are usually made, N_x is so close to linear in its temperature dependence ($N_x = C(T - T_0)$), that equation (12) is indistinguishable from the empirical Fulcher equation

$$\tau = \tau_0 \exp [B/(T - T_0)] \tag{13}$$

which is frequently utilized to describe the viscosity of supercooled liquids in the range $10 - 10^5$ cp. Since the linear relation implies $N_x \rightarrow 0$ at $T = T_0$, this model provides a reasonable account of the correspondence of $T_{0(\tau)}$ with the T_0 determined from calorimetric data. Some results showing how the temperature dependence of viscosity, η, shear relaxation time, τ_S, and dielectric relaxation time, τ_D, may be correlated with the configurational heat capacity through equations (1) and (12), are shown in Figure 4.

It should be noted that equation (12) contains no activation energy in the normal sense of the term, in this respect being similar to the free volume transport equation of Cohen and Turnbull (1959). This implies that de-

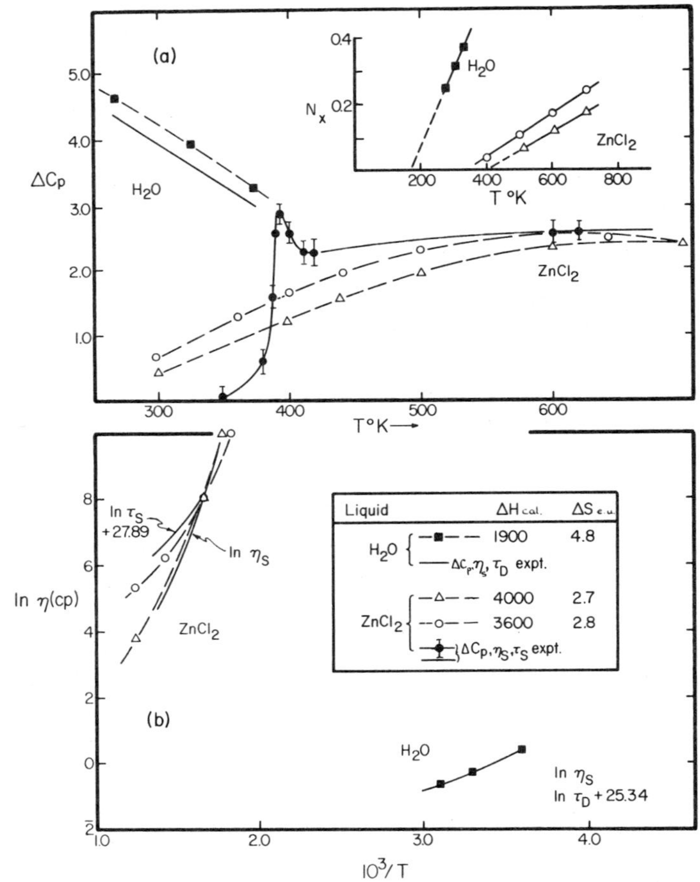

Figure 4. Fit of excess heat capacity, and comparisons of theoretical viscosity temperature dependence and shear and dielectric relaxation times for H_2O and $ZnCl_2$. Data for τ_S, η_S ($ZnCl_2$) taken from G. J. Gruber and T. A. Litovitz (1964). *J. chem. Phys.*, **40**, 13; for η (H_2O) from *Handbook of Physics and Chemistry*; τ_D (H_2O) from C. H. Collie, J. B. Hasted, and D. M. Riston (1948). *Proc. phys. Soc.*, **60**, 145

creases in 'activation energy' on addition of 'network modifying' components may be discussed in terms of changes in equilibrium thermodynamic properties. Indeed dramatic increases in ΔC_p may be observed to accompany such network breaking (Rao and Angell, 1971).

When ΔS of equation (1) is small, and ΔH is large, a steep slope to the Arrhenius plot ln τ *vs.* $1/\tau$, is predicted by equation (12), and curvature will not readily be detected. This accords qualitatively with the observations for BeF_2, GeO_2 and SiO_2.

We have not attempted to discuss transport properties in the cooperative region. Cooperative effects are hard to assess even qualitatively since clustering will make critical groups more probable, tending to decrease the curvature of the Arrhenius plot, while conversely the accelerated temperature dependence of N_x in this region will have the opposite effect. An explanation of the now-familiar return to Arrhenius behaviour for some liquids in this temperature region probably has its origin in the cooperative behaviour, but a quantitative treatment may prove difficult. It should be noted that it is in the temperature region immediately below the melting point, in which Figure 3 suggests cooperative effects need not be invoked to gain accord with experiment, that the first four orders of magnitude of transport constants are observed, and it is from data in this region that T_0 (transport) values are usually obtained.

REFERENCES

Adam, G., and J. H. Gibbs (1965). *J. chem. Phys.*, **43**, 139.

Angell, C. A. (1971). *J. Phys. Chem.*, to be published.

Angell, C. A., and C. T. Moynihan (1969). In G. Mamantov (Ed.), *Molten Salts: Characterization and Analysis.* Marcel Dekker Inc., New York.

Angell, C. A., and J. Wong. (1970). *J. chem. Phys.*, **53**, 2053.

Bell, R. J., M. F. Bird, and P. Dean (1968). *J. Phys.*, C, (Proc. Phys. Soc.) ser 2, **1**, 299.

Bueche, F. (1959). *J. chem. Phys.*, **30**, 748

Capps, W., P. B. Macedo, and T. A. Litovitz (1966). *J. chem. Phys.*, **44**, 3357.

Cohen, M. H., and D. Turnbull (1959). *J. chem. Phys.*, **31**, 1164.

De Neufville, J., and D. Turnbull (1970). *Disc. Faraday Soc.*

Dexter, A. R., and A. J. Matheson (1968). *Trans. Faraday Soc.*, **64**, 2632.

Gibbs, J. H., and E. A. Dimarzio (1958). *J. chem. Phys.*, **28**, 373.

Kauzmann, W. (1948). *Chem. Rev.*, **43**, 219.

Leidecker, H. W., P. B. Macedo, and T. A. Litovitz (1969). *O.N.R. Technical Report No.* 1, N00014-68-A-0506-00021.

Rao, K. J., and C. A. Angell (1971). To be published.

Schottky, W. (1921). *Phys. Z.*, **22**, 1. (1922). *Phys. Z.*, **23**, 9.

Taylor, A. R., and Gardner. U.S. Department of Interior, Bureau of Mines, Report of Investigations No. 6664.

19

The Description of Glasses by Generalized Equations of State

B. ECKSTEIN

Technische Hochschule, Aachen, Germany

The arbitrary control of non-equilibrium materials is of great practical importance. Especially, the sensitive dependence of the properties of amorphous semiconductors on their thermal history gives fresh interest to the thermodynamics of non-equilibrium materials.

As is well known from the stabilization of glasses, their volume for constant temperature and pressure is time-dependent, in contrast to equilibrium materials with $V(p, T)$ totally fixed. $\mathrm{d}V/\mathrm{d}t|_{p,T} \neq 0$, however, implies the existence of additional 'inner parameters' φ_i in the equation of state. Assuming uniform pressure (non-uniform stresses causing complications), the equation of state may be generalized to

$$f\,[p, V, T_{\mathrm{ph}}; \varphi_1(t), \ldots, \varphi_n(t)] = 0$$

with the 'phenomenological temperature' T_{ph} interpreted as the temperature of the material, and the φ_i describing the detailed deviations from equilibrium for the various inner degrees of freedom, and thus the detailed properties of the material. For the simplest case of a monocomponent glass,

$$f(p, V, T_{\mathrm{ph}}; T_{\mathrm{c}}^{(1)}, T_{\mathrm{c}}^{(2)}) = 0$$

with the 'configurational temperatures' $T_{\mathrm{c}}^{(1)}$ connected with the concentration and $T_{\mathrm{c}}^{(2)}$ with the distribution of frozen-in structural defects (Eckstein, 1968).

Configurational freedoms generally are connected with relaxation times $\tau_i = \tau_{0i} \exp{(E_i/kT_{\mathrm{ph}})}$ with energetic barriers $E_i = E_i(p, T_{\mathrm{ph}}; \varphi_1, \ldots, \varphi_n)$.

The spectrum of the $\tau_i(T_{\mathrm{ph}})$ governs the range of arbitrary variability of the properties of materials accessible by variations of the production process (Eckstein, 1970). In varying the value of one inner parameter by thermal treatment, generally the values of the others shift simultaneously. Within a certain range, they may be fixed independently of each other during a cooling programme, if and when the connected relaxation times $\tau_i(T_{ph})$ reach the value of about 10 minutes at clearly distinguished (phenomenological) temperatures, and every τ_i then decreases steeply with further decrease in T_{ph}.

To obtain a special set of φ_i, rather complicated cooling programmes may be necessary, and a slight deviation from the programme (even in the previous molten state) may seriously alter the properties of the material.

The numerous discrepancies in the observed properties of amorphous solids do not usually indicate erroneous measurements, but different thermal histories of the material, resulting in different values of the inner parameters.

REFERENCES

Eckstein, B. (1968). *Mater. Res. Bull.*, **3**, 199.
Eckstein, B. (1970). *J. non-cryst. Solids*, **4**, 366.

VISCOSITY AND DIFFUSION PROCESSES

20

Atomic Transport in Simple Fluids

R. C. BROWN and N. H. MARCH

Department of Physics, The University, Sheffield, Great Britain

---

The purpose of this paper is to discuss atomic transport properties in simple fluids from the standpoint of the dynamical behaviour of atoms, and to enquire whether any of the concepts used throw light on atomic motions in glassy materials.

In a simple monatomic liquid, like Na or Ar, one frequency, the high energy end of the frequency spectrum, or the Debye frequency ω_d provides us with an extremely useful semiquantitative description of atomic motions.

However, we shall argue in this paper, that, in general, there is another frequency of central importance in liquids and amorphous materials; namely that frequency below which it is not possible to transmit transverse waves. We shall argue that an essential difference between the atomic dynamics of simple liquids like Ar and Na, and that of glasses, resides in the fact that this frequency is essentially the same as ω_d in the former cases, but a very low frequency compared with ω_d in the latter. Thus, the spectral density for viscosity, which we define below, has a crucially different shape for simple liquids and for glasses.

Rather than attempting to give a theory of the viscosity of glasses, we shall ask what we can learn about the spectral density from experimental measurements of viscosity.

MODERN THEORY OF ATOMIC TRANSPORT

Older theories of diffusion and viscosity focused on the non-equilibrium distribution function in a fluid. Modern theory has shifted the emphasis to a discussion of time-dependent correlation functions in liquids, which are intimately related to neutron scattering. This is discussed in the paper by Dr. Leadbetter at this conference, and so we shall only make occasional reference to it.

Microscopic theories of fluids must, of course, in the appropriate long-wavelength–low-frequency limit, reduce to the equations of hydrodynamics, and thus must relate to atomic transport coefficients. It is this relation that is our prime interest here.

To illustrate the approach, let us suppose that we follow the motion of a chosen atom through the fluid, assuming it to be at the origin $\boldsymbol{r} = 0$ at time $t = 0$. Then we define the correlation function $G_s(\boldsymbol{r}, t)$ such that it gives us the probability of finding this chosen atom at $\boldsymbol{r}$ at time t.

Now the simplest theory we can write down for $G_s(\boldsymbol{r}, t)$ is to suppose it satisfies the diffusion equation

$$D\nabla^2 G_s(\boldsymbol{r}, t) = \frac{\partial G_s}{\partial t}. \tag{1}$$

The appropriate solution is then readily obtained as

$$G_s(\boldsymbol{r}, t) = \left[\frac{1}{(4\pi D|t|)^{\frac{3}{2}}}\right] \exp\left(\frac{-r^2}{4D|t|}\right). \tag{2}$$

We now take the Fourier transform of $G_s(\boldsymbol{r}, t)$ with respect to $\boldsymbol{r}$ and t, the resulting function being $S_s(\boldsymbol{k}\omega)$ given by

$$S_s(\boldsymbol{k}, \omega) = \frac{Dk^2}{\pi[\omega^2 + (Dk^2)^2]}. \tag{3}$$

Now let us view this as a way to relate D to the time-dependent correlation functions. The macroscopic argument we have given above is valid only for long time and large $\boldsymbol{r}$, which means that we have the right to use equation (3) only for small $\boldsymbol{k}$ and ω. We note now that

$$\frac{D}{\pi\omega^2} = \lim_{k \to 0} \frac{S_s(k, \omega)}{k^2}. \tag{4}$$

This equation focuses attention on the frequency function $g(\omega)$ defined by

$$g(\omega) = \omega^2 \lim_{k \to 0} \frac{S_s(k, \omega)}{k^2}. \tag{5}$$

In general, equation (4) is true only as $\omega \to 0$, and thus we see that

$$\frac{D}{\pi} = g(0) \tag{6}$$

and the diffusion constant, apart from the factor π, is given as a low frequency limit of $g(\omega)$.

If we evaluate equation (5) explicitly for a Debye solid, we find

$$g(\omega) \propto \omega^2, \qquad \omega < \omega_d, \tag{7}$$

i.e. $g(\omega)$ is the usual frequency spectrum.

In the liquid, $g(\omega)$ is not only reflecting the vibrational motions, but, as we can see from equation (6), it is describing the diffusive modes at low frequencies.

A similar, though slightly more complicated argument, from the hydrodynamic equations, shows that the sound wave attenuation coefficient, proportional to $\frac{4}{3}\eta + \zeta$, where η and ζ are respectively the shear and bulk viscosities, can be related to $S(\mathbf{k}, \omega)$, the usual scattering function, which is essentially the probability that, in neutron scattering, there is momentum transfer $\hbar\mathbf{k}$ and energy transfer $\hbar\omega$, from the neutron to the liquid, (for a general reference, see, for example, March, 1968) by

$$\tfrac{4}{3}\eta + \zeta = \frac{\pi M^2 \rho}{k_B T} \lim_{\omega \to 0} \omega^4 \lim_{k \to 0} \frac{S(k, \omega)}{k^4}, \tag{8}$$

where ρ is the number of atoms per unit volume. Before the $\omega \to 0$ limit, we have again the important frequency function

$$s(\omega) = \omega^4 \lim_{k \to 0} \frac{S(k, \omega)}{k^4}. \tag{9}$$

This function is the focal point of interest in the present paper. In the absence, to date, of detailed neutron measurements of $S(k, \omega)$, we might ask what information about $s(\omega)$ we can infer from our knowledge of atomic transport.

Finally, we remark that, since experiments on ζ are not at present very reliable even in simple liquids, there is a formula like equation (8) for the shear viscosity η, involving a frequency function $s_\eta(\omega)$, rather than $s(\omega)$ of equation (9).

FORMS OF FREQUENCY FUNCTIONS IN SIMPLE LIQUIDS

Diffusion

We have information about the frequency spectrum $g(\omega)$ from machine calculations for one or two liquids, and in particular for Na and Ar.

However, we are interested here in relating properties of a variety of liquids, and we shall therefore use a much cruder approach, which is applicable only at the melting temperature T_m.

We argue that the high frequency limit of the frequency spectrum is the Debye frequency ω_d, and we can then write down the area $\int_{-\infty}^{\infty} g(\omega)\mathrm{d}\omega$ under the $g(\omega)$ curve as

$$\int_{-\infty}^{\infty} g(\omega)\mathrm{d}\omega \simeq \text{Height} \times \text{width} \simeq (D/\pi)2\omega_\mathrm{d}. \tag{10}$$

But for classical monatomic fluids, it follows quite generally that

$$\int_{-\infty}^{\infty} g(\omega)\mathrm{d}\omega = \frac{k_\mathrm{B}T}{M}. \tag{11}$$

From equations (10) and (11) we then obtain at the melting temperature T_m

$$D \sim \frac{k_\mathrm{B}T_\mathrm{m}}{M\omega_\mathrm{d}}. \tag{12}$$

This 'rough and ready' theory is compared with experiment in Table 1 below.

Table 1. Diffusion constants at melting temperature

Liquid	$D_\mathrm{expt} \times 10^5\ \mathrm{cm^2\ sec^{-1}}$	$\frac{k_\mathrm{B}T_\mathrm{m}}{M\omega_\mathrm{d}} = D_\mathrm{calc} \times 10^5$	$D_\mathrm{calc}/D_\mathrm{expt}$
Li	6·5	12·3	1·9
Na	4·1	6·4	1·6
K	3·9	5·5	1·4
Cu	4·0	4·0	1·0
Ag	2·6	3·4	1·3
Zn	2·0	2·0	1·0
Mg	1·2	1·0	0·9
In	1·7	2·1	1·3
Sn	2·0	1·0	0·5
Pb	2·2	2·1	0·9
Ar	1·5	1·7	1·1

It will be seen that, within a group, say the alkali metals, the ratio $D_\mathrm{calc}/D_\mathrm{expt}$ is reasonably constant. The factor 1·4–1·9 in $D_\mathrm{calc}/D_\mathrm{expt}$ for the alkali metals must be due to the wrong shape assumed for $g(\omega)$: our argument is essentially taking a rectangular distribution: for the alkali metals $g(0)$ must be less than the average value to agree with experiment. But the shape of $g(\omega)$ is not, it would seem, too different for metals in the same group.

Actually, although this theory hinges on the role of the Debye frequency, we can use the Lindemann law of melting,

$$\omega_\mathrm{d} = \text{const}\ T_\mathrm{m}^{\frac{1}{2}}/M^{\frac{1}{2}}\Omega^{\frac{1}{3}} \tag{13}$$

where Ω is the atomic volume. This shows, eliminating ω_d in equation (12), that $DM^{\frac{1}{2}}/T_\mathrm{m}^{\frac{1}{2}}\,\Omega^{\frac{1}{3}}$ is a constant and this result has also been obtained from a hard sphere model (Ascarelli and Paskin, 1968).

However, their arguments are probably best suited to liquid argon, with generally short-range forces, whereas ours are more appropriate for treating the long-range forces in liquid metals.

Viscosity

The above crude discussion is more interesting when applied to relate the viscosities of a variety of liquid metals at the melting point.

The frequency function analogous to equation (5) which plays a fundamental role in theories of viscosity is given by equation (9).

The most primitive theory of $S(k, \omega)$ builds it from the structure factor $S(k)$ and the self-correlation function $S_s(k, \omega)$; and according to Vineyard (1958) it has the form

$$S(k, \omega) = S(k)S_s(k, \omega). \tag{14}$$

This form does not have the correct behaviour for small k and ω, for the limit in the formula (see equations (8) and (9),

$$\tfrac{4}{3}\eta + \zeta = \frac{\pi M^2 \rho}{k_B T} \lim_{\omega \to 0} s(\omega) \tag{15}$$

to be properly defined. However, it suggests that $s(\omega)$ has the same frequency range as $S_s(k, \omega)$, and we find, using the area argument at the melting temperature

$$\int_{-\infty}^{\infty} s(\omega)\,d\omega \doteqdot s(0)\,2\omega_d \tag{16}$$

But we have an exact formula for the area, if we can assume the atoms interact via a pair potential

$$\int_{-\infty}^{\infty} s(\omega)d\omega = 3\left(\frac{k_B T}{M}\right)^2 + \frac{\rho k_B T}{2M^2}\int d\mathbf{r}\; g(r)x^2\frac{\partial^2 \phi}{\partial x^2}. \tag{17}$$

Actually, at the melting point, we can approximate the structure dependent term (see Brown and March, 1968) in terms of the structure factor $S(k)$ in the long wavelength limit.

A somewhat oversimplified version of the above argument is as follows. We require

$$\int_{-\infty}^{\infty} s(\omega)\,d\omega \equiv \int_{-\infty}^{\infty} \frac{\omega^4}{k^4} S(k, \omega)\,d\omega \tag{18}$$

Remembering that we want the limit $\omega \to 0$ and $k \to 0$, let us replace ω^4/k^4 by v_s^4, where v_s is the velocity of sound, and recall that $S(k)$ is the integral of $S(k, \omega)$ over all energy transfers, i.e.

$$\int_{-\infty}^{\infty} S(k, \omega)\,d\omega = S(k). \tag{19}$$

Thus, it suggests that

$$\int_{-\infty}^{\infty} s(\omega)\,\mathrm{d}\omega \simeq v_s^4\, S(0). \tag{20}$$

Applying this again at the melting temperature T_m, we find

$$(\tfrac{4}{3}\eta + \zeta)\frac{k_B T}{\pi M^2 \rho} 2\omega_d \simeq S_{T_m}(0)\, v_s^4. \tag{21}$$

But

$$S_{T_m}(0) = \frac{C_p\, k_B\, T_m}{C_v\, M v_s^2} \tag{22}$$

where C_p/C_v is the ratio of specific heats, which is about unity for liquid metals at the melting point, and we obtain

$$\tfrac{4}{3}\eta + \zeta \,|_{T_m} \simeq \frac{\rho\, k_B T_m}{\omega_d\, S_{T_m}(0)} \tag{23}$$

Actually, since ζ is not very reliably known experimentally, an argument similar to the above can be given for η, and again, in order of magnitude

$$\eta \sim \frac{\rho k_B T_m}{\omega_d S_{T_m}(0)} \tag{24}$$

If we neglect the spread in $S_{T_m}(0)$, and use Lindemann's law for ω_d, then we find

$$\eta = \frac{\text{const}\; T_m^{\frac{1}{2}} M^{\frac{1}{2}}}{\Omega^{\frac{2}{3}}} \tag{25}$$

Table 2. Shear viscosities of liquids at melting point

Liquid	Expt (poise)	Theory $\propto (MT_m)^{\frac{1}{2}}\,\Omega^{-\frac{2}{3}}$
Li	0·0060	0·0056
Na	0·0069	0·0062
K	0·0054	0·0050
Rb	0·0067	0·0062
Cs	0·0069	0·0066
Cu	0·041	0·042
Ag	0·039	0·041
Au	0·054	0·058
In	0·019	0·020
Sn	0·021	0·021
Ne	0·0016	0·0019
Ar	0·0025	0·0036
Kr	0·0039	0·0055
Xe	0·0052	0·0068

and this formula was, in fact, given earlier by Andrade (1934), from a kinetic theory argument. His constant is used in comparing the formula with experiment in Table 2.

Our constant is an underestimate by a factor between 2 and 3, and this tells us, in contrast to the frequency spectrum $g(\omega)$, that $s(\omega)$ probably has a quite pronounced maximum at the origin, before falling to zero around ω_d.

TEMPERATURE DEPENDENCE OF VISCOSITY

Simple liquids

The Andrade form for the temperature dependence of viscosity must presumably be connected with the temperature dependence of $g(r)$, as was first proposed, to our knowledge, by Green (1952). We can write in a classical fluid

$$g(r) = e^{-U(r)/k_B T} \tag{26}$$

where $U(r)$ is a potential of mean force. $U(r)$, in a liquid metal, does not appear to be too different from the pair potential $\phi(r)$, and if we assume that the temperature dependence of η is dominated by the first minimum of $\phi(r)$, we might expect

$$\eta \sim e^{(|\phi_{min}|/k_B T)} \tag{27}$$

very crudely.

Table 3 gives a table of activation energies for metals, and there is evidence that these are of the order of the depth of the potential. The activation energies are quite well correlated with the melting temperatures, which again are presumably related grossly to the depth of $\phi(r)$. Thus, in a very rough way, we can give some microscopic interpretation of the viscosity of simple monatomic liquids.

We want now to compare and contrast the situation outlined above with that in some simple glasses.

Table 3. Activation energies for viscosity of metals (based on Menz and Sauerwald, 1966)

Metal	T_m °K	E_η(eV)	1st Min (eV)
Hg	234	0·027	
Bi	544	0·069	
Pb	599	0·086	
Sn	505	0·060	
Zn	692	0·141	
Cd	594	0·089	
Sb	903	0·136	
Cu	1356	0·154	0·24
Ag	1233	0·211	
Na	371	0·065	0·024–0·048
K	337	0·050	

Glasses

The viscosity of glasses is several orders of magnitude higher than that of the liquid metals we have been discussing. The qualitative reason for this is quite clear in, say, silica glass. In viscous flow one would need to break and reform Si–O bonds, and this would take a lot of energy. We see, in Tables 4(a) and 4(b), how the activation energies relate to the bond energies. The

Table 4(a). Activation energies for viscosity

	Energy eV	Reference
SiO_2	6·01	Bacon, J. F., *et al.*
GeO_2	3·22	Kurkjian, C. R., and Douglas, R. W.

Table 4(b). Bond energies in glasses

	Energy eV	Reference
Si–Si	1·8	Kittel, C.
	2·3	Wells, A. F.
Si–O	3·8	Syrkin, Y. K., and Dyatkina, M. E.
	4·6	Wells, A. F.
O–O	1·4	Kittel, C.
	1·5	Wells, A. F.
Ge–Ge	1·6	Rawson, H.

orders of magnitude are the same and the activation energies are greater than the bond energies. We must add that even the concept of activation energy is not sharp; it varies with temperature and, in the end, it will probably turn out to involve, at least weakly, structure and forces, as we have seen in the case of simple liquids. No one, to our knowledge has, so far, been successful in giving a theory of viscosity in systems with strong directional bonding.

However, since the empirical facts on viscosity are now well established, it is worth asking whether we can learn more about the basic atomic dynamics in glasses. We shall try to do this, semiempirically, by assuming that we can still talk of a frequency function for viscosity in a multicomponent system. Thus, we shall try to take cognizance of the very high viscosity, that is the very high value of $s_\eta(\omega)$ at the origin, and at the same time give due attention to the fact that a Debye frequency can still be usefully discussed, as is shown in Table 5. We shall separate $s_\eta(\omega)$ into a low frequency peak and a part which extends out to the Debye frequency, ω_d, by writing the following expression, which gives only a very general description of the shape:

$$s_\eta(\omega) = \frac{A}{\omega_0} \exp\left(-\frac{\omega^2}{\omega_0^2}\right) + B\alpha \exp(-\alpha|\omega|). \tag{28}$$

Table 5. Debye temperature of silica (due to Anderson, 1966)

T °K	Debye temperature		°K
	Cristobalite	Vitreous SiO_2 1308 °C quench	Vitreous SiO_2 1070 °C quench
15	294·0	282·2	276·4
20	312·0	320·4	313·6
30	420·0	370·3	367·6
40	459·3	423·1	421·6
60	521·8	518·5	517·9
100	665·7	664·1	662·8

Our aim now is to see what we can learn about the parameters in this expression, and, in particular, about the width of the low frequency peak. We shall see that this can be related to the so-called Maxwell relaxation time.

Thus we find that, if we require

(i) $s_\eta(0) = (k_B T/\pi)\eta$

(ii) The sum rule of viscosity theory, approximated in terms of the isothermal compressibility K_T as above.

(iii) A fraction of the area, $1/\gamma$, is in the low frequency peak.

(iv) A criterion that 90% of the area is within the Debye frequency ω_d,

the parameters come out, with a tiny mathematical approximation of no consequence in a glass, to be

$$A = \frac{3k_B T}{5\sqrt{\pi}\gamma K_T}$$

$$B = \frac{(\gamma - 1)\, 3k_B T}{10\gamma K_T}$$

$$\alpha = \ln\left(\frac{10\gamma - 10}{\gamma}\right)\Big/\omega_d \tag{29}$$

$$\frac{1}{\omega_0} = \frac{5\sqrt{\pi}\gamma}{3}\left[\frac{\eta K_T}{\pi} - \frac{3(\gamma - 1)}{10\gamma\omega_d}\ln\left(\frac{10\gamma - 10}{\gamma}\right)\right]$$

The interesting parameter from our point of view is the quantity ω_0 which is related to the half-width of the low-frequency peak by

$$\omega_{\frac{1}{2}} = \omega_0\sqrt{\ln 2} \simeq \frac{3\sqrt{\pi}\sqrt{\ln 2}}{5\gamma\eta K_T} \tag{30}$$

since the first term in the expression for $1/\omega_0$ is several orders of magnitude larger than the second.
Then

$$\omega_{\frac{1}{2}} \sim \exp\left(-\frac{E_\eta}{k_\mathrm{B}T}\right) \tag{31}$$

if we assume that the temperature dependence of η is dominant. We then find, if we use the viscoelastic theory summarized below, a quite interesting situation. The idea of Maxwell was that we had to combine an equation for an elastic solid with one for viscous flow. This can be done as follows.

VISCOELASTIC THEORY AND MAXWELL'S RELAXATION TIME

Suppose we apply a shearing stress P_{xy} and that we have viscous flow along the x axis. Then, if v_v is the velocity for viscous flow, we can write immediately

$$\frac{\partial v_\mathrm{v}}{\partial y} = \frac{1}{\eta}P_{xy} \tag{32}$$

Alternatively, for a solid elastic body, with the same shearing stress, we have

$$\frac{\partial v_\mathrm{e}}{\partial y} = \frac{1}{G}\frac{\partial P_{xy}}{\partial t} \tag{33}$$

since

$$\frac{\partial u_x}{\partial y} = \frac{1}{G}P_{xy} \tag{34}$$

where u_x is the elastic displacement, v_e is the corresponding velocity, and G the rigidity modulus. If we combine elastic and viscous effects, then the total velocity, $v = v_\mathrm{v} + v_\mathrm{e}$, is given by

$$\frac{\partial v}{\partial y} = \frac{1}{G}\frac{\partial P_{xy}}{\partial t} + \frac{1}{\eta}P_{xy} \tag{35}$$

and this is the basis of Maxwell's theory of viscoelasticity.

When motion comes suddenly to a standstill, we have

$$\frac{1}{G}\frac{\partial P_{xy}}{\partial t} + \frac{1}{\eta}P_{xy} = 0 \tag{36}$$

as the equation determining how the stress relaxes. The solution is

$$P = P_0 \exp(-t/\tau). \tag{37}$$

where P_0 is the initial stress at time $t = 0$. The relaxation time τ is evidently given by $\tau = \eta/G$.

This theory yields, for $s_\eta(\omega)$, the result, obtained essentially by Fourier transforming equation (37)

$$s_\eta(\omega) = \frac{k_B T}{\pi} G \frac{\tau}{1+\omega^2\tau^2} \tag{38}$$

with area

$$\int_{-\infty}^{\infty} s_\eta(\omega)d\omega = k_B TG \tag{39}$$

If we use the results of the viscoelastic theory for the area under $s_\eta(\omega)$, which is just a slight variant on the procedure adopted above, then we can see immediately that, treating the area under the peak crudely as the height × width again, then, since we have assumed the peak to have fractional area $1/\gamma$,

$$\frac{1}{\gamma} k_B TG \simeq \omega_{\frac{1}{2}} \frac{k_B T}{\pi} \eta \tag{40}$$

Thus we see that the half-width $\omega_{\frac{1}{2}}$ can be written in the alternative form

$$\omega_{\frac{1}{2}} \simeq \frac{\pi}{\gamma} \frac{G}{\eta} \tag{41}$$

and this is clearly related to Maxwell's relaxation time. Indeed, we see that the halfwidth is directly related to the frequency $1/\tau$, and though we do not know what to take for γ, we can conclude, since it is obviously greater than 1, perhaps 10, that a frequency lower than $1/\tau$ is involved in the frequency function for viscosity.

Perhaps the concept of $s_\eta(\omega)$ in a glass will prove too crude, but, in vitreous silica at the melting temperature of cristobalite $1/\tau$ is less than 10^{-6} sec^{-1}, very much less than the Debye frequency and this new element, the Maxwell relaxation time, enters crucially. As discussed by Frenkel and others, this has a very direct physical significance. τ^{-1} is the lowest frequency at which transverse waves can propagate through the medium without great attenuation. It is the fact that in liquid metals this frequency comes out as the same order of magnitude as the Debye frequency that allows the general picture, outlined at the beginning of the paper, to emerge. In the glass, it is crucial to distinguish the two frequencies, and the viscosity is determined largely by $\omega_{\frac{1}{2}}$ and not so much be ω_0, though the complete separation we make is, no doubt, an oversimplification.

We might conclude, then, by asking what experiments can be done to test the conclusion we put forward here. It seems, on the whole, that inelastic neutron scattering experiments, for small momentum transfer, will be the most directly relevant but apparently present experimental resolution will be a problem in the interesting range of ω. However, some further studies of, say, the temperature dependence of the frequency at which transverse sound waves cease to be heavily attenuated would also seem of interest. Needless to say, the theorist has a great deal to do in refining the crude starting point above, and we are trying to put a more detailed account of the structure into the theory at the present time.

REFERENCES

Anderson, O. L. (1966). *Structure of Glass*, Consultants Bureau, New York, **6**, 25.
Andrade, E. N. da C. (1934). *Phil. Mag.*, **17**, 497
Ascarelli, P., and A. Paskin (1968). *Phys. Rev.*, **165**, 222.
Bacon, J. F., A. A. Hasapis, and J. W. Wholley. (1960). *Physics Chem. Glasses*, **1**, 90.
Brown, R. C., and N. H. March (1968). *Physics Chem. Liquids*, **1**, 141.
Egelstaff, P. A. (1967). *An Introduction to the Liquid State*. Academic Press, London.
Frenkel, J. (1946). *Kinetic Theory of Liquids*. Clarendon Press, Oxford.
Green, H. S. (1952). *Molecular Theory of Fluids*. North Holland Press, Amsterdam.
Kittel, C. (1966). *Introduction to Solid State Physics*. Wiley, New York.
Kurkjian, C. R., and R. W. Douglas (1960). *Physics Chem. Glasses*, **1**, 19.
March, N. H. (1968). *Theory of Condensed Matter*. International Atomic Agency, Vienna, p. 93.
Menz, W. and F. Sauerwald (1966). *Acta metall.*, **14**, 1617.
Rawson, H. (1967). *Inorganic Glass Forming Systems*. Academic Press, London.
Syrkin, Y. K., and M. E. Dyatkina (1950). *Structure of Molecules and the Chemical Bond*. Butterworth, London.
Wells, A. F. (1962). *Structural Inorganic Chemistry*. O.U.P., Oxford.

21

The Behaviour of Viscosity in Semiconducting and Metallic Melts of the System Ge–As–Se

M. FISCHER and H. KREBS

Institut für Anorganische Chemie der Universität Stuttgart, Germany

In chalcogenide glasses attention has so far been mainly focused on electrical effects. However, these cannot be understood satisfactorily without taking into account the mobility of the atoms. Therefore, we have undertaken viscosity experiments in addition to X-ray and conductivity investigations.

The main problems we had to cope with were:

(1) The viscometer should measure viscosities varying over many orders of magnitude, because we deal with viscous and metallic melts.

(2) Melts of chalcogenide glasses are fairly aggressive and therefore, mostly quartz-glass is used.

(3) The melts react with air; therefore, vacuum conditions or an inert gas atmosphere are necessary.

(4) The vapour-pressures of Se and As increase rapidly at higher temperatures.

(5) Economic reasons make small sample-sizes necessary.

For some or all of these reasons the usual methods and commercial instruments (see 1961 reference) are not applicable.

THE APPARATUS

Figure 1 explains the principle of the viscometer. Gravity causes the liquid to flow through the capillary of length L and diameter $2r$. Capillary flow can be treated theoretically fairly easy and reliably. It is governed by the Hagen–Poiseuille equation, cited at the top of Figure 1. As the melt flows from the upper container to the lower one, the difference in levels, $H_{(t)}$, and therefore the pressure difference, diminishes. In consequence the volume V flowing through the capillary per unit time-interval becomes smaller. The solution of the corresponding differential equation leads to an exponential function. If the viscometer is turned through 90° the relative height of the two containers is changed. The liquid flows back to the first container and after a turn in the opposite direction the starting position is reached again.

The two containers are also connected by a tube to guarantee equal gas pressure on both surfaces. Finally, there is a tube to admit the sample, to evacuate the viscometer and to seal it hermetically by melting-off. This tube also serves as a mould for the melt after measurements are finished. The solidifying melt could easily damage the viscometer if it remained in the capillary. The viscometer is made of quartz-glass to withstand temperatures up to 1000 °C. Viscometers made of appropriate metals like platinum or molybdenum would be suitable for melts of oxide glasses at much higher temperatures.

The sample volume required is only 0·8 cm^3, that is about 3 gm of chalcogenide glass. The viscometer is housed in a furnace with 20 cm open width. In order to operate the viscometer and also to trace the flow when the furnace is closed, the viscometer is fixed to an axis of rotation perpendicular to the plane of the viscometer.

Figure 2 shows the arrangement as a block-diagram. The viscometer is represented by the capillary in the middle of the furnace. The axis of rotation is a quartz tube. This tube is supported by two bearings outside the furnace and it also houses the thermocouple.

Firstly, as already mentioned, the tube makes it possible to operate the viscometer inside the furnace, secondly, it can be connected with the balance and the dash-pot. The flow of the melt from the left upper container to the lower container on the right hand side causes a change of the torque. This torque acting on the balance is measured by means of a tension and compression sensing transducer. The X–Y recorder plots a signal proportional to the torque, against time. Two PID temperature regulators control the upper and the lower half of the furnace separately to eliminate temperature gradients.

EVALUATION OF OBSERVATIONS

The X–Y recorder plots the exponential function expected according to the solution of the differential equation. Only the exponent of this function is needed. Therefore, we determine the slopes s_1 and s_2 of the plotted curve at two times t_1 and t_2. Dividing s_1 by s_2 results in the fourth equation in Figure 1. The next line shows the logarithms of this equation. All symbols represent known values (ν is the kinetic viscosity, g is the acceleration due to gravity).

Therefore, the equation is solvable with respect to the kinetic viscosity

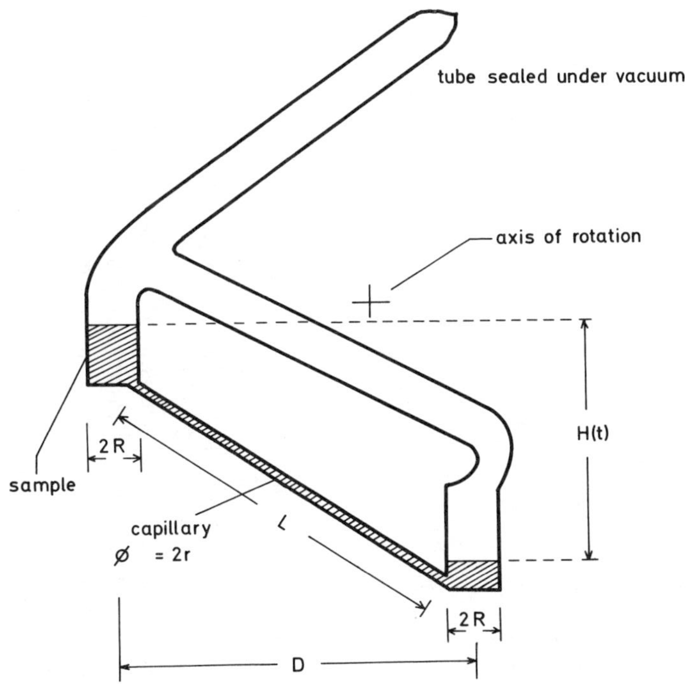

Hagen-Poiseuille Equation

$$V = \frac{\pi}{8} \cdot \frac{r^4(p_1 - p_2)\,t}{\nu\,\rho\,L}$$

pressure difference $p_1 - p_2 = H_{(t)} \cdot g\,\rho$

Resulting torque of viscometer

$$T_{(t)} = g\,\rho\,D\,V_{(t)}$$

Solution of the problem

$$T_{(t)} = g\rho D H_o \frac{\pi R^2}{2} \cdot \exp\left(-\frac{r^4\,g\,t}{4R^2 L \nu}\right) + \text{const.}$$

$$\left(\frac{dT_{(t)}}{dt}\right)_{t_1} \Big/ \left(\frac{dT}{dt}\right)_{t_2} = \frac{s_1}{s_2} = \exp\frac{r^4 g}{4R^2 L\nu}\,(t_2 - t_1)$$

$$\ln s_1 - \ln s_2 = \frac{r^4 g}{4R^2 L\nu} \cdot (t_2 - t_1)$$

Kinetic Viscosity

$$\nu = \frac{g\,r^4}{4R^2 L} \cdot \frac{t_2 - t_1}{\ln s_1 - \ln s_2} = k \cdot \frac{t_2 - t_1}{\ln s_1 - \ln s_2}$$

Figure 1. Quartz viscometer

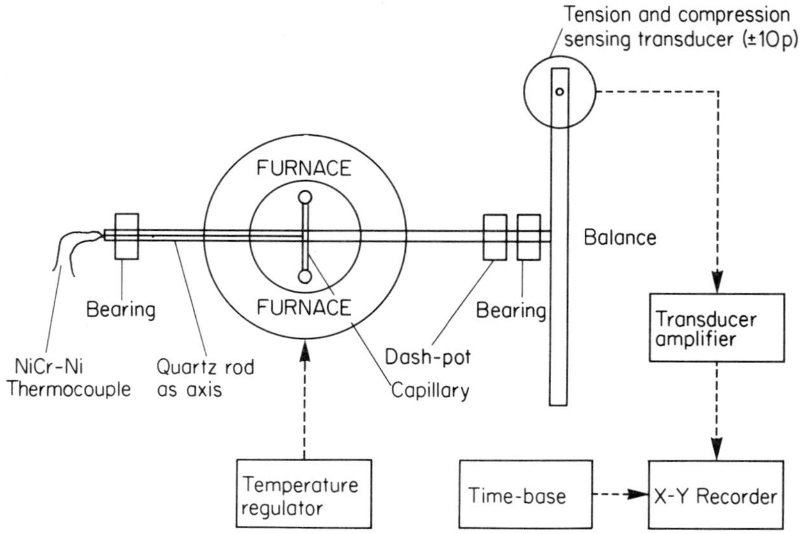

Figure 2. Block-diagram of apparatus

v as is done in the last line. The constant k contains no values of the recording system, nor of the balance, but solely values of the viscometer, that is to say, of the radii of capillary and containers and the length of the capillary. k is easily calculated and coincides pleasantly with the calibration constant within margins of error. The viscometer was calibrated by comparing the results with those obtained from Schott's very precise Ubbelohde viscometer in the temperature range between room temperature and 100 °C.

RESULTS

The results of viscosity measurements on Se_{50} As_{30} Ge_{20} and Se_{15} As_{30} Ge_{55} are presented in Figure 3 together with measurements of electrical resistivity on the same compositions by R. W. Haisty (1969). Log v and log ρ are plotted versus $1000/T$. The viscosity of Se_{15} As_{30} Ge_{55} (a composition that is situated well outside the glassforming region) is about 1 cSt at 750 °C. This value lies by a factor of 20 below the viscosity of Se_{50} As_{30} Ge_{20} at the same temperature and is characteristic of metals. The latter composition is situated in the centre of the glassforming region and exhibits typical viscous behaviour. Although there is not a straight line, but rather a slightly bent line, in accordance with resistivity measurements in this temperature region, an estimation of the activation energy can be made. At temperatures of about

750 °C this results in 74 Kcal/mole for viscosity and resistivity. This amazingly high value together with the non-linear dependence on $1000/T$ suggest that we are not concerned only with normal flow processes, but that in addition some structural change in the short range order of the melt takes place (Krebs

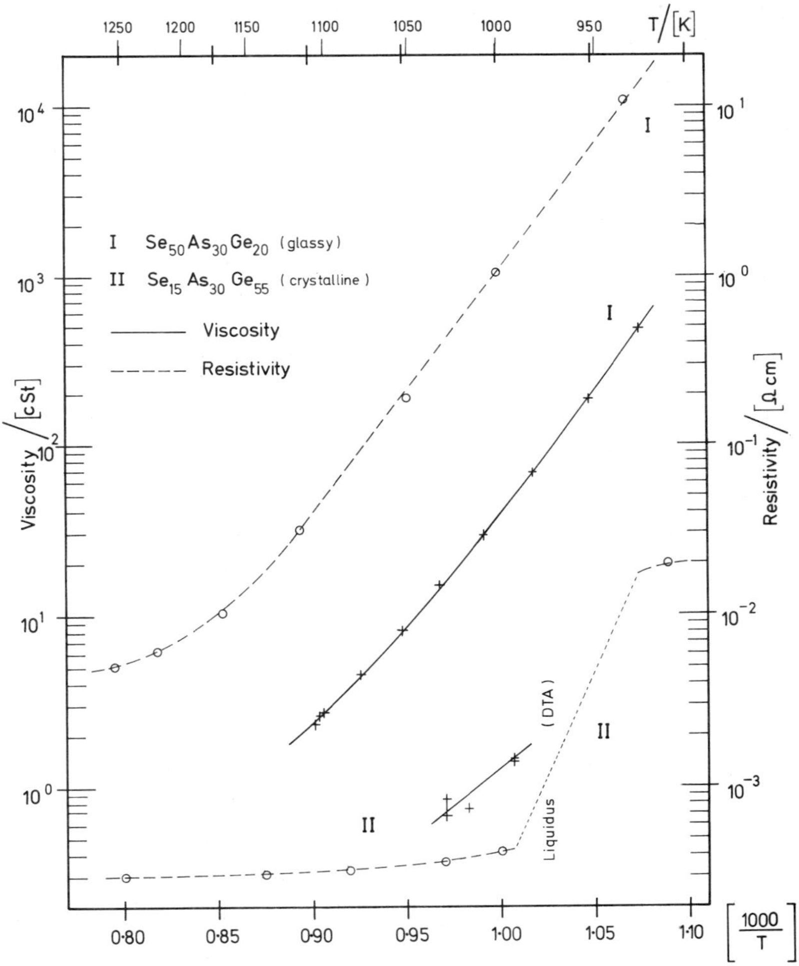

Figure 3. Viscosity and electrical resistivity of two melts

and Fischer, 1970). Radial distribution curves of the same samples (Ackermann and Krebs, 1970) give more support to such ideas. Certainly, more examples and a wider temperature range are needed and, therefore, the work is going on.

REFERENCES

Ackermann, F., and H. Krebs (1970). This volume, Chapter 32.

Haisty, R. W., and H. Krebs (1969). *J. non-cryst. Solids*, **1**, 399 and 427.

Krebs, H., and P. Fischer (1970). Paper presented to the general discussion on the vitreous state of the *Faraday Society* in Bristol, September 22–24, 1970.

Ullmanns Encyklopädie der technischen Chemie, (1961). Urban und Schwarzenberg, München–Berlin.

22

Viscous Flow in Glass-Forming Liquids

D. R. UHLMANN

Department of Metallurgy and Materials Science, Center for Materials Science and Engineering, Massachusetts Institute of Technology, Cambridge, Massachusetts, U.S.A.

A number of theoretical models have been advanced to describe the process of viscous flow in liquids. Of these, three have received widest attention; and to provide perspective for the discussion to follow, these will be briefly summarized.

Absolute rate theory

Viscosity is viewed as an example of a rate process dominated by a transition state of high energy. The standard treatment (Glasstone *et al.*, 1941) yields an expression for the viscosity:

$$\eta = \frac{\tau \exp(\Delta E/kT)}{a_0 \sinh(\tau V_a/2kT)} \approx \eta_0 \exp(\Delta E/kT) \tag{1}$$

where τ is the shear stress; a_0 is the jump distance; ΔE is the height of the energy barrier in the absence of stress; and V_a is the flow volume.

Linear relations between $\log \eta$ and $1/T$ were observed for a number of liquids, generally in the range of high fluidity. In the range of somewhat higher viscosities, however, curvature is generally seen in the $\log \eta - 1/T$ relations, with the slope (apparent activation energy) increasing with falling temperature. In oxide glass-forming systems, $\Delta E_{\text{LOW TEMPERATURE}}$ is typically larger by a factor of 2 to 3 than $\Delta E_{\text{HIGH TEMPERATURE}}$ while in organic systems, $\Delta E_{\text{LOW } T}$ often exceeds $\Delta E_{\text{HIGH } T}$ by an order of magnitude.

Free volume model

This approach to liquid viscosity was adopted by Doolittle (1951), who described the flow of hydrocarbon liquids in the range of high fluidity by the empirical relation:

$$\eta = \eta_0' \exp(\gamma V_0/V_f) \tag{2}$$

where V_0 is the van der Waals volume of the molecule; γ is a constant (of the order unity); and V_f is the free volume ($V_f = V - V_0$, where V is the actual molecular volume). The temperature dependence of η is represented here by the temperature dependence of V_f.

Subsequently, Williams, Landel and Ferry (1955) suggested an empirical relation to describe relaxation behavior in the vicinity of the glass transition (T_g), suggesting that V_f goes to some small value in the vicinity of T_g. The relation applied to viscosity is:

$$\eta = \eta_0' \exp\left[\frac{b}{f_g + \Delta\alpha(T - T_g)}\right] \tag{3}$$

where $b \sim 1{\cdot}0$; $f_g \sim 0{\cdot}025$; and $\Delta\alpha \sim 5 \times 10^{-4}\ {}^\circ\mathrm{K}^{-1}$ for a variety of organic materials.

Descriptions of flow in terms of free volume were provided by a number of workers (e.g. Fox and Flory, 1951; Bueche, 1959; Cohen and Turnbull, 1959; Turnbull and Cohen, 1961), who derived expressions of the WLF form. These models suggested that the critical step in flow is opening a void of some critical volume to permit molecular (or segmental) motion. The void was treated as being opened by the redistribution of free volume in the system.

Excess entropy model

This approach to flow and relaxation (e.g. Gibbs and DiMarzio, 1958; Adams and Gibbs, 1965) is based on the concept that the decrease in configurational entropy with falling temperature results in a diminished ease of deformation. In a recent development (Adam and Gibbs, 1965), attention was directed to the size of the smallest region which can change to a new configuration without a simultaneous external change. This was related to the configurational entropy, and an expression was obtained for the average transition probability. The corresponding expression for the viscosity is:

$$\eta = \eta_0'' \exp(C/TS_c) \tag{4}$$

Here S_c is the configurational entropy of the macroscopic sample; and C should be nearly constant, proportional to the potential energy barrier to molecular rearrangement. Over a limited range of temperature near T_g, this expression is hardly distinguishable from the WLF relation.

The predictions of the free volume and excess entropy models can be expressed in the form of the empirical Vogel–Fulcher relation:

$$\eta = \eta_0''' \exp\left[\frac{B}{T - T_0}\right] \tag{5}$$

Depending upon the magnitudes of the respective constants, this expression can be equivalent to the WLF relation. When this relation is compared with experimental data on the temperature dependence of viscosity, reasonable agreement in form is generally obtained, Unfortunately, most experimental data have been obtained in the range of low to intermediate viscosities. In the high viscosity range near the glass transition, where the models were generally assumed to be the most appropriate, relatively little viscosity data have been available; and of the available data, a large fraction were obtained on inorganic oxide liquids, which are sufficiently complex that the direct application of simple models might well be questioned. The experimental results to be described in the present paper were obtained to provide additional information on the flow of relatively simple organic liquids in the range of high viscosity.

VISCOUS FLOW AT HIGH VISCOSITIES

The experimental data described below were obtained by Dr. Laughlin in our laboratory; the details of the experimental techniques and a more elaborate discussion of the results will be published elsewhere (Laughlin and Uhlmann, 1970). The liquids (see page 208) are: *o*-terphenyl, salol, α-phenyl *o*-cresol and glycerin. Viscosity data were obtained using the falling sphere and beam-bending techniques; thermal expansion coefficients of liquids and glasses were obtained from volume measurements as a function of temperature; and heat capacities were determined with a differential scanning calorimeter. All the viscosity data to be reported here were obtained on stabilized liquids whose measured viscosities were independent of time.

The results obtained on *o*-terphenyl are shown in Figure 1, where the data of Laughlin are combined with those of previous investigators. As shown there, the high temperature data can be described well by equation (5) over more than 8 orders of magnitude in viscosity. At lower temperatures, however, the relation significantly overestimates the viscosity (keeping the same set of constants). The low temperature data also exhibit curvature on a log η *vs.* $1/T$ plot and can be described by a relation of the Vogel–Fulcher form, but with quite different constants from those used to describe the high temperature data. The Vogel–Fulcher relation fit to the low-temperature viscosity data significantly overestimates the high-temperature data.

The suggestion that more than a single set of Vogel–Fulcher constants are required to describe the viscosity data on this material is similar to the previous suggestion of other workers (Plazek and Magill, 1968; Barlow *et al.*, 1966) investigating the flow behaviour of organic liquids.

The results obtained on salol and α-phenyl *o*-cresol are shown in Figure 1. Again, the relation of equation (5) provides a good description of the high-temperature viscosity data, but grossly overestimates the viscosity in the

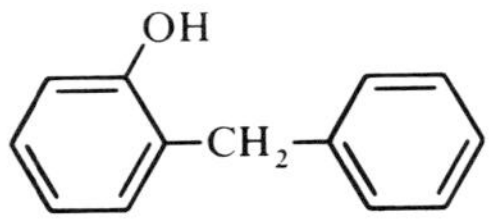

α-Phenyl *o*-cresol

I

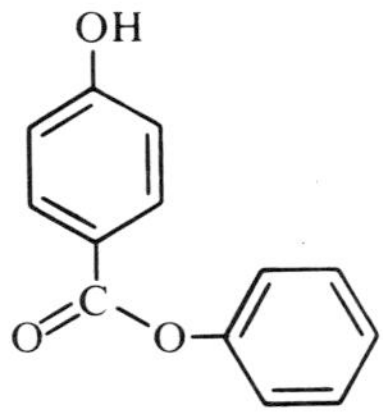

Phenyl salicylate
(Salol)

II

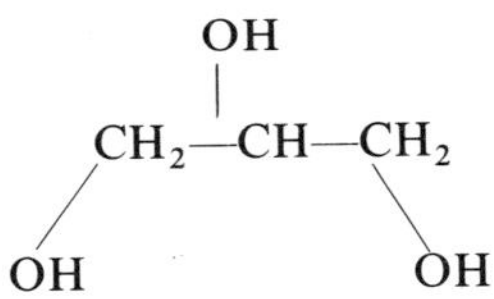

Glycerin

III

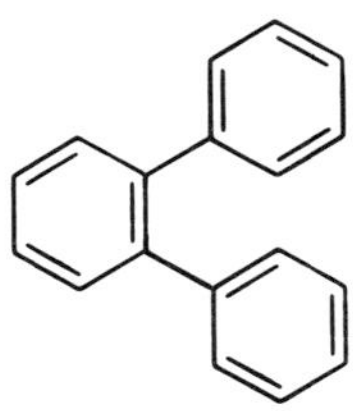

o-Terphenyl

IV

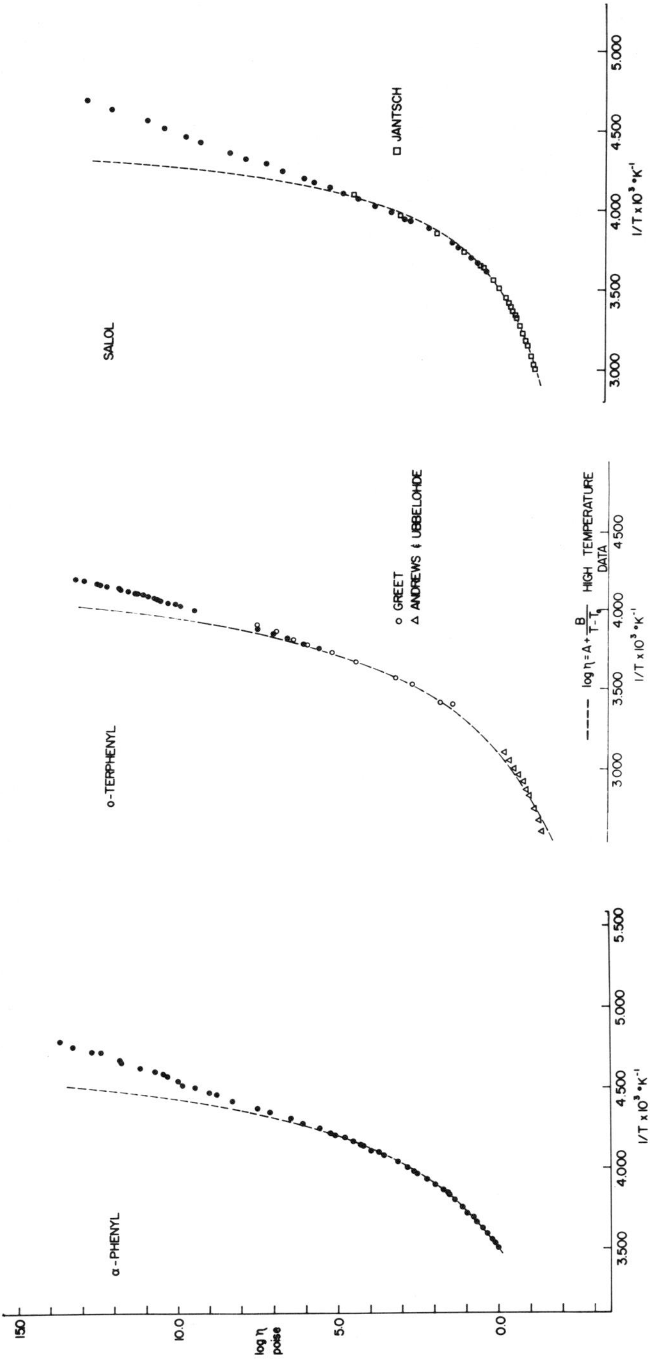

Figure 1. Variation of viscosity with temperature for *o*-terphenyl. Data from (Laughlin and Uhlmann, 1970). (Greet and Turnbull, 1967), and (Andrews and Ubbelohde, 1955)

low-temperature region. Unlike *o*-terphenyl, however, the low-temperature viscosity data are not describable by a Vogel–Fulcher relation save with $T_0 = 0$. That is, over the range of viscosity from about 10^5 to 10^{13} poise, an Arrhenius temperature dependence is exhibited by both salol and α-phenyl *o*-cresol. For both these materials, the apparent activation energies, estimated from the slopes of the low temperature Arrhenius regions, are larger by a factor of 6 to 7 than the respective heats of vaporization.

The data on salol are compared with the predictions of the excess entropy model in Figure 2. As shown in the figure, a curve of sigmoid shape is observed, rather than a straight line as predicted by the theory (equation 4). A similar sigmoid relation is exhibited by α-phenyl *o*-cresol.

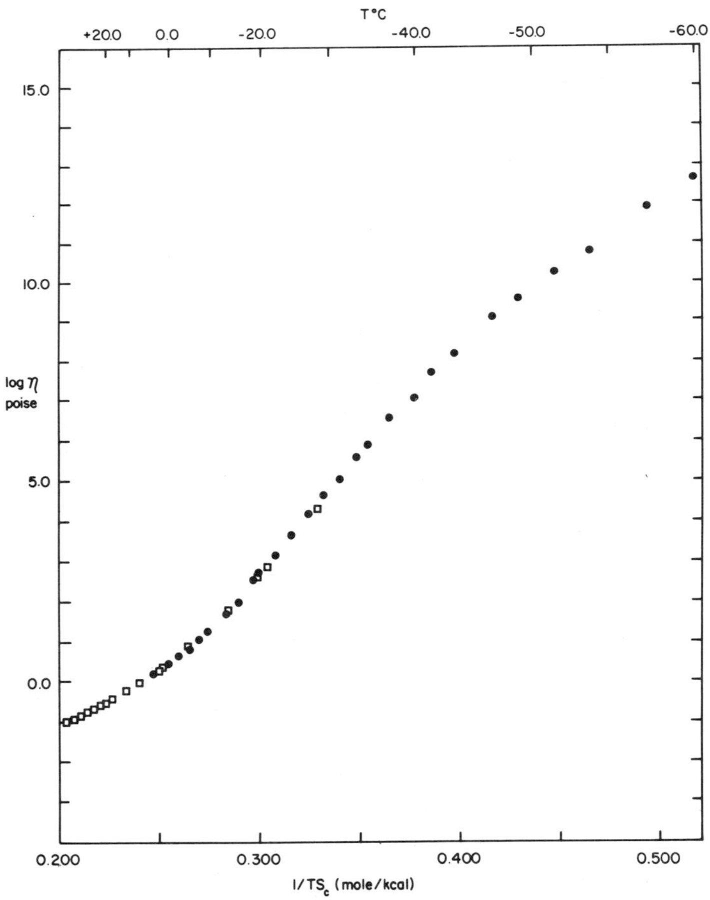

Figure 2. Variation of viscosity with TS_c for salol. Data from (Laughlin and Uhlmann, 1970)

The viscosity data on glycerin, not shown, are described reasonably well by the WLF relation with the original WLF constants over the full range of viscosity from 10^2 to 10^{13} poise. The measured $\Delta\alpha$ ($\Delta\alpha = \alpha_{\text{LIQUID}} - \alpha_{\text{GLASS}}$) for this material is, however, about half that of most organic compounds and when the WLF constant is modified accordingly, the predicted viscosity is about three orders of magnitude high over a range of low-to-intermediate viscosities, and the fit at high viscosities leaves much to be desired. The log η *vs.* $1/TS_c$ relation provides a much better description of the data than the WLF relation; but the potential energy barrier to molecular rearrangement estimated from the slope of the relation is larger than the heat of vaporization by about a factor of 5.

Finally, when the high-temperature data on *o*-terphenyl, α-phenyl *o*-cresol and salol are represented by the Vogel–Fulcher relation, the results shown in Table 1 are obtained. As shown there, there appears to be no simple relation

Table 1. High temperature viscosity data represented by the equation $\log \eta = A + B/(T - T_0)$

	o-Terphenyl	α-Phenyl	Salol
log A	−4·18	−3·43	−2·93
B (°K)	388·0	277·6	177·0
T_0 (°K)	227·2	204·6	221·0
T_g (°K)	239	210	212

$T_g \equiv$ temperature where $\eta = 10^{13}$ poise.

between the T_0 values derived from the high-temperature data and the measured T_g values—save that both are in the same general range of temperature.

DISCUSSION

It is apparent from the results considered in the preceding section, obtained on relatively simple organic liquids, that none of the standard models for flow are adequate for describing the observed behaviour over the full range of measured viscosity. It is also apparent that significant differences exist in detailed flow behaviour among the various materials.

When the range of materials considered is broadened to include inorganic liquids, the complexity of observed behaviour becomes even more apparent. For some materials, e.g. *o*-terphenyl and tri-α-naphthyl benzene, curvature in the log η *vs.* $1/T$ relation is observed over the full range of viscosity, with the curvature being small at high and low temperatures and large at intermediate temperatures; for others, e.g. glycerin, more gradual curvature is observed over the entire range of temperature; for others, e.g. salol and α-phenyl *o*-cresol, curvature is exhibited at high temperatures and intermediate temperatures, and Arrhenius behaviour is found at low temperatures; for others, e.g. SiO_2 and GeO_2 (Bruckner, 1964; Fontana and Plummer, 1966), Arrhenius or nearly-Arrhenius behaviour is exhibited over the full range of temperature; for others, e.g. B_2O_3 and possibly the alkali silicates (Lillie, 1939; Taylor and Doran, 1941; Poole, 1949; Shartsis *et al.*, 1965; Macedo and Napolitano, 1968), Arrhenius behaviour is found at high temperatures and low temperatures, with an intermediate region of curvature.

It is clear that such a complexity in even the form of the temperature dependence of viscosity is beyond description by any of the standard theoretical models, each of which is based on a relatively simple picture of the flow process. Recent developments, based on molecular dynamics studies of transport in hard sphere fluids (Alder and Wainwright, 1967, e.g.) and extensions of the free volume theory (Turnbull and Cohen, 1970), have provided significant new information; and when they are combined with the experimental results described above, our understanding of the flow process can be clarified (at least in the case of relatively simple liquids).

The molecular dynamics calculations have indicated self-diffusion coefficients for hard-sphere fluids which are in good agreement with the Enskog values for low densities, but which fall far below the Enskog predictions at the highest fluid densities. That is, while the fluids exhibited gas-like behaviour at relatively low densities, at high average densities, diffusive motion was cancelled to an appreciable extent by backscattering in the locally-more-dense parts of the system.

The extension of the free volume theory for hard sphere fluids has included consideration of a variable diffusive displacement and association of a correlation factor with each magnitude of the displacement. Approximating the correlation factor by a step function, the modified free volume theory was found to provide a good description of the results of the molecular dynamics calculations in the high density range.

Considering the good fit between the modified free volume theory and the results of molecular dynamics calculations in the range of density corresponding to relatively high fluidity, and the utility of equation (5) in describing the flow behaviour of the relatively simple organic liquids in the high temperature range, it is suggested that free volume theories should properly be applied to describe flow behaviour in this range and not in the vicinity of the glass transition.

This suggestion is in consonance with the original use of the free volume concept to describe the flow of hydrocarbons at high fluidities (Doolittle, 1951) and the satisfactory description of the temperature and pressure dependences of the self-diffusion coefficients of rare-gas liquids by a modified free volume treatment (Naghizadeh, 1964). It also avoids the problem posed to free volume theories by Goldstein (1963), who noted that such theories apparently provide a poor description of the pressure dependence of the glass transition temperature. It also suggests that the original form of the simple free volume theories which led to expressions of the WLF form require modification, since the constants appropriate for describing the high temperature data are, as we have seen, quite different from the WLF constants.

It is further suggested that the essential postulates of the free volume and excess entropy models concerning the glass transition—that it is determined by the free volume or the configurational entropy falling to some low values—must be regarded with some caution. As we have seen, neither of the models as presently developed is capable of describing the flow behaviour in the high viscosity region (as the glass transition is approached).

Using the free volume model to describe flow in the high temperature region, as suggested previously, the results shown in Figure 1 indicate that the model grossly overestimates the difficulty of flow in the high viscosity region, and that an easier mechanism of flow becomes dominant in this region.

The characteristics of this low temperature flow mechanism remain to be elucidated satisfactorily. It seems attractive to consider flow in this region from the viewpoint of transport over potential energy barriers; but it is clear from the data in Figure 1 that the transport cannot be regarded as a simply-activated process. Rather, some extensive degree of cooperative motion is indicated, and the description of this cooperation within the framework of a potential energy model constitutes a noteworthy objective of future theoretical developments.

Even if such developments were successful, however, the evaluation of a truly general theory for liquid viscosity could still present a formidable challenge; and the possible importance in flow of various factors, such as type of bonding, molecular shape and flexibility of molecular linkages, indicates that a general theory will also be a complex theory—and very likely will lack the simplicity and physical appeal of the models discussed above.

ACKNOWLEDGEMENTS

The author is happy to acknowledge stimulating discussions with Professor D. Turnbull of Harvard and Dr. W. T. Laughlin of M.I.T. The financial support of the Advanced Research Projects Agency is also gratefully acknowledged.

REFERENCES

Adam, G., and J. H. Gibbs (1965). *J. chem. Phys.*, **43**, 139.

Alder, B. J., and T. E. Wainwright (1967). *Phys. Rev. Lett.*, **18**, 988. See also results cited in Turnbull and Cohen, 1970.

Andrews, J. N., and A. R. Ubbelohde (1955). *Proc. R. Soc.*, **A228**, 435.

Barlow, A. J., J. Lamb, and A. J. Matheson (1966). *Proc. R. Soc.*, **A292**, 322.

Bruckner, R. (1964). *Glastech. Ber.*, **37**, 413.

Bueche, F. (1959). *J. chem. Phys.*, **30**, 748.

Cohen, H. H., and D. Turnbull, (1959) *J. chem. Phys.*, **31**, 1164.

Doolittle, A. K. (1951). *J. appl. Phys.*, **22**, 1471.

Fontana, E. H., and W. H. Plummer (1966). *Physics Chem. Glasses*, **7**, 139.

Fox, T. G., and P. J. Flory (1951). *J. appl. Phys.*, **21**, 581.

Gibbs, J. H., and E. A. DiMarzio (1958). *J. chem. Phys.*, **28**, 373.

Glasstone, S., K. J. Laidler, and H. Eyring (1941). *The Theory of Rate Processes.* McGraw–Hill, New York.

Goldstein, M. (1963). *J. chem. Phys.*, **39**, 3369.

Greet, R. J., and D. Turnbull (1967). *J. chem. Phys.*, **46**, 1243.

Jantsch, O. (1956). *Z. Kristallogr.*, **108**, 185.

Laughlin, W. T., and D. R. Uhlmann (1970). To be published. See Laughlin, W. T. and D. R. Uhlmann (1968). *Bull. Am. Ceram. Soc.*, **47**, 402. See also Laughlin W. T. (1969). Viscous flow and volume relaxation in simple glass-forming liquids. *Sc.D. Thesis*, M.I.T.

Lillie, H. R. (1939). *J. Am. Ceram. Soc.*, **22**, 367.

Macedo, P. B., and A. Napolitano (1968). *J. chem. Phys.*, **49**, 1887.

Naghizadeh, J. (1964). *J. appl. Phys.*, **35**, 1162.

Plazek, D. J., and J. H. Magill (1968). *J. chem. Phys.*, **49**, 3678.

Poole, J. P. (1949). *J. Am. Ceram. Soc.*, **32**, 230.

Shartsis, L., S. Spinner, and W. Capps (1965). *J. Am. Ceram. Soc.*, **35**, 155.

Taylor, N. W., and R. F. Doran (1941). *J. Am. Ceram. Soc.*, **24**, 103.

Turnbull, D., and M. H. Cohen (1961). *J. chem. Phys.*, **34**, 120.

Turnbull, D., and M. H. Cohen (1970). *J. chem. Phys.*, **52**, 3038.

Williams, M. F., R. F. Landel, and J. D. Ferry (1955). *J. Am. chem. Soc.*, **77**, 3701.

23

Which Viscosity Theory? An Indistinguishability Condition

J. F. HUTTON and M. C. PHILLIPS

Shell Research Ltd., Thornton Research Centre, Chester, Great Britain

Many equations have been proposed to describe the variation of viscosity with temperature. Some, such as the one based on the configurational-entropy theory of Adam and Gibbs (1965) have been derived from models or theories of how flow occurs. Others have been proposed empirically and later shown to be derivable on the basis of a theory: an example is the Vogel (1921) equation which Cohen and Turnbull (1959) derived from their free-volume equation under the condition that free volume varies linearly with temperature. A final type is the purely empirical equation that has, as yet, no theoretical basis; an example is the T^n equation proposed by Litovitz (1952) which is shown below together with the previous two equations.

Adam–Gibbs (AG): $$\ln \eta = A + B[T \ln (T/T_2)]^{-1} \quad (1)$$

Vogel (V): $$\ln \eta = A + B(T - T_0)^{-1} \quad (2)$$

Empirical (T^n): $$\ln \eta = A + BT^{-n} \quad (3)$$

Experimentally the current problem is to gather data of sufficient accuracy to test these equations, and to tabulate the values of the parameters (and, in particular, the 'third parameter', T_2, T_0 or n) which are presumed to be correlated with other molecular parameters.

The authors have measured the viscosity–temperature relationship for three supercooled non-polymeric liquids and, in attempting to fit their results to the above three equations, made the following three observations:

(1) All three equations fitted the experimental results about equally well but not within the limits of experimental accuracy.

(2) The residuals (errors remaining after the equations were fitted to the results) showed a series correlation that was cubic in form.

(3) The T_0 and T_2 values for each liquid were equal to within 7 degrees (or 5%) but with $T_2 < T_0$ always.

The complete understanding of the significance of these three observations was not possible at the time and led the authors to construct a theory to describe the fitting of data to an equation. The theory explained the observations and led to predictions that have since been verified.

To provide a critical test of the theory the laboratory viscosity data were supplemented by two synthetic data-sets which simulated viscosity data with no scatter.

EXPERIMENTAL

Three liquids, di(2-ethylhexyl) phthalate, cedrene and a mineral oil, LVI 260, were used. Viscosity was measured in the range $10^{-0\cdot7}$ P to $10^{6\cdot3}$ P with suspended-level glass-capillary viscometers, a Ferranti–Shirley cone-and-plate rotational viscometer, and a plunger viscometer with axial flow through the annulus between long concentric cylinders. The results showed agreement in the range of viscosities $10^2\text{P} < \eta < 10^4\text{P}$, a range in which all three techniques could be used.

Two synthetic data sets were also computed. Set I consisted of number-pairs which exactly fitted the equation:

$$\ln \eta = 20{\cdot}48\,(250/T)^6$$

whilst set II exactly fitted the equation:

$$\ln \eta = 1000/(T\text{-}200)$$

By taking various numbers of data-pairs from a given data-set, data-ranges of varying extents (measured in decades of η) could be constructed.

The results of data-fitting to viscosity equations (1), (2) and (3) are presented in Tables 1, 2 and 3 and in Figure 1, and are discussed below. In Tables 1, 2 and 3 the final two columns present parameters calculated from the theory of the following section. The preceding columns present the parameters obtained from a least-squares procedure in which the Spiral algorithm (Jones, 1970) was used on a digital computer. From columns 3 and 4 of Table 1 it is seen that $T_2 \simeq T_0$ but with $T_2 < T_0$ always, (observation 3 of the introduction).

In Figure 1, the strong series correlation of residuals remaining after the least-squares procedure has been applied to experimental results for di(2-ethylhexyl) phthalate can be observed. (For the synthetic data-sets this correlation is even more obvious because it is not basically obscured by experimental scatter.) It is cubic in form as stated in Observation 2 of the introduction. The scatter of the points from a smooth curve drawn through the residuals gives the experimental scatter, σ_L, which is of the order of 0·02.

Table 1. Parameters of viscosity equations for three liquids

Experimental details		Parameters from least-squares procedure			Parameters calculated from equations (15) and (13)	
Liquid	Data-range midpoint, T_d, K	T_2, K	T_0, K	n	T_2, K	T_0, K
Di(2-ethylhexyl) phthalate	233	127	134	3·4	129	127
Cedrene	206	135	137	5·1	139	139
LVI 260 oil	249	172	175	5·2	169	169
Error		2	2	0·1	2	3

Table 2. Parameters of viscosity equations for two synthetic data-sets

Details of synthetic data-set			Parameters from least-squares procedure		Parameters calculated from equation (13)	
Data-set number	Important parameter	Data-range midpoint, T_d, K	T_0, K	n	T_0, K	n
I	$n = 6$	300	216	—	214	—
		290	209	—	207	—
		283	204	—	202	—
		276	199	—	197	—
II	$T_0 = 200$	300	—	4·9	—	5·0
		290	—	5·3	—	5·4
		283	—	5·7	—	5·8
		276	—	6·1	—	6·3

Table 3. Errors in fitting viscosity equations to three liquids

Liquid, L	Fitting errors from least-squares procedure			Fitting errors calculated from equations (10) and (11) and the value of $\sigma(L, V)$	
	$\sigma(L, AG)$	$\sigma(L, V)$	$\sigma(L, T'')$	$\sigma(L, AG)$	$\sigma(L, T'')$
Di(2-ethylhexyl) phthalate	0·075	0·082	0·048	0·[illegible]	0·028
Cedrene	0·034	0·035	0·027	0·031	0·003
LVI 260 oil	0·059	0·056	0·075	0·062	0·113

Similar plots were obtained from the fitting of other liquids and equations and provided experimental scatter values, σ_L, of 0·025 and 0·04 for cedrene and LVI 260 oil respectively. For each liquid the σ_L value is less than the fitting

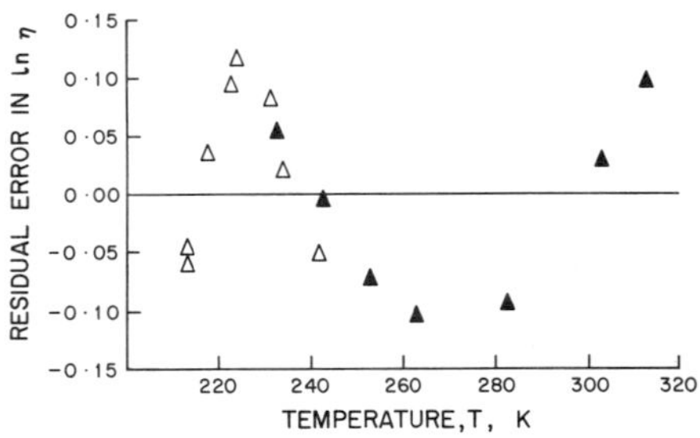

Figure 1. Residual errors for di(2-ethylhexyl) phthalate fitted to the T^n equation by the least-squares procedure (▲ suspended level-viscometer; △ plunger viscometer)

error calculated from the least-squares procedure, as can be observed from Table 3 (Observation 1 of the introduction).

THEORY

A description of the theory is now presented, and is illustrated by means of a specific example. Consider a set of data (η, T) which *exactly* fits the equation:

$$y_1 = \ln \eta = A_1 + B_1(T - T_0)^{-1} \tag{4}$$

when suitable values of A_1, B_1 and T_0 are chosen. This data-set is now to be fitted to the equation:

$$y_2 = \ln \eta = A_2 + B_2 T^{-n} \tag{5}$$

This is performed by the least-squares procedure, that is, by adjusting the values of A_2, B_2 and n, so that the sum of the values of $(y_1 - y_2)^2$ at the various values of T are minimized. To do this, y_1 must be made almost equal to y_2, i.e.

$$y_1 \simeq y_2 \tag{6}$$

However, from equation (5), y_2 is linear in T^{-n} as shown in Figure 2. (The range of values of T^{-n} has been chosen so as just to include the greatest and least values of T in the original data-set.) Thus, to ensure that $y_1 \simeq y_2$, y_1

must also be made linear in T^{-n} and so the graph of y_1 against T^{-n} must have a point of inflection, i.e.

$$\partial^2 y_1/\partial(T^{-n})^2 = 0 \tag{7}$$

With this condition the graph of y_1 can be as shown in Figure 2. It is evident that $y_1 \simeq y_2$ over the range of values of T considered, so that a good fit of the data-set (η_1, T) to the curve y_2 has been achieved. It is also evident from Figure 2 that the residual error after fitting is cubic in form and of the form shown in Figure 1.

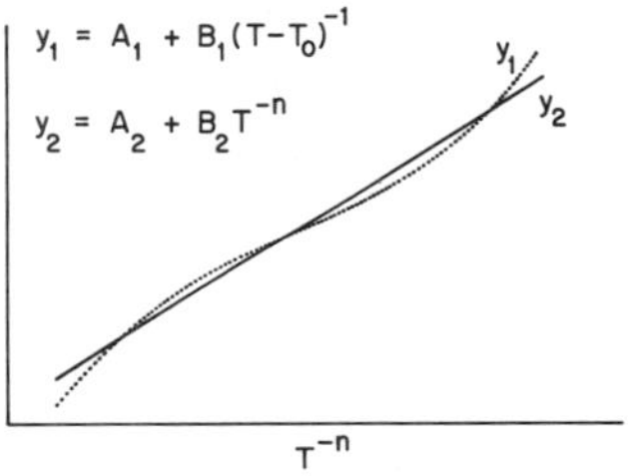

Figure 2. Graphs of the functions y_1 and y_2 against T^{-n}

A theory has been developed (Hutton and Phillips, unpublished work) to describe this procedure logically and mathematically. It has been formulated so as to apply to any two non-linear equations of the form:

$$y = A + Bx(T, C) \tag{8}$$

where x is a non-linear function of both T and C. Equations (4) and (5) are both of this form. To proceed with this example, the theory starts by writing $(y_1 - y_2)$ as a Taylor expansion about a point $T^{-n} = c$. It is assumed that successive terms in the expansion are of decreasing importance so that those after the term $(\frac{1}{6})(T^{-n} - c)^3(\partial^3 y_1/\partial(T^{-n})^3)$ may be neglected. The theory then proves that for minimization of the fitting-error σ, equation (7) is valid at the midpoint y_{1d} of the range of values of y_1:

i.e.
$$|\partial^2 y_1/\partial y_2^2|_d = 0 \tag{9}$$

The minimum value of the fitting-error produced by fitting the data-set η_1, T (which lies exactly on the function y_1) to the function y_2 will be denoted by σ (1, 2), and correspondingly for other data-sets and functions.

The theory enables the value of σ to be expressed as a function of the extent of the data range, Δy, of y_1:

$$\sigma(1, 2) = \nu\Delta y_3\,[\partial^3 y_1/\partial(T^{-n})^3]\,[\partial y_1/\partial(T^{-n})]^{-3} \tag{10}$$

where the value of v depends slightly on the number of data-points, N, e.g. for $N = 10$, $v = 0{\cdot}0051$ and as $N \to \infty$, $v \to 0{\cdot}0031$. (The N data-points have been assumed to be uniformly spaced in y_1 values). Finally, the residual error, $(y_1 - y_2)$, is shown to have a specific cubic form.

APPLICATION OF THEORY

The application of the theory requires the assumption that any given set of experimental results for a liquid L, can be exactly fitted (to within their experimental scatter, σ_L) by some unspecified and unknown three-parameter non-linear function, y_L, of the form of equation (8). This is a reasonable assumption since one may invent new non-linear functions of the series type with specified coefficients, to fit almost any data-set. The assumption is also justified by the success of the results which follow from it. With this assumption it follows from the theory that:

$$\sigma(L, 1) = \sigma(L, 2) \mp \sigma(1, 2) \text{ if } \sigma \quad = 0 \tag{11}$$

(Equation (11) states that, for systematic errors of a specific form, σ values are additive. By contrast, for random errors σ^2 values are additive.)

For $\sigma_L \neq 0$, the formula is more complicated. However application of the F-test for ratio of variances allows the formula to be expressed as:

$$\sigma(L, 1) = \sigma(L, 2) \qquad \text{if } \sigma(1, 2)/\sigma_L \leqslant E \tag{12}$$

In the range $7 \leqslant N \leqslant 30$, E can be taken as $(10/N)$.

Equation (12) will be called an indistinguishability condition, as it states that the quantities $\sigma(L, 1)$ and $\sigma(L, 2)$ are not significantly different (at the 95% level) if the inequality is obeyed.

Originality is not claimed for equation (12); what is important is that the equation is easy to apply as $\sigma(1, 2)$ can be calculated from the theory.

A least-squares computer-fitting of experimental results to the function y_2 provides a value of $\sigma(L, 2)$ and also of the parameters A_2, B_2, C_2 appearing in equation y_2. Also equation (10) permits $\sigma(1, 2)$ to be calculated. (For a numerical evaluation, the number of original data-pairs, N, and the data-range, Δy, are required, as are the values of the parameters B_2 and C_2.) Hence $\sigma(L, 1)$ can be calculated using equation (11). If desired this result can be tested by comparing it with the results of a second least-squares computer-fitting of the experimental results to the function y_1. Such tests are applied in the following section. (The ambiguity in sign in equation (11) arises because standard error values σ, are by definition positive. However, the values of, e.g., $\sigma(1, 2)$ can be (by convention) given the same sign as $y_1 - y_2$ at T_{max}. If such a convention is adopted the negative sign in equation (11) always applies.)

Equation (9) can be written explicitly as:

$$\partial^2 y_1(T_d, C_1)/\partial y_2^2(T_d, C_2) = 0$$

It permits the parameter C_2 to be calculated from C_1 and the data-range midpoint, T_d. Thus for y_1 and y_2 given by equations (2) and (3), $C_1 \equiv T_0$ and $C_2 \equiv n$ and equation (9) gives:

$$T_0 = T_d(n - 1)/(n + 1) \tag{13}$$

Similarly for y_1 and y_2 given by equations (1) and (3), equation (9) gives:

$$n = (2/t) + t/(1 + t), \text{ where } t = \ln(T_d/T_2) \tag{14}$$

$$\therefore\ T_2/T_d = \exp(-t) \simeq \exp(-2/n) \tag{15}$$

From equations (13) and (14) it can be shown that $T_2 \leqslant T_0$ for all $n > 1$, and that, for the data-range of interest, $T_2 \simeq T_0$ to within 4%.

COMPARISON OF THEORY WITH LEAST-SQUARE PROCEDURE RESULTS

The agreement between least-squares procedure results and values calculated from the theory is good and is illustrated in Tables 1, 2 and 3. In each table the final two columns show values calculated using the theory.

Table 3 also shows that $\sigma(L, AG)$ values are very similar to the $\sigma(L, V)$ values from which they are calculated. This arises from the fact that $\sigma(V, AG)$ values as calculated from equation (10) are very small. The Table also shows that the calculated $\sigma(L, T^n)$ values are rather different from the $\sigma(L, V)$ values. The theory consistently correctly predicts the sign of this difference and only in the case of cedrene is the predicted value incorrect by a factor of more than 2. But for cedrene the predicted value is $\sigma(L, T^n) = 0{\cdot}003$ which is very much less than the experimental scatter, $\sigma_L = 0{\cdot}025$. The least squares procedure value of 0·027 agrees well with this.

COMPARISON WITH OTHER WORK

Greet (1966) has come nearest to producing results similar to those of this paper. He noted 'that for many glass-forming systems the difference' between the W.L.F. equation (Williams *et al.* 1955) (which is equivalent to the Vogel equation used herein) and the Adam–Gibbs equation 'cannot be distinguished with current experimental resolution'. He also noted 'that a correlation exists in the numbers cited by Adam and Gibbs which was not emphasized by them'. Greet explained these effects qualitatively by expanding the equations in a power series about the W.L.F. reference temperature, T_s, and by letting $T_0 = T_2$.

He noted that 'if the restriction $T_2 = T_0$ were not imposed, then the second-order coefficients could be exactly equated and differences in the expressions would not begin to show up until the third-order terms'. However, using the assumption that $T_2 = T_0$, he showed how the correlations that he had observed in the paper of Adam and Gibbs could be rationalized. Thus Greet's analysis was limited to a particular pair of equations, and was an attempt at a treatment correct to the second order.

Eisenberg and Saito (1966) noted that the parameters of the Gibbs–DiMarzio (1958) theory (a forerunner of the Adam–Gibbs equation) and the free-volume theory were related for a set of uncrosslinked homopolymers that they investigated. By enquiring into the nature of the two theories, they decided that it was reasonable that the parameters of the two theories should be related, although they did not derive any equation to describe this relationship. They concluded that the predictions of the two theories were identical and that this is because the parameters are related.

Finally, Bartenev *et al.* (1969) have derived an equation relating the parameters appearing in free-volume theory and a relaxation theory of molecular mobility. This is another attempt to show that superficially different theories are essentially equivalent, if not identical, owing to certain physical relationships.

By contrast, this paper demonstrates that the experimental equivalence of various theories (as described by their particular viscosity–temperature equations) is a direct consequence of the form of the equations used. The theories *may* have common origins or even be identical, but it is not necessary to search for this similarity in order to explain their experimental equivalence. The theory presented applies quite generally to any non-linear equations of a certain form, and is quantitatively correct to terms of the third order.

DISCUSSION AND CONCLUSIONS

The first achievements of the theory are that it directly explains Observations (2) and (3) at the beginning. It also predicts further correlations, e.g. between the T_0 and n parameters, that had not been previously noted.

The theory also indicates that certain non-linear equations fit each other so well that they cannot be resolved with existing experimental data. For example, the Adam–Gibbs and Vogel equations both fit the T^n equation to better than 3% with a data-range of four decades. Consequently, according to the indistinguishability condition of equation (12), typical sets of experimental results ($N \simeq 12$) with as little as 4% scatter ($\sigma_L \geqslant 0{\cdot}04$) will be fitted equally well by all three equations. Thus Observation (1) is explained. The moral is that caution must be exercised in concluding that because a particular equation describes a set of results, the model on which it is based is necessarily valid.

The indistinguishability condition of the theory can be used to predict the usefulness of providing data of a given accuracy over a given range in an attempt to distinguish between two equations. In particular, the Vogel and Adam–Gibbs equations are so similar that data of 1% accuracy over ten decades in viscosity would be required to discriminate or distinguish between them (on the assumption of typical experimental values of B and T_0 in the Vogel equation).

The theory can be especially useful when a complete set of experimental results is not available, but the fit to a given equation over a given range of values is given, together with the parameter values for the equation. The theory then permits the accuracy of the fit of the results to another equation to be estimated, together with the value of parameters (e.g. T_2 or n) appearing in the new equation.

In general, the theory is useful when data-fitting to non-linear equations.

REFERENCES

Adam, G., and J. H. Gibbs (1965). *J. chem. Phys.*, **43**, 139.
Bartenev, G. M., I. V. Razumovskaya, D. S. Sanditov, and I. A. Lukyanov (1969). *J. Polym Sci. A*1, **7**, 2147.
Cohen, M. H., and D. Turnbull (1959). *J. chem. Phys.*, **31**, 1164.
Eisenberg, A., and S. Saito (1966). *J. chem. Phys.*, **45**, 1673.
Gibbs, J. H., and E. A. DiMarzio (1958). *J. chem. Phys.*, **28**, 373.
Greet, R. J. (1966). *J. chem. Phys.*, **45**, 2479.
Jones, A. (1970). *Comput. J.*, **13**, 301.
Litovitz, T. A. (1952). *J. chem. Phys.*, **20**, 1088.
Vogel, H. (1921). *Phys. Z.*, **22**, 645.
Williams, M. L., R. F. Landel, and J. D. Ferry (1955). *J. Am. chem. Soc.*, **77**, 3701.

24

Viscosity of Sodium Borosilicate Glasses in the Transformation Range

M. PROD'HOMME

Laboratoire des Verres, C.N.R.S., Boulevard Pasteur, Paris, France

In a previous paper (Prod'homme, 1968) the isothermal change in viscosity of homogeneous optical glasses was studied over their transformation range and the stabilization of this property was shown to occur at temperatures within this range, while the viscous flow was found to be newtonian.

The question arises now as to what is the behaviour of viscosity in phase separated glasses where two vitreous networks are present either discontinuous (one dispersed phase), or continuous (two interconnected phases) (Vogel, 1965). Few papers have been devoted to this subject till now (Li and Uhlmann, 1970); thus it will be of interest to investigate sodium borosilicate glasses in which phase separation has already been studied by other methods

EXPERIMENTAL PROCEDURE

Determination of viscosity was made in the range from 10^{10} to 10^{15} poises using a rod under uniaxial pressure, as previously described (Prod'homme, 1968).

The viscosity is derived from the rate of strain of the rod under stress.

$$\eta = \frac{\sigma}{3\dfrac{dl}{l dt}}$$

where $\sigma = mg/S$ denotes the stress, l and S are respectively the length and cross-sectional area of the rod, dl/dt is the strain rate of the rod.

Four phase separated borosilicate glasses were chosen for study. Their mole percent compositions were as follows:

	SiO_2	B_2O_3	Na_2O	ZnO
371	68	27	5	—
299	73	19	8	—
1435	77	13	3	7
1209	60	30	10	—

Before studying the rheological behaviour of these glasses over the transformation range, it is necessary to describe that of a homogeneous optical crown glass. At a given temperature the structure of a glass changes in the transformation range as a function of time, in order to reach an equilibrium stucture proper to that temperature. This stabilization effect appears in all physical properties. Specifically, the apparent viscosity of a chilled glass increases as a function of time up to an equilibrium value (Figure 1). The stabilization times become longer and longer and the overall changes in viscosity are larger and larger when the temperature is lowered. In contrast, when a glass is already stabilized at the temperature of the experiment, its viscosity does not change at all.

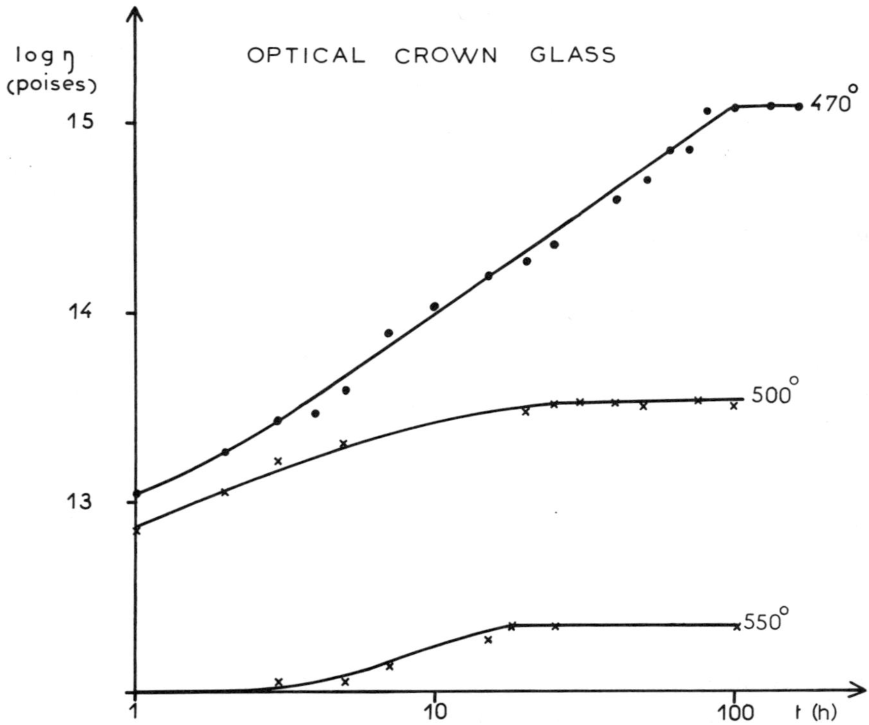

Figure 1. Variation of viscosity with time for an optical crown glass

Furthermore it was shown (Prod'homme, 1968) that the viscous flow of a stabilized glass* was newtonian and the non-linear part of the viscosity curve was due to a stabilization of density. This is illustrated by Figure 2 where log (dl/dt) is plotted against the logarithm of the stress in an experiment carried out at 480 °C on the ordinary crown sample. After a 100 hours

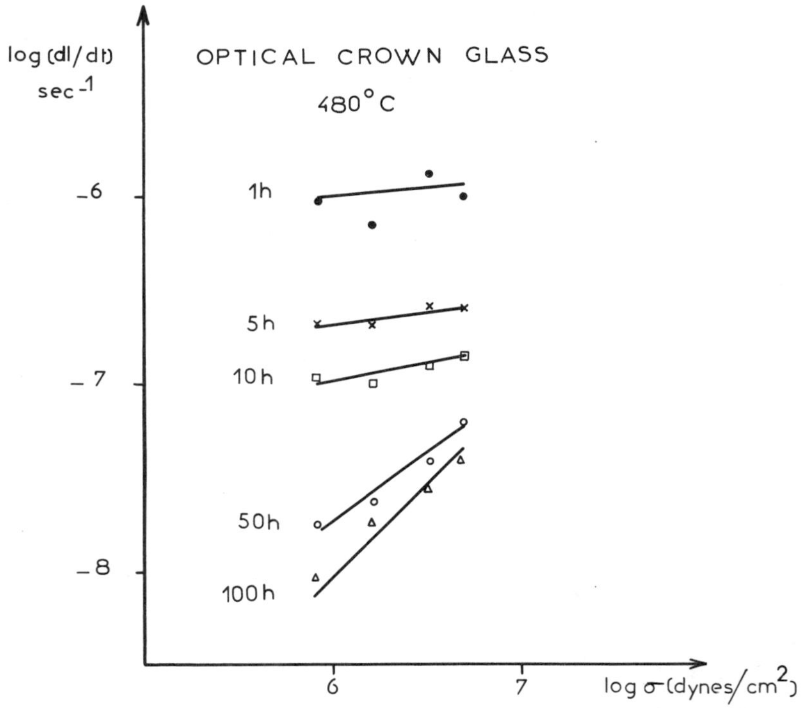

Figure 2. Strain rate versus stress relation for an optical crown glass tested at 480 °C

testing time the change of log (dl/dt) in terms of log σ becomes linear with a slope equal to unity, which is typical of a newtonian fluid.

EXPERIMENTAL RESULTS

BSC 371

Experiments were carried out from 440 °C to 640 °C on untreated samples and on samples heated for 72 hours at 440 °C in order to reduce the stabilization phenomena before testing.

* Maximum stress used = $9{\cdot}4 \times 10^6$ dynes/cm^2.

A comparison of the stabilization tests carried out on treated and non-treated rod samples shows that both total change in viscosity and stabilization time significantly decrease during the test conducted at 440 °C. At higher temperatures these variations are smaller.

BSC 299

The viscosity was measured in succession on untreated samples, on samples heated for 170 hours at 500 °C, and on samples heated at the testing temperatures.

The temperature range extending from 450 °C to 660 °C corresponds to an initial viscosity of the untreated glass ranging from 10^{15} down to 10^{10}

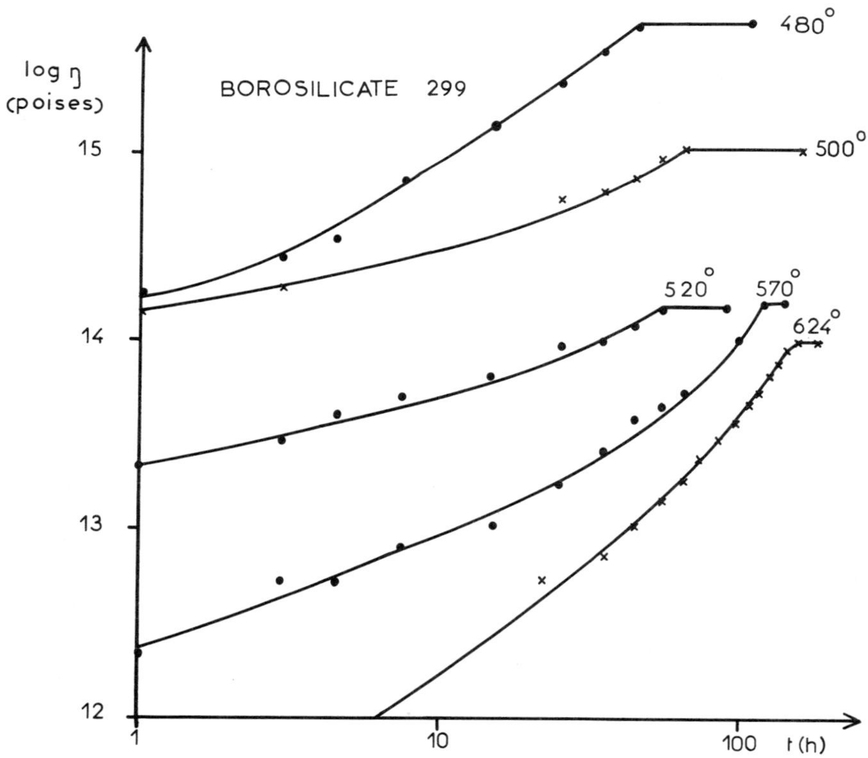

Figure 3. Variation of viscosity with time for borosilicate 299

poises. Tests for stabilization have shown that at every temperature up to 520 °C, the viscosity reaches a constant value which decreases when the temperature is raised (Figure 3). From 520 °C upwards at any temperature, the viscosity always reaches the same approximate stabilization value of about 10^{14} poises. The time necessary to reach this limit is longer and longer, in proportion, as the temperature rises.

When rod samples, previously heated at the test temperatures were measured, the overall change in viscosity, as well as the stabilization time, was considerably reduced but the final viscosity (linear flow) was the same.

BSC 1435

Tests were made on untreated samples and on samples heated for 72 hours at 600 °C.

The initial viscosity of the untreated glass was observed to decrease from $10^{13.9}$ poises at 560 °C to 10^{11} poises at 730 °C. Stabilization of viscosity was obtained during tests conducted at temperatures lower than 650 °C. At temperatures higher than 650 °C, the viscosity steadily increased and never reached a constant value, even after the 135 hours at 660 °C (Figure 4). The same kind of behaviour was also found in the glass treated at 600 °C, but the overall change in viscosity during stabilization was significantly reduced.

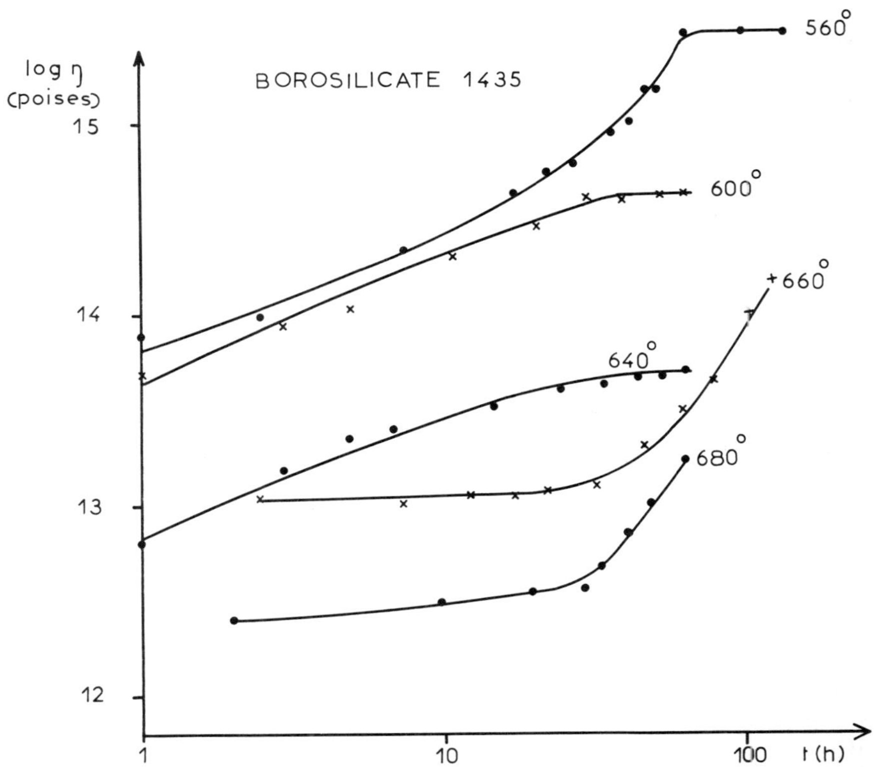

Figure 4. Variation of viscosity with time for borosilicate 1435

For the tests conducted at temperatures higher than 640 °C, no evidence of linear flow was found.

BSC 1209

Tests were made on untreated samples and on samples heated at the temperature (710 °C) where opalescence is a maximum.

For the untreated samples the viscosity was found to decrease from $10^{14\cdot6}$ poises to 10^{11} poises in the 450–550 °C temperature range. At every temperature lower than 500 °C stabilization of viscosity was obtained (Figure 5). Specifically, at 490 °C, the flow is linear at the beginning of the test. But

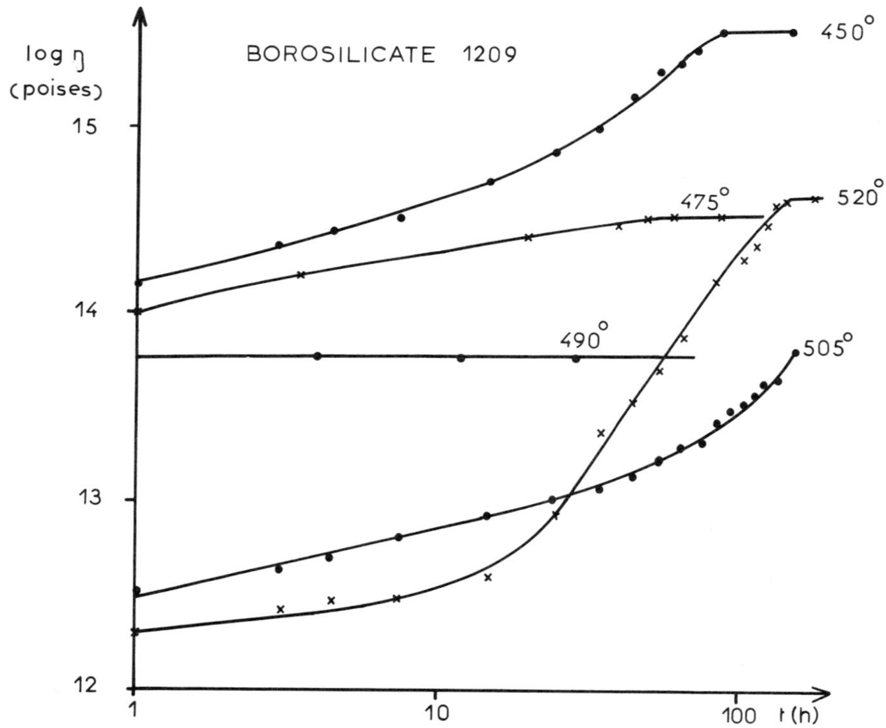

Figure 5. Variation of viscosity with time for borosilicate 1209

above 500 °C, there is again a significant increase in viscosity with time. As the temperature increases this rise is larger and steeper. So at 520 °C, after a 30 hours testing, the viscosity is higher than the value at 505 °C and after a 60 hours testing it is higher than the value at 490 °C.

The same kind of behaviour was also found in the rod samples previously heated at 710 °C. Stabilization of viscosity was obtained when the temperature did not exceed 500 °C, but further on a constant rise was observed. The initial viscosity is always lower than that of an untreated glass.

In an attempt to explain the nature of viscous flow in this glass, the rate of strain of untreated rods was measured at three temperatures (475 °C, 505 °C and 520 °C) and under several loads ($7{\cdot}8 \times 10^5$ to $6{\cdot}8 \times 10^6$ dynes/cm^2) (Figure 6). Although measurements were inaccurate, evidence of a newtonian

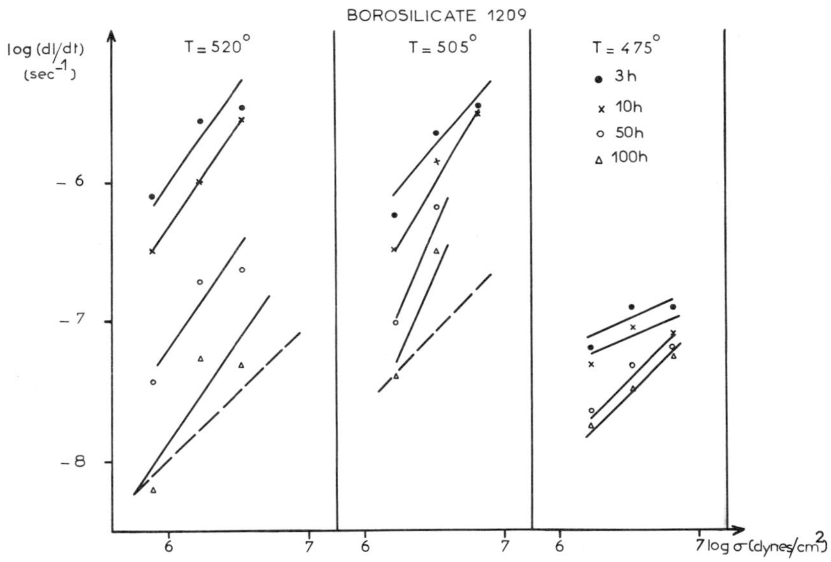

Figure 6. Rate of strain *vs.* stress relation for borosilicate 1209 tested at 520 °C, 505 °C and 475 °C

flow (slope of unity) was shown at 475 °C after a testing duration of about 50 hours. When the duration was shorter than 50 hours, the apparent viscosity $\sigma/(dl/dt)$ was an increasing function of the strain rate (slope smaller than unity). A similar result was obtained in the case of an homogeneous glass because of the stabilization of density.

At 505 °C and 520 °C with times up to 100 hours a newtonian flow was not observed. Furthermore the apparent viscosity decreased as the strain rate increased (slope larger than unity). This abnormal viscosity, sometimes defined as structural viscosity, is typical of the behaviour of heterogeneous systems such as suspensions or emulsions (De Vries, 1960).

DISCUSSION

The preceding experiments have shown that the change in viscosity as a function of time for phase separated glasses is quite different from that of homogeneous optical glasses. Above a definite temperature a linear flow can no longer be obtained, at least for testing times shorter than 120 or 150 hours depending upon samples.

However this limiting temperature corresponds, for all glasses under test, to an initial viscosity which is lower than 10^{13} poises, that is, to a viscosity where stabilization of the physical properties occurs without delay in the case of an homogeneous glass. For the last two glasses investigated, this increase in viscosity becomes more and more rapid when the tests are conducted at higher temperatures, so that, after a definite period, the viscosity becomes higher than that observed at lower temperatures. Li and Uhlmann (1970) have reported a similar result with a soda-silica glass.

The phase separation was investigated in some of the same borosilicate glasses by several other methods: electron microscopy (Cordelier, 1970), light scattering (Prod'homme 1968), small angle X-ray scattering (Jeanmaire, 1970) and differential thermal analysis (Prod'homme, 1970). It is well known that one of the separated phases essentially consists of B_2O_3, while the other is essentially made of silica.

If both vitreous networks are interconnected, as is the case for the first three glasses investigated, the stabilization process of viscosity is a rather complicated one. This fact may account for nearly similar stabilization times for viscosity, whatever the testing temperature.

If both phases are discontinuous and their respective transformation temperatures are not too close to one another, as is the case for BSC 1209, it is suggested that stabilization for both networks occurs in succession. One is stabilized at temperatures lower than 500 °C; the other network is then stabilized over a significantly longer period, as a result of its very much higher viscosity.

To this stabilization process is now added the phenomenon of coalescence of the domains (droplets) of the discontinuous phase, which is very much faster at higher temperatures and for this reason the viscosity at 520 °C rapidly becomes higher than at lower temperatures.

When treated for 120 hours at 710 °C, the phase domains have largely developed, the diameter of the silica rich droplets reaching 4000 Å in a boron oxide rich continuum. The initial viscosity for this treated glass is less than that of the untreated one, because the volume percentage of the boron oxide rich phase is higher in the treated than in the untreated sample (Haller *et al.*, 1970).

The stabilization of rod samples treated at 710 °C is similar to that of untreated samples. When the samples have been heated previously at the testing temperatures, the initial viscosity is higher than for the untreated glass; but after a definite time its value differs slightly from that of the untreated glass. This supports the idea that the increase in viscosity, during tests carried out above 500 °C, is due to the stabilization process of the silica rich network and eventually to a coalescence of the phase domains.

Finally it was shown that even for the compressive loads used, which are always lower than 10^7 dynes/cm^2, the viscous flow is not newtonian at temperatures above 500 °C. Other cases of a non-newtonian behaviour have been reported by several authors (Bartenev, 1965; Li and Uhlmann, 1970; Stein and Stevels, 1965; Watanabe and Koyama, 1957) for systems exhibiting a liquid–liquid immiscibility. Further tests would therefore be necessary for a better understanding of the influence of phase separation on the relationship existing between stress and viscosity. It is suggested that, when the apparent viscosity is an increasing function of the strain rate, the deviation from a newtonian behaviour is due to a stabilization of the glass network, but when it is a decreasing function of the strain rate, this deviation is due mainly to a coalescence of the phase domains.

REFERENCES

Bartenev, G. M. (1965). In J. A. Prins (Ed.), *Physics of Non-crystalline Solids*. North-Holland, Amsterdam. p. 461.

Cordelier, J. F. (1970). *Verres Réfract.*, **24**, 113.

De Vries, A. J. (1960). In B. Persoz (Ed.), *Introduction à l' Étude de la Rhéologie*. Dunod, Paris. p. 199.

Haller, W., D. H. Blackburn, F. E. Wagstaff, and R. J. Charles (1970). *J. Am. Ceram. Soc.*, **53**, 34.

Jeanmaire, M. (1970). *Verres Réfract.*, **24**, 3.

Li, J. H., and D. R. Uhlmann (1970). *J. non-cryst. Solids*, **3**, 205.

Prod'homme, L. (1968). *Verres Réfract.*, **22**, 604.

Prod'homme, M. (1968). *Verres Réfract.*, **22**, 614.

Prod'homme, M. (1970). *Verres Réfract.*, **24**, 151.

Stein, H. N., and J. M. Stevels (1965). In *Comptes Rendus VII Congres International du Verre*, Bruxelles. No. 25.

Vogel, W. (1965). In *VEB Deutscher Verlag für Grundstoffindustrie*. Leipzig. p. 74.

Watanabe, M., and R. Koyama (1957). *J.S.G.T.*, **41**, 137, 296.

25

Influence of Microstructure on Diffusion Processes in Glasses

G. H. FRISCHAT *

Max-Planck-Institut für Silikatforschung, Würzburg, Germany

Only perfect single crystals containing structurally equal lattice sites show tracer diffusion processes ideally describable by a unique diffusion coefficient. Deviations from this simple behaviour may occur if in addition to the lattice properties dislocations or grain-boundaries are of influence.

Normal homogeneous glasses behave like perfect single crystals in this respect. Recently more complex diffusion profiles have been found for ^{110}Ag in Pyrex glass by Doremus (1968) and for ^{22}Na in silica glass and in mixed-alkali glasses by Frischat (1970 a, b). As a reason for this behaviour, phase separation of Pyrex glass on a fine scale, a non-homogeneous microstructure of SiO_2 glass, and a loss of structural homogeneity of a glass after addition of a second alkali have been suggested.

As already known the binary glass system Na_2O–SiO_2 forms two immiscible liquids at high temperatures in the composition range between about 4 and 22 mol % Na_2O. Both metastable and unstable decomposition may occur, see, for example, Redwine and Field (1968). As a consequence the resulting glasses display a glass-in-glass separation. This paper is concerned with the investigation of the influence of such microstructure on ^{22}Na diffusion in glasses of this system.

EXPERIMENTAL

Samples

Several glasses of the Na_2O–SiO_2 system with Na_2O contents between 12·6 and 33·3 mol % were investigated. As a typical example a glass con-

* Now at the Lehrstuhl für Steine und Erden (Glas and Keramik), Technische Universität Clausthal, Germany.

taining 14·8 % Na_2O is discussed here. For preparing the glass the melt was held for 1 h in a Pt crucible at 1500 °C and was stirred for homogenization. The resulting homogeneous glass then was annealed for 0·5 h at 480 °C. In order to attain phase separation the glass was heat-treated for 8 h at 600 °C. According to Redwine and Field (1968) and Hähnert and Hinz (1967), and other authors, this treatment yields a microstructure in the glass consisting of two phases in which both phases are independently interconnected in three dimensions. In electron micrographs these two interconnected phases are directly visible. ^{22}Na diffusion experiments were carried out both on homogeneous and on phase-separated glasses.

Tracer experiment

One drop of an aqueous solution of ^{22}NaCl was put on the surface of a polished glass specimen (diameter 20 mm, thickness 2 mm). This was covered with an identical disc to prevent loss of active material and was heated in a furnace. After this treatment the specimens were cleaned and sectioned by grinding off thin layers parallel to the surface. The residual activity of the sample was measured with a scintillation counter after removal of each layer, using a discriminator for the 1·28 MeV γ-rays of the ^{22}Na emission spectrum, the specimen being placed in a lead chamber to reduce background. Further experimental details may be taken from Frischat and Oel (1966).

RESULTS

Figure 1 shows diffusion profiles for ^{22}Na in homogeneous and phase separated glass after holding for 4·5 h at 245 °C. Both curves look similar except for the insignificant difference in the values for the initial activities at $x = 0$. For a diffusion process with a single diffusion coefficient, Fick's second diffusion law holds. With the initial condition of an instantaneous source one obtains a ^{22}Na tracer distribution in the glass describable by a Gaussian function.

The residual activity, $I'(x, t)$, measured after grinding away a thickness, x, of sample consists of two terms, the constant radioactive background, I_∞, and a term, $I(x, t)$, which is due to the diffusion effect. I_∞ was determined experimentally and had a value of about 30 counts/min for the experimental conditions employed.

In the present case of hard γ-rays the residual activity may be written as (Frischat and Oel, 1966)

$$I(x, t) = I'(x, t) - I_\infty = I_0[1\text{-erf}\,(x/2\sqrt{(Dt)})], \qquad (1)$$

where I_0 is the initial activity, D the diffusion coefficient, and t the diffusion time. As may be seen from this equation D is independent of I_0. Substituting $q = x/2\sqrt{(Dt)}$ into equation (1) gives

$$\frac{I(x, t)}{I_0} \equiv F(q) = 1 - \operatorname{erf}(q). \tag{2}$$

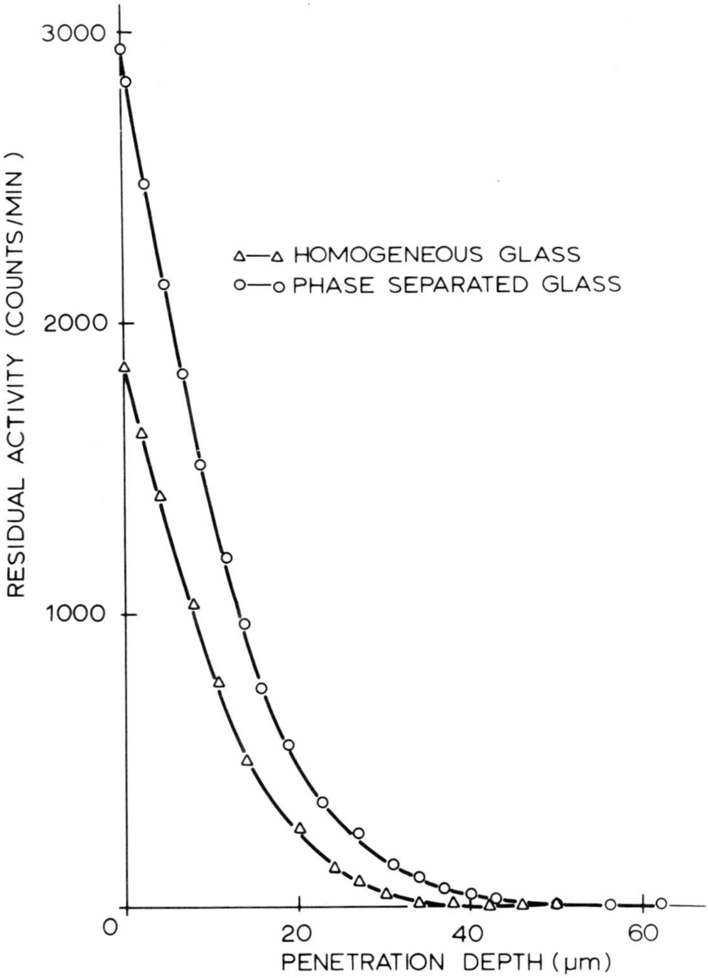

Figure 1. Residual γ-ray activity of ^{22}Na following diffusion in a homogeneous and in a phase separated glass, $T = 245\ ^\circ$C, $t = 4{\cdot}5$ h

As $F(q)$ is known from the experimental data for a series of depths, q may be calculated for these depths from equation (2). If there is indeed only one single diffusion process and if D is independent of experimental conditions, a plot of q against x is a straight line passing through the origin, thus D is given by

$$D = x^2/4q^2t. \tag{3}$$

Figure 2 illustrates a graphical attempt to evaluate the diffusion profiles

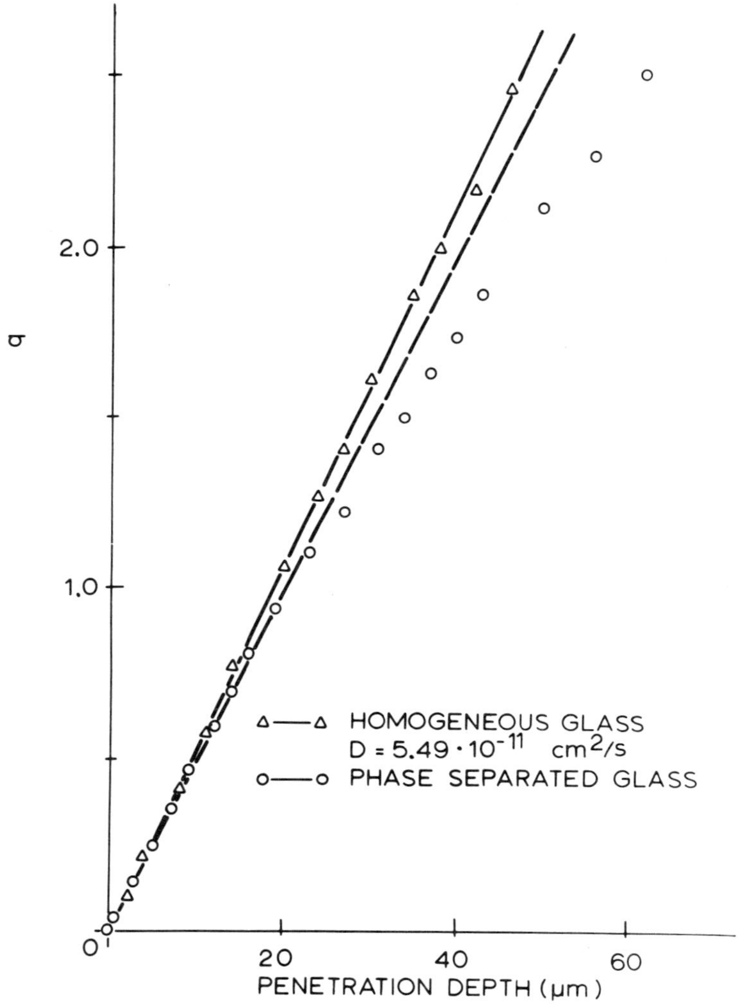

Figure 2. Attempted graphical evaluation of diffusion profiles from Figure 1

from Figure 1. Although the diffusion curve for the homogeneous glass can be described for the total activity range by a constant D value of $5{\cdot}49 \times 10^{-11}$ cm^2/s according to equations (2) and (3), strong deviations from linearity occur for penetration depths greater than 20 μm in the case of the phase separated glass. This simple method of evaluation no longer holds.

For a more appropriate description of the observations it is assumed that the diffusion profile reflects a superposition of two simple diffusion processes in the two phases of the glass with diffusion coefficients D_1 and D_2 ($D_1 < D_2$). This is in accordance with the method of evaluation of the diffusion profiles of ^{110}Ag in Pyrex glass (Doremus, 1968) and of ^{22}Na in SiO_2 glass and in mixed-alkali glasses (Frischat, 1970 a, b).

The tracer concentration in the glass will always be an additive superposition of two Gaussian functions and the measured residual activity may be described by

$$I(x, t) = I'(x, t) - I_\infty = I_{01}\,[1 - \operatorname{erf}(x/2\sqrt{(D_1 t)})] + I_{02}\,[1 - \operatorname{erf}(x/2\sqrt{(D_2 t)})] \quad (4)$$

where the I_{0i} are the respective initial activities of both processes.

Figure 3 illustrates an attempt to evaluate the diffusion experiment for the phase separated glass according to equation (4) by means of a computer least squares fit. The parameters of this fit are

$$I_\infty = 29{\cdot}7 \pm 3{\cdot}9 \text{ counts/min},$$
$$I_{01} = 1385 \pm 36 \text{ counts/min},$$
$$I_{02} = 1565 \pm 27 \text{ counts/min},$$
$$D_1 = (3{\cdot}55 \pm 0{\cdot}21) \times 10^{-11} \text{ cm}^2/\text{s},$$
$$D_2 = (1{\cdot}01 \pm 0{\cdot}03) \times 10^{-10} \text{ cm}^2/\text{s}.$$

As may be seen, at least at this stage, the assumption of two such processes is justified: the computed I_∞ is the same as that determined experimentally, and the deviations of the other variables are within the limit.

According to this simple model a portion of the active particles originally present at the surface of the glass diffuses *via* sites of one phase, the other portion *via* sites of the other phase. The ratio of these different diffusing particles may be obtained from the areas of both activity distributions. Integration of equation (1) gives

$$Z = I_0 \int_0^\infty [1 - \operatorname{erf}(x/2\sqrt{(Dt)})]\,\mathrm{d}x = \frac{2}{\sqrt{\pi}} I_0 \sqrt{(Dt)}. \quad (5)$$

In the case of two processes equation (4) has to be integrated. The total activity, Z, of the sample may be divided into two parts Z_1 and Z_2, corresponding to the two processes. The percentage of diffusing particles involved in the first process is thus

I

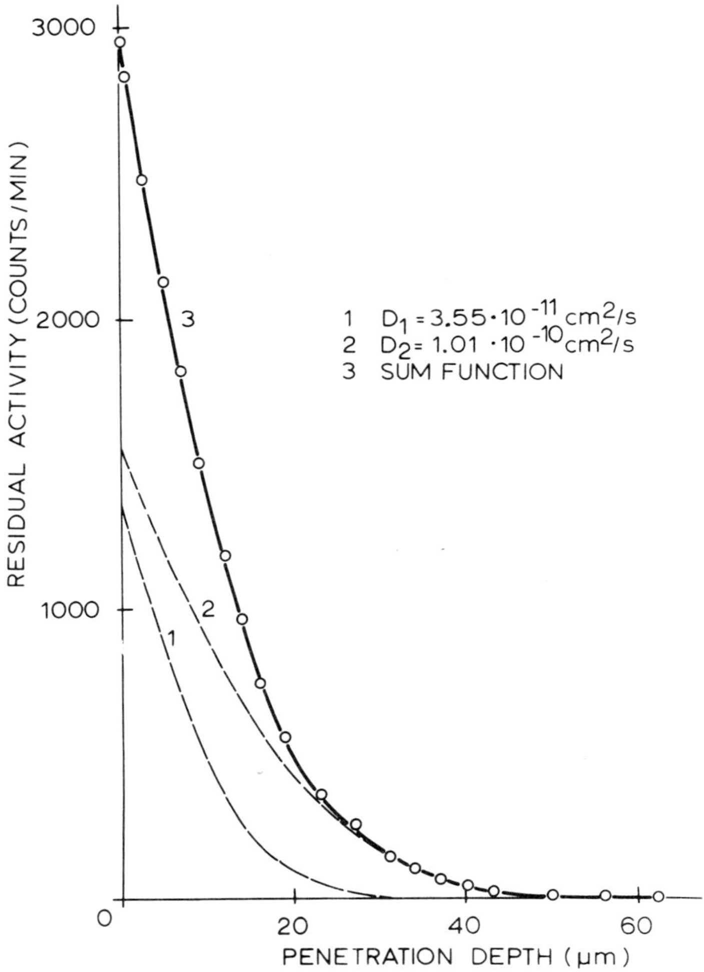

Figure 3. Computer least squares fit of the diffusion profile in the phase separated glass assuming two superimposing processes

$$z_1 = \frac{100}{1 + (I_{02}/I_{01})\sqrt{(D_2/D_1)}}[\%]. \tag{6}$$

Calculation of the separated diffusion profile of Figure 3 gave a value of $z_1 = 34{\cdot}5\ \%$.

Similar results were obtained for different glasses and temperatures.

DISCUSSION

Homogeneous glasses of the binary Na_2O–SiO_2 system always show simple diffusion behaviour. A comparison of the D value at 245 °C with a result of Johnson *et al.* (1951) for a similar binary glass (14·7 % Na_2O) gives good agreement. On the other hand the splitting of the diffusion profile for the phase separated glass clearly demonstrates the influence of such microstructure on the process of sodium self-diffusion. In the present case the diffusivities differ by a factor of 2·9 only. Greater differences are possible depending on the composition of the glass and on arrangement of the microstructure. The diffusion coefficient of the homogeneous glass lies between the coefficients for the two phases of the phase separated glass. One can suggest that the phase with the higher diffusivity is Na_2O-rich, whereas the other is SiO_2-rich. Qualitatively this tendency agrees with the course of coefficients for Na self-diffusion in glasses of higher (> 30 % Na_2O, Frischat, 1967) and of lower (SiO_2 glass, 8 ppm Na, Frischat, 1968) Na_2O contents.

Redwine and Field (1969) investigated the effect of microstructure on other transport properties (internal friction and DC electrical conductivity) of phase separated glasses of the binary Na_2O-SiO_2 system. The influence of the two three-dimensionally interconnected phases on internal friction peaks was of minor importance only, whereas electrical conductivity resulted in a significant temperature dependence of activation energy. From their results these authors stated that transport properties which depend on long-range order or average over the sample volume such as electrical conductivity are influenced by the two-phase structure. Nevertheless there was an influence on internal friction, that is transport of Na^+ ions between nearest neighbours. For the 14·8 % Na_2O glass this resulted in an asymmetrical low-temperature peak and a tail at lower temperatures. The authors speculated that the asymmetry might be caused by the overlapping of a further peak due to a contribution from the moving sodium ions in the silica rich phase.

This paper has shown the existence of at least two diffusion path possibilities, one of them attributable to the Na_2O-rich, the other to the SiO_2-rich phase. It is believed that these results give direct evidence for the assumption that the asymmetrical internal friction peak also demonstrates a superposition of at least two peaks, attributed to the stress-induced movement of Na^+ ions in both phases.

Haven and Verkerk (1965) discussed the mechanism of Na self-diffusion in binary Na_2O-SiO_2 glasses in terms of a defect structure of the glass. From the value of the experimental correlation factor and from geometrical structure models for such glasses they concluded that an interstitialcy mechanism would be the most probable one. There is no reason why similar conclusions should not be valid for each phase of an amorphously decomposed glass. But up to now it seems to be difficult to give direct experimental evidence for that statement.

Evaluation of these non-uniform diffusion profiles is based on the simple two-phase model. According to this model the overall concentration, $c(x, t)$, is described by the diffusion equation

$$\frac{\partial c}{\partial t} = D_1 \cdot \frac{\partial^2 c_1}{\partial x^2} + D_2 \frac{\partial^2 c_2}{\partial x^2}, \tag{7}$$

where the $c_i(x, t)$ are the respective tracer concentrations in the two phases. Although profiles may be described according to equation (7), considering D_1 and D_2 as effective diffusion coefficients, obviously this equation is too simple for a real substance, as it does not contain possible interactions between both phases. A more realistic diffusion model, therefore, should contain such interactions, that is equation (7) has to be replaced by equations containing both diffusion into, and interaction between, the two phases. Currently such a model is under investigation.

For the moment it must be stressed that the proposed evaluation is based on experimental arguments only. Nevertheless, in all cases investigated the hypothesis of two superimposed processes was adequately fulfilled, demonstrating that the description of Na self-diffusion in a phase separated glass, where both phases are three-dimensionally interconnected, by the effective coefficients D_1 and D_2 is already more realistic than a description by one D value only. The most important condition for finding such a superposition is accurate activity measure ment to $< 1\%$ of I_0; otherwise quite different diffusion behaviour may be simulated.

REFERENCES

Doremus, R. H. (1968). *Physics Chem. Glasses*, **9**, 128.
Frischat, G. H., and H. J. Oel (1966). *Z. angew. Phys.*, **20**, 195.
Frischat, G. H. (1967). *Glastech. Ber.*, **40**, 382.
Frischat, G. H. (1968). *J. Am. Ceram. Soc.*, **51**, 528.
Frischat, G. H. (1970 a). *Physics Chem. Glasses*, **11**, 25.
Frischat, G. H. (1970b). *J. Am. Ceram. Soc.*, **53**, 285.
Hähnert, M., and W. Hinz (1967). *Silikattechnik*, **18**, 377.
Haven, Y., and B. Verkerk (1965). *Physics Chem. Glasses*, **6**, 38.
Johnson, J. R., R. H. Bristow, and H. H. Blau (1951). *J. Am. Ceram. Soc.*, **34**, 165.
Redwine, R. H., and M. B. Field (1968). *J. Mater. Sci.*, **3**, 380.
Redwine, R. H., and M. B. Field (1969). *J. Mater. Sci.*, **4**, 713.

26

Diffusion Processes in Sodium Silicate Glasses

L. W. BARR
Department of Physics, Paisley College of Technology, Paisley, Renfrewshire, Great Britain

J. N. MUNDY
Argonne National Laboratory, Argonne, Illinois, U.S.A.

and A. H. ROWE
Solid State Division, AERE, Harwell, Great Britain

The purpose of this paper is to describe the application of isotope effect measurements to the determination of mechanisms of sodium diffusion in sodium silicate glass. In this technique the relative diffusion rates of two isotopes of the same element, differing only in mass, are used to obtain information about possible diffusion mechanisms. The technique and the underlying theory have been discussed in several places (Barr and Le Claire, 1964; Le Claire, 1970).

While isotope effect measurements have been used to determine diffusion mechanisms in crystalline solids they have not hitherto been applied to non-crystalline solids. Because of this, theoretical discussions on the method have been confined to crystalline solids and this paper must, of necessity, be written within such a framework.

Sodium silicate glasses are ionic conductors, the sodium ions being the current carriers, and comparison of sodium diffusion coefficients with ionic conductivity provides additional, and closely related information, about diffusion mechanisms. Such measurements have already been reported (Haven and Verkerk, 1965) and the measurements reported in this paper were made on a glass of the same composition (17·4 mol % Na_2O) to enable results to be compared.

DIFFUSION MECHANISMS

Possible diffusion mechanisms operating in sodium silicate glasses are limited by the condition that they also account for ionic conduction. For this reason only mechanisms involving charge carrying defects need be considered.

Such mechanisms are:

(a) the vacancy mechanism

(b) the interstitial migrating from interstitial site to interstitial site (the direct interstitial mechanism)

(c) the interstitialcy mechanism where an interstitial moves by jumping on to a lattice site pushing the occupying atom on to another interstitial site.

We shall consider each of these mechanisms in turn.

In the vacancy mechanism atom diffusion occurs by jumps of adjacent atoms into the vacancy as it moves through the lattice. While the vacancy moves at random the diffusing atom does not (Bardeen and Herring, 1952). When an atom has just exchanged with a vacancy there is a higher than random probability that the next jump of that atom will be back into the same vacancy. The atoms moving by the vacancy execute a correlated random walk.

This is taken into account by writing the well known expression for the diffusion coefficient D in terms of the number of atom jumps per second, Γ, and the jump distance, r, as

$$D = \tfrac{1}{6}\Gamma\, r^2 f \tag{1}$$

Here f is the correlation factor and is, for self-diffusion, a constant which is a function of mechanism and lattice geometry. A good approximation for f is given, for the vacancy mechanism, by the equation (Barr and Le Claire, 1964).

$$f = 1 - \frac{2}{Z} \tag{2}$$

where Z is the coordination number of the lattice. This equation reflects the fact that a return jump of an atom into a vacancy restores the original position and that the probability of such a return jump is $1/Z$.

Consider now the relative jump rates to two isotopes of the same solute. The energy barrier to be surmounted by these isotopes will be the same since it depends on chemical properties only. Since their vibration frequencies, and hence their jump frequencies, are proportional to $m^{-\frac{1}{2}}$, the lighter isotope will tend to exchange more frequently with the vacancy than the average atom. Similarly a heavier isotope will tend to exchange less frequently. The correlation factor will thus enter into the diffusion of such isotopes in slightly different ways. Using this Schoen (1958) showed that

$$\left(\frac{D^\alpha}{D^\beta} - 1\right) = f\left(\frac{\omega^\alpha}{\omega^\beta} - 1\right) = f\left[\left(\frac{m^\beta}{m^\alpha}\right)^{\frac{1}{2}} - 1\right] \tag{3}$$

where ω^α and ω^β are the jump frequencies of the isotopes. A determination of the ratio of the isotopic diffusion coefficients D^α/D^β thus enables f to be found.

The same equation holds for interstitial motion by the direct interstitial mechanism but since such motion is uncorrelated, f is equal to 1.

The interstitialcy case is more complex since two atoms take part in the diffusion step. In this case

$$\left(\frac{\omega^\alpha}{\omega^\beta} - 1\right) = \left[\left(\frac{m^\beta + m}{m^\alpha + m}\right)^{\frac{1}{2}} - 1\right] \tag{4}$$

(Vineyard, 1957), where m is the average mass of the non-tracer atoms.

For interstitialcy diffusion f is given approximately by the expression (Haven and Verkerk, 1965)

$$f = 1 - \frac{1}{Z'} \tag{5}$$

where Z' is the number of neighbouring interstitial sites to which an atom from a lattice site can jump.

Experimental evidence has accumulated from a wide range of crystalline solids that equation (3) is correct to within $\sim 10\%$. It has been shown to hold, for example, for NaCl (Barr and Le Claire, 1964), Pd (Peterson, 1964), Zn (Peterson and Rothman, 1967), Cu (Rothman and Peterson, 1969) and Ag (Peterson, Barr and Le Claire, 1970).

Similarly equation (4) has been found to hold for the collinear interstitialcy diffusion of Ag in AgBr (Peterson, Barr and Le Claire, 1970).

For interstitial diffusion in quartz equation (3) has been shown to hold to a high degree of accuracy (Frank, Swets and Lee, 1961; Jones, 1953).

The correlation factor f applies to the atoms moved by defects and does not apply to the defects themselves. Hence, ionic conductivity, which can be discussed purely in terms of the defect movements, does not contain the correlation factor. For this reason comparison of the tracer diffusion coefficient and the ionic conductivity, σ, can also be used to obtain an experimental value of the correlation factor. This relationship can be written

$$\frac{D}{\sigma} = \frac{kTf_\sigma}{Ne^2} \quad \text{or} \quad \frac{D}{D_\sigma} = f_\sigma \tag{6}$$

(Le Claire, 1970) where N is the number of change carrying ions per unit volume. k, T and e have their usual meaning. For the particular cases of the vacancy and direct intersticial mechanism.

$$f_\sigma = f$$

but this equality does not hold generally. For this reason the quantity f_σ should *not* be called a correlation factor.

For example, in the case of the interstitialcy mechanism (Le Claire, 1970)

$$f_\sigma = 2(r_T/r_\sigma)^2 f \tag{7}$$

The factor 2 appears because 2 atoms are involved in each interstitialcy jump and the factor (r_T/r_σ) because the distance moved by the charge, r_σ differs from that moved by a tracer atom, r_T. For the collinear interstitialcy jump $r_T = r_\sigma/2$.

This brief discussion shows that each of the two techniques of isotope effect and diffusion–ionic conductivity measurements can yield information as to the diffusion mechanism. From this point of view sodium silicate glasses are ideal. Not only are these glasses ionic conductors with the sodium ions carrying the current but there are two radioactive isotopes of sodium ^{22}Na and ^{24}Na having one of the largest mass differences and also very convenient radioactive properties.

EXPERIMENTAL METHOD

The sodium silicate glass used in this work contained 17·4 mol % Na_2O and was prepared in the Department of Glass Technology, Sheffield University.

The diffusion coefficients of ^{22}Na and ^{24}Na were found to the required precision by diffusing these tracers simultaneously from a thin surface layer. Under these conditions the solution of the diffusion equation for either isotope is

$$C_x = C_0 \exp\left(\frac{-x^2}{4Dt}\right) \tag{8}$$

where C_x is the concentration of either isotope a distance x from the surface after an anneal of time t. This concentration profile was found by standard radiotracer techniques using acid sectioning.

For the case of simultaneous diffusion of both tracers

$$\begin{aligned} \ln(C^\alpha/C^\beta) &= \text{constant} + \frac{x^2}{4D^\alpha t}\left(\frac{D^\alpha}{D^\beta} - 1\right) \\ &= \text{constant} - \ln C^\alpha \left(\frac{D^\alpha}{D^\beta} - 1\right) \end{aligned} \tag{9}$$

(Barr and Le Claire, 1964). The concentration of each tracer separately was found using half-life separation of the isotopes. This method eliminates errors due to time and temperature of anneal and sectioning.

The ionic conductivities were measured on the same glass samples using conventional a.c. bridge techniques.

RESULTS

The results of the measurements are summarized in Table 1. An Arrhenius plot of the diffusion coefficients is shown in Figure 1 together with the data of Haven and Verkerk (1965). The results fit the equation

Table 1. Diffusion coefficients, isotope effects and the diffusion-conductivity ratio as a function of temperature in sodium silicate glass

T°C	D cm^2 sec^{-1}	$[(D^\alpha/D^\beta) - 1]$	f_σ
446	$9{\cdot}19 \times 10^{-9}$	—	$0{\cdot}44 \pm 0{\cdot}02$
446	$8{\cdot}8 \times 10^{-9}$	—	—
432	$7{\cdot}6 \times 10^{-9}$	$0{\cdot}011 \pm 0{\cdot}002$	$0{\cdot}37 \pm 0{\cdot}01$
411	$5{\cdot}7 \times 10^{-9}$	$0{\cdot}011 \pm 0{\cdot}002$	$0{\cdot}50 \pm 0{\cdot}1$
388	$3{\cdot}5 \times 10^{-9}$	—	—
354	$1{\cdot}64 \times 10^{-9}$	$0{\cdot}012 \pm 0{\cdot}002$	$0{\cdot}39 \pm 0{\cdot}02$
348	$1{\cdot}56 \times 10^{-9}$	—	$0{\cdot}39 \pm 0{\cdot}05$

$[(m^\beta/m^\alpha)^{\frac{1}{2}} - 1] = [(24/22)^{\frac{1}{2}} - 1] = 0{\cdot}044.$

$$D = (8 \pm 3) \times 10^{-4} \exp\left(\frac{-16{\cdot}3 \pm 0{\cdot}5}{RT}\right) \text{cm}^2 \text{ sec}^{-1}$$

where the activation energy is in kcal mol^{-1}.

A typical plot, using equation (9), of the isotope effect is shown in Figure 2 together with a zero isotope effect line and a theoretical line drawn using equation (3) with f equal to one.

Neither $(D^\alpha/D^\beta - 1)$ nor D/D_σ were found to vary with temperature.

DISCUSSION

For the vacancy mechanism $f_\sigma = f$ and the value of f can be deduced from both the isotope effect measurements and the diffusion-conductivity measurements. From the isotope effect f is found to be 0·25 and from the conductivity $f = 0{\cdot}40$. These values are clearly incompatible and the vacancy mechanism must be excluded.

Since the value of f obtained by either method is very much smaller than one the direct interstitial mechanism must also be excluded.

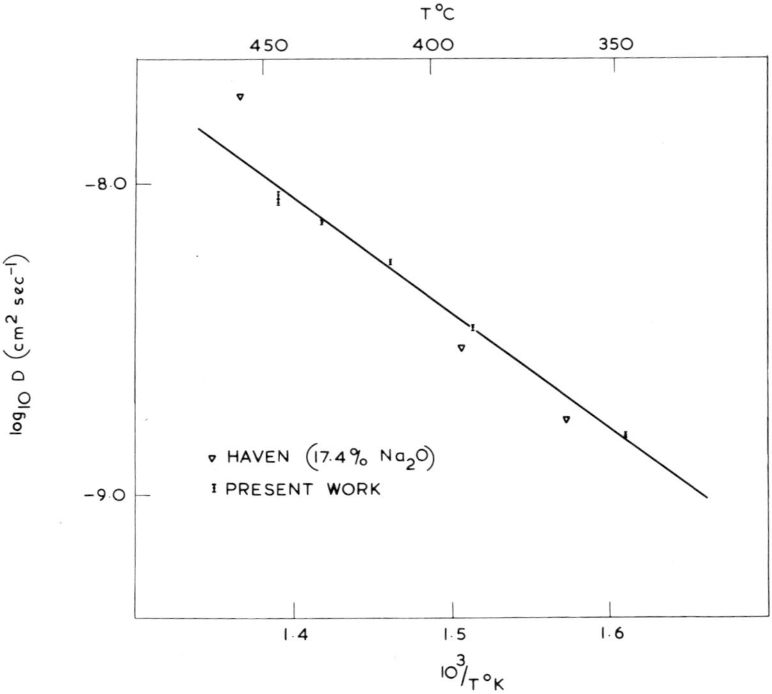

Figure 1. ^{22}Na diffusion as a function of temperature in sodium silicate glass

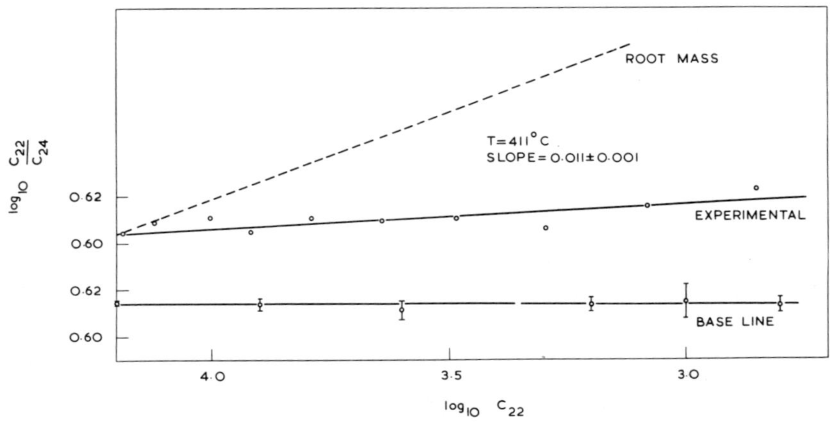

Figure 2. The isotope effect for ^{22}Na and ^{24}Na diffusion in sodium silicate glass

Turning now to the interstitialcy mechanism it is instructive to compare isotope effects and diffusion-conductivity results in AgBr and sodium silicate glass. This is done in Table 2. The similarity is striking and strongly suggests that the diffusion mechanism is the same in both cases. In AgBr, of course, it is well known that Ag diffusion is by the interstitialcy mechanism (Friauf, 1957). On this basis, the low isotope effect arises because of the two atoms involved in the jump process: the low value of f_σ largely because of the difference between the tracer and charge jump differences.

Table 2. Comparison of D/D_σ and the normalized isotope effect,

$$\left(\frac{D^\alpha}{D^\beta} - 1\right) \Big/ \left[\left(\frac{m^\beta}{m^\alpha}\right)^{\frac{1}{2}} - 1\right],$$

for Ag in AgBr and Na in sodium silicate glass. These quantities vary with temperature in AgBr. In the temperature region chosen the interstitialcy mechanism predominates

Mechanism of diffusion	AgBr interstitialcy	Sodium silicate glass
$\dfrac{D}{D_\sigma} = f_\sigma$	0·46–0·53[a] (140–200 °C)	0·37–0·50 (350–450 °C)
$\left(\dfrac{D^\alpha}{D^\beta} - 1\right) \Big/ \left[\left(\dfrac{m^\beta}{m^\alpha}\right)^{\frac{1}{2}} - 1\right]$	0·29[b] (175–200 °C)	0·25 (350–430 °C)

[a] Friauf (1957).
[b] Peterson, Barr and Le Claire (1970).

If it is assumed that equation (4) is exact then $f = 0{\cdot}5$. Such a low value of f implies about 2 interstitial sites/sodium ion. Applying equation (7) we find

$$(r_T/r_\sigma) \sim 0{\cdot}6$$

These numerical estimates, however, must be treated with great caution. Almost all deviations reported from equations (3) and (4) are in the direction of weaker isotope effects. If this is the case f could be larger and (r_T/r_σ) smaller than calculated above.

Since Haven and Verkerk (1965) found f_σ varied only slightly with soda content in a wide range of sodium silicate glasses and values of f_σ similar in magnitude to those reported in this work and the work of Haven and Verkerk, have been reported in sodium alumino-silicate glasses (Terai, 1969), it seems likely that the interstitialcy type mechanism occurs in a range of glasses.

CONCLUDING REMARKS

As was remarked earlier the treatment in this paper has been in terms of a crystalline lattice. While equation (3) has been established experimentally and theoretically (Tharmalingham and Lidiard, 1959) for a wide range of lattices its extension to non-crystalline solids has not been examined. Because f is largely determined by local structure, as indicated by equations (2) and (5), it is plausible to argue that equation (3) will hold. However, a more detailed analysis is required.

Finally, caution must be exercised in visualizing the diffusion step in terms of an interstitialcy in a crystalline lattice. All that can be said on the basis of the present evidence is that *two atoms* are involved and that $r_\sigma > r_T$.

Clearly the work reported here represents only a beginning. Several pairs of isotopes are available which would be suitable for studying diffusion in other glass systems and such measurements and the extension of the measurements reported here over a wider range of soda content and temperature are obviously called for.

ACKNOWLEDGEMENTS

We are grateful to Dr. J. O. Isard of the Department of Glass Technology of the University of Sheffield both for supplying the glass used in this work and for many interesting discussions and wise advice.

REFERENCES

Bardeen, J., and C. Herring (1952). In W. Shockley (Ed.), *Imperfections in Nearly Perfect Crystals.* Wiley, New York. p. 261.
Barr, L. W., and A. D. Le Claire (1964). *Proc. Br. Ceram. Soc.*, **1**, 109
Frank, R. C., D. E. Swets, and R. W. Lee (1961). *J. chem. Phys.*, **35**, 1451.
Friauf, R. J. (1957). *Phys. Rev.*, **105**, 843.
Haven, Y., and B. Verkerk (1965). *Physics Chem. Glasses*, **6**, 38.
Jones, W. M. (1953). *J. Am. chem. Soc.*, **75**, 3093.
Le Claire, A. D. (1970). In H. Eyring, D. Henderson and W. Jost (Eds.), *Physical Chemistry*, Vol. X. Academic Press, New York. p. 261.
Peterson, N. L. (1964). *Phys. Rev.*, **136A**, A586.
Peterson, N. L., L. W. Barr, and A. D. Le Claire (1970). To be published.
Peterson, N. L., and S. J. Rothman (1967). *Phys. Rev.*, **163**, 645.
Rothman, S. J., and N. L. Peterson (1969). *Phys. Stat. Sol.*, **35**, 305.
Schoen, A. (1958). *Phys. Rev. Lett.* **1**, 138.
Terai, R. (1969). *Physics Chem. Glasses*, **10**, 146.
Tharmalingham, K. and A. B. Lidiard (1959). *Phil. Mag.*, **4**, 899.

27

Alkali Ion Conduction in Silicate Glass *

P. B. MACEDO, R. BOSE, V. PROVENZANO and T. A. LITOVITZ

Vitreous State Laboratory, The Catholic University of America, Washington, D.C., U.S.A.

There have been many models which have proposed that conduction occurs in a series of steps. We have shown that this assumption demands that one adds dielectric 'moduli' rather than susceptibilities. This has not been done in the past. When this is done it is easy to show that a distribution of jump times leads to a frequency dependent conductivity and to an additional polarizability. These theoretical considerations have been applied to sodium trisilicate glass and have allowed us to separate clearly the conduction jump processes from the trapped ionic motion. An important by-product of this work is the demonstration that Eyring's Formula for conductivity is in error.

Of the large number of conductivity measurements that have been reported in the literature, only a few authors have concerned themselves with measurements of the frequency dependence of the conductivity; and in even fewer systems (Volger, 1960) has there been any understanding of this frequency dependence. The fundamental problem in such analysis is to separate and understand those ionic processes causing dielectric relaxations from those related to conductivity. One of the most important efforts in relating these two ionic effects is due to Taylor (1957, 1959). He studied the dielectric properties of a large number of glasses and proposed that the frequency dependence of the conductivity, σ, and the permittivity, ϵ, could be accounted for by assuming a distribution of barrier heights over which the mobile ions in the medium must move. However, Taylor did not attempt any theoretical treatment of this model.

Following Taylor, Charles (1961) proposed an elegant model associating the dielectric relaxation in silicate glasses with the migration of alkali ions

* This work was supported by ONR contract N00014-68-A-0506-0002 and the Air Force Office of Scientific Research, Grant No. AFOSR 68-1376.

about a given oxygen site and relating the D.C. conductivity to the migration of the ions between different oxygen sites. This model is very useful in relating dielectric relaxations to the structure of glasses. In explaining the D.C. conductivity, Charles suggests that the conduction process takes place through a *series* of alkali ionic jumps, first about a given oxygen site, and then between two such sites. His theoretical treatment, however, is inconsistent with his own assumptions, since he assumes a distribution of jump times (obtained from the dielectric relaxation data) and then adds them in *parallel*.

A further defect in Charles' treatment is that he neglects the excess polarization effect that arises from a distribution of ionic jump times. This latter process has been well understood for two phase systems (Volger, 1960) but never considered for single phase systems with a distribution of barrier heights, until our recent theoretical paper on the 'dielectric modulus' (Macedo, Bose and Litovitz, 1970). The purpose of this paper is to show how one can understand better the motions of sodium ions in silicate glasses by the use of the 'dielectric modulus' picture.

GENERALIZED SUSCEPTIBILITY

When describing a lossy conducting dielectric medium, it is useful to define a quantity known as the 'generalized electrical susceptibility,' ε^*, given by

$$\varepsilon^* = \varepsilon_1 - i\varepsilon_2 = \epsilon_1 - i\epsilon_2 - i\sigma_0/\omega \tag{1}$$

where ε_1 and ε_2 are the real and the imaginary parts of the complex permittivity, ε^*, and σ_0 is the static or D.C. conductance of the medium dependent permittivity. It has been demonstrated that there can be a non-trivial frequency dependence of the conductive response term and therefore the analysis of the generalized susceptibility ε^* is more useful than that of the permittivity ϵ^*.

FREQUENCY DEPENDENCE OF ε^* IN A NON-CONDUCTING DIELECTRIC ($\sigma_0 = 0$). INTRODUCTION OF THE 'ELECTRICAL MODULUS'

In the absence of contributions from conduction currents, the susceptibility function ε^* is identical to the permittivity ϵ^* of the medium. Assuming a single relaxation time τ_E, one obtains the familiar equation

$$\varepsilon^*_{(\omega)} = \epsilon_\infty + \frac{\epsilon_S - \epsilon_\infty}{1 + i\omega\tau_E} \tag{2}$$

where ϵ_S and ϵ_∞ are the limiting low- and high-frequency values of the measured permittivity associated with the relaxation process.

Below we shall show that it is of interest to define a 'generalized electrical modulus' for a system as

$$M^*(\omega) = \varepsilon^*(\omega)^{-1} = \frac{\varepsilon_1}{\varepsilon_1^2 + \varepsilon_2^2} + \frac{i\varepsilon_2}{\varepsilon_1^2 + \varepsilon_2^2} \tag{3}$$

In terms of this newly defined quantity, equation (2) can be rewritten as

$$M^*(\omega) = M_S + (M_\infty - M_S)\left[\frac{\omega^2\tau_D^2 + i\,\omega\tau_D}{1 + \omega^2\tau_D^2}\right] \tag{4}$$

where $M_S = 1/\epsilon_S$, $M_\infty = 1/\epsilon_\infty$ and $\tau_D = (M_S/M_\infty)\tau_E$. The relaxation time for the susceptibility τ_E, and the relaxation time for the modulus, τ_D, are *not equal* but proportional to each other. The susceptibility relaxation time corresponds to the time taken by the electrical strain $\vec{D}$ in the medium to decay to e^{-1} of its initial value while the applied stress $\vec{E}$ is held constant, and the modulus relaxation time τ_D corresponds to the time taken by the stress $\vec{E}$ to decay by the same ratio while the strain $\vec{D}$ is kept constant.

FREQUENCY DEPENDENCE OF ε^* AND M^* IN A CONDUCTING MEDIUM WITH NO DIELECTRIC LOSS ($\sigma_0 \neq 0, \epsilon_2 = 0$)

In the absence of all dielectric loss, the permittivity of the medium is real, dispersionless and given by a frequency independent constant ϵ_S and ϵ_2 is zero. One can see from equation (1) that the imaginary part of $\varepsilon^*(\omega)$ now describes a conductive loss which decreases as ω^{-1}. Equation (1) can be inverted to find the generalized modulus.

$$M^*(\omega) = \frac{1}{\epsilon_S}\left[\frac{\omega^2\tau_\sigma^2 + i\,\omega\tau_\sigma}{1 + \omega^2\tau_\sigma^2}\right] \tag{5}$$

where $\tau_\sigma = \epsilon_S/\sigma_0$. As pointed out above, this modulus relaxation time τ_σ has the physical significance of being the time for an initial electric stress $\vec{E}$ to decay to e^{-1} of its value, when the strain $\vec{D}$ is held constant. This is the well known Maxwell relaxation time.

We can now go back and analyse equation (5) for the modulus relaxation spectrum. Evaluating the frequency limits of $M^*(\omega)$ we obtain for the real part

$$\lim_{\omega\tau_\sigma \ll 1} M'(\omega) = M_S = 0$$

and

$$\lim_{\tau_\sigma \omega \gg 1} M'(\omega) = M_\infty = 1/\epsilon_S.$$

In the low frequency limit the real part of the conduction modulus goes to zero immediately suggesting an analogy with the shear flow process, where the real part of the shear modulus also goes to zero at low frequencies thereby allowing the liquid to flow. The lack of a finite zero frequency electrical modulus implies the absence of any restoring force in the medium permitting irreversible flow of ions. This is in striking contrast with the dielectric relaxation case.

FREQUENCY DEPENDENCE OF ε^* AND M^* FOR A CONDUCTING MEDIUM WITH A DISTRIBUTION OF MODULUS RELAXATION TIMES BUT NO DIELECTRIC LOSS

Since the relaxational characteristics involved in a conduction process are only apparent in the modulus picture, the next logical step is to investigate the effect of a distribution of relaxation times on the spectrum of the generalized modulus, as well as on the conductivity of the system. Denoting the distribution function for τ_σ as $g(\tau_\sigma)$, equation (5) for the modulus is now replaced by

$$M^*(\omega) = M_\infty \int_0^\infty \mathrm{d}\tau_\sigma g(\tau_\sigma) \left[\frac{\omega^2\tau_\sigma^2 + \mathrm{i}\,\omega\tau_\sigma}{1 + \omega^2\tau_\sigma^2}\right] \tag{6}$$

As expected, the only effect of the distribution of relaxation times is to broaden the spectrum of $M'(\omega)$ and depress the maximum of $M''(\omega)$. We emphasize once again the fact that characteristically for a conduction process, the low frequency limit of $M'(\omega)$ goes to zero.

The most significant effect of the distribution of relaxation times becomes evident when we calculate the conductivity of the material. From equation (3) the conductivity $\sigma(\omega)$ is defined as

$$\sigma(\omega) = \omega\varepsilon_2 = \omega \frac{M''}{M'^2 + M''^2} \quad . \tag{7}$$

It is not possible to find an analytic expression for the conductivity from the above $g(\tau_\sigma)$ unless a specific form of the distribution function is given. On the other hand, the zero and infinite frequency limits of equation (6) can be easily evaluated. Thus we obtain

$$\operatorname*{Lim}_{\omega\tau_\sigma \ll 1} \sigma(\omega) = \sigma_0 = \frac{1}{M_\infty \langle\tau_\sigma\rangle} = \epsilon_S/\langle\tau_\sigma\rangle \tag{8}$$

and

$$\lim_{\omega\tau_\sigma \gg 1} \sigma(\omega) = \sigma_\infty = \frac{1}{M_\alpha}\left\langle\frac{1}{\tau_\sigma}\right\rangle = \epsilon_S\left\langle\frac{1}{\tau_\sigma}\right\rangle \tag{9}$$

the angular brackets in both cases denoting statistical average. Since $\langle 1/\tau_\sigma\rangle \neq (1/\langle\tau_\sigma\rangle)$, we come upon the conclusion that $\sigma_0 \neq \sigma_\infty$, and that the conductivity itself undergoes a relaxation.

The important features of this 'conductivity relaxation' are as follows. The zero mobility of an ion is determined by the probability that a series of many jumps will occur. In this process the ion samples all the time constants and the probability of a jump is proportional to $(1/\langle\tau_\sigma\rangle)$. At high frequencies only a few ions perform more than one jump. and the probability that a single jump will occur is now proportional to $\langle 1/\tau_\sigma\rangle$.

The total dispersion of the conductivity 'relaxation', $\sigma_r = \sigma_\infty - \sigma_0$ is given by

$$\sigma_r = \epsilon_S\left[\left\langle\frac{1}{\tau_\sigma}\right\rangle - \frac{1}{\langle\tau_\sigma\rangle}\right] \tag{10}$$

Thus the conductivity dispersion is very much dependent on the nature of $g(\tau_\sigma)$, the contrast with ordinary relaxation mechanisms where the magnitude of the dispersion is independent of the nature of the distribution of relaxation times. Thus, in the previous case of a single relaxation time where $g(\tau_\sigma) = \delta\,(\tau_\sigma - \tau_0)$, we get $\langle 1/\tau_\sigma\rangle = (1/\langle\tau_\sigma\rangle)$. so that σ_r vanishes. As the distribution function becomes broader, the magnitude of the conductivity dispersion increases.

It is interesting now to find out how the corresponding dielectric susceptibility function ε^* behaves in the presence of this distribution of relaxation times. Once again, analytical forms being dependent on the specific nature of distribution functions, we shall only evaluate the limiting values of ε^*. Thus

$$\varepsilon_S = \epsilon_S\frac{\langle\tau_\sigma^2\rangle}{\langle\tau_\sigma\rangle^2} - \mathrm{i}\,\sigma_0/\omega = \epsilon_S - \mathrm{i}\,\sigma_0/\omega + \epsilon_S\frac{\langle\tau_\sigma^2\rangle - \langle\tau_\sigma\rangle^2}{\langle\tau_\sigma\rangle^2} \tag{11}$$

$$\varepsilon_\infty = \epsilon_S - \mathrm{i}\,\sigma_0/\omega = \epsilon_S - \mathrm{i}\,\sigma_0/\omega + \frac{\mathrm{i}\epsilon_S}{\omega}\left[\left\langle\frac{1}{\tau_\sigma}\right\rangle - \frac{1}{\langle\tau_\sigma\rangle}\right] \tag{12}$$

Interpolating between limits, we have in general

$$\varepsilon^*(\omega) = \epsilon_S - \mathrm{i}\,\sigma_0/\omega + \Phi^*(\omega) \tag{13}$$

where $\Phi^*(\omega)$ is a frequency dependent term which is determined by the given distribution. Notice that the dispersion of the real part of ε^* is now determined by the second moment of the relaxation time and in contrast to normal dielectric relaxations, depends again on the specific form of the distribution function $g(\tau_\sigma)$.

It is tempting at this point to calculate a conductivity 'jump distance' as is customarily done in the literature. The usual procedure is to use Eyring's rate theory to calculate a mean square jump distance $\langle l^2 \rangle$ from the formula

$$\sigma_0 = \frac{F|Z|N\langle l^2 \rangle}{CkT\langle \tau \rangle} \tag{14}$$

where F is the Faraday constant, Z is the average charge per ion, N is the number density, C is a geometrical constant usually set at 6, and $\langle \tau \rangle$ is an average relaxation time associated with the conductivity process. Such a calculation should be entirely feasible except for the fact that the average time $\langle \tau \rangle$ appearing in the above formula is a relaxation time at constant stress, i.e. $\langle \tau_E \rangle$ in our notation. However, the measured value of the conductivity is related to a relaxation time at constant strain, i.e. $\sigma_0 = \epsilon_S / \langle \tau_D \rangle$. Although the stress and strain relaxation times are related by

$$\tau_E = (M_\infty / M_S)\tau_D \tag{15}$$

in the case of conduction, where $M_S = 0$, τ_E becomes infinite and its significance in the original Eyring formula becomes unclear. It appears to us that it is not possible to calculate a conductivity jump distance from Eyring's formula. In fact Eyring's equation is in error.

APPLICATION OF A MATHEMATICAL FRAMEWORK TO SODIUM TRISILICATE GLASS

The model ofCharles and Taylor describe the sodium ions performing a series of jumps about an oxygen ion, then between oxygen ions. Since these two processes occur in a series one must add their contributions to the 'dielectric modulus,' rather than to the dielectric susceptibility. We will therefore use equation (6) to describe the sodium ion jumps between oxygen sites (conduction process). Then we must add the further effect of the ions jumping about an oxygen ion (trapped ionic motion).

Since in this latter process there is a maximum possible displacement of the ions in the direction of the electric field, there is a non-zero M_S. $M^*(\omega)$ has the form given by equation (4) modified to include a distribution of relaxation times. Thus the total effect in a sodium silicate glass can be described by

$$M^*(\omega) = M_1 \int_0^\infty g_\sigma(\tau) \left[\frac{(\omega\tau)^2 + \mathrm{i}\,\omega\tau}{1 + \omega^2\tau^2} \right] \mathrm{d}\tau + (M_\infty - M_1) \int_0^\infty g_{\mathrm{TI}}(\tau) \left[\frac{(\omega^2\tau^2) + \mathrm{i}\,\omega\tau}{1 + (\omega\tau)^2} \right] \mathrm{d}\tau \tag{16}$$

where M_1 is the zero frequency limit of the trapped ionic motion which of course coincides with the infinite frequency limit of the conductivity. M_∞ is measured at frequencies so high that the ions have no time to make either type of jump. The distribution of conduction times and of trapped ionic times are given respectively by $g_\sigma(\tau_\sigma)$ and $g_{TI}(\tau_D)$.

As described in the previous section $g_\sigma(\tau_\sigma)$ causes a very complicated frequency dependence of the susceptibility. The added presence of the trapped ionic motion makes the behaviour even more complicated. The data of

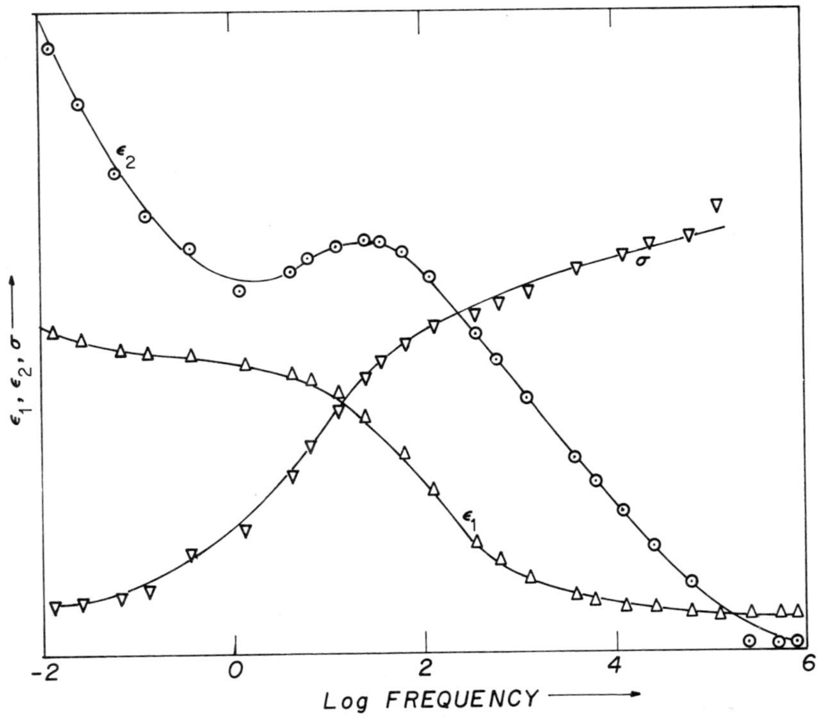

Figure 1. The logarithm of the real part (ε_1) and the imaginary part (ε_2) of the generalized susceptibility as well as the conductivity (σ) are plotted versus logarithm of the frequency. The points represent actual data on sodium trisilicate glass and the solid lines are smooth curves without theoretical implications

Provenzano, Macedo and Voltera (1970) on sodium trisilicate glass, shown in Figure 1 demonstrates this complexity. It is exceedingly difficult to analyse the data in this form. However, converting the same data to the modulus representation (as was done in Figure 2) allows a clear representation of the effects of the two types of ion motions. The data is very well fitted by assuming

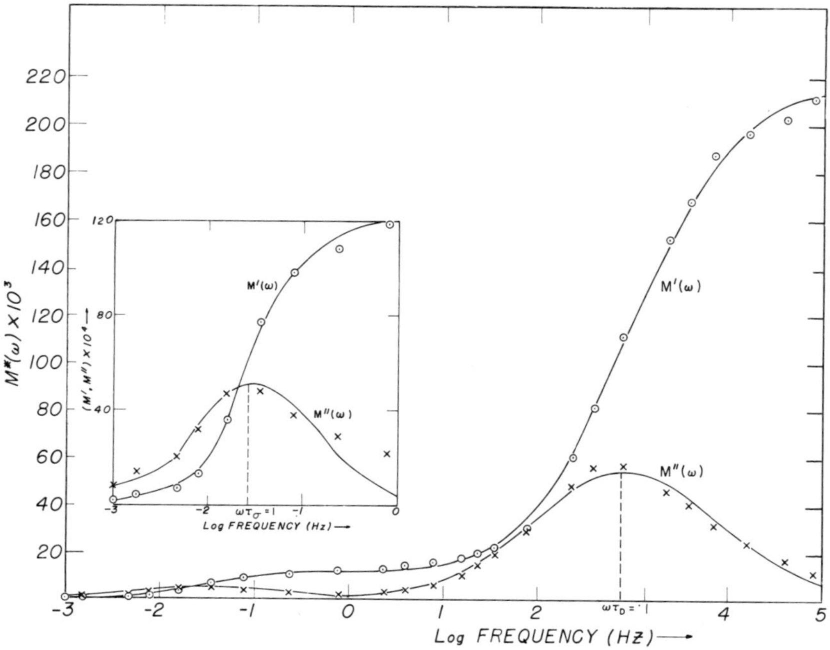

Figure 2. The points are the same data as Figure 1 replotted in the modulus representation. The solid lines represent the theoretical values given in equation (16) with the following values for the parameters. $M_1 = 0{\cdot}012 \pm 0{\cdot}001$, $\tau'_\sigma = 3{\cdot}04207 \pm 0{\cdot}00001$, $b_\sigma = 0{\cdot}785 \pm 0{\cdot}002$, $M_\infty = 0{\cdot}215 \pm 0{\cdot}001$, $\tau'_{TI} = 1{\cdot}395 \times 10^{-4} \pm 2 \times 10^{-7}$, $b_{TI} = 0{\cdot}3875 \pm 0{\cdot}0004$

that each type of ionic motion is associated with a log-Gaussian distribution of relaxation times.

CONCLUSIONS

(1) The present mathematical framework has shown clearly that Eyring's formula for ionic conduction is in error.

(2) It was shown that both effects, originally proposed by Taylor and then further developed by Charles, can be found in the dielectric data when analyzed in the modulus representation.

(3) The present work shows that Charles confused the trapped ionic motion with the conduction jumps in his treatment of the data (even though his physical model was correct). Charles obtained his distribution of trapped ionic times from the peak in the ϵ_2 curve (see Figure 1 above) and then inserted the average of these times in Eyring's equation in an attempt to fit the low frequency conductivity. This approach is faulty for three reasons:

(a) Eyring's equation is not valid; (b) the two ionic processes were added in parallel rather than in series (i.e. susceptibilities were added, not moduli) and (c) assuming that the trapped ionic relaxation times were the same as the conduction jump times neglects the fact that if these times were the same the ions would not be trapped.

REFERENCES

Charles, R. J. (1961). *J. appl. Phys.*, **32**, 1115.

Glasstone, S., H. Eyring, and K. Laidler, *Theory of Rate Processes.* (1941). McGraw–Hill, New York.

Macedo, P. B., R. Bose, and T. A. Litovitz. Paper submitted to *J. chem. Phys.*, 1970.

Provenzano, V., P. B. Macedo, and V. Voltera (1970) (unpublished work).

Taylor, H. E. (1957). *J.S.G.T.*, **41**, 350T.

(1959). *J.S.G.T.*, **43**, 124T.

Volger, J. In A. F. Gibson (Ed.), *Progress in Semiconductors*, Vol. 4, 1960. John Wiley & Sons, New York. p. 205.

STRUCTURE

28

X-Ray Structural Determination of Glasses

B. E. WARREN
Massachusetts Institute of Technology, Cambridge, Massachusetts, U.S.A.

The application of X-ray diffraction methods to the study of glass structures started some 40 years ago (Randall, Rooksby and Cooper, 1930). Two years later, Zachariasen's important random network model for glass structures was presented (Zachariasen, 1932). Largely as a result of these two papers, an intensive X-ray diffraction study of glass structures began. In this early period, the experimental methods were crude, and the theory used in the interpretation involved serious approximations. It was not possible to do much more than to obtain rough values for interatomic distances and number of surrounding atoms for the first few neighbours, and to show that the diffraction pattern could be interpreted on the basis of a random network model. The methods were too rough to obtain any of the fine details in the structure, and after a few years the X-ray diffraction studies were largely discontinued.

In recent years the interest has revived, partly as a result of new and challenging problems to be studied, and partly as a result of important improvements in both theory and experimentation. To have good resolution in the final distribution curves, it is necessary to have accurate measurements of the intensity of unmodified scattering out to very large values of sin (θ/λ). But at these large angles of scattering and for elements of low atomic number, the unmodified intensity is buried under a Compton modified intensity which may be five or six times as strong. This was a major experimental limitation in all of the early measurements.

To obtain an accurate measurement of the unmodified intensity, it is necessary to eliminate the Compton scattering. A satisfactory scheme for doing this is now available in the method of fluorescence excitation (Warren and Mavel, 1965). Monochromatic $\mathbf{Rh}K_\alpha$ radiation from a Rh target tube fitted with a crystal monochromator falls upon the glass sample producing unmodified and modified scattering. The total scattering passes through the receiving slit and falls upon a sheet of Mo at 45° to the scattering direction. The fluorescence MoK radiation which is produced is recorded by a counter placed close to the Mo sheet. Since $\lambda(\mathrm{Rh}K_\alpha) = 0{\cdot}615$ Å and λ(MoK abs. edge) $= 0{\cdot}620$ Å, the unmodified scattering is ideally suited to excite strongly the MoK fluorescence radiation, whereas most of the Compton scattering is of wavelength longer than 0·620 Å, and hence not capable of exciting fluorescence radiation. With this arrangement, the major part of the counter reading is due to the unmodified scattering, and the Compton component has been largely eliminated.

Most of the early X-ray studies were based upon a Fourier inversion which involved the approximation that the different atomic scattering factors could be set proportional to one another. By the use of 'pair functions', a concept first introduced in the Norwegian school of Finbak, a completely rigorous and exact treatment is now available (Warren, 1969). It is convenient to introduce the term unit of composition (u.c.), where for vitreous silica, the obvious unit would be 1 silicon and 2 oxygens. With $k = 4\pi \sin(\theta/\lambda)$, a sharpening factor $g^2(k)$, and a convergence factor $\exp[-\alpha^2 k^2]$, the pair function for a pair of atoms i and j at separation r_{ij} is given by

$$P_{ij}(r) = \int_0^{k_m} \frac{f_i f_j}{g^2(k)} \mathrm{e}^{-\alpha^2 k^2} \sin k\, r_{ij} \sin kr\, \mathrm{d}k \tag{1}$$

In terms of I_{eu}/N the intensity of unmodified scattering per unit of composition, we define an intensity function $i(k)$ by

$$i(k) = \frac{I_{\mathrm{eu}}/N - \sum_{\mathrm{uc}} f_j^2}{g^2(k)} \tag{2}$$

The pair function distribution curve is then given by

$$\sum_{\mathrm{uc}} \sum_{i} \frac{N_{ij}}{r_{ij}} P_{ij}(r) = 2\pi^2 r\, \rho_{\mathrm{e}} \sum_{\mathrm{uc}} Z_j + \int_0^{k_m} ki(k) \mathrm{e}^{-\alpha^2 k^2} \sin rk\, \mathrm{d}k \tag{3}$$

The right hand side of equation (3) is readily evaluated for a series of values of r from known quantities such as ρ_{e} the average electron density in the sample, and the experimental intensity function $i(k)$. When plotted as a function of r, the abscissae values of the peaks give directly the interatomic distances in the glass. The left hand side of equation (3) represents the computed pair function distribution curve, where r_{ij} is the distance between an atom j and the surrounding atoms in shell i, N_{ij} is the number of surrounding atoms in shell i, and $P_{ij}(r)$ is the pair function for atoms i and j evaluated from equation (1). The number of surrounding atoms is obtained by finding the values of N_{ij} necessary to make the computed left hand side of equation (3) agree with the experimentally obtained right hand side.

As a first example, we shall summarize the results of a recent determination of the structure of vitreous silica (Mozzi and Warren, 1969). For the details, reference should be made to the original publication. The sample was optical quality fused quartz under the trade name Amersil, with a density of 2·20 gm cm^{-3}. The intensity curve was obtained by combining measurements made with crystal monochromated CuK_α ($\lambda = 1{\cdot}542$ Å) and with RhK_α ($\lambda = 0{\cdot}6147$ Å) by the method of fluorescence excitation.

In crystalline silicates, each silicon atom is tetrahedrally surrounded by 4 oxygen atoms with a Si–O distance of about 1·62 Å, and an O–O distance of about 2·65 Å. With this information, the first peak on the pair function distribution curve occurring at $r = 1{\cdot}62$ Å is readily identified as a Si–O peak. Similarly the second peak at $r = 2{\cdot}65$ Å is identified as an O–O peak. We start with a model in which each silicon is tetrahedrally bonded to 4 oxygens, and each oxygen is bonded to 2 silicons. The first peak is due to oxygen neighbours of a silicon and silicon neighbours of an oxygen, and in the unit of composition there are 1 silicon and 2 oxygen atoms. For the first peak

$$\sum_{uc}\sum_{i} N_{ij}/r_{ij} = (1 \times 4 + 2 \times 2)/1{\cdot}62 = 8/1{\cdot}62$$

If the computed Si–O pair function $P_{ij}(r)$ is multiplied by 8/1·62 and plotted on the first peak of the distribution curve, it is found that the areas of the two peaks are closely the same, but that the measured peak is a little lower and broader than the computed peak. This indicates that there is a small variation in the Si–O distance of about $\pm 0{\cdot}05$ Å around the mean value 1·62 Å. Part of this variation is of course due to temperature vibration. The second peak at $r = 2{\cdot}65$ Å is an O–O contribution, and here the comparison of peak widths indicates a variation of the order of $\pm$ 0·10 Å about the mean separation 2·65 Å. This variation may be due partly to temperature vibration, and partly to a small variation in the O–Si–O bond angle around the tetrahedral value.

The third peak occurring at $r = 3{\cdot}12$ Å is a Si–Si contribution. The fact that $3{\cdot}12 < 2 \times 1{\cdot}62$ indicates that the average Si–O–Si bond angle is significantly less than 180°. In addition the third peak is very much broader than the Si–Si pair function, indicating a wide variation in this angle. The matching of computed and experimental curves for the Si–Si contribution leads to a distribution function for the Si–O–Si bond angle which goes all the way from 120° to 180° with a maximum at 144°. The fourth peak is a silicon-second oxygen contribution and the fifth peak is a superposition of silicon-second silicon and oxygen-second oxygen contributions. To obtain satisfactory agreement for these peaks, it was necessary to assume that there is a randomness in orientation about the Si–O bond direction.

The interpretation leads to a random network structure in which each silicon is tetrahedrally surrounded by 4 oxygens. Each oxygen is bonded to 2 silicons, and the Si–O–Si bond angle varies from 120° to 180° with a maximum at 144°. There is essentially random orientation about the SiO bond direction. The wide variation in the Si–O–Si bond angle clearly distinguishes vitreous silica from the crystalline forms of SiO_2.

As a second example, we shall consider the recent determination of the structure of vitreous B_2O_3 (Mozzi and Warren, 1970). Following the methods already described for vitreous SiO_2, a pair function distribution curve was obtained. In crystalline borates, the boron atom is sometimes found in triangular coordination with an average distance B–O = 1·37 Å, and sometimes in tetrahedral coordination with an average distance B–O = 1·48 Å. The first peak on the distribution curve occurs at r = 1·37 Å· and this clearly indicates that in vitreous B_2O_3 the boron atom is predominantly in triangular coordination. The second peak at r = 2·40 Å is an O–O contribution. Both the positions and areas of the first two peaks indicate that in vitreous B_2O_3 each boron is triangularly surrounded by 3 oxygens. The simplest model which is in agreement with the first two peaks is a random network consisting of BO_3 groups linked by shared oxygens, with random orientation about the B–O bond directions. The calculated distribution curve for this model gives good agreement for the first two peaks, but complete disagreement beyond the second peak. It is quite incapable of explaining the sharp peaks and dips which extend out to about r = 6 Å, and a larger unit of structure in the glass is clearly indicated.

It has been suggested that boroxol groups exist in vitreous B_2O_3 (Krogh-Moe, 1965, 1969). The boroxol group B_3O_6 is a hexagonal ring of 3 borons and 3 oxygens with 3 corner oxygens outside of the ring so that each of the borons is triangularly bonded to 3 oxygens. For our next model of vitreous B_2O_3, we take a random network of boroxol groups which are linked by a sharing of the corner oxygens, with randomness in orientation about the B–O bond directions at the shared oxygens. The computed distribution curve for this model is in very satisfactory agreement with the experimental curve, each peak and dip in the curve being closely reproduced. No other model gave even approximate agreement, and we conclude that in vitreous B_2O_3 we have a random network of boroxol groups linked by a sharing of the corner oxygens, with randomness in orientation about the B–O bond directions at the corner oxygens. Since crystalline B_2O_3 does not contain boroxol groups, it is evident that in forming crystalline B_2O_3 from the glass it is necessary to break the boroxol rings and rejoin the BO_3 groups. This should be a difficult process and it seems very likely that this is at least part of the explanation for the difficulty in crystallizing vitreous B_2O_3.

With the new X-ray methods, it is possible to obtain much of the fine detail in the glass, and hence to obtain a fairly comprehensive picture of the glass structure. It will be well worthwhile to repeat all of the earlier X-ray studies of glass samples.

REFERENCES

Krogh-Moe, J. (1965). *Physics. Chem. Glasses*, **6**, 46.
Krogh-Moe, J. (1969). *J. non-cryst. Solids*, **1**, 269.
Mozzi, R. L. and B. E. Warren (1969). *J. appl. Cryst.*, **2**, 164.
Mozzi, R. L. and B. E. Warren (1970). *J. appl. Cryst.*, **3**, 251.
Randall, J. T., H. P. Rooksby, and B. S. Cooper (1930). *Z. Kristallogr.*, **75**, 196.
Warren, B. E. and G. Marvel (1965). *Rev. scient. Instrum.*, **36**, 196.
Warren, B. E. (1969). *X-Ray Diffraction.* Addison–Wesley, Reading, Mass., U.S.A. pp. 135–142.
Zachariasen, W. H. (1932). *J. Am. chem. Soc.*, **54**, 3841.

29

The Structon Theory: Stabilities of Different Structon Types

M. L. HUGGINS

Arcadia Institute for Scientific Research, Northridge Lane, Woodside, California, U.S.A.

This paper is an amplification of part of a paper presented at the recent Bragg Symposium (Huggins, 1970b). It is concerned with the relative stabilities of different types of close neighbour arrangements (structons; see Huggins, 1968a, b) and of combinations of these types. Consideration will be limited to structon types known to be present in anhydrous borates.

Although some knowledge of close-neighbour arrangements has been obtained from glasses, especially by X-ray diffraction and nuclear magnetic resonance techniques, much more has come from X-ray studies of the structures of crystals. In a recent paper (Huggins, 1971a) I have collected, summarized, and discussed these results. That paper forms the basis for much of what I shall present now.

One must of course be cautious about transferring knowledge of crystal structures to the structures of glasses, especially with regard to arrangements of non-closest-neighbour atoms, as these arrangements in crystals are often determined by the requirement for 3-dimensional repetition of a small unit structure motif, and since certain combinations of structon types, that may be present in the glass or in the liquid from which the crystal was grown, may not be capable of combining to give a crystalline arrangement having stable interatomic distances and bond angles. Nevertheless, careful study of the structon types in stable crystals can lead to some definite and useful conclusions with regard to the structures of glasses of the same or similar compositions.

STRUCTONS IN BORATE CRYSTALS

Table 1 lists the types of boron structons that have been found in boric oxide and the anhydrous borates of known structure. Note that no tetrahedral boron atoms (B_4) have non-bridging oxygen (O′) neighbours. This is not unreasonable, considering the fact that both B_4 and O′ have formal charges of −1. (Of course, the charge on a non-bridging oxygen is partly—but only partly—compensated by the opposite charges on the cations that are invariably close to it.)

Table 1. Boron structon types

$\langle B_3(3\cdot O'')\rangle$	$\langle B_3(O', 2\cdot O'')\rangle$	$\langle B_3(2\cdot O', O'')\rangle$	$\langle B_3(3\cdot O')\rangle$
$\langle B_4(4\cdot O'')\rangle$			
	$\langle B_4(O'', 3\cdot O''')\rangle$	$\langle B_4(2\cdot O'', 2\cdot O''')\rangle$	$\langle B_4(3\cdot O'', O''')\rangle$

Table 2 lists the oxygen structons (neglecting their cation neighbours) found in the crystalline borates. The occurrence of oxygens (O‴) with three boron neighbours is noteworthy, since oxygens with three silicon neighbours are unknown in the silicates. In the latter case the formal charge on a tribonded oxygen would be +1 and none of its neighbours would have a negative formal charge. A tribonded oxygen in a borate, also with a formal charge of +1, can be stabilized by having three B_4 neighbours, each with a formal charge of −1. I know of no O‴ oxygens with any B_3 neighbours.

Table 2. Oxygen structon types, neglecting cations

$\langle O'(B_3)\rangle$		
$\langle O''(2B_3)\rangle$	$\langle O''(B_3, B_4)\rangle$	$\langle O''(2B_4)\rangle$
$\langle O'''(3B_4)\rangle$		

The cations in the crystals, and presumably in the glasses, are invariably surrounded by oxygens. The numbers of oxygen neighbours, observed in the crystals, are shown in Table 3. The energy of each cation-to-oxygen contact is much smaller than that of a boron-to-oxygen contact. The total energy of interaction between a cation and its oxygen neighbours depends on both the number of neighbours and their effective negative charges. The numbers and types of geometrically available oxygens vary from compound to compound, hence the variation in coordination number is to be expected.

Since the distribution of contacts of a cation to the different types of oxygen depends on the effective negative charges on these oxygens, the distribution of cations around these oxygens also depends on these charges. A thorough discussion would necessarily be lengthy and will not be attempted here. It is pertinent, however, to compare the relative effective charges (Huggins, 1970a) on the oxygens of the different types, computed, without considering

Table 3. Coordination numbers of cations

Li	(4), 5				
Na	7, 8	Mg	6	Al	6
K	7	Ca	8, (10), (12)	Ga	6
Rb	7	Sr	8, 9, (10), (12)	Cr	6
Cs	10	Ba	(6), 9, 10	Ln	6
Ag	7, 8	Cd	7		
		Co	6		

The numbers in parentheses have been observed only in unstable phases.

the cation neighbours, as is done in applying Pauling's (1939) electrostatic valence rule. These charges, given in Table 4, should be in the same order as the attraction energies of the oxygens for a given cation. A study of the crystal structure data shows general agreement with this assumption, and agreement can also be expected in glass systems.

Table 4. Effective negative charges

$\langle O'(B_4)\rangle$	−1·25
$\langle O'(B_3)\rangle$	−1·00
$\langle O''(2B_4)\rangle$	−0·50
$\langle O''(B_3, B_4)\rangle$	−0·25
$\langle O''(2B_3)\rangle$	0
$\langle O'''(3B_4)\rangle$	+0·25

The number of cation neighbours for any oxygen of a given type is usually, in stable crystals, the number that will make the effective charges (calculated without neglect of these cations) either zero or close to zero. The number is limited, however, by cation-to-cation repulsions. Non-bridging oxygens (O′) have been found with as many as five monovalent cation, or three bivalent cation, neighbours. Bridging oxygens (O″) have not been found with more than two (monovalent or bivalent) cation neighbours. All observed tribonded oxygens (O‴) have no cation neighbours.

STRUCTONS IN BORATE GLASSES

Alkali oxides and alkaline earth oxides are essentially assemblages of ions. The initial change occurring when such an oxide is added to boric oxide can be expressed (again temporarily neglecting the cations) by the simple equation:

$$B_3 + O^{-2} \rightarrow B_4 + O' \qquad (1)$$

If the amount of added oxide is relatively small, X-ray diffraction and nuclear magnetic resonance studies of glasses and X-ray studies of crystals all show that the change represented by the following equation takes place nearly quantitatively:

$$B_3 + O' \rightarrow B_4 + O'' \qquad (2)$$

Apparently, tribonded boron and non-bridging oxygen atoms do not coexist stably (at least in large amounts, at the temperatures at which the boron–oxygen bonding becomes frozen), unless cations are present in sufficient quantity to stabilize the non-bridging oxygens. The boron–oxygen bonds are partially polar and coulombic energies between these atoms favour B_4 over B_3.

Of course, the changes indicated by equations 1 and 2 must be accompanied by other changes. In my opinion, the best way to consider these changes is to determine, if possible, the set of structon types that is (or seems likely to be) present at a given composition or in a given composition range and then to see how the relative numbers of the different types vary with the composition. If two or more sets seem possible, then one can consider the factors affecting their relative stability—such factors as those I have already discussed. Eventually, perhaps, we can assign relative energies or free energies to the sets being compared.

It is convenient, for this purpose and others, to use a formula for a borate composition, or for the polyanion it contains, that has been normalized to give the numbers of each kind of atom per atom of boron. (These numbers are, in general, not integers.) The overall composition is then conveniently represented by the (negative) charge on the anion, designated as ρ. The numbers of the different structon types, per boron atom, are then functions of ρ. If the number of structon types in a set is the minimum number required to give the overall composition (i.e. if the set is a 'minimum set'), these functions are readily determined.

For example, the only minimum sets (neglecting cations) having all oxygens of the O'' type and containing no structon types other than those found in borate crystals (Tables 1 and 2) are those of Class A, given in Table 5 with their composition ranges. The number of each type, per boron atom in the material, is given before its structon formula.

The stable structures of crystalline boric oxide, all crystalline borates of

Table 5. Minimum structon sets, neglecting cations

	ρ	Structons
Class A. All O atoms bridging between borons.		
1	0	$1\langle B_3(3\cdot O'')\rangle$ $(\frac{3}{2})\langle O''(2B_3)\rangle$
2	0 to 0·4286	$(1-\rho)\langle B_3(3\cdot O'')\rangle$ $\rho\langle B_4(4\cdot O'')\rangle$ $(\frac{3}{2}-7\rho/2)\langle O''(2B_3)\rangle$ $4\rho\langle O''(B_3, B_4)\rangle$
3	0·4286	$(\frac{4}{7})\langle B_3(3\cdot O'')\rangle$ $(\frac{3}{7})\langle B_4(4\cdot O'')\rangle$ $(\frac{12}{7})\langle O''(B_3, B_4)\rangle$
4	0·4286 to 1	$(1-\rho)\langle B_3(3\cdot O'')\rangle$ $\rho\langle B_4(4\cdot O'')\rangle$ $(3-3\rho)\langle O''(B_3, B_4)\rangle$ $(7\rho/2-\frac{3}{2})\langle O''(2B_4)\rangle$
5	1	$1\langle B_4(4\cdot O'')\rangle$ $2\langle O''(2B_4)\rangle$
Class B. All B atoms tribonded to oxygens.		
1	As above	
6	0 to 1	$(1-\rho)\langle B_3(3\cdot O'')\rangle$ $\rho\langle B_3(O', 2\cdot 0'')\rangle$ $(\frac{3}{2}-\rho/2)\langle O''(2B_3)\rangle$ $\rho\langle O'(B_3)\rangle$
7	1	$1\langle B_3(O', 2\cdot 0'')\rangle$ $1\langle O''(2B_3)\rangle$ $1\langle O'(B_3)\rangle$
8	1 to 2	$(2-\rho)\langle B_3(O', 2\cdot O'')\rangle$ $(\rho-1)\langle B_3(2\cdot O', O'')\rangle$ $(\frac{3}{2}-\rho/2)\langle O''(2B_3)\rangle$ $\rho\langle O'(B_3)\rangle$
9	2	$1\langle B_3(2\cdot O', O'')\rangle$ $\frac{1}{2}\langle O''(2B_3)\rangle$ $2\langle O'(B_3)\rangle$
10	2 to 3	$(3-\rho)\langle B_3(2\cdot O', O'')\rangle$ $(\rho-2)\langle B_3(3\cdot O')\rangle$ $(\frac{3}{2}-\rho/2)\langle O''(2B_3)\rangle$ $\rho\langle O'(B_3)\rangle$
11	3	$1\langle B_3(3\cdot O'')\rangle$ $3\langle O'(B_3)\rangle$
Class C. Containing some oxygens tribonded to borons.		
12	0	$1\langle B_4(O'', 3\cdot O''')\rangle$ $(\frac{1}{2})\langle O''(2B_4)\rangle$ $1\langle O'''(3B_4)\rangle$
13	0 to 0·3333	$3\rho\langle B_4(2\cdot O'', 2\cdot O''')\rangle$ $(1-3\rho)\langle B_4(O'', 3\cdot O''')\rangle$ $(\frac{1}{2}+3\rho/2)\langle O''(2B_4)\rangle$ $(1-\rho)\langle O'''(3B_4)\rangle$
14	0·3333	$1\langle B_4(2\cdot O'', 2\cdot O''')\rangle$ $1\langle O''(2B_4)\rangle$ $(\frac{2}{3})\langle O'''(3B_4)\rangle$
15	0·3333 to 0·6667	$(3\rho-1)\langle B_4(3\cdot O'', O''')\rangle$ $(2-3\rho)\langle B_4(2\cdot O'', 2\cdot O''')\rangle$ $(\frac{1}{2}-3\rho/2)\langle O''(2B_4)\rangle$ $(1-\rho)\langle O'''(3B_4)\rangle$
16	0·6667	$1\langle B_4(3\cdot O'', O''')\rangle$ $(\frac{3}{2})\langle O''(2B_4)\rangle$ $(\frac{1}{3})\langle O'''(3B_4)\rangle$
17	0·6667 to 1	$(3\rho-2)\langle B_4(4\cdot O'')\rangle$ $(3-3\rho)\langle B_4(3\cdot O'', O''')\rangle$ $(\frac{1}{2}+3\rho/2)\langle O''(2B_4)\rangle$ $(1-\rho)\langle O'''(3B_4)\rangle$
5	As above	
Class D. Containing both 3- and 4-coordinated borons and both bridging and non-bridging oxygens.		
18	0·4286 to 1	$(1-\rho)\langle B_3(3\cdot O'')\rangle$ $(7\rho/6-\frac{1}{2})\langle B_3(O', 2\,O'')\rangle$ $(\frac{1}{2}-\rho/6)\langle B_4(4\cdot O'')\rangle$ $(7\rho/6-\frac{1}{2})\langle O'(B_3)\rangle$ $(2-2\rho/3)\langle O''(B_3, B_4)\rangle$
19	1	$(\frac{2}{3})\langle B_3(O', 2\cdot O'')\rangle$ $(\frac{1}{3})\langle B_4(4\cdot O'')\rangle$ $(\frac{2}{3})\langle O'(B_3)\rangle$ $(\frac{4}{3})\langle O''(B_3, B_4)\rangle$
20	1 to 1·800	$(\frac{3}{2}-5\rho/6)\langle B_3(O', 2\cdot O'')\rangle$ $(\rho-1)\langle B_3(2\cdot O', O'')\rangle$ $(\frac{1}{2}-\rho/6)\langle B_4(4\cdot O'')\rangle$ $(7\rho/6-\frac{1}{2})\langle O'(B_3)\rangle$ $(2-2\rho/3)\langle O''(B_3, B_4)\rangle$
21	1·800	$(\frac{4}{5})\langle B_3(2\cdot O', O'')\rangle$ $(\frac{1}{5})\langle B_4(4\cdot O'')\rangle$ $(\frac{8}{5})\langle O'(B_3)\rangle$ $(\frac{4}{5})\langle O''(B_3, B_4)\rangle$
22	1·800 to 3	$(2-2\rho/3)\langle B_3(2\cdot O', O'')\rangle$ $(5\rho/6-\frac{3}{2})\langle B_3(3\cdot O')\rangle$ $(\frac{1}{2}-\rho/6)\langle B_4(4\cdot O'')\rangle$ $(7\rho/6-\frac{1}{2})\langle O'(B_3)\rangle$ $(2-2\rho/3)\langle O''(B_3, B_4)\rangle$

known structure having ρ less than one-half, and one crystalline compound (CdB_4O_7) having ρ equal to one-half have structures conforming to the minimum sets of Class A. In the glassy state, B_2O_3 has a structure (Warren, 1970) conforming to Set 1 of this class, and nuclear magnetic resonance evidence (e.g., Bray and O'Keefe, 1963; Svanson, Forslind and Krogh-Moe, 1962) shows that the proportion of boron of the B_4 type in borates of monovalent and bivalent cations, for concentrations below $\rho = \frac{3}{7}$, is in agreement with the assumption that Set 2 is predominant in this range.

BORATES CONTAINING BOTH NON-BRIDGING (O′) AND BRIDGING (O″) OXYGENS

At higher concentrations the relative amount of B_4 departs more and more from that required by a Class A structure. To fit the data one must assume that the reaction of Equation 2 has taken place to some extent, but not completely. It is reasonable to assume that, at a temperature just above that at which the boron–oxygen bond structure becomes fixed, an equilibrium exists between a minimum set of structons of the Class A type and another minimum set containing non-bridging oxygen atoms.

Typical of the data mentioned are those for lithium borates, shown in Figure 1. The full line corresponds to Set 2. The upper dashed line represents Set 4; the lower dashed line represents Set 18 (for $\rho < 1$) and other similar sets, containing both O′ and O″, at higher concentrations.

For the concentration range in which ρ is between $\frac{3}{7}$ and 1, the only minimum set possible, containing non-bridging oxygens and only structon types found in the crystalline borates, is that of Set 18:

$$\text{Set 18} \quad \rho = 0{\cdot}4286 \text{ to } 1 \quad (1 - \rho)\langle B_3(3\cdot O'')\rangle \quad (\tfrac{7}{6}\rho - \tfrac{1}{2})\langle B_3(O', 2\cdot O'')\rangle$$

$$(\tfrac{1}{2} - (\rho/6))\langle B_4(4\cdot O'')\rangle \quad (\tfrac{7}{6}\rho - \tfrac{1}{2})\langle O'(B_3)\rangle \quad (2 - \tfrac{2}{3}\rho)\langle O''(B_3, B_4)\rangle$$

As will be shown in another paper (Huggins, 1971b), the assumption of equilibrium between Sets 4 and 18 in this composition range leads to agreement with the NMR results, within the apparent probable error of the latter.

Favouring Set 18 over Set 4 is the fact that O′ oxygens can hold more cation neighbours and hold each more strongly than O″ oxygens. Also, with only Set 4, some of the O″ oxygens would necessarily have three cation neighbours or else some cations would have coordination numbers less than four. Both of these structural alternatives would decrease the stability.

At higher values of ρ, equilibria between other pairs of minimum sets appear to be required by the meagre amount of NMR data available. These equilibria will also be discussed in the paper just referred to.

The foregoing treatment of glass structure in this composition range is fundamentally similar to an earlier treatment by Huggins and Abe (1957).

It differs from that treatment primarily with regard to the compositions at which significant amounts of non-bridging oxygens appear. Evidence then available from expansion coefficients seemed to indicate that the break occurs at considerably lower proportions of alkali oxide than is indicated by the more recent NMR measurements.

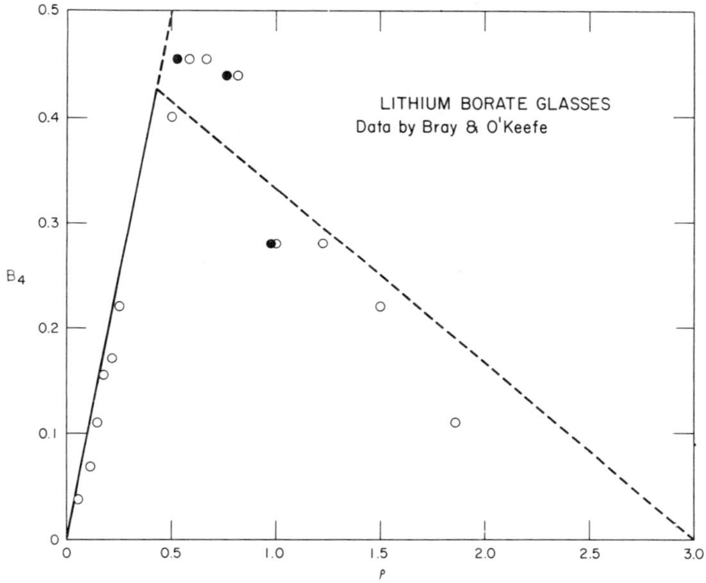

Figure 1. Dependence of the fraction of boron that is 4-coordinated (B_4) on the concentration (ρ) for lithium borate glasses. The circles are experimental values from NMR measurements by Bray and O'Keefe (1963). The compositions plotted are batch compositions for the open circles, analytical compositions for the filled circles. The lines are theoretical; see text.

My pseudo-equilibrium treatment of borate glass structure in the transition region is similar to that of Beekenkamp (1965), but more specific with regard to the types (and amounts of each type) of structons.

BORATES CONTAINING TRIBONDED (O‴) OXYGENS

The existence of two known crystalline borates (SrB_4O_7 and PbB_4O_7) with structures containing tribonded oxygen (O‴) and one (CdB_4O_7) in which all the oxygen is of the O″ type, all with $\rho = \frac{1}{2}$, suggests that a corresponding equilibrium between structon sets may exist in glasses, at high temperatures, at and near this composition.

Minimum structon sets, containing O″ and O‴, with no structon types other than those found in crystalline borates, are listed in Table 5, under Class C. SrB_4O_7 and PbB_4O_7 have set 15: CdB_4O_7 has Set 4 of Class A.

Let us consider the relative stabilities of Class A and Class C structures. To make the comparison more definite, let us consider only borates of bivalent cations, having a coordination number of 9, as in SrB_4O_7 and PbB_4O_7.

The problem may be divided into two parts: (1) a comparison of the relative stabilities of the cation–oxygen interactions, and (2) a comparison of the relative energies of the boron–oxygen interactions.

With regard to the first of these parts, it seems reasonable to assume that, other things being equal, the most stable structure is that having the lowest summation of the magnitudes of the effective charges on the oxygen atoms. The effective charge magnitudes measure, roughly, the departures from a structure in which the charges at each atom are balanced, which structure would presumably have maximum stability as regards coulombic interactions. The effective charge sums for Class A and Class C structon sets, for a series of ρ values, are given in Table 6, with the differences between these sums. From

Table 6. Effective charge sums and fractions of boron atoms that are tetrahedrally coordinated, for structon sets A and C

	Class A Sets		Class C Sets		Differences	
ρ	$\Sigma\|c_0\|$	B_4	$\Sigma\|c_0\|$	B_4	$\Delta\Sigma\|c_0\|$	ΔB_4
0	0	0	0·500	1	0·500	1·000
0·111	0·022	0·111	0·444	1	0·422	0·889
0·200	0·039	0·200	0·400	1	0·361	0·800
0·250	0·049	0·250	0·375	1	0·326	0·750
0·429	0·083	0·429	0·286	1	0·203	0·571
0·500	(0·083) 0·097	0·500	0·250	1	0·153 (0·167)	0·500
0·667	(0·111) 0·130	0·667	0·167	1	0·037 (0·056)	0·333
0·750	(0·125) 0·146	0·750	(0·167)	1	(0·042)	0·250
0·800	(0·133) 0·156	0·800	(0·167)	1	(0·033)	0·200
0·900	(0·144)	0·900	(0·167)	1	(0·023)	0·100

The effective charge sums in parentheses are for structon sets that include ⟨O″ (2B_4, 3M)⟩. Otherwise, the only oxygen structon types are those in borate crystals of known structure.

these differences one can conclude that, if no other factors were involved, the Class A sets should be more stable than the Class C sets over the whole range of composition for which these sets are valid, but that the difference in stability decreases as ρ increases, reaching a moderately low value at $\rho = \frac{1}{2}$ and a very low value at $\rho = \frac{2}{3}$. When ρ is larger than $\frac{2}{3}$, Class C structures are only possible if the structon type O″ (2B_4, 3M) is allowable. This type has not been observed in any stable crystal phase; it may require too much crowding of electropositive atoms around the oxygen in question.

The boron atoms in Class C sets all have four close oxygen neighbours. In the Class A sets some have three and some have four oxygen neighbours. If these atoms were all ions, if the oxygen-to-oxygen repulsions were negligible, and if the B–O distances were all equal, the ratio of the boron-to-oxygen attraction energy for each B_4 boron to that for each B_3 boron would be $\frac{4}{3}$. Of course these provisos are not valid, but, on the average, the magnitude of the attraction energy between a boron and the oxygens surrounding it would be expected to be somewhat greater for a B_4 boron than for a B_3 boron. Other things being equal, then, the larger the fraction of tetrahedral boron atoms in the structure, the greater should be the stability. This fraction is given in the columns headed B_4 in Table 6. From the differences between these values (in the last column), one sees that the boron–oxygen interaction energy contributions favour the Class C structures at all values of ρ, but decreasingly as ρ increases. At $\rho = \frac{1}{2}$, for which the Sr and Pb borates are known to have a Class C structure, B_4 is $\frac{1}{2}$. This is apparently enough to overcome the small positive value of $\Sigma|c_0|$, favouring a Class A structure. (It is possible, of course, that the C structure occurs in the crystal because it is favoured by geometric factors.) It seems likely that a glass or glass melt, having ρ at or near $\frac{1}{2}$, contains considerable amounts of tribonded oxygen.

Up to now, the effect of the density of packing of the atoms in the structure has been neglected. Increasing the number of close neighbours around an atom obviously increases the density of packing. If other structural factors are the same, Class C structures, with all borons of the B_4 type and with some tribonded oxygens, should thus have higher densities than Class A structures of the same overall compositions. Changing the density changes the interaction energy between non-closest-neighbour atoms. Except in very unusual cases, the van der Waals interactions between non-closest-neighbours are all attractions, increasing as the interatomic distances decrease, hence increasing as the density increases. The coulombic energies between non-closest-neighbours, which would be expected to be fairly large in the borates, also increase in magnitude as the distances decrease. These are both attractions and repulsions, and it seems impossible to evaluate the net effect in any simple way.

Neglecting a possible opposing effect of these non-closest-neighbour coulombic interactions, increasing pressure should favour Class C structures. This is in line with the fact that the stable structures for crystalline and glassy boric oxide at normal pressures are of the Class A type, whereas a high pressure crystalline phase has a Class C structure.

REFERENCES

Beekenkamp, P. (1965). *Proc. Int. Conf. Physics of Non-crystalline Solids*, Delft, July 1964, 512.

Bray, P. J., and J. G. O'Keefe (1963). *Physics Chem. Glasses*, **4**, 37.

Huggins, M. L. (1968a). *Macromolecules*, **1**, 184.

Huggins, M. L. (1968b). *Inorg. Chem.*, **7**, 2108.

Huggins, M. L. (1970a). *Acta Crystallogr.*, **B26**, 219.

Huggins, M. L. (1970b). Paper presented at *Bragg Symposium*, London, 1970.

Huggins, M. L. (1971a). *Inorg. Chem.*, **10**, 791.

Huggins, M. L. (1971b). Paper in preparation.

Huggins, M. L., and T. Abe (1957). *J. Am. Ceram. Soc.*, **40**, 287.

Pauling, L. (1939). *The Nature of the Chemical Bond.* Cornell University Press, Ithaca, N.Y.

Svanson, S.-E., E. Forslind, and J. Krogh-Moe (1962). *J. phys. Chem.*, **66**, 174.

Warren, B. E. (1970). Lecture at *Bragg Symposium*, London, 1970.

30

Statistical Mechanics of Polymerized Materials

S. F. EDWARDS

Schuster Laboratory, University of Manchester, Great Britain

The kind of problem arising in the statistical mechanics of non crystalline solids is seen at once if one tries to calculate the free energy by means of the Gibbs formula

$$e^{-F/\kappa T} = \int e^{-H/\kappa T}\, d\,(\text{phase space}) \tag{1}$$

At a sufficiently low temperature a glass will result in systems which cannot crystallize (a situation readily available in some polymerized systems whose irregularity precludes crystallization), and even if the system can in principle crystallize, there is good reason to believe that a calculation which in some way excludes the crystallized state in the integration of (1) will give correct correlation functions. Thus if one calculates the two-body correlation function

$$a(\boldsymbol{r}, \boldsymbol{r}') = \sum_{\alpha,\beta} \langle \delta(\boldsymbol{r} - \boldsymbol{R}_\alpha)\, \delta(\boldsymbol{r}' - \boldsymbol{R}_\beta) \rangle \tag{2}$$

$$= \int e^{(F-H)/\kappa T} \sum_{\alpha,\beta} \delta(\boldsymbol{r} - \boldsymbol{R}_\alpha)\, \delta(\boldsymbol{r}' - \boldsymbol{R}_\beta) \tag{3}$$

(where F is defined by (1)), one can expect an answer in some accord with experiment, indeed under some conditions (to be discussed) it will be a correct formula. But F itself from (1) is a function solely of the volume, for there is no macroscopic order in a glass. With $F = F(V, T)$ one has a pressure $P = -(\partial F/\partial V)$ and a compressibility $(\partial^2 F/\partial V^2)$, but no shear ie.

$$\frac{\partial^2 F}{\partial \lambda_1^2} = \frac{\partial^2 F}{\partial \lambda_2^2} \tag{4}$$

where λ_1 is an extension in the x direction and λ_2 in the y direction. Clearly F is the free energy of a liquid and as such has $F = F(V)$ which is characteristic of a system which cannot sustain shear. But glasses can and do sustain shear, so (1) is inadequate. For a crystalline solid one could also get this paradox if it is assumed that the integral over phase space includes the crystal axes, for then one has

$$F(\text{of }(1)) = \int F\,(\text{customary, containing axes})\,d\Omega$$

where Ω are angular variables. It is easy to put this right by keeping the axes in, but what is their analogue in a glass? For a normal, say SiO_2 glass, the answer is that any particular Si will have a certain coordination, and its neighbours also, and these coordinations and correlations between them are permanent. Whereas a liquid would change these coordinations, and flow, the glass will not, or at least only very slowly—or it would not be a glass. Let us consider the totality of information required to specify an environment in a glass, in our SiO_2 glass a set of coordinations and *their* coordinations starting with one atom out to a point where it is no longer important. This is still a microscopic region. Call the specification of the region, and all the other regions 'm'. Call the free energy with this specification F_m. Then taking a macroscopic body, there will be a probability of an 'm' occurring, p_m. At this point it is valid to take an ensemble average and to introduce

$$\mathscr{F} = \sum_m p_m F_m \tag{5}$$

which will be called the effective free energy. (The entropy of forming the ensemble plays no interesting role and will be ignored.) Now suppose some change is made, like shearing the system. Label the new variables by a tilde

$$\tilde{\mathscr{F}} = \sum p_m \tilde{F}_m\,. \tag{6}$$

Notice that p_m does *not* change. To give a specific example, suppose a glass is formed in an idealized experiment in which a fluid at temperature T_0 is instantaneously dropped to T. Since it was in thermal equilibrium at T_0, the probability of finding 'm' is

$$p_m = \exp\left((F_0 - F_m)/\kappa T_0\right) \tag{7}$$

where

$$e^{-F_m/\kappa T_0} = \int_m e^{-H/\kappa T_0} \tag{8}$$

the integral meaning integrate over all configurations with the specification 'm', and

$$e^{-F_0/\kappa T_0} = \sum_m e^{-F_m/\kappa T_0} \tag{9}$$

Thus

$$\tilde{\mathscr{F}} = \sum_m e^{(F_0 - F_m)/\kappa T_0}\,\tilde{F}_m \tag{10}$$

where

$$e^{-\tilde{F}_m} = \tilde{\int}_m e^{-H/\kappa T} \tag{11}$$

so that

$$\tilde{\mathscr{F}} = \tilde{\mathscr{F}}(T_0, T). \tag{12}$$

To illustrate the formula, suppose one had an extremely viscous liquid in which the creep response to shear is so slow that it can be ignored, but nevertheless exists so that the system can have a quasi-thermal equilibrium (i.e. without crystallizing). Then one can set $T_0 = T$. This simplifies the algebra a bit, and one finds that, for an infinitesimal change

$$\Delta\tilde{\mathscr{F}} = \Delta\tilde{F} \tag{13}$$

where

$$e^{-\tilde{F}/\kappa T} = \tilde{\int} e^{-H/\kappa T}. \tag{14}$$

i.e. the pressure from our effective free energy is identical to that leading from (1). But the second derivative

$$\Delta^2\tilde{\mathscr{F}} = \Delta^2 F + (1/\kappa T)\sum_m e^{(F-F_m)/\kappa T}\{\Delta\tilde{F}_m\Delta\tilde{F}_m - (\Delta\tilde{F})(\Delta\tilde{F})\}. \tag{15}$$

This last term gives the shear modulus, and indeed is Hooke's Law.

It is possible to contract these basic formulae in a useful way. Suppose one considers the initial state at its initial temperature in its initial volume, and further considers n other states at the final temperature and final volume, (sides $\lambda_1, \lambda_2, \lambda_3$ of the original, permitting an arbitrary strain). These n other systems have the same 'm' constraint as the initial. Then consider a partition function defined jointly for the $n + 1$ systems

$$Z = \sum_m \int \exp(-H/\kappa T_0)\tilde{\int}\dots\tilde{\int}\exp(-\sum H^{(\alpha)}/\kappa T) \tag{16}$$

$$= Z(T_0, T, \rho_0; \lambda_1, \lambda_2, \lambda_3; n). \tag{17}$$

At this point one makes a mathematical assumption which, *a posteriori*, is justifiable, that n can be treated as a continuous variable. At $n = 0$ one clearly has

$$Z_0 = \exp(-F^{(0)}/\kappa T_0) \tag{18}$$

and from their definitions

$$Z = \sum_m \exp[(F^{(0)} - F_m^{(0)})/\kappa T_0 - n\tilde{F}_m/\kappa T]. \tag{19}$$

Hence expanding in n

$$Z = Z_0(1 - n\tilde{F}_{\text{effective}}/\kappa T) \tag{20}$$

or

$$Z = Z_0 \exp(-n\tilde{F}_{\text{eff}}/\kappa T - O(n)]. \tag{21}$$

A method of calculating is then to consider the solid in $3n + 3$ dimensions, 3 of initial dimension and $3n$ in the strained dimensions, with temperatures T_0 and T respectively, but the same constraints in all dimensions. A solution of this problem will provide the required free energy.

At this point the reader will probably be asking 'but what *is* "m"', for the coordination numbers which were quoted above are only rather loosely defined in say an SiO_2 glass. The point of the present paper is that although an 'm' specification must exist for amorphous SiO_2, it *is* difficult to specify, but in polymerized materials it is much easier, so that, in spite of the formidable look of (16) that formula can be made to yield the elastic constants of a rubber.

The specification will be now given.

THE SPECIFICATION OF POLYMERIZED MATERIAL

Polymer aggregates can exist in at least three forms, in solution, in a rubber and in a glass. These forms roughly parallel the gas, liquid and solid phases of normal substances, for in solution the polymer molecules can move more or less freely, in a glass they are tightly constrained by their neighbours and the basic motion is vibration, whilst in the rubber, although segments of a chain move freely, the chain as a whole moves very slowly. If a vulcanization takes place, permanently linking pairs of chains, the rubber loses its ability to creep and becomes a true solid. However even without vulcanization the entanglement of the chains is such that, although on a long enough time scale the material will creep, on a shorter time scale there are well defined elastic moduli. A sketch of rubber will then look like the figure below,

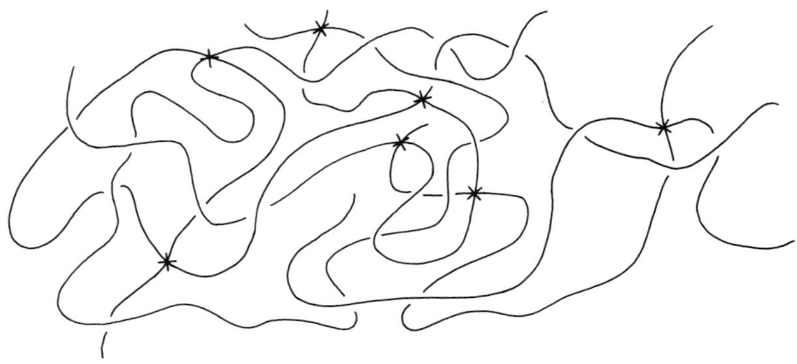

or rather a much more random, kinked set of curves. The crosses denote permanent inter chain links and it will be noted that one finds situations which can be symbolically abbreviated to (a) and (b). The shaded regions mean areas of other entanglements and cross links; the two situations (a) and (b) are different. It is assumed that the shaded areas keep them so, i.e. (a) and (b) are analogues of the topological difference between (c) and (d).

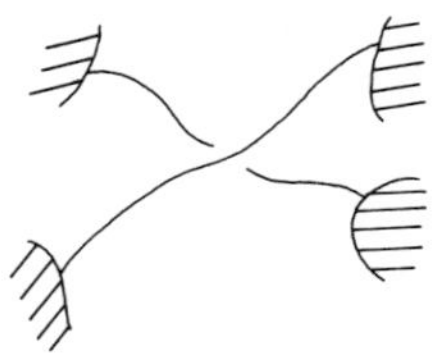

(a)

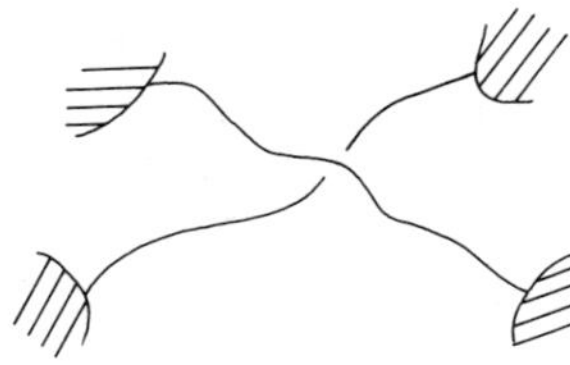

(b)

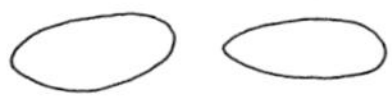

(c)

(d)

There is of course no difference between (e) and (f). for the chain, having a free end, can make its way from (e) to (f). The simplest assumption to start a theory is to take the cross links sufficiently numerous and the chains sufficiently long, that situation (e), (f) is of negligible occurrence compared

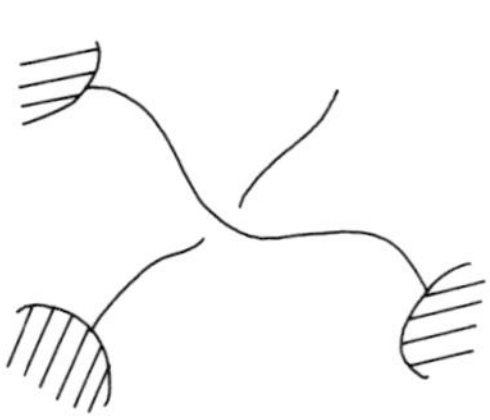

(e)

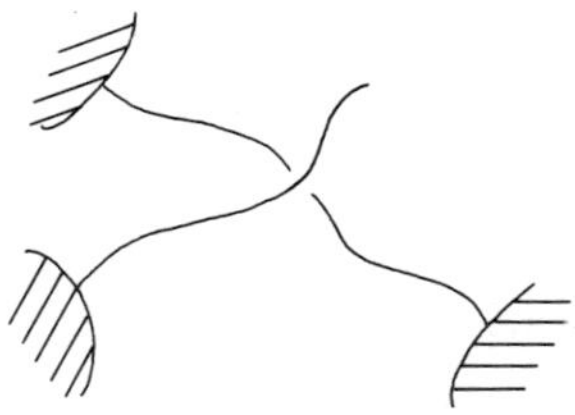

(f)

with (a), (b), and this will be done here. (The literature has a considerable discussion of chain ends, they are ignored here to make the analysis as simple as possible.) This does not mean however that situations (g) and (h) will be counted as a constraint, the calculation to be presented will properly discount such configurations.

(g)

(h)

Now one needs to translate these pictures into mathematics.

Firstly the cross links which are easy. The constraint is that if the S_ath monomer of polymer a meets the S_bth monomer of polymer b with a cross link, it is the S_ath monomer, etc., for all the other polymers. Suppose it is possible to describe the polymer by a function $\boldsymbol{R}(s)$, being the arc length along the polymer, $\boldsymbol{R}(s)$ its position in space. Label the initial configuration $\boldsymbol{R}^{(0)}(s)$, n subsequent positions $\boldsymbol{R}^{(\alpha)}(s)$. (Since the end effects are to be ignored, it will simplify the analysis if the polymers are considered part of one enormous chain, and a label saying which polymer molecule is under discussion will be suppressed unless it specifically matters.) Then a cross-link at S_1, S_2 imposes a constraint

$$\delta(\boldsymbol{R}^{(0)}(s_1) - \boldsymbol{R}^{(0)}(s_2))\,\delta(\boldsymbol{R}^{(1)}(s_1) - \boldsymbol{R}^{(2)}(s_2))\,\delta(\ldots\ldots \tag{22}$$

There will be N such terms if there are N cross links, and they appear at random along the chain, so in (22) the cross link 'm' is represented by

$$\Big[\int \prod_{q=0}^{q=n} \delta(\boldsymbol{R}^{(q)}(s) - \boldsymbol{R}^{(q)}(s'))\,\mathrm{d}s\,\mathrm{d}s'\Big]^N \tag{23}$$

That the cross-links are at random is of course an assumption, and a vulcanizing process extending over some period of time, or cross linking in a solution, will almost certainly cause bunching of the links (for one link will increase the chain density around it and enhance the chance of a second settling near). However the chemistry can be arranged so that the links are truly random (as in a forthcoming paper by Allen, Burgess, Edwards and Walsh), or alternatively one can argue that randomness is surely the simplest state so one should solve this problem at first.

Now for the entanglements: these can be described by a series of invariants, which are the generalization of angle. In two dimensions one can distinguish between the configurations by evaluating the angle (*not* modulo 2π). If one

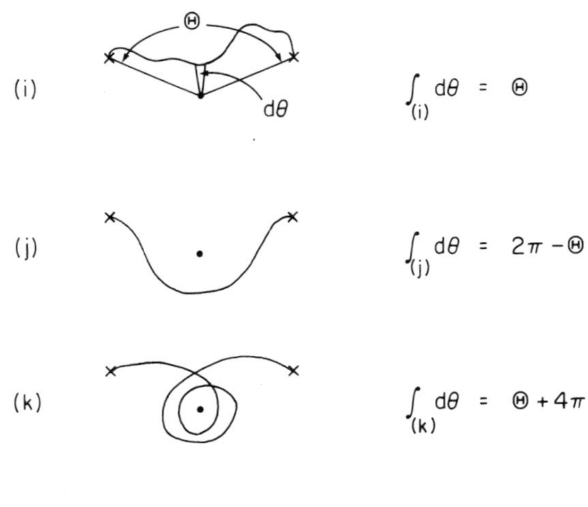

imagines an infinite wire at right angles to the plane carrying a current, these integrals represent the work done in transporting a magnetic pole from one end point to the other. The work done is readily extended from the work of Ampere and Gauss to three dimensions, and the basic invariant between two curves $\boldsymbol{A}(s)$ and $\boldsymbol{B}(s')$is

$$\iint \frac{(\mathrm{d}\boldsymbol{A}(s) \times \mathrm{d}\boldsymbol{B}(s')) \cdot (\boldsymbol{A}(s) - \boldsymbol{B}(s'))}{|\boldsymbol{A}(s) - \boldsymbol{B}(s')|^3} \tag{24}$$

For example if the two rings in (c), (d) are labelled $R_1(s_1)$ and $R_2(s_2)$, then

$$\begin{aligned} I_2 &= \oint\oint \frac{(\mathrm{d}\boldsymbol{R}_1 \times \mathrm{d}\boldsymbol{R}_2) \cdot (\boldsymbol{R}_1 - \boldsymbol{R}_2)}{|\boldsymbol{R}_1 - \boldsymbol{R}_2|^3} \\ &\left.\begin{aligned} &= 0 \text{ for (c)} \\ &= 4\pi \text{ for (d).} \end{aligned}\right\} \end{aligned} \tag{25}$$

Similarly a self integral I_{11} will have values differing by integral multiples of 4π according to its knottedness. However notice an unpleasant feature: $\boldsymbol{R}(s)$ is a vector and points along the direction of increasing s. The value of I_{12}, I_{11} depends on the senses chosen, whereas physically they are not significant i.e. if $I_{12} = \vartheta$, then by altering the sense of $\boldsymbol{R}_1(s)$ as in Figure (*l*) one can make

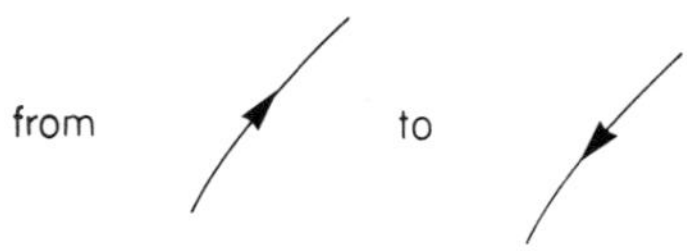

(l)

$I_{12} = -\vartheta$. Altering (2) also, brings it back. There seems no escape from this redundancy which appears essential for a full classification. The simple invariant I_{12} does not give a complete specification because, to look at a single chain, there are positive and negative knots, and these cannot be cancelled even though $I_{11} = 0$. A clear example appears with three rings, which have

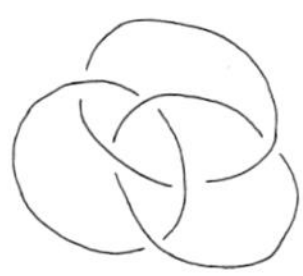

(m)

$I_{12} = I_{23} = I_{31} = 0$ whereas they clearly cannot be separated. There must be (and indeed are) higher invariants I_{123} etc. which specify such situations. These higher invariants are so involved that, in this paper attention will be confined to I_{12}, I_{11} for that already is difficult and its analysis incomplete, but it does handle the vital situation (a), (b). One must just hope at this point that the configuration (n) which has the same I_{12} as (a), has a low probability.

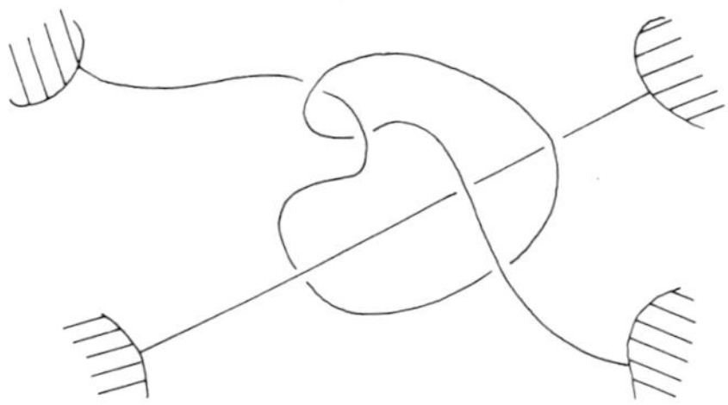

(n)

The constraint from the entanglements is then

$$\delta(I^{(0)}, \tilde{I}^{(1)})\,\delta(I^{(0)}, \tilde{I}^{(2)})\ldots\delta(I^{(0)}, \tilde{I}^{(n)}) \tag{27}$$

where $\delta(I^{(0)}, \tilde{I}^{(1)})$ is unity when $I^{(0)} = \tilde{I}^{(1)}$ and zero otherwise. It is a Kronecker rather than Dirac δ, because the values of I differ by integers, rather than continuously.

Putting it all together then, one has

$$Z = \int\tilde{\int}\cdots\tilde{\int}\exp\left(-\frac{H^{(0)}}{\kappa T_0} - \sum\frac{H^{(\alpha)}}{\kappa T}\right)\prod_{\alpha=1}^{n}\delta(I^{(0)}, I^{(\alpha)}) \times\left[\int\prod_{\beta=0}^{n}\delta(R^{(\beta)}(s) - R^{(\beta)}(s'))\mathrm{d}s\mathrm{d}s'\right]^N. \tag{28}$$

It is convenient to express the N power in terms of a chemical potential, by the identity

$$\frac{1}{2\pi}\oint\frac{\mathrm{d}\mu}{\mu^{N+1}}N!\exp x\mu = x^N \tag{29}$$

so finally

$$Z = \frac{1}{2\pi}\oint\frac{N!\,\mathrm{d}\mu}{\mu^{N+1}}\int_{(0)}\tilde{\int}_{(1)}\cdots\tilde{\int}_{(n)}\exp\left[-H^{(0)}/\kappa T_0 - \sum H^{(\alpha)}/\kappa T - \mu\int\int\mathrm{d}s\mathrm{d}s'\prod_{\beta=0}^{n}\delta(R^{\beta}(s) - R^{\beta}(s'))\right]\prod_{\alpha=1}^{n}\delta(I^{(0)}, I^{(\alpha)}). \tag{30}$$

This integral will be evaluated in the next section.

EVALUATION: Phantom Chains

To evaluate one must firstly decide on the mathematical specification of the polymer. Since the analysis is to be kept as simple as possible the first model is that of a freely hinged random flight molecule, which can further be simplified into a diffusion equation description. Thus if a chain is made up of monomers of length l, over large distances, the number of configurations between $\boldsymbol{r}$, $\boldsymbol{r}'$, using s, s' for the arc length (i.e. no. of monomers $\times$ l) is $\mathcal{N}_0 G(\boldsymbol{r}, \boldsymbol{r}'; s, s')$ where $\mathcal{N}_0$ is the total number of configurations without the restriction that $\boldsymbol{R}(s) = \boldsymbol{r}$, $\boldsymbol{R}(s') = \boldsymbol{r}'$, and

$$\left(\frac{\partial}{\partial s} + \frac{l}{6}\nabla^2\right) G(\boldsymbol{r}, \boldsymbol{r}'; s, s') = \delta(r - r')\,\delta(s - s') \tag{31}$$

This differential equation is obtained by approximating the true random configuration (o) by smooth curves (p). It introduces much simplification to

(o)

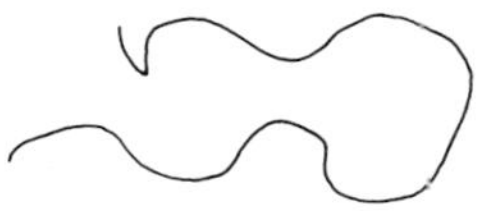
(p)

do this, but is not quite adequate. A better approximation is to stiffen the smooth curves by introducing a penalty against curvature. If the derivative $\partial \boldsymbol{R}/\partial s$ is denoted by $\boldsymbol{v}/s)$ and a weight factor $\exp(-1/2q \int (\partial \boldsymbol{v}/\partial s)^2 \, ds)$ introduced G becomes modified to

$$\left(\frac{\partial}{\partial s} + v\frac{\partial}{\partial r} + \frac{\eta}{2}\frac{\partial^2}{\partial v^2} + \frac{3}{2l}v^2\right) G = \delta(\boldsymbol{r} - \boldsymbol{r}')\,\delta\,(\boldsymbol{v} - \boldsymbol{v}')\,\delta(s - s') \tag{32}$$

Similarly a further extension is to introduce torsion, but this will not be pursued.

To handle the effect of constraints it is worth giving a method of deriving (31) and (32) directly. Suppose that the number of ways a link, having one end at $\boldsymbol{R}_q$, will arrive at $\boldsymbol{R}_{q+1}$ is $g(\boldsymbol{R}_q, \boldsymbol{R}_{q+1})$. Then the numbers of configurations of M links starting at R_1 ending at R_M is

$$\mathcal{N}_0(G(R_1, R_M) = \int g(\boldsymbol{R}_1, \boldsymbol{R}_2)\, g(\boldsymbol{R}_2, \boldsymbol{R}_3) \ldots\ldots g(\boldsymbol{R}_{M-1}, \boldsymbol{R}_M)\, d\boldsymbol{R}_2 \ldots\ldots d\boldsymbol{R}_{M-1} \tag{33}$$

In the simplest case g is $g(\boldsymbol{R}_1 - \boldsymbol{R}_2)$, so that if the Fourier transform of $g(\boldsymbol{R})$ $\gamma(\boldsymbol{k})$ and of G is $G(\boldsymbol{k})$, since the integrals are all convolutions

$$\mathcal{N}_0\, G(\boldsymbol{k}) = \gamma(\boldsymbol{k})^M \tag{34}$$

$$= [\gamma(\boldsymbol{O})]^M \exp(M \log[\gamma/\boldsymbol{k})/\gamma(\boldsymbol{O})]) \tag{35}$$

$$= [\gamma(\boldsymbol{O})]^M \exp(M\gamma''(\mathrm{O})\boldsymbol{k}^2 + \mathrm{O}(k^4)\, M + \ldots) \tag{36}$$

For $M \gg 1$, the terms in k^4 are negligible, and will be dropped. For a rod link of length l, γ'' comes out to be $-l^2/6$, so that

$$\mathcal{N}_0\, G(\boldsymbol{R}_1\, \boldsymbol{R}_M) = [\gamma(\mathrm{O})]^M \left(\frac{2\pi l^2 M}{3}\right)^{-\frac{3}{2}} \exp\left\{-\frac{3}{2}\frac{(\boldsymbol{R}_1 - \boldsymbol{R}_M)^2}{Ml^2}\right\} \tag{37}$$

or if one writes

$$R_1 = r, R_M = r', s - s' = Ml, [\gamma(\mathrm{O})]^M = \mathcal{N}_0 \tag{38}$$

(for this is what it is)

$$G(\boldsymbol{r}, \boldsymbol{r}'; s - s') = \left(\frac{3}{2\pi l^2 M}\right)^{\frac{3}{2}} \exp\left\{-\frac{3}{2l}\frac{(\boldsymbol{r} - \boldsymbol{r}')^2}{|s - s'|}\right\}. \tag{39}$$

$$(s > s')$$

Notice that G can be broken at an arbitrary number of points, not just at the $\boldsymbol{R}_q$:

$$G(\boldsymbol{r}, \boldsymbol{r}'; s, s') = \int \ldots \int G(\boldsymbol{r}, \boldsymbol{r}_1; s, s_1)\, G(\boldsymbol{r}_1\, \boldsymbol{r}_2; s_1, s_2) \ldots\ldots G(\boldsymbol{r}_n, \boldsymbol{r}'; s_n, s')\, \mathrm{d}^3 r_1 \ldots\ldots \mathrm{d}^3 r_n, \tag{40}$$

this being the characteristic of a Markov chain. The set of points $r, r_1 \ldots r_n, r'$ can now be made numerous so that they define a curve in space $\boldsymbol{R}(s)$ and the integral $\mathrm{d}^3 r_1\, \mathrm{d}^3 r_2 \ldots \to \mathrm{d}$ (curve), which one can write $\delta\boldsymbol{R}(s)$ say. Over a small interval

$$\left.\begin{aligned} s_i - s_{i+1} &\to \mathrm{d}s \\ \boldsymbol{R}_i - \boldsymbol{R}_{i+1} &\to \dot{\boldsymbol{R}}(s_i)\, \mathrm{d}s \end{aligned}\right\} \tag{41}$$

and

$$G(\boldsymbol{r}_i, \boldsymbol{r}_{i+1}; s_i, s_{i+1}) \to \left(\frac{3}{2l\pi \mathrm{d}s}\right)^{\frac{3}{2}} \exp\left\{-\frac{3}{2l}\dot{\boldsymbol{R}}^2(s_i)\, \mathrm{d}s\right\} \tag{42}$$

so that

$$G(\boldsymbol{r}, \boldsymbol{r}'; s, s') = \left\{\prod_s \left(\frac{3}{2l\pi \mathrm{d}s}\right)^{\frac{3}{4}}\right\} \int \delta R \exp\left\{-\frac{3}{2l}\int_{s'}^{s} \dot{\boldsymbol{R}}^2(\sigma)\, \mathrm{d}\sigma\right\}. \tag{43}$$

This formula is of course now very close to the original (33), but it may be objected that whereas one started with a finite number of links, one now has a continuum. The point is this: by approximating a polymer chain to a continuous curve one is able to use a simple diffusion equation to describe it. This produces an enormous simplification in the analysis which will become clear as we go on. At small distances, along the chain the Wiener integral (43) cannot correctly represent the function, but one can for example include a curvature term to simulate this to some extent. For consider (32) which also has the convolution property

$$G(\boldsymbol{r}, \boldsymbol{r}'; \boldsymbol{v}, \boldsymbol{v}'; s, s') = \int \ldots \int \Pi \, \mathrm{d}\boldsymbol{r} \, \Pi \, \mathrm{d}\boldsymbol{v} \, G \cdot G \cdot G \ldots\ldots G \tag{44}$$

Over a small $(s_1 - s_2)$ separation one can write the solution of (32) as

$$G(\boldsymbol{r}_1, \boldsymbol{r}_2; \boldsymbol{v}_1, \boldsymbol{v}_2; s_1, s_2) = \int \mathrm{d}^3k_1 \, \mathrm{d}^3q_1 \exp\{\mathrm{i}\boldsymbol{k}_1(\boldsymbol{r}_1 - \boldsymbol{r}_2) + \mathrm{i}\boldsymbol{q}_1(\boldsymbol{v}_1 - \boldsymbol{v}_2) - (s_1 - s_2)(\mathrm{i}\boldsymbol{v}_1 \cdot \boldsymbol{k}_1 - \boldsymbol{q}_1^2 + \boldsymbol{v}_1^2)\} \tag{45}$$

where it is immaterial whether one writes $\boldsymbol{v}_1 \cdot \boldsymbol{k}_1$ or $\frac{1}{2}(\boldsymbol{v}_1 + \boldsymbol{v}_2) \cdot \boldsymbol{k}_1$ etc., since the limit $s_1 = s_2 + \mathrm{d}s$ will be taken shortly. This leads from (32) to

$$G = \int \prod \mathrm{d}^3k_i \, \mathrm{d}^3q_i \, \mathrm{d}^3r_i \, \mathrm{d}^3v_i \exp\{\mathrm{i} \sum \boldsymbol{k}_\alpha(\boldsymbol{r}_\alpha - \boldsymbol{r}_{\alpha+1}) + \mathrm{i} \sum \boldsymbol{q}_\alpha(\boldsymbol{v}_\alpha - \boldsymbol{v}_{\alpha+1}) - \sum (s_\alpha - s_{\alpha+1}) [\mathrm{i}\boldsymbol{v}_\alpha \boldsymbol{k}_\alpha - (\eta \boldsymbol{q}_\alpha^2/2) + (3\boldsymbol{v}_\alpha^2/2l)]\} \tag{46}$$

$$= \int \delta\boldsymbol{k}(s) \, \delta\boldsymbol{q}(s) \, \mathrm{d}\boldsymbol{R}(s) \, \delta\boldsymbol{v}(s) \exp\left\{\mathrm{i}\int \boldsymbol{k} \cdot \dot{\boldsymbol{R}} \, \mathrm{d}\sigma - \mathrm{i}\int \boldsymbol{k} \cdot \boldsymbol{V} \, \mathrm{d}\sigma + \mathrm{i}\int \boldsymbol{q}\dot{\boldsymbol{V}} \, \mathrm{d}\sigma \frac{\eta}{2} \int \boldsymbol{q}^2 \, \mathrm{d}\sigma + \frac{3}{2l} \int \boldsymbol{V}^2 \, \mathrm{d}\sigma \right\} \tag{47}$$

$$= \mathcal{N} \int \prod \delta(\boldsymbol{V} - \dot{\boldsymbol{R}}) \exp\left[-\frac{1}{2\eta} \int \dot{V}^2 \, \mathrm{d}\sigma - \frac{3}{2l} \int V^2 \, \mathrm{d}\sigma \right] \delta R \delta V \tag{48}$$

$$= \mathcal{N} \int \delta\boldsymbol{R} \exp\left[-\frac{1}{2\eta} \int \ddot{\boldsymbol{R}}^2 \, \mathrm{d}\sigma - \frac{3}{2l} \int \dot{\boldsymbol{R}}^2 \, \mathrm{d}\sigma \right]. \tag{49}$$

(The symbol $\mathcal{N}$ is here used for any normalization that arises.) This now is the same form as (43) with an additional term. This is just the form a Boltzmann factor for flexural energy has, but alternately the modification can be regarded as a device for using differential equations to smaller distances than are justified by using (43). One can readily calculate

$$\left.\begin{aligned} \langle(\boldsymbol{R}(s) - \boldsymbol{R}(s'))^2\rangle &= l(s - s') \text{ using the } G \text{ of } (31) \\ &= l(s - s') - a(1 - \mathrm{e}^{-l(s-s')/a} \text{ using the } G \text{ of } (32) \\ &= l(s - s') \text{ for large } s - s' \\ &= \frac{l}{a}(s - s')^2 \text{ for small } s - s'. \end{aligned}\right\} \tag{50}$$

This method of analysis can clearly be extended to any differential equation of the type

$$\left(\frac{\partial}{\partial s} + \mathscr{D}\left(x_i, \frac{\partial}{\partial x_i}, \frac{\partial^2}{\partial x_i^2} \ldots\right)\right) G = \prod_k \delta(X_k - X'_k) \, \delta(s - s') \tag{51}$$

but can only be sensibly reduced to

$$\int F([\boldsymbol{R}])\,\delta R$$

in special circumstances, otherwise it will be

$$\int \prod \delta q_k\,\delta X_k\,\mathscr{F}([q],[X])(X_0 \equiv R). \tag{52}$$

At this point one may formulate the constraints explicitly. The chain structure contributes exp $(-\ U([R]))$ say, where

$$U = \frac{3}{2l}\int \dot{\boldsymbol{R}}^2\,\mathrm{d}\sigma$$

or something more complicated, so one has

$$Z = \mathscr{N}\oint_0 \frac{(N+1)!\,\mathrm{d}\mu}{\mu^{N+1}}\int \prod_0^n \delta \boldsymbol{R}^{(\beta)}$$

$$\times \exp\left[-\sum_0^n U([\boldsymbol{R}^{(0)}]) - H([\boldsymbol{R}^{(0)}]/\kappa T_0 - \sum_0^n H(\boldsymbol{R}^{(\alpha)})/\kappa T\right.$$

$$\left. - \mu \prod_{\alpha=0}^{n}\int\int \delta(\boldsymbol{R}^{(\alpha)}(s_1) - \boldsymbol{R}^{(\alpha)}(s_2))\,\mathrm{d}s_1\,\mathrm{d}s_2\right]\prod_{\gamma=1}^{n}\delta(I([\boldsymbol{R}^0]) - I([\boldsymbol{R}^{(\gamma)}]))\ . \tag{53}$$

To make a start on this, the crudest approximation is that of phantom cross linked chains. In these one drops the energy and the invariants altogether. In particular circumstances it may be possible to justify the first approximation in that at some temperature the attractions and repulsions cancel, or that the effect of the forces is to make the system incompressible but that is all. The topology however can never be ignored in a real situation, hence Flory's description of them as phantoms. If one takes the phantom chains to satisfy the diffusion equation, they are described by

$$\oint \frac{\mathrm{d}\mu(N+1)!}{\mu^{N+1}}\int \delta\boldsymbol{R}^{(0)}\,\delta\boldsymbol{R}^{(n)} \exp\left\{-\frac{3}{2l}\int_0^L \sum_\alpha \dot{\boldsymbol{R}}^{(\alpha)2}\,\mathrm{d}s\right.$$

$$\left. + \mu \prod_\alpha \int_0^L\int_0^L \mathrm{d}s_1\,\mathrm{d}s_2\,\delta(\boldsymbol{R}^{(\alpha)}(s_1) - \boldsymbol{R}^{(\alpha)}(s_2)\right\}. \tag{54}$$

This expression on its own however cannot describe the essential property that the material fills a box uniformly. That property is a consequence of the intermolecular forces, and without building it in, in some new way, the integral will represent quite a different system. It is interesting to see what this system is. It is assumed that there are going to be many cross links, several per individual polymer. This means that one may treat the whole of the material as belonging to one enormous chain, simply for simplicity. What happens if a very long chain (without cross links) is put in a box? The boundary condition of a box means that the chain does not get outside, and with a second order differential equation like (31), this means that the solution vanishes at the walls of the box. Thus the solution (39) is modified by walls to the bilinear combination of sines

$$G = \left(\frac{2}{V^{\frac{1}{3}}}\right)^3 \sum_{n_1 n_2 n_3} \sin\frac{\pi n_1 x}{V^{\frac{1}{3}}} \sin\frac{\pi n_1 x'}{V^{\frac{1}{3}}} \sin\frac{\pi n_2 y}{V^{\frac{1}{3}}} \sin\frac{\pi n_2 y'}{V^{\frac{1}{3}}}$$

$$\sin\frac{\pi n_3 z}{V^{\frac{1}{3}}} \sin\frac{\pi n_3 z'}{V^{\frac{1}{3}}} \exp\left[-\frac{\pi^2 l^2}{6V^{\frac{2}{3}}}(n_1^2 + n_2^2 + n_3^2)(s - s')\right] \tag{55}$$

and if $s - s'$ is long enough, to the lowest eigenfunctions

$$n_1 = n_2 = n_3 = 1.$$

Now although this kind of solution may be meaningful for a segment of chain boxed in locally by its neighbours, it cannot apply generally because the density should be uniform. One can get a uniform density by applying instead *cyclic* boundary conditions, i.e. the cosines are also introduced and

$$\left(\frac{3}{V^{\frac{1}{3}}}\right)^3 \sum \left(\frac{\delta_{n_1 . 0}}{2}\left(\frac{L}{l}\right)^{\frac{1}{3}} + \exp(2\pi i n_1 x/V^{\frac{1}{3}} - 2\pi i n_1 x'/V^{\frac{1}{3}})\right)\begin{pmatrix}\text{do.}\\ y\end{pmatrix}\begin{pmatrix}\text{do.}\\ z\end{pmatrix}$$
$$\times \exp\left[-\frac{2}{3}\frac{\pi^2 l^2}{V^{\frac{2}{3}}}(n_1^2 + n_2^2 + n_3^2)(s - s')\right] \tag{56}$$

which is very much the same as

$$\frac{L}{lV} + \left(\frac{32\pi l(s - s')}{3}\right)^{\frac{3}{2}} \exp\left(-\frac{3}{2l}\frac{(\boldsymbol{r} - \boldsymbol{r}')^2}{(s - s')}\right) \tag{57}$$

i.e. when a chain goes through $\boldsymbol{r}'$ and $\boldsymbol{r}$ either they are points on the same chain, corresponding to the second term in (57), or the chain through $\boldsymbol{r}'$ leaves the box before passing through $\boldsymbol{r}$. If the cyclic condition is chosen however, the theory cannot be expected to give sensible information about the bulk modulus; but that anyway appears from the forces. These arguments also apply when the cross links are present. A free chain will shrink when cross linked. One can estimate how much by considering the $\boldsymbol{R}^{(0)}$ variable alone, and assuming that as before for large L only the lowest eigenfunction appears. If this is $\phi(\boldsymbol{r})\,e^{\sigma L}$ one has the probability of finding the chain at $\boldsymbol{r}$ to be $\phi^2(\boldsymbol{r})$ so that one can replace

$$\iint ds_1\, ds_2\, \delta(\boldsymbol{R}(s_1) - \boldsymbol{R}(s_2))$$

by

$$L\int \delta(\boldsymbol{R}/s) - \boldsymbol{r})\,\phi^2(\boldsymbol{r})\,d^3r \tag{58}$$

which gives the differential equation

$$\left(\frac{\partial}{\partial s} + \frac{l^2}{6}\nabla^2 + \mu L\phi^2(\boldsymbol{r})\right) G(\boldsymbol{r}, \boldsymbol{r}') = \delta(r - r') \tag{59}$$

dominated by

$$G \sim \phi(\boldsymbol{r})\,\phi(\boldsymbol{r}')\,\mathrm{e}^{\sigma(s-s')} \tag{60}$$

i.e.

$$\left(\sigma + \frac{l}{6}\nabla^2 + \mu L\phi^2(\boldsymbol{r})\right)\phi(\boldsymbol{r}) = 0. \tag{61}$$

The entropy per unit length is σ, and this equation represents diffusion in a potential well $\mu L\phi^2$. This is not a recognized equation but its solution is evidently a monotonically decreasing function, and if one solves it by the Rayleigh–Ritz method, ϕ is approximately given by

$$\left(\frac{\omega}{2\pi}\right)^{\frac{3}{4}} \exp\left(-\tfrac{1}{2}\omega \boldsymbol{r}^2\right) \tag{62}$$

where

$$\omega \sim \frac{N}{Ll} \qquad \text{and} \qquad \sigma \sim \frac{N}{L}. \tag{63}$$

The precise calculation is similar in structure to one to be presented below for the free energy, so this result is just quoted.

Thus the figures below represent the situations discussed. (α) is a chain with cyclic boundary conditions with its density (α'). (β) is a chain with true boundary conditions, density (β'). (γ) and (δ) represent the consequences of cross linking, which do not disturb the density (γ'), and reduce it to an absurdity of radius $\sqrt{Ll/N}$ as in (δ'), and we must clearly live with situation (γ). The discussion of case (δ) has value even when our picture of the situation is (γ), for now return to the full integral (54). Physically the cross links ensure that each of the chains $\boldsymbol{R}^{(1)} \ldots \boldsymbol{R}^{(n)}$ do not stray far from $\boldsymbol{R}^{(0)}$. Thus if one considers a centre of mass coordinate $X^{(0)}$ and a set of other relative co-ordinates, $X^{(0)}$ behaves as in picture (γ), but the others are kept at small values i.e. the *relative* coordinates behave as in picture (δ). One has to apply the cyclic space filling condition to $X^{(0)}$, but not to the rest. If the system is now deformed, the mean positions of the cross links can be expected to undergo the same transformation as the macroscopic body to some approximation, and it turns out that for gaussian chains, as James and Guth showed long ago, that this is exactly true. But for mean positions only. It is natural therefore to attempt to evaluate the integral by applying an affine transformation to the whole set of coordinates, most conveniently by first taking the c.m.

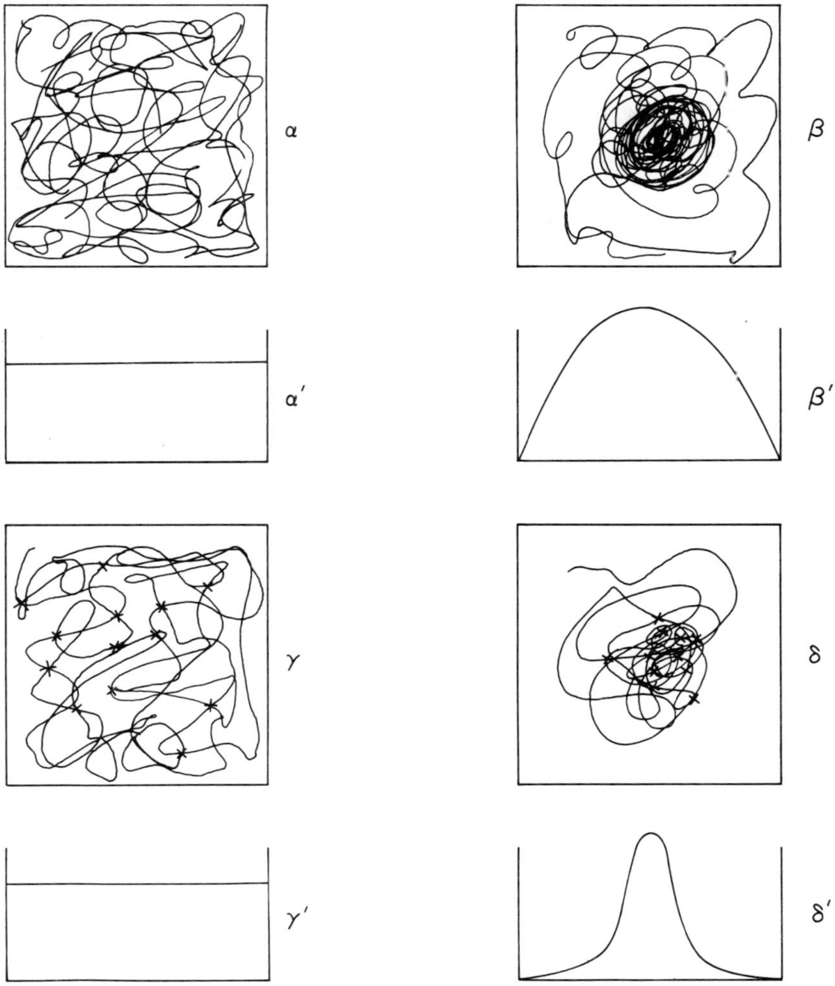

transformation for the $1 \ldots n$ coordinates, then taking a combination of their c.m. with $\boldsymbol{R}^{(0)}$. The form is (i being a cartesian index)

$$X_i^{(0)} = \frac{R_i^{(0)} + \lambda_i \sum_1^n R_i^{(\alpha)}}{\sqrt{(1 + n\lambda_i^2)}} \qquad X_i^{(1)} = \frac{R_i^{(0)} - (1/\lambda_i) \sum_1^n R^{(\alpha)}}{\sqrt{[1 + (1/n\lambda_i^2)]}}$$

$$Y_i^{(m)} = \frac{1}{\sqrt{N}} \sum_1^n R_i^{(\alpha)} \exp{(2\pi i m\alpha/n)} \tag{64}$$

Although it may look curious, the complex form for the Y's is most convenient, but not essential. This transformation now takes

$$\prod_{\alpha=0}^{n} \delta(\boldsymbol{R}^{(\alpha)}(s_1) - \boldsymbol{R}^{(\alpha)}(s_2)) \rightarrow$$

$$\prod_{-(n-1)/2}^{(n-1)/2} \delta(\boldsymbol{Y}^{(\alpha)}(s_1) - \boldsymbol{Y}^{(\alpha)}(s_2))\, \delta(\boldsymbol{X}^{(0)}(s_1) - \boldsymbol{X}^{(0)}(s_2)\, \delta(\boldsymbol{X}^{(1)}(s\ ) - \boldsymbol{X}^{(1)}(s_2) \tag{65}$$

and

$$\sum_{\alpha=0}^{n} \dot{\boldsymbol{R}}^{(\alpha)2} \rightarrow \dot{\boldsymbol{X}}^{(0)2} + \boldsymbol{X}^{(1)2} + \sum_{-(n-1)/2}^{(n-1)/2} \dot{\boldsymbol{Y}}^{2(\alpha)} \tag{66}$$

The original box, which is deformed by λ_1, λ_2, λ_3 is now replaced by the deformation of $X^{(0)}$ which (by Pythagoras in $n+1$ dimensions) is $\sqrt{(1+n\lambda_1^2)}$, $\sqrt{(1+n\lambda_2^2)}$, $\sqrt{(1+n\lambda_3^2)}$. The deformation of $X^{(1)}$ and the Y is not significant, because, as in (δ), they have an exponentially small chance of reaching their boundary limits, as will appear in explicit calculation.

The problem is now to evaluate

$$\int \frac{d\mu(N+1)!}{\mu^{N+1}} \int \exp\left\{-\frac{3}{2l}\int \dot{\boldsymbol{X}}^{(0)2} - \frac{3}{2l}\int \dot{\boldsymbol{X}}^{(1)2} - \frac{3}{2l}\sum_{\alpha}\int \dot{\boldsymbol{Y}}^{(\alpha)}\dot{\boldsymbol{Y}}^{(\alpha)*}\right.$$

$$\left. + \mu\int\!\!\int \delta(\boldsymbol{X}^{(0)}(s_1) - \boldsymbol{X}^{(0)}(s_2)\, \delta(\boldsymbol{X}^{)1(}(s_1) - \boldsymbol{X}^{(1)}(s_2)) \quad \delta(\boldsymbol{Y}^{(\alpha)}(s_1) - \boldsymbol{Y}^{(\alpha)}(s_2))\right\}$$

$$\delta\boldsymbol{X}^{(0)}\,\delta\boldsymbol{X}^{(1)}\prod \delta\boldsymbol{Y} \tag{67}$$

This is tackled by choosing an integrable model, which is taken to be

$$-\left(\omega^2\frac{l^2}{6}\int \boldsymbol{X}^{(1)2} + \omega^2\frac{l^2}{6}\sum\int \boldsymbol{Y}\boldsymbol{Y}^*\right) \text{replacing the } \mu x \text{ delta functions} \tag{68}$$

Physically this means that one is trying to simulate the constraint felt by a chain due to cross-links, by the çonstraint of a harmonic well i.e. a kind of pipe.

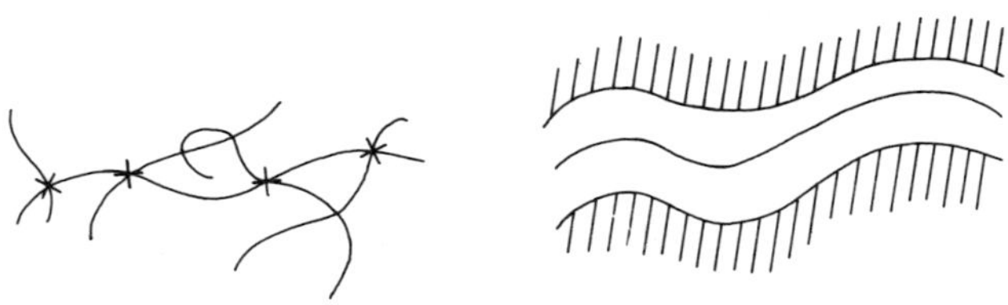

It is possible to make a variational principle by using the inequality

$$e^{Q} > 1 + Q \tag{69}$$

so if one takes for Q

$$
\begin{aligned}
Q &= \delta(\boldsymbol{X}^{(0)} - \boldsymbol{X}^{(0)})\,\delta(\boldsymbol{X}^{(1)} - \boldsymbol{X}^{(1)}) \prod \delta(\boldsymbol{Y} - \boldsymbol{Y}) \\
&\quad - \frac{\omega^2 l^2}{6} \int \dot{\boldsymbol{X}}^{(1)2} - \frac{\omega^2 l^2}{6} \sum \int \dot{\boldsymbol{Y}} \dot{\boldsymbol{Y}}^* - < \text{same} > \\
&= P - \langle P \rangle
\end{aligned} \tag{70}
$$

where

$$\langle f \rangle = \int f \mathrm{e}^{A}\, \delta(\text{all}),$$

one has

$$\exp(\mathscr{F}/\kappa T) < \int \exp(A + \langle P \rangle). \tag{71}$$

The integrals are easily worked out in the case of large L, for $X^{(1)}$, $Y^{(\alpha)}$ are governed by Hermite's equation

$$\left(\frac{\partial}{\partial s} + \frac{l}{6} \frac{\partial^2}{\partial \boldsymbol{x}^{(1)2}} + \frac{\omega^2 l}{6} \boldsymbol{x}^{(1)2} \right) G = \delta(s - s_1)\, \delta(\boldsymbol{x}^{(1)} - \boldsymbol{x}^{(1)\prime}) \tag{72}$$

which is dominated by the ground state eigenfunction

$$\left(\frac{\omega}{2\pi} \right)^{\frac{3}{4}} \exp\left(-\frac{l}{2} \omega (s - s') \right) \exp\left(-\tfrac{1}{2} \omega \boldsymbol{x}^{(1)2} \right) \tag{73}$$

whereas $X^{(0)}$ is dominated by the constant term of (70).

Hence

$$\langle \delta(\boldsymbol{Y} - \boldsymbol{Y}) \rangle = \omega^{\frac{3}{2}} / (8\pi^3)^{\frac{1}{2}}$$

and

$$\langle \delta(\boldsymbol{x}^{(0)} - \boldsymbol{x}^{(0)}) \rangle = \frac{L^2}{V} \prod_{1,2,3} (1 + n\lambda_k^2)^{-\frac{1}{2}}. \tag{74}$$

Thus

$$\langle P \rangle = -(N + 1) \log \mu + \frac{\mu L^2}{V} \left(\frac{\omega}{2\pi} \right)^{3n/2} \prod_i (1 + n\lambda_i^2)^{-\frac{1}{2}} + \frac{nl}{12} (3\omega L)$$

and the ground state eigenfunction itself, coming from

$\int \mathrm{e}^{A}$ alone gives

$$-\frac{nl}{6}(3\omega)L,$$

so that the total exponent is

$$-(N+1)\log\mu + \frac{\mu L^2}{V}\left(\frac{\omega}{2\pi}\right)^{3n/2} \prod_{i=1,2,3} (1+n\lambda_i^2)^{-\frac{1}{2}} - \frac{nl}{12}(3\omega)L. \quad (75)$$

The values of ω, μ are now determined by the variational principle, and the method of steepest descent, and

$$\omega = 6N/Ll \quad (76)$$

$$\mu = \frac{NV^2}{L^2}\left(\frac{2\pi}{\omega}\right)^{3n/2} \prod_i (1+n\lambda_i^2)^{\frac{1}{2}} \quad (77)$$

so that expanding $\mathscr{F}(n)$ one has

$$\tilde{F}/\kappa T = -3N\log\left(\frac{\omega}{2\pi}\right) + \frac{N}{2}\sum \lambda_i^2 \quad (78)$$

and the variational principle suggests that this $\tilde{F}$ is greater than the true value.

There are several different aspects to this expansion which are discussed elsewhere (Edwards 1971), but the key point is that it should be valid provided that the number of cross links exceeds a certain fraction of the number of monomers (L/l), and of course that there are a large number of polymers so that the space is filled i.e. $(Ll)^{\frac{3}{2}} \gg V$. The key claim here is that although there is nothing new in itself in the quadratic for $\tilde{F}$, it has been derived from a quite general expression (54) and this can be expanded further and approximated in different ways. One can for example extend the present treatment down to the gel point, and one can also estimate the effects of various particular configurations, which have all been included in an average way above, the average only being valid for large numbers of cross links.

Before leaving this problem it is worth comparing the above result with the simpler arguments in the literature. Suppose one characterized the problem by looking at one piece of polymer between two cross-link end points. Suppose initially these are distant $\boldsymbol{R}$, and suppose the probability of finding this configuration in the first place is the normal gaussian chain form

$$\left(\frac{3}{2Ll\pi}\right)^{\frac{3}{2}} e^{-(3/2Ll)\boldsymbol{R}^2},$$

the number of configurations being

$$N_0\left(\frac{3}{2Ll\pi}\right)^{\frac{3}{2}} e^{-(3/2Ll)\boldsymbol{R}^2}$$

where N_0 is the total number of configurations of the chain. If we take '$\boldsymbol{R}$' as our characteristic of the constraint and suppose the chain ends affinely deformed, the number of configurations alters to

$$N_0 \left(\frac{3}{2Ll\pi}\right)^{\frac{3}{2}} \exp\left(-\frac{3}{2Ll\pi}(\lambda_1^2 x^2 + \lambda_2^2 y^2 + \lambda_3^2 z^2)\right) \tag{79}$$

and

$$\tilde{F}/\kappa T = \sum \int d^3R \exp\left\{-\left(\frac{3}{2Ll}\right)\boldsymbol{R}^2\right\}\left(\frac{3}{2Ll\pi}\right)^{\frac{3}{2}}$$
$$\times \log\left(\frac{3}{2Ll\pi}\right)^{\frac{3}{2}} \exp\left\{-\frac{3}{2Ll\pi}(\lambda_1^2 x^2 + \lambda_2^2 y^2 + \lambda_3^2 z^2)\right\} \tag{80}$$

$$= \frac{\mathscr{M}}{2}\sum \lambda_i^2 \tag{81}$$

where $\mathscr{M}$ is number of chains (i.e. $2N$). This is a result due to James and Guth (without invoking their discussion of the cure of the rubber). But it can be studied another way. If one considers only entropy, the general formula is

$$\tilde{S}_{\text{eff}} = \kappa T \sum \frac{\Omega_m}{\Omega} \log \tilde{\Omega}_m \tag{82}$$

where Ω_m is number of initial configurations, $\tilde{\Omega}_m$ the final and $\Omega = \sum_m \Omega_m$. This can be rewritten using probabilities

$$p_m = \frac{\Omega_m}{\Omega}, \tilde{p}_m = \frac{\tilde{\Omega}_m}{\tilde{\Omega}}$$

as

$$\tilde{S}_{\text{eff}} = \kappa T \sum p_m \log \tilde{\Omega}\tilde{p}_m = \kappa T \log \tilde{\Omega} + \kappa T \sum p_m \log \tilde{p}_m$$
$$\tilde{S}_{\text{eff}} = \tilde{S} + \kappa T \sum_m p_m \log \tilde{p}_m \tag{83}$$

where $\tilde{S}$ is the normal entropy of the system, without constraints. If the calculation above is now repeated, $\tilde{p}_m$ differs by the normalization

$$\tilde{p}_m = \left(\frac{3}{2lL\pi}\right)^{\frac{3}{2}} (\lambda_1 \lambda_2 \lambda_3)^{\frac{3}{2}} \exp\left(-3\left(\frac{\lambda_1^2 x^2 + \lambda_2^2 y^2 + \lambda_3^2 z^2}{2Ll}\right)\right) \tag{84}$$

and

$$\tilde{S}_{\text{eff}} = \tilde{S} - \frac{\mathscr{M}}{\kappa T}\left(\sum_i (\lambda_i^2 - \log \lambda_i)\right) \tag{85}$$

which satisfies the condition that

$$\partial \tilde{S}_{\text{eff}}/\partial \lambda_1 = \partial \tilde{S}/\partial \lambda_1 \text{ at } \lambda_1 = 1, \quad \text{as it must.}$$

This form is derived by Flory and Wall. The logarithmic term has caused much argument, which cannot be discussed in this paper, since in the present discussion it is tied up with boundary conditions and hence with the molecular forces so far ignored. But these two discussions differ by a factor of $\frac{1}{2}$ from that offered here. This is presumably due to the concentration of attention on the affinely deformed single-chain, whilst the present method is based on treating many cross links on a single chain. It has the advantage of the very general formalism and of being able to incorporate entanglements as will be discussed in the next section.

ENTANGLEMENTS

The entanglements involve the additional constraint of

$$\prod_{\alpha} \delta(I^{(0)} - I^{(\alpha)}).$$

This is a *Kronecker* delta, and can be parametrized as

$$\prod_{\alpha} \frac{1}{2\pi} \int_0^{2\pi} \exp\left[i\xi_\alpha (I^{(0)} - I^{(\alpha)})\right] d\xi_\alpha . \tag{86}$$

Since I is made up of many pieces, this expression can be approximated by expansion and resummation i.e. if the integral for Z is written as

$$Z = \int e^{Q} \prod \frac{1}{2\pi} \int \exp\left[i\xi_\alpha (I^{(0)} - I^{(\alpha)})\right] d\xi_\alpha \tag{87}$$

$$\cong e^{Q}\left(1 + i\sum_{\alpha} \frac{\xi_\alpha}{2\pi}(I^{(0)} - I^{(\alpha)}) - \sum \frac{\xi_\alpha \xi_\beta}{4\pi^2}(I^{(0)} - I^{(\alpha)})(I^{(0)} - I^{(\alpha)}\right) + \ldots \tag{88}$$

$$= \exp\left\{Q - \frac{\pi^2}{6}\langle \sum_{\alpha} (I^{(0)} - I^{(\alpha)})^2 \rangle\right\} \tag{89}$$

in an obvious notation, the linear terms vanishing.

In this approximation one can now do a direct evaluation of $\langle (I^{(0)} - I^{(\alpha)})^2 \rangle$ in terms of the approximation to Q, with a resulting modification of the equations to ω, μ and the final value of the free energy. The details of this calculation are given elsewhere (Edwards 1971); it is based on the crude assumption that one may write

$$\begin{aligned} &(I([\boldsymbol{R}^{(0)}]) - I([\boldsymbol{R}^{(\alpha)}]))^2 \\ &= (I([\boldsymbol{X}^{(0)}]\,[\boldsymbol{X}^{(1)}], [\boldsymbol{Y}] - I([\boldsymbol{X}^{(0)}]\,[\boldsymbol{X}^{(1)}]\,[\boldsymbol{Y}], \lambda))^2 \\ &= \left(I^{(0)}([\boldsymbol{X}^{(0)}]) - I^{(\alpha)}([\boldsymbol{X}_0]) + \boldsymbol{X}^{(1)} \frac{\partial I}{\partial \boldsymbol{x}^{(1)}} + \ldots \right. \\ &\qquad \left. + \sum_i (\lambda_i - 1)\left(\frac{\partial I}{\partial \lambda_i}\right)_{\lambda = 1} + \ldots \right)^2 \end{aligned} \tag{90}$$

so that the free energy will be modified by terms in ω and $(\lambda_i - 1)$. The final form comes out to be

$$\omega = \frac{6N}{lL}\left[\frac{1}{2} + \sqrt{\left(\frac{1}{4} + \beta\frac{\mathscr{B}lL^2}{3N^2}\right)}\right] \tag{91}$$

and

$$\tilde{F} = \tfrac{1}{2}N\kappa T\sum(\lambda_i - 1)^2 + L\rho\kappa T\mathscr{B}\sum(\lambda_i - 1)^2 \tag{92}$$

where ρ is the density of polymer, and $\mathscr{B}$ a constant depending on the specification of a chain i.e. thickness, effect of curvature and torsion etc. The free energy is given for small deformations only and it is assumed that

$$\lambda_1 + \lambda_2 + \lambda_3 = 3. \tag{93}$$

The crudity of this approach is a consequence of the complexity of the full equations. Roughly speaking the inclusion of entanglements is like adding currents down the phantom chains, solving the statistical mechanics and then averaging the currents out. The above argument though not perturbation theory, assumes reasonable mathematical behaviour, but the strength of magnetic forces $1/r^3$ casts doubt on this. One can always get finite answers by giving the chains stiffness and thickness, but I suspect that some of these sledgehammer tactics are unnecessary in getting finite answers. In two dimensions one can go very far and solve some problems analytically, but experience in quantum field theory (which, rather implausibly, this problem is an exact analogue to) suggests that there is no solution to a pure random walk entanglement problem, i.e., chains without any curvature energy or thickness are always infinitely entangled—this is not unreasonable intuitively, for when one permits infinitesimal steps in a random walk, the knottedness could well be unbounded. Clearly much more work is required on this problem.

CONCLUSION

This paper has done little more than set out the mathematical problem of polymerized material. The object has been to derive for sych systems with permanent entanglements the analogue of the Gibbs equation for systems in which all phase space is accessible. Having the formula, its evaluation still presents great difficulty; nevertheless, it is formulated and can be attacked.

Similar problems arise when dynamics are introduced, but this would take at least as long again to discuss.

ACKNOWLEDGEMENTS

I have had many helpful discussions with the Manchester Polymer Group, Professors Gee, Allen and Treloar, and with my coworkers Professor Freed and Dr. R. Alexander-Katz.

REFERENCES

The problem of networks is the subject of a symposium on *Highly Cross-linked Networks*, Plenum Press, 1971. The standard reference is Flory (1958), *Principles of Polymer Chemistry*, Cornell U.P.

References to topological constraints are:
Edwards, S. F. (1967). *Proc. Phys. Soc.*, **91**, 513.
Edwards, S. F. (1968). *J. Phys.*, A, **1**, 15.

31

Mathematical Models of One- and Two-Dimensional Amorphous Structures

A. R. COOPER
Case Western Reserve University, Cleveland, Ohio, U.S.A.

and P. F. AUBOURG
Paris, France

To represent the structure of a crystalline material, one needs only a small amount of information. All its configurational features are known when its unit cell and the symmetries of its space group are specified. It is also possible to describe statistically the structure of a dilute gas, simply by its density. For high density non-crystalline materials, no simple description has yet been proposed as there is neither perfect randomness nor perfect order. For such a structure, one requires functions such as correlation functions or distribution functions to describe statistically the atomic configuration of the material. Knowledge of the correlation functions for all temperatures and of the interatomic forces allows one to compute the thermodynamic variables of the system (Hill, 1956).

There seems no *a priori* reason why an irregular structure cannot be described statistically by a set of a few numbers. These of course may be the moments of an appropriate distribution function. Our purpose is to consider simple amorphous structures which can be readily conceived and then to calculate the pair correlation function and/or the distribution of 'bond' angles.

Besides the interest in looking for simple descriptions of amorphous structures, we also recognize that the interpretations of X-ray scattering from liquids benefits from an idea of the correlation function at long distances.

Our concern will be with the correlations (or lack of randomness) that result from the simplest of considerations such as the requirement of non-overlap of atoms in one dimension, or from the requirement of closure or continuity of a network in two or three dimensions.

Following Bernal (1959, 1960), many investigators have considered the irregular packing of spheres (Scott, 1960, 1962). Others (Ordway, 1964; Evans and King, 1966; Bell and Dean, 1966) have made models of amorphous structures in three dimensions, some of which have a close correspondence to the density and vibrational properties of silica glass. However, there is a degree of arbitrariness in the construction of these models. Further, they all depend on decisions made by the constructor that cannot be completely objective or rigorously described. Finally, they are tedious to build and study.

By comparison, models can be built following a set of preconceived rules, using Monte Carlo techniques, in a totally objective manner. Time for construction is short ($\sim$3 seconds on a UNIVAC 1108). Their difficulty lies in the choice of a set of rules which are both self-consistent and sufficiently general. This limited the initial efforts described here to two dimensions and even here we were unable to achieve as much generality as desired.

The terminology used in the literature has not always been consistent. We shall, therefore, first define the quantities we shall use. These definitions, which follow closely the ones used by J. Kirkwood (1968), are given below.

The probability of finding an atom centre within a volume element dv at position $\vec{r}_0$ is written $p^{(1)}(\vec{r}_0)\,dv$; then $p^{(1)}(\vec{r}_0)$ is termed the single particle distribution function. Extending to n atoms the probability of finding a centre within dv of $\vec{r}_0$, another within dv of $\vec{r}_1 \ldots.,$ and another within dv of $\vec{r}_{n-1}$ is written $p^{(n)}(\vec{r}_0, \vec{r}_1 \ldots., \vec{r}_{n-1})\,dv^n$; then $p^{(n)}(\vec{r}_0, \vec{r}_1 \ldots.. \vec{r}_{n-1})$ is the n-particle distribution function.

As the origin of atomic coordinates is arbitary, $p^{(1)}(\vec{r}_0)$ is just the probability density for finding an atom centre, i.e. the atomic density, ρ. By definition of conditional probability $p^{(2)}(\vec{r}_0, \vec{r}_1)\,dv^2$ then is the product of ρdv and the conditional probability, $p[\vec{r}_1|\vec{r}_0]\,dv$, of finding an atom centre in an incremental volume dv at a position $\vec{r}_1$ when another atom centre is within dv of $\vec{r}_0$. As all atoms are equivalent, we may think of $p[\vec{r}_1|\vec{r}_0]\,dv$ as the simple probability $p(\vec{r}_{01})\,dv$ of finding an atom at a position $\vec{r}_{01} \equiv (\vec{r}_1 - \vec{r}_0)$ away from another atom centre. Putting the origin of the coordinate system at r_0 gives $\vec{r}_{01} = \vec{r}$. The assignment of direction, i.e. the rotation of a coordinate system remains arbitrary even after the origin of coordinates is fixed at an atom centre. Thus $p(\vec{r})$ depends only on the magnitude of r, i.e. $p(\vec{r}) = p(r)$.

The third order distribution function, $p^{(3)}(\vec{r}_0, \vec{r}_1, \vec{r}_2)$, can be decomposed into the product of ρ, $p(r_{01})$ and the conditional probability density, $p[\vec{r}_2|\vec{r}_1, \vec{r}_0]$, of finding an atom centre within an incremental volume dv at $\vec{r}_2$ when other centres are at $\vec{r}_0$ and $\vec{r}_1$. The conditional probability density $p[\vec{r}_2|\vec{r}_1, \vec{r}_0]$ depends on the angle, $\theta(0, 1, 2)$ between $\vec{r}_{01}$ and $\vec{r}_{12}$, and hence the third order distribution function is a function of two scalar distances and an angle and its simplest representation is $p^{(3)}(r_{01}, r_{12}, \theta_{0,1,2})$.

The n th order correlation function $g^{(n)}(\vec{r}_0, \vec{r}_1 \ldots \vec{r}_{n-1})$ is equal to the n th order distribution function normalized by dividing by the n th power of the atomic density,

$$g^{(n)}(\vec{r}_0, \vec{r}_1 \ldots \vec{r}_{n-1}) = \frac{p^{(n)}(\vec{r}_0, \vec{r}_1 \ldots . \vec{r}_{n-1})}{\rho^n}$$

From what has preceded, it should be clear that in its simplest representation, the pair correlation function $g^{(2)}$ is given by

$$g^{(2)}(r) = \frac{p(r)}{\rho} \tag{1}$$

Non-overlap of atoms of diameter, d, requires

$$g^{(2)}(r) = 0, r < d$$

For amorphous structures under normal conditions, correlations tend to vanish at long distances,

$$\text{i.e.} \qquad \lim_{r \to \infty} g^{(n)}(r) = 1 \qquad \text{for all n.}$$

Thus a generalized correlation function, $h^{(n)}(r) \equiv g^{(n)}(r) - 1$, which vanishes when correlations are absent, is frequently used.

NON-OVERLAPPING 'ATOMS' IN ONE DIMENSION

Let us consider the case of uniform length, one-dimensional, non-overlapping atoms placed at random on a line.

The density will always be the atomic density, i.e. the number of atoms per unit length. We choose a scale such that the length of the atoms is equal to unity.

Call 'a' the average distance between the edges of the nearest neighbour atoms. Then atomic density, the quotient of the atomic length by the average separation between atom centres $\rho = 1/(1 + a)$.

The term random non-overlapping line segments does not unambiguously define a configuration as there are two distinct interpretations. The first, known among statisticians as the 'random car parking procedures' involves placing equal size 'atoms' on a line at random and rejecting those events which cause overlap. If carried to completion, this process results in a density of 0·7476 (Solomon, 1954–55).

The second interpretation which we follow starts from the premise that the probability, $p_{01}(r)\,\mathrm{d}r$, of finding a nearest neighbour atom centre within dr of a distance, r, is only influenced by the non-overlap condition, i.e.

$$p_{01}(r)\,\mathrm{d}r = 0 : r < 1 \qquad \text{(a)}$$
$$p_{01}(r)\,\mathrm{d}r = \frac{1}{a}\exp -((r-1)/a)\,\mathrm{d}r : 1 \leqslant r \qquad \text{(b)} \tag{2}$$

The R.H.S. of equation 2b can be recognized as the product of two factors: (i) exp $-((r-1)/a)$, the probability that the nearest neighbour does not lie closer than r; and (ii) dr/a, the probability that the centre lies within dr of r when the average spacing is a.

While there are other derivations of the calculations of this one dimensional problem, e.g. Frenkel (1946) and Kirkwood (1968), the one below which recognizes the direct application of Poisson distribution is perhaps most succinct.

The probability density, $p(r)$, of finding any atom in dr of a distance of 'r' from another one can be written as a sum of probability densities of finding the nearest neighbour, p_{01}, the second nearest neighbour, p_{02}, etc.

$$p(r) = \sum_{i=1}^{v} p_{0i}(r)$$

where 'v' is an integer such that $v \leqslant r < v + 1$ and $p_{0i}(r)$ is the probability density that if a segment of length 'r' has its origin in the centre of the 0 th atom, its end will be on the centre of the i th atom.

To compute $p_{0i}(r)$, let us consider the simpler problem of points placed at random on a line such that their average separation is a. We place the origin at a point and recognize that the probability density, $p_{0i}(L)$, that (in direction of L) the i th nearest neighbour is at L is given by a Poisson distribution (Girault, 1966), i.e.

$$p_{0i}(L) = \frac{1}{a}\frac{\exp -(L/a)}{(i-1)!}\left(\frac{L}{a}\right)^{i-1}$$

Now, if we associate each point with an atom centre (atom 0 at the origin) and L with the unoccupied distance between atom 0 and atom i, then the relation between r and $L$is easily expressed, i.e.

$$r = L + (i+1) - 2 + \frac{1}{2} + \frac{1}{2} = L + i$$

and the probability density, $p_{0i}(r)$ then becomes

$$p_{0i}(r) = 1/a\,\frac{\exp(-(r-i)/a)}{(i-1)!}\cdot\left(\frac{r-i}{a}\right)^{i-1} \quad \text{for } r \geqslant i$$

$$p_{0i} = 0 \text{ for } r < i$$

Introducing this result in the initial formula

$$p(r) = 1/a \sum_{i=1}^{v} \frac{\exp(-(r-i)/a)}{(i-1)\,!} \cdot \left(\frac{r-i}{a}\right)^{i-1}$$

Therefore, the generalized pair correlation function is equal to

$$h^{(2)}(r) = \frac{p(r)}{\rho} - 1 = \frac{a+1}{a} \sum_{i=1}^{v} \frac{\exp(-(r-i)/a)}{(i-1)\,!} \cdot \left(\frac{r-i}{a}\right)^{i-1} - 1 : r \geqslant 1$$

$$h^{(2)}(r) = 0 : r < 1$$

One can check that when r tends toward infinity, $h^{(2)}(r)$ tends toward zero. Also $h^{(2)}(1) = 1/a$.

We drew the generalized correlation function, $h^{(2)}(r)$, for various densities (Figure 1), 1, $\sim$ 0·75, 0·5, 0·1. In contrast to the car parking problem, this approach permits the correlation function to be drawn for all densities. It is clear that the order in the system increases monotonically with density.

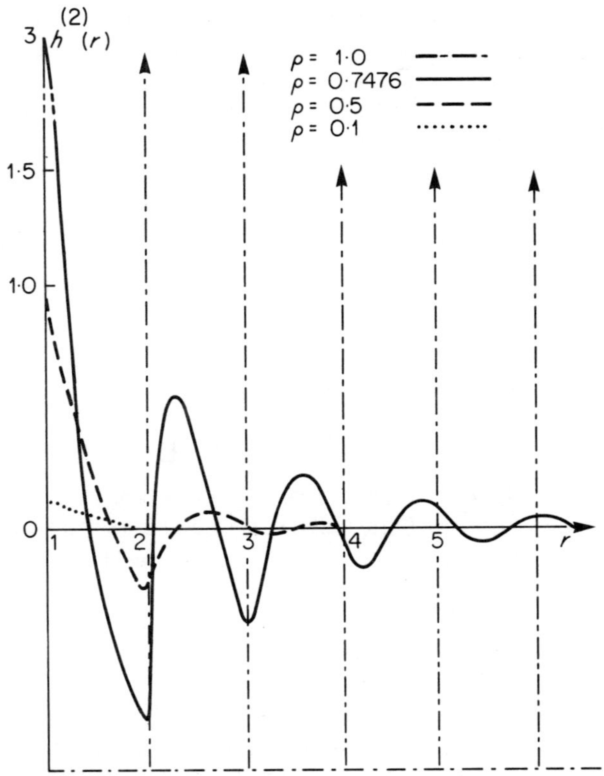

Figure 1. Pair correlation function $h^{(2)}(r)$ for random distribution of equal length segments with no overlap; density ρ is a parameter

While we show the correlation function for the density equivalent to the maximum 'car parking density', we do not expect the correlation functions to be equivalent because in our case the probability of an empty space larger than unity is $\sim \exp(-3) \simeq \frac{1}{20}$ while in car parking this probability is exactly zero. Thus, in this sense the Poisson distribution is, as expected, more random.

If at density 0·5 the structure were perfectly ordered, the distance between peaks would be 2. For the random structure the peaks are greatly diminished and the distances reduced to approximately $1{\cdot}4 \pm 0{\cdot}1$ as we have a peak at 1, 2·63, 3·93, 5·35, 6·7, and 8·1.

The case of a low density corresponds to a gas. One sees that the $h^{(2)}$ tends very quickly toward zero.

The correlation function obtained from a Poisson distribution corresponds to a hard core potential energy distribution between atoms. This is the simplest imaginable arrangement for non-overlapping atoms, and it is not surprising that the correlation function depends on density alone as there is no way to introduce another parameter into such a distribution.

It is possible to obtain information about all higher order correlation functions from this derivation because the angle $\theta_{i,\,i+1,\,i+2}$ for $r_{i+2} > r_{i+1} > r_i$ is always equal to π and thus higher order correlations can ignore any dependence on θ. Also for a Poisson distribution the probability density of finding a third particle is in no way influenced by the position of the first. Thus for this case

$$p[r_2 | r_0, r_1] = p[r_2 | r_1] = p(r_{12})$$

and

$$p^{(3)}(r_0, r_1, r_2) = \rho \,.\, p(r_{01})\, p(r_{12})$$

$$h^{(3)}(r_{01}, r_{12}) = g^{(2)}(r_{01})\, g^{(2)}(r_{12}) - 1$$

$$h^{(3)}(r_{01}, r_{12}) = h^{(2)}(r_{01})\, h^{(2)}(r_{12}) + h^{(2)}(r_{01}) + h^{(2)}(r_{12})$$

Less ideal nearest neighbour probability densities, p_{01}, i.e. those with clustering or anti-clustering tendencies, can be converted to pair correlation functions following a similar procedure to that above. However, the calculation becomes cumbersome.

TWO-DIMENSIONAL MODELS OF NETWORKS

Our first idea in studying two-dimensional models was to compute directly $h^{(2)}(r)$ as we had done in the one-dimensional case by decomposing the probability density, $p(r)$, into a sum. The difficulty of the direct computation increases quickly with the number of atoms and there is no simple recurrence formula. Furthermore our main interest is in modelling random networks which have no direct analogy in one dimension. Therefore, we studied by iterative methods the bond angle, θ, distribution and $h^{(2)}(r)$ of two dimensional models of networks constructed from circular atoms of unit diameter.

Because three-fold coordination gives the greatest freedom for network formation in two dimensions, we imposed the restriction that each atom has exactly three touching atoms and hence the average bond angle is $2\pi/3$.

The atoms, therefore, form polygonal loops* (on the average hexagons). Due to the difficulty of defining in a non-ambiguous way the construction of structures of more than one type of polygon, the study was limited to random structures of hexagons. We started from a central atom, atom zero, and added atoms progressively in a given order. 1 . . . 2 . . ., etc. In this way we completed first a hexagon and then shells of hexagons surrounding the initial one, as shown schematically in Figure 2.

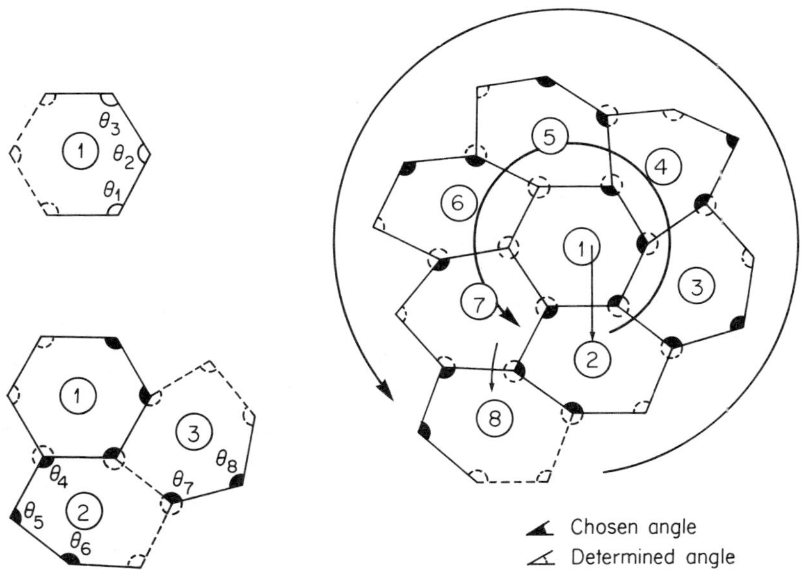

Figure 2. Method of construction of a model

For each hexagon, one, two or three bond angles had to be chosen (the shaded ones on Figure 2). It can be shown that for a structure of n shells containing $3n^2 + 3n + 1$ hexagons it is necessary to choose $3n^2 + 9n + 3$ angles to determine the structures. Thus in the limit of large structures, only one angle per hexagon is required.

* For convenience of language and representation, we often substitute equal length lines joining the atom centres for the atoms themselves.

The angles were chosen from an *a priori* given distribution of angles. The maximum interval of variation was limited to $(\pi/3, \pi)$ to avoid overlap and re-entrant angles. For each angle, the choice was often further limited by conditions which assured that the structure could be extended (Aubourg, 1969). These conditions arise from the fact that the sum of the angles at each vertex must be 2π and from the necessity to close the hexagonal loops and the rings of loops. For example, it is obvious that if the first angle of an hexagon is equal to π, the second cannot be chosen equal also to π as one would have a completely flattened hexagon. (It was not always possible to take into account all the conditions, and from time to time, it was not possible to extend a structure. In this case the structure was not used.)

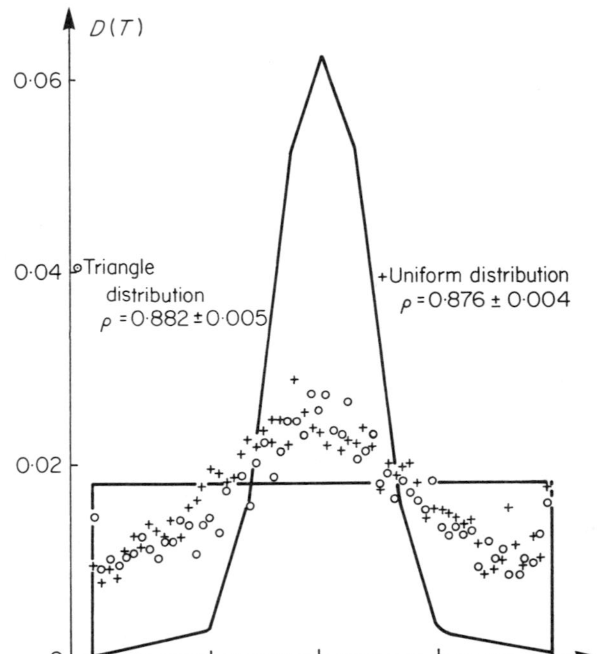

Figure 3. Comparison of final angular distribution for two different *a priori* distributions. Models of 37 hexagons. Results averaged over 30 (0) and 40 (+) models

By this method, structures of 19 and 37 hexagons were built, in approximately 3 seconds on a UNIVAC 1108. After construction, the distribution of angles was determined on at least 12 models and more often on 30 or 50 models.

Since the structures were expected to depend on the choice of the initial angular distribution, our goal was to find by trial and error a self-consistent structure in which the empirically determined distribution of angles matched the *a priori* distribution.

We found that the density and the determined distribution of angles were surprisingly insensitive to the *a priori* distribution and depended almost entirely on the bounds for θ. For example, five different *a priori* distributions ($\pi/3 < \theta < \pi$) give final distributions that are hardly distinguishable. As an illustration, we show in Figure 3 the final distributions for two very different initial distributions.

Figure 4 shows that we were able to achieve the goal of a matching initial and final distribution.

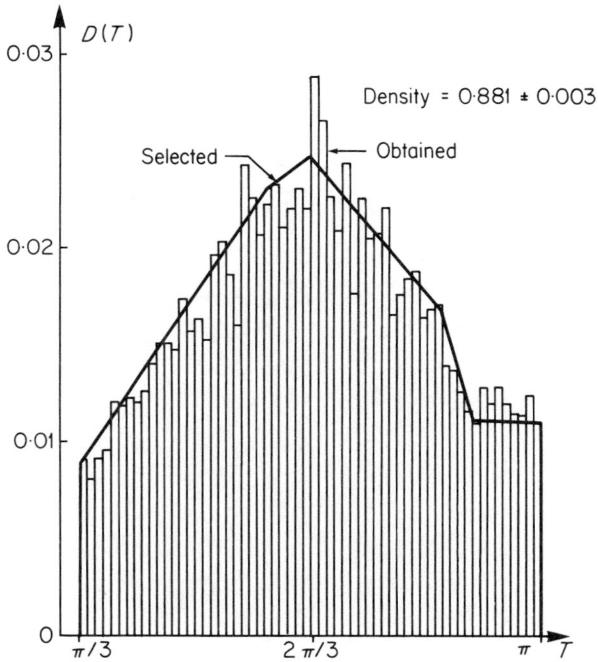

Figure 4. Comparison of *a priori* and obtained angular distribution. Model of 37 hexagons. Results averaged over 20 models

Pair correlation functions were determined for the various distributions in the following manner.

An atom of the first hexagon was chosen as central atom. The number, $S(r)$, of atom centres (vertices) within a ring between $r - \Delta r/2$ and $r + \Delta r/2$ for $1 \leqslant r \leqslant 4{\cdot}5$ was counted. The procedure was repeated for each of six central atoms in 16 structures, i.e. 96 central atoms. The probability density,

$$p(r) = \frac{1}{2{\cdot}96} \sum^{96} \frac{S(r)}{\pi r \Delta r}$$

The density, ρ, was determined by calculating the area of each hexagon. Together, ρ and $p(r)$ allow $h^{(2)}(r)$ to be obtained. A value of $\frac{1}{10}$ for Δr was found to be appropriate; the choice depends mainly on the number of cases considered and the accuracy of the data. The smaller Δr, the better the resolution of the resulting curve. The larger Δr the lower the uncertainty of $S(r)$ and hence of $h^{(2)}(r)$. For our case of 96 central atoms standard deviation from counting statistics $\sigma(h^{(2)}(r)) \simeq 0{\cdot}04/\sqrt{r\Delta r}$. Thus the choice of $\Delta r = 0{\cdot}1$ leads to reasonable values of σ. The error of measurement of the distance between the atoms increases with distance but this factor was not important under our conditions.

$h^{(2)}(r)$ for the case of a uniform initial distribution is shown in Figure 5. From the height of the first peak one can compute the coordination number, CN, since $\text{CN} = 2\pi\Delta r\rho\,(h(r) + 1)$. It is higher than 3. This is due to the fact

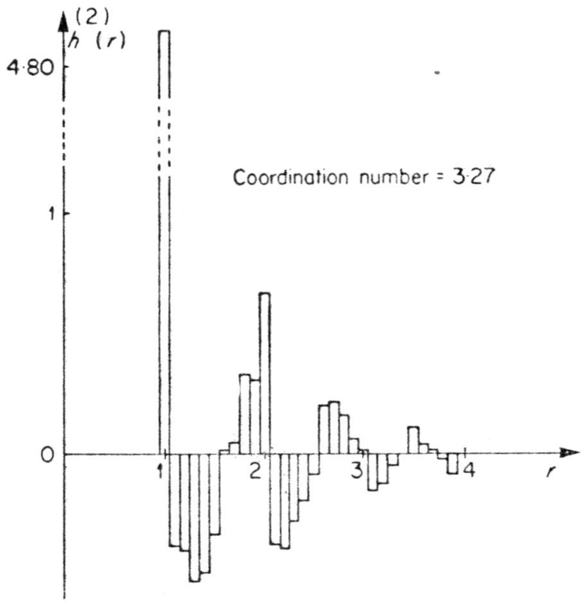

Figure 5. Correlation function $h^{(2)}(r)$ when angles are selected from a uniform distribution (results averaged over 100 central atoms)

that all distances up to 1·05 contribute to the first coordination shell in this calculation, as hexagons with a nearly rectangular shape or angles nearly equal to $\pi/3$ allow atoms to approach to within 1·05 of each other.

The curve shows maxima at 1, 2, 2·75 and $h^{(2)}(r)$ decreases rapidly with r. For $r = 4$, the values are of the same order as the uncertainty in the measurements.

We conclude from our study of hexagonal network models that the conditions of non-overlap and closure are powerful in fixing the density, the angular distribution, and the pair correlation functions of these structures once the limits of maximum angles are set.

SUMMARY

In one dimension the idea of a network is not well defined. However, the idea of non-overlap of atoms is well understood and fortunately a simple transformation allows the pair correlation to be determined by a sum of Poisson distributions. These results show clearly the effect of non-overlap in producing correlations that persist over distances which depend entirely on the density. Such structures represent the most random arrangements that can be achieved at a given density under the restraint of no overlap. Tendencies for clustering or anti-clustering introduced by atom interactions would reduce the randomness.

Two dimensional structures permit network formation. It was shown that the requirement for network formation is a powerful influence in determining the pair correlation function, the angular distribution and naturally the density, for structures of hexagonal rings once the range of angles permitted was established. The results are quite independent of the details of the initial distribution of angles.

The most serious limitation is that we could not devise a procedure which permitted the additional randomness that would have been associated with allowing pentagonal, septagonal, etc. rings to occur.

CONCLUSIONS

In the two simple cases we have considered amorphous structures are described by a small set of numbers. For random non-overlapping lines the structure is completely described by the density (and the size of the line segment). For equi-diameter circles arranged randomly in 3 coordinations in hexagonal loops the structure appears, within the limits of our uncertainties, to be well described once the range of angles is established. We recognize that these are especially simple amorphous structures, but believe that it may be fruitful to describe more complicated structures in terms of their deviation from simpler ones.

ACKNOWLEDGEMENT

We gratefully acknowledge the financial support of the American Optical Corporation through the Fellowship awarded to one of us (P.A.).

REFERENCES

Aubourg, P. F. (1969). *M.Sc. Thesis*, Case Western Reserve Univ., Cleve., Ohio.
Bell, R. J. and P. Dean (1966). *Nature*, **212**, 1354.
Bernal, J. D. (1959). *Nature*, **183**, 141.
Bernal, J. D. (1960). *Nature*, **185**, 69.
Bernal, J. D., and J. Mason (1960). *Nature*, **188**, 910.
Evans, D. L., and S. V. King (1966). *Nature*, **212**, 1353.
Frenkel, Y. I. (1946). *Kinetic Theory of Liquids*. Oxford University Press. p. 126.
Girault, M. (1966). *Stochastic Processes*. Springer Verlag. p. 14.
Hill, T. H. (1956). *Statistical Mechanics*. McGraw-Hill. pp. 172–193.
Kirkwood, J. G. (1968). *Theory of Liquids*. Gordon and Breach.
Ordway, F. (1964). *Science*, **142**, 800.
Scott, D. G. (1960). *Nature*, **188**, 908.
Scott, D. G. (1962). *Nature*, **194**, 956.
Solomon, H. (1954–55). *Fifth Berkeley Symposium on Mathematical Statistics and Probability*, Vol. 3. Univ. California Press. pp. 119–134.

32

Short Range Order of the Atoms at Higher Temperatures in Semiconducting and Metallic Melts of the System Ge–As–Se

F. ACKERMAN and H. KREBS

Institut für Anorganische Chemie, Universität Stuttgart, Germany

---

Atomic distribution curves of $Ge_xAs_{30}Se_{(70-x)}$ glasses and melts were determined by X-ray measurements with a focusing goniometer especially developed by Gruber and Krebs (1964) for the investigation of liquids at high temperatures.

Measurements were made on three specimens of composition

I: $Ge_{0\cdot20}As_{0\cdot30}Se_{0\cdot50}$ at 20 °C and 620 °C
II: $Ge_{0\cdot45}As_{0\cdot30}Se_{0\cdot25}$ at 750 °C
III: $Ge_{0\cdot55}As_{0\cdot30}Se_{0\cdot15}$ at 750 °C.

The results are shown in figure 1. The density of the melts which it is necessary to know for the determination of the coordination number has been measured with an accuracy of about 3%. More accurate density measurements are in preparation.

Sample I lies in the middle of the glass-forming region. The radial distribution curves of the glassy and the liquid state of this sample are very similar, though the curve at 620 °C is smeared out to a certain extent by the thermal movement of the atoms. In both states the mean coordination number does not reach the weighted mean value for the typical coordination number (Se = 2, As = 3, Ge = 4). Therefore we believe that most of the Ge-atoms have only two neighbours. The interatomic distances correspond to summed covalent bonding radii. We find 2·3 next nearest neighbours as the mean coordination number with a maximum at 2·41 Å. The space between the first and the second coordination shell is not populated by atoms.

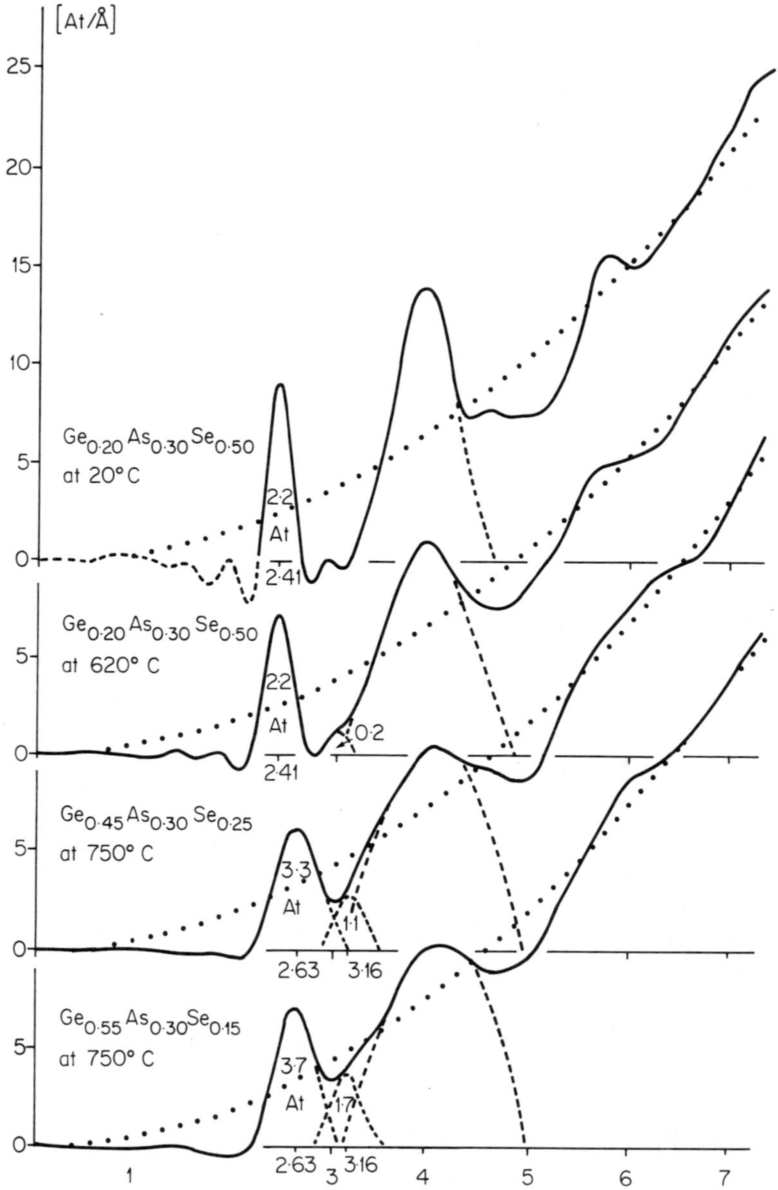

Figure 1. Radial distribution functions of chalcogenide melts and glasses

In the melts of sample II which lies near the border, and sample III which is situated well outside the glass forming region, the distances of nearest neighbours increase from the calculated values (assuming localized bonds) of 2·45 Å and 2·46 Å respectively to the experimental value of 2·63 Å. Atoms of the second peak of the radial distribution function begin to move into the unpopulated space situated at about 3 Å from the central atom between the first and the second coordination shell. We find about 3·3 neighbours at a distance of 2·63 Å and about 1·1 more neighbours at a distance 20% larger, i.e. at 3·16 Å for sample II. The tendency mentioned above is particularly evident for sample III. Here we find about 3·7 neighbours at 2·63 Å and about 1·7 neighbours at 3·16 Å. It is suggested that these observations are consistent with the development of a pσ-bonding system, similar to that attributed to liquid Ge just above the melting point by Krebs, Lazarev and Winkler (1967).

REFERENCES

Gruber, H. U., and H. Krebs (1969). *Z. anorg. allg. Chem.*, **369**, 184.
Krebs, H., V. B. Lazarev, and L. Winkler (1967). *Z. anorg. allg. Chem.*, **352**, 277.

33

Crystallization of Lead Silicate Glass as Studied by Trimethylsilylation and Chromatographic Separation of the Anionic Constituents*

J. GÖTZ, C. R. MASSON
Atlantic Regional Laboratory, National Research Council of Canada, Halifax, Nova Scotia, Canada

and L. M. CASTELLIZ
Atlantic Industrial Research Institute, Nova Scotia Technical College, Halifax, Nova Scotia, Canada

---

It has become increasingly apparent in recent years that silicate melts and glasses may be regarded as polymeric materials which contain a wide variety of constituents. Recent experimental evidence supports the view that liquid silicates may be classed as ionic condensation polymers, considered to be derived from the 'monomeric' ion $SiO_4{}^{4-}$. These ions can undergo self-condensation, with elimination of free oxide ions O^{2-}, to yield linear and branched polyions of formulae $Si_2O_7{}^{6-}$, $Si_3O_{10}{}^{8-}$, $Si_4O_{13}{}^{10-}$ and higher species. These polyions are in a state of dynamic equilibrium in the melt.

At high silica contents, in the range commonly associated with glass formation, the proportion of discrete anions in silicate melts is very small. Indeed, such ions have not so far been detected in ordinary commercial glasses of high silica content. Their presence can, however, be shown experimentally and theoretically in melts and glasses more basic than the metasilicate composition.

Theoretical evidence for the presence of discrete silicate anions in basic melts has been obtained from interpretation of the thermodynamic properties of these systems (Masson, 1965, 1968). More recently, direct application of polymer theory (Whiteway, Smith and Masson, 1970; Masson, Smith and

* Issued as NRCC 12035.

Whiteway, 1970a, b) has shown that for all binary systems MO–SiO_2 so far investigated, the activity of MO can be expressed as a function of composition in terms of an adjustable parameter k_{11}. This is the equilibrium constant for the self-condensation of $SiO_4{}^{4-}$ ions:

$$2SiO_4{}^{4-} \rightleftharpoons Si_2O_7{}^{6-} + O^{2-}$$

$$k_{11} = N_{Si_2O_7{}^{6-}} N_{O^{2-}} / N^2_{SiO_4{}^{4-}}$$

where N is the ion fraction of the species indicated. The value of k_{11} is specific for a given system and is determined by the nature of the cation. Knowledge of its value allows the distribution of silicate ions and their average chain length to be calculated as a function of the silica content.

APPLICATION OF POLYMER THEORY TO PbO–SiO_2 MELTS

The value of k_{11} for this system as determined from the experimental activity *vs.* composition data of Richardson and Webb (1955), Sridhar and Jeffes (1967) and Charette and Flengas (1968) is 0·196. Ionic distributions in PbO–SiO_2 melts at 1000 °C as calculated from this value are shown in Figure 1. In their general form, these curves resemble the distribution curves for all binary

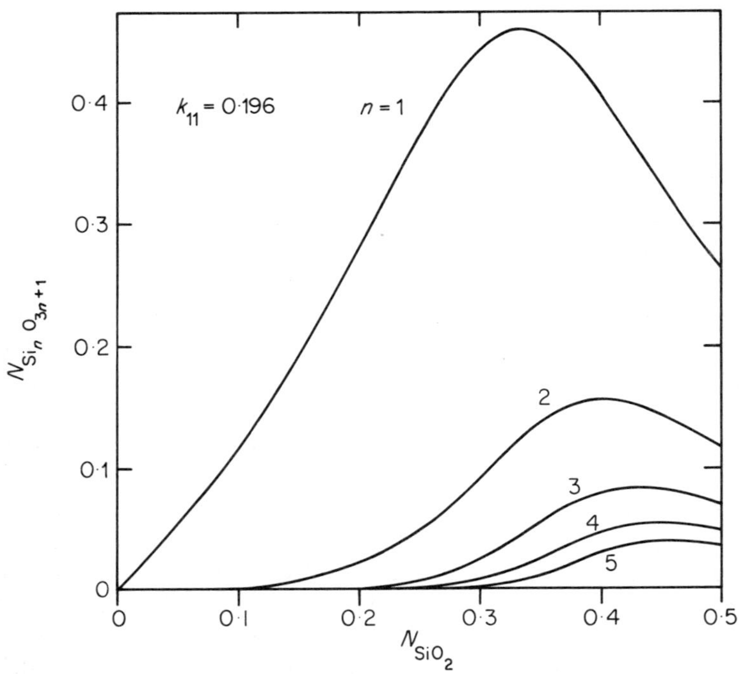

Figure 1. Ionic distributions in PbO–SiO_2 melts at 1000 °C as calculated from polymer theory

silicate melts at high basicities. They show that, at any given silica content, the ion $SiO_4{}^{4-}$ is always the most abundant species of discrete silicate ion followed, in turn, by the ions $Si_2O_7{}^{6-}$, $Si_3O_{10}{}^{8-}$, etc.

The assumptions made in deriving these curves are identical with those made in conventional polymer theory. In particular it should be noted that only linear and branched chains are considered to be present in the system, and that no allowance has been made for the possible formation of ring structures or of networks. This is a limitation which is, at present, inherent in polymer theory itself. The other main assumption is that the principle of equal reactivity of functional groups is applicable, i.e. that all 0^- groups on the polyionic chains are chemically equivalent, regardless of the size of the polyions. Experience in polymer chemistry suggests that this assumption may not be strictly valid for the smallest species but should become increasingly valid for the larger chains.

Distribution curves calculated in this way from thermodynamic data should thus be considered as approximate, and indicative only of the broad features of the ionic constitution. They cannot be regarded as applicable with high precision, particularly for the smallest species.

EXPERIMENTAL DETERMINATION OF IONIC DISTRIBUTIONS IN SILICATE GLASSES

Until recently it has been difficult to demonstrate conclusively the presence of discrete anions in silicate glasses, although the analogous species $PO_4{}^{3-}$, $P_2O_7{}^{4-}$, $P_3O_{10}{}^{5-}$, etc. have been demonstrated in phosphate glasses by chromatographic techniques (Westman, 1960). Application of similar techniques to the silicates has proved more difficult due to the sparingly soluble nature of many silicate glasses and, for the soluble silicates, to the labile nature of the silicate anion in aqueous solution.

Lentz (1964) showed that discrete silicate anions could be extracted from certain minerals and aqueous silicate solutions as their trimethylsilyl (TMS) derivatives and that these stable derivatives could be separated by gas–liquid chromatography. The method was applied by Lentz (1967) to the study of portland cement and by Kolb and Hansen (1965) to borate glasses. Although the method of trimethylsilylation described by Lentz is not quantitative it represents a valuable contribution to the study of silicate structures.

In a preliminary study of this method as applied to silicate glasses (Götz, Jamieson and Masson, 1971) the influence of all possible reaction conditions on the products was investigated. It was concluded that side reactions occur to a significant extent and a direct method of trimethylsilylation was subsequently developed and tested with minerals of known structure (Götz and Masson, 1970, 1971). In this procedure, trimethylsilyl chloride was used as the trimethylsilylating reagent. The products of trimethylsilylation were separated by gas–liquid chromatography and characterized by high-resolution mass spectrometry (Wu, Götz, Jamieson and Masson, 1970).

RESULTS FOR LEAD ORTHOSILICATE GLASS

Figure 2 shows the chromatogram obtained by this technique for a glass of composition corresponding to the formula $2PbO \cdot SiO_2$. The glass was prepared by melting Pb_3O_4 and SiO_2 in a platinum crucible, crushing, sieving, remelting and subsequently quenching in an ice–water mixture.

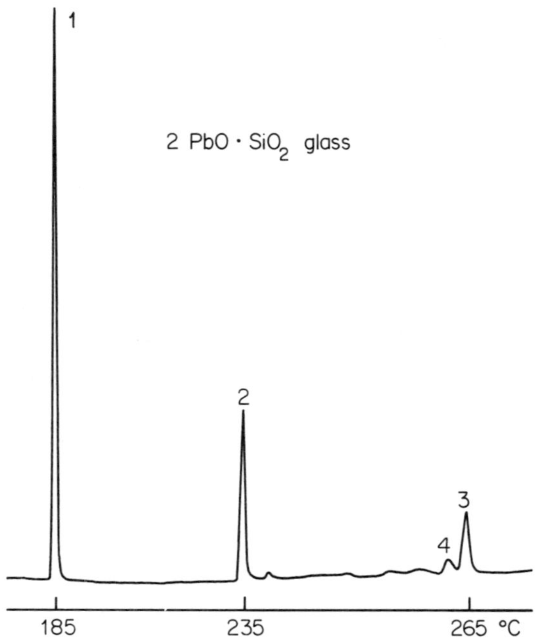

Figure 2. Chromatographic pattern for lead orthosilicate glass. The peaks correspond to the TMS derivatives of the anions (1) $SiO_4{}^{4-}$, (2) $Si_2O_7{}^{6-}$, (3) $Si_3O_{10}{}^{8-}$ and (4) $Si_4O_{12}{}^{8-}$

The largest peak is that due to the monomeric derivative $SiO_4(SiMe_3)_4$, which accounts for 61·9% of the total area in the chromatogram. Other peaks clearly identified, with the corresponding percentages of the total area, are due to the derivatives of the chain ions $Si_2O_7{}^{6-}$ (22·9%) and $Si_3O_{10}{}^{8-}$ (12·2%) and the cyclic ion $Si_4O_{12}{}^{8-}$ (3·0%).

These results are qualitatively in line with the predictions of Figure 1 which indicates ion fractions for $SiO_4{}^{4-}$, $Si_2O_7{}^{6-}$ and $Si_3O_{10}{}^{8-}$ of 0·461,

0·123 and 0·044 respectively at the orthosilicate composition. At this composition also the calculated ion fraction of free oxide ion O^{2-} is 0·339 so that in theory these four constituents account for 96·7% of the total anions on a molar basis, the remaining 3·3% being attributable to longer chains. This is again in line with the chromatographic results, which show little or no evidence of discrete anions larger than $Si_4O_{12}{}^{8-}$. The small proportion of tetrameric ring structure in the glass is not accounted for by the theory.

CRYSTALLIZATION OF LEAD ORTHOSILICATE GLASS

Small quantities (approx. 7 gm) of the orthosilicate glass, ground to pass 325 mesh, were heat-treated in a platinum vessel at 498–500 °C for various times, quenched at room temperature, and their constitution studied by the above technique.

After heat-treatment for 7 min there was an increase in the proportion of the dimeric derivative $Si_2O_7{}^{6-}$, as compared with the original glass. This was accompanied by a decrease in the proportion of the monomeric and trimeric chain derivatives and an increase in the tetrameric ring. After 25 min the dimeric derivative had increased to over 50% of the total area of the peaks and the monomeric derivative had dropped to about 25%. Further annealing resulted in a decrease in the dimeric derivative and a rapid build-up of the tetrameric ring structure, so that, after 45 min, the tetrameric ring derivative accounted for 52·7% of the total peak area. The corresponding figures for the $Si_2O_7{}^{6-}$ and $SiO_4{}^{4-}$ derivatives were 25·6% and 17·2% respectively. Finally, after 24 hr at 500 °C, the dominant constituent was the tetrameric ring derivative, which represented 83·3% of the total area in the chromatogram from the crystallized material. The remaining contributions due to monomer (9·8%), dimer (4·3%) and trimer (2·6%) may have represented a small proportion of glass still present after crystallization.

These results are summarized in Figure 3 which shows the percentages of the total peak area due to monomer, dimer and tetrameric ring derivatives as functions of time during crystallization. The main features are the smooth decrease in the monomeric derivative, the maximum in the dimeric derivative after a short interval and the establishment of a plateau for the tetrameric ring structure.

Relations exist between the variation in proportions of the three TMS derivatives with increasing devitrification and the changes in X-ray diffraction pattern of the crystallization products. The most conspicuous changes are summarized in Figure 4 which represents a low-angle section of diffractometer charts taken from $2PbO \cdot SiO_2$ in various stages of crystallization.

The glass phase is characterized by two broad maxima, one of which (curve a) is shown in Figure 4. With Cu*K* radiation the low-angle halo extends from 26 to 36° in 2θ, covering the range of Bragg angles of the crystal

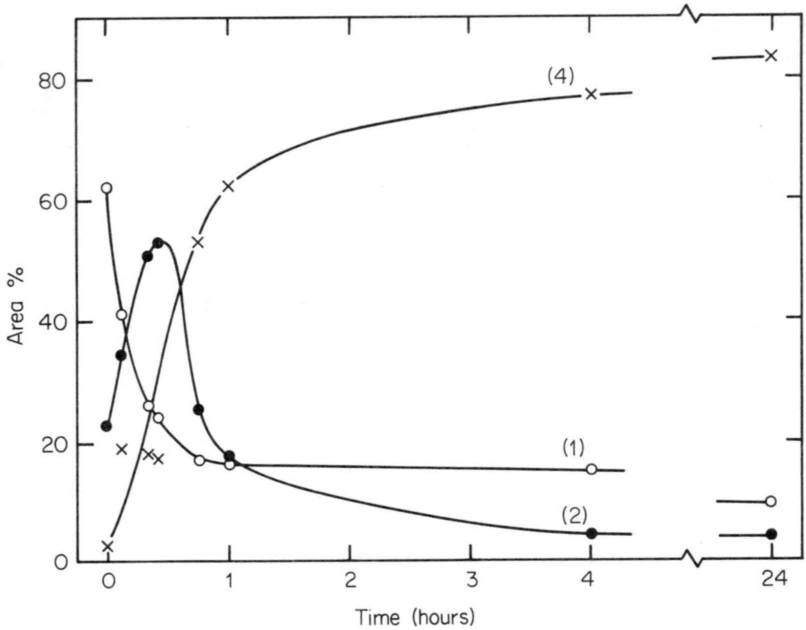

Figure 3. Change in constitution of lead orthosilicate glass on crystallization at 500 °C. The curves represent the percentages of the total area in the chromatograms as (1) SiO_4^{4-}, (2) $Si_2O_7^{6-}$ and (4) $Si_4O_{12}^{8-}$ derivatives

planes reflecting with the highest intensities. The second maximum, not shown in Figure 4, is flatter and broader, hardly discernible from the background, and extends over a width of 20° up to an angle of 70° in 2θ.

After 7 min annealing time (curve b) the low angle halo of the glass state is seen to be broken up into essential, but still broadened, reflections of crystallized $2PbO \cdot SiO_2$. After 20 min (curve c) these reflections have assumed the sharpness of the final crystallization product, but on further heating (curves d–h) there occur still further changes in the relative intensities of certain reflections which parallel the changes in the relative amounts of the discrete polyions.

In particular, the appearance and gradual increase of the reflections at approx. 24·3 and 28·9°, corresponding to interplanar spacings of 3·66 and 3·10 Å respectively, and the steady increase of two other peaks at 32·8 and 35·0° (d = 2·73 and 2·56) might be considered to be in causal connection with the steady increase of the relative amount of the tetrameric ring derivative.

A second group of reflections, corresponding to the peaks at 23·2, 30·8 and 33·8° (d = 3·83, 2·90 and 2·65) assume their maximum relative intensities at about 25 min annealing time, which corresponds to the time at which the relative abundance curve of the $Si_2O_7{}^{6-}$ ion passes through a maximum. The reflection at 26·3° (d = 3·39) may also be in this category.

Finally, the decreasing relative intensity of the well-defined peak at

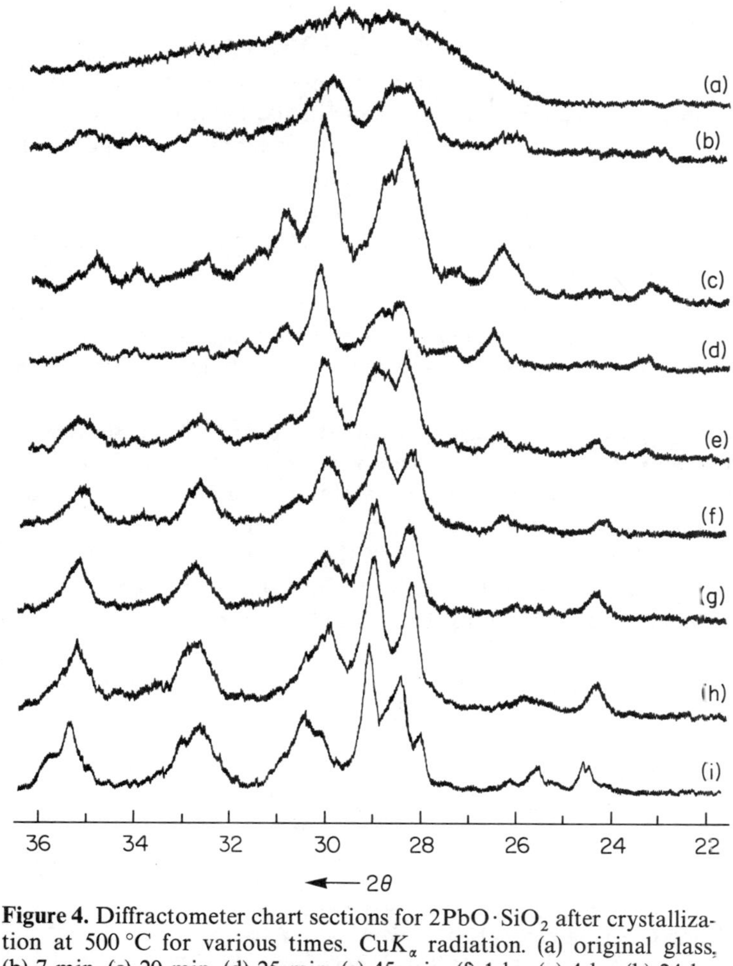

Figure 4. Diffractometer chart sections for $2PbO \cdot SiO_2$ after crystallization at 500 °C for various times. CuK_α radiation. (a) original glass, (b) 7 min, (c) 20 min, (d) 25 min, (e) 45 min, (f) 1 hr, (g) 4 hr, (h) 24 hr, (i) after crystallization for 24 hr at 725 °C

30·0° (d = 2·98), and possibly also the gradual decrease in intensity of the reflection at 26·3° may be related to the gradual decrease of the monomeric ion $SiO_4{}^{4-}$, which levels off at about the same annealing time as these intensities.

The final pattern (curve i) in Figure 4 is for the fully crystalline $2PbO \cdot SiO_2$, after heat treatment at 725 °C for 24 hr. It is identical with that reported by

Argyle and Hummel (1960) for crystalline $2PbO \cdot SiO_2$ fired at 725 °C and by Smirnova, Tikhomirov and Alekseev (1967) for material fired at 620 °C, except for one additional low-angle reflection close to 28°.

CONCLUSIONS

Although the technique of trimethylsilylation is at an early stage of development and further improvements in the method are to be expected, it offers a useful procedure for the structural elucidation of silicates. It would appear that structures can be identified in silicate glasses by this method which are difficult to detect by other means. The method also promises to be a useful adjunct to X-ray procedures for the study of glasses and crystalline silicates with discrete anions.

Our results for the lead orthosilicate glass confirm the view that a distribution of polyanions is present in this material. In this respect silicate glasses are analogous to the phosphates, although there are important differences in detail.

The dominance of the tetrameric ring structure in the crystalline orthosilicate was unexpected. For stoichiometric reasons, there must be four O^{2-} ions for each $Si_4O_{12}{}^{8-}$ complex in the crystalline material. The changes in the diffraction pattern cannot be interpreted in terms of changes in atomic position until the detailed structure of the crystalline orthosilicate is known. However it seems reasonably safe to conclude that, during crystallization, the $SiO_4{}^{4-}$ groups which predominate in the glass undergo dimerization initially, with elimination of free O^{2-} ions, before finally transforming into the stable structure which is characterized by the Si_4O_{12} ring complexes.

REFERENCES

Argyle, J. F. and F. A. Hummel (1960). *J. Am. Ceram. Soc.*, **43**, 452.

Charette, G. G. and S. N. Flengas (1968). *Can. Met. Quart.*, **7**, 191.

Götz, J., W. D. Jamieson, and C. R. Masson (1971). *Proceedings of the Annual Meeting, International Commission on Glass*, Toronto. Sept. 3–6, 1969 (to be published).

Götz, J. and C. R. Masson (1970). *J. chem. Soc., A*, 2683.

Götz, J. and C. R. Masson (1971). *J. chem. Soc., A*, 686

Kolb, K. E. and K. W. Hansen (1965). *J. Am. Ceram. Soc.*, **48**, 439.

Lentz, C. W. (1964). *Inorg. Chem.*, **3**, 574.

Lentz, C. W. (1967). *Industrie chim. belge*, **2**, Special Number (Part 2), 487.

Masson, C. R. (1965). *Proc. R. Soc.* **287A**, 201.

Masson, C. R. (1968). *J. Am. Ceram. Soc.*, **51**, 134.

Masson, C. R., I. B. Smith, and S. G. Whiteway (1970a). *Can. J. Chem.*, **48**, 201.

Masson, C. R., I. B. Smith, and S. G. Whiteway (1970b). *Can. J. Chem.*, **48**, 1456.

Richardson, F. D. and L. E. Webb (1955). *Trans. Inst. Mining Met.*, **64**, 529.

Smirnova, E. V., G. P. Tikhomirov, and A. G. Alekseev (1967). *Izv. Akad. Nauk. SSSR*, Neorg. Mat. **3**, 1230.

Sridhar, R. and J. H. E. Jeffes (1967). *Trans. Inst. Min. Met.* **76**, C44.

Westman, A. E. R. (1960). In J. D. MacKenzie (Ed.), *Modern Aspects of the Vitreous State*, Vol. I. Butterworths, London. pp. 63–91.
Whiteway, S. G., I. B. Smith, and C. R. Masson (1970). *Can. J. Chem.*, **48**, 33.
Wu, F. F. H., J. Götz, W. D. Jamieson, and C. R. Masson (1970). *J. Chromat.* **48**, 515.

34

On the Existence of Submicrostructure in Fused Silica

T. P. SEWARD, III

Research and Development Laboratories, Corning Glass Works, Corning, New York, U.S.A.

and D. R. UHLMANN

Department of Metallurgy and Materials Science, Center for Materials Science and Engineering, Massachusetts Institute of Technology, Cambridge, Massachusetts, U.S.A.

The structure of glass has been the subject of considerable attention in recent years. Particular emphasis has been directed to two scales of glass structure: local atomic arrangements, on a scale of about 1 to 10 Å; and submicrostructural features on a scale of about 30 to a few thousand Å. On the larger scale, it has been convincingly demonstrated that liquid–liquid immiscibility is widespread in binary and more complex glass-forming systems. Such immiscibility often gives rise to heterogeneous submicrostructures in these complex glasses. However, for single component glasses, such as SiO_2 and GeO_2, the picture of glass structure which has been evolving is not one composed of discrete regions having definite boundaries and breaks, but rather a continuous medium.

At variance with this general picture of glass structure are a number of observations carried out in the electron microscope. These observations, e.g. (Prebus and Michener, 1954; Zarzycki and Mezard, 1962; Oberlies, 1965), have indicated the presence of submicrostructure in single-component glasses such as SiO_2 and GeO_2. Some of these observations were interpreted in terms of paracrystal or micelle models of glass structure, and they have been cited as evidence of other 'granular' models, e.g. (Warshausky and Robinson, 1968).

In the present paper, we shall consider at some length the possible existence of submicrostructure on a scale of 30 to a few thousand Å in silica glass.

ELECTRON MICROSCOPE OBSERVATIONS

Prebus and Michener (1954), using direct transmission electron microscopy of fractured thin sections, reported a granular structure in fused silica which ranged in scale from less than 30 Å to about 100 Å. They interpreted the granularity as evidence of both an increased density and increased degree of order within so-called micellar regions.

Zarzycki and Mezard (1962), in another transmission electron microscope study, showed similar evidence of a granular structure in both SiO_2 and GeO_2 glasses, with a scale of inhomogeneity between 50 Å and 100 Å. The granularity was interpreted in terms of a paracrystalline model of the glass, with the image contrast being a phase contrast effect due to mutual misorientation of the paracrystalline domains.

Oberlies (1966) has also reported the presence of submicrostructure in fused silica, observed in electron microscope studies of ultra-microtome sections. The submicrostructure was observable only in relatively thick specimens, and when seen was typically on a scale of 500–2000 Å.

The present authors have carried out an extensive electron microscope study of this material. Our principal source of silica glass was General Electric type 204, which is produced by the fusion of natural quartz. Samples of Spectrasil and I. R. Vitreosil were also used. Sections were prepared for viewing in the electron microscope by mechanical thinning and fracturing techniques. In addition, thin films of amorphous silica were formed from the fused natural quartz by the combined techniques of electron beam evaporation and vacuum vapour deposition in a partial atmosphere of oxygen (see Seward, 1967).

The absence of any observable submicrostructure in SiO_2 was found as a general rule for all starting materials and all techniques used in preparing sections for electron microscope viewing.

However, when evaporated carbon films were used to support the glass sections, substructure on a scale of 50–100 Å was often observed (see Plate I). This structure was shown to be an artifact caused by the carbon film, and was not observed when the carbon film was eliminated (see Plate II).

Further, in some (but not all) of the thinnest sections of bulk glass and thinnest vapour deposited films, apparent irregularities were observed. These could not be seen by direct visual examination of the viewing screen, but only in the developed photographic plates. For bulk glass, the scale of these irregularities—when found—was typically in the range 30–50 Å for mechanically-thinned samples and 50–100 Å for fractured specimens *of the same glass.* No difference in the scale or form of the irregularities—when observed—was found after heat treating the glass at 1050 °C for three weeks.*

* Prior to preparing the electron microscope samples, the surfaces of the bulk glass were polished to remove any contaminants or devitrification products.

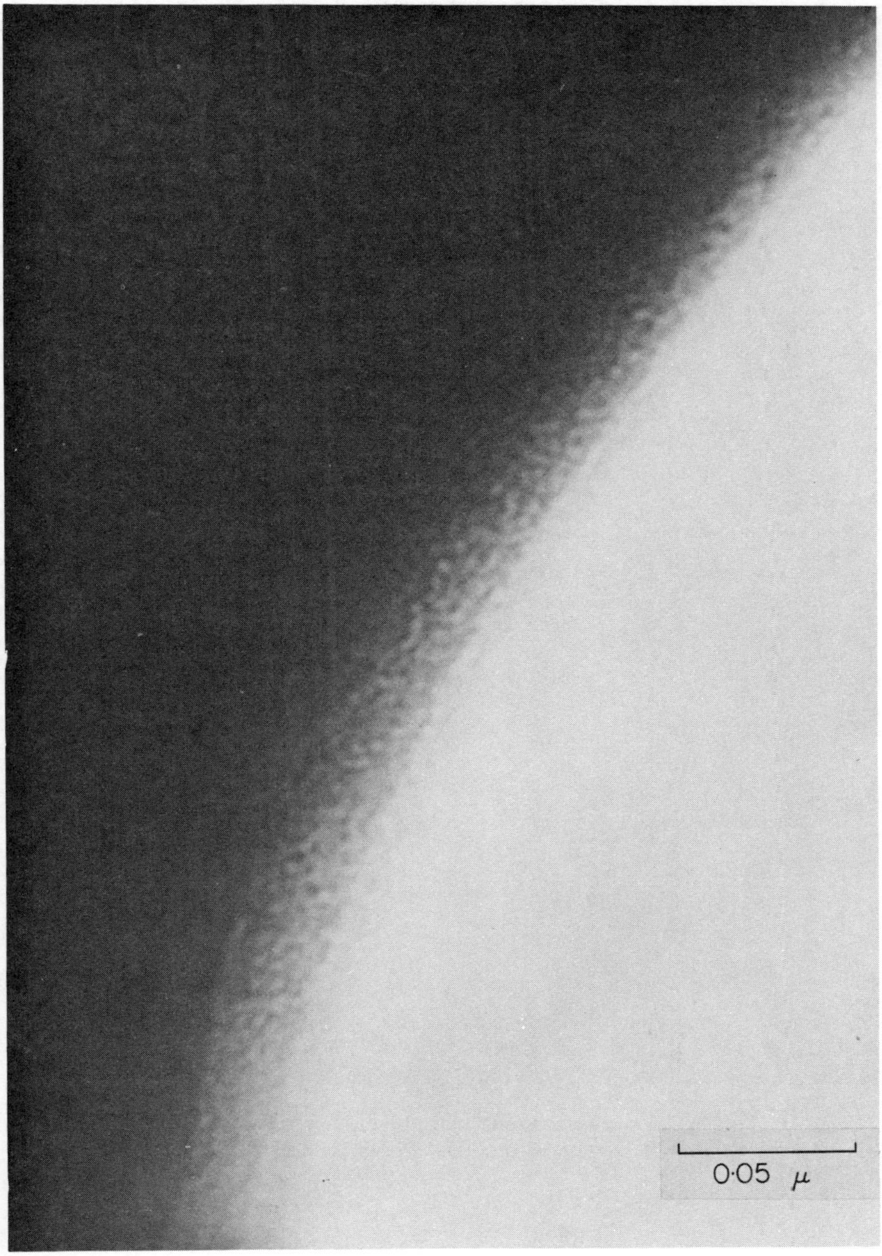

Plate I. Transmission electron micrograph of silica glass. Sample prepared by mechanical thinning technique developed by Doherty and Leombruno (1964). Submicrostructure due to evaporated-carbon support film.

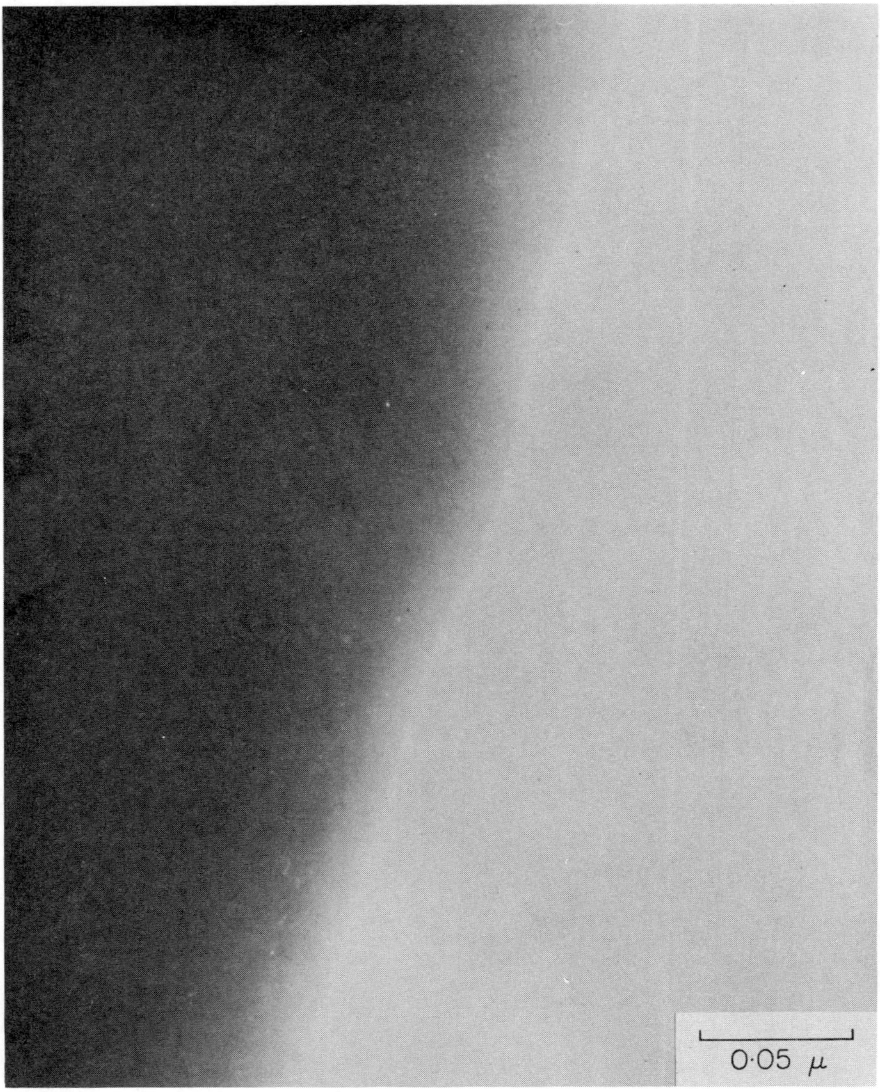

Plate II. Transmission electron micrograph of silica glass. Sample prepared by mechanical thinning technique as modified by Seward, *et al.* (1967). No support film present.

Vapour-deposited films of thickness between 600 and 1500 Å showed no detectable submicrostructure, except at fractured edges where surface irregularities on a scale of 50–150 Å were sometimes seen. These were similar to those observed in fractured bulk silica glasses. Very thin vapour deposited films of thickness between 150 and 300 Å often showed what looked like a surface roughness on a scale of 50 to 100 Å.

In no case did we observe a well-defined granular structure similar to those reported elsewhere.

There are three compelling reasons for concluding that such irregularities —when observed—do not reflect submicrostructural features intrinsic to silica glass, but rather are surface artifacts introduced in the sample preparation: The fact that irregularities were only sometimes observed in some of the thinnest sections; the fact that the scale of the irregularities—when observed—was different in samples of the same material prepared by different techniques; and the fact that the observed irregularities did not change on prolonged heat treatment at temperatures near the glass transition region where changes in the density fluctuations would be expected.

A similar absence of well-defined heterogeneities observable in the electron microscope was noted by Doherty and Leombruno (1965) in a large number of observations on mechanically thinned samples. A faint waviness on a scale of about 70 Å was noted by de Neufville (1969) in SiO_2, GeO_2, and a number of GeO_2–Ge glasses. This waviness—when observed—was found to be independent of sample thickness and to be present in dark field as well as bright field photographs, suggesting that it is associated with a surface diffraction effect and does not reflect the presence of submicrostructure as a characteristic feature of the glasses.

It is appropriate at this point to consider the origin of image contrast in the electron microscope. This will be useful in elucidating the origin of inhomogeneities which are sometimes observed and in evaluating whether such inhomogeneities, if characteristic of the glass, are consistent with data provided by other techniques.

IMAGE CONTRAST IN ELECTRON MICROSCOPY

Contrast in transmission electron microscopy arises from either amplitude or phase modulations of the electron beam, and not from variations in the absorption of electrons by the sample, provided its thickness is less than the electron mean free path. (Samples are usually thinner than this to minimize fogging effects due to multiple scattering.)

Amplitude contrast

Amplitude contrast arises from variations in the degree to which incident electrons are scattered outside the acceptance angle of the objective aperture.

For amorphous materials, such scattering may result from variations in the density (or electron density) of the sample or from thickness variations. Recent analysis by the present authors* has indicated that for samples of constant thickness, t, divided into two regions of density ρ_1 and ρ_2 (Figure 1a), the density difference ($\Delta\rho = \rho_2 - \rho_1$) needed for the minimum observable contrast (5%) on photographic plates† is, assuming 50 kV electrons and an objective aperture angle of 5×10^{-3} radians:

$$\Delta\rho \approx 5 \times 10^{-7}\, t^{-1} \tag{1}$$

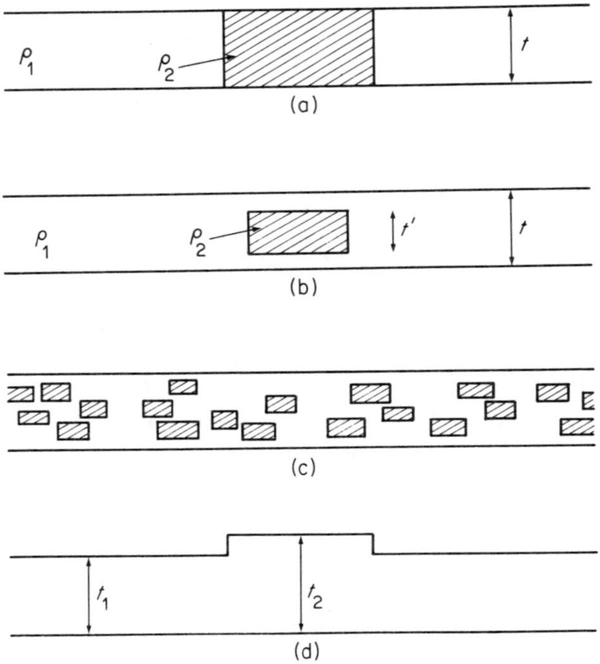

Figure 1. Schematic representation of thin sample configurations showing sources of amplitude contrast. (a) A higher density region extending through the sample. (b) A higher density region within the sample. (c) Many regions of higher density within the sample. (d) A protuberance on the sample surface

* A detailed report of this analysis is being prepared for publication by the authors. See (Seward and Uhlmann, 1970).

† The density differences required for visual observation of contrast on the viewing screen are larger by about a factor of 4.

Hence, for thicknesses of 100, 500, and 1000 Å, density difference of 0·5, 0·1, and 0·05 gm cm^{-3} would be required for the regions to be discerned in the electron microscope. An accelerating voltage of 100 kV would require density differences of twice this value.

If the regions of different density do not extend through the sample (Figure 1b), the density requirements for observability will be independent of sample thickness (provided t is less than or comparable to the electron mean free path so that fogging by multiple scattering does not result). For regions of density ρ_2 and thickness $t' < t$, the requirements on $\Delta\rho$ for $t' = 100$, 500, and 1000 Å are again 0·5, 0·1, and 0·05 gm cm^{-3}.

If the heterogeneities in the sample occur with a sufficient packing density that their images overlap (Figure 1c), the contrast calculations become more involved. For the extreme case where the sample thickness is much greater than the size of the inhomogeneities and the latter occupy a sufficient fraction of the volume, all contrast will be lost in the image, which will be characteristic of a uniform medium of density equal to the average density of the sample.

For the case of a sample of uniform density but varying thickness (Figure 1d), the contrast condition is:

$$\Delta t \approx 5 \times 10^{-7} \rho^{-1} \tag{2}$$

Applying equations (1) and (2) to the question of submicrostructure sometimes observable in SiO_2, we shall take $t = 100$ Å (typical of the reported heterogeneities) and $\rho = 2{\cdot}2$ gm cm^{-3}. For density differences to account for the contrast, these differences between the granules and the intergranular regions must exceed 0·5 gm cm^{-3} (approximately 25% of the bulk density). For thickness variations to account for the contrast, the sample must have variations in thickness exceeding about 23 Å.

Phase contrast

Phase contrast arises from interference effects associated with variations in phase shifts experienced by electrons passing through each region of the sample. Such interference effects, for amorphous materials, can arise from two sources: first Fresnel fringes or edge contours may appear at discontinuities in thickness or refractive index (internal potential) of the sample in a defocused image. Second, phase contrast can result from interference effects when the elastically scattered radiation entering the objective aperture recombines constructively or destructively with the transmitted beam.

Considering the Fresnel fringe case for SiO_2, our analysis has shown that for a sample with uniform density but variable thickness, just-observable contrast would be produced by steps approximately 10 Å high on the surfaces of the samples. Alternatively, if the samples were of uniform thickness but varied in internal potential from one region to the next, differences in internal potential of about 1 volt would be required for observable contrast.

Our analysis has shown that for inhomogeneities 30 Å or more in size, we may ignore contrast produced both by interference effects between elastically scattered and transmitted waves and by Bragg type diffraction.*

Discussion of contrast effects

The diffraction calculations of the preceding sections were based on a minimum observable contrast of 5%. The contrasts in the micrographs reported elsewhere are much greater than this, perhaps even in excess of 50% contrast. Consequently, differences in density, internal potential, or thickness an order of magnitude larger than those calculated may be needed to account for the observed contrast. That is, the more-than-minimal contrast in some of the published electron micrographs and the typical heterogeneity size of 50–100 Å indicate that density variations of 1–2 gm cm^{-3} (or more) and variations in internal potential of 2–4 volts (or more) would be required to account for the observed contrast. Considering that the density of fused silica is about 2·2 gm cm^{-3} and its internal potential is about 10 volts, variations of the required magnitude between the 'grains' and the intergranular regions seem highly unlikely in both cases. As will be discussed in the following sections, such variations would also be inconsistent with the observed small angle X-ray scattering from the material.

We believe, therefore, that the heterogeneities sometimes observed in electron microscope studies of silica glass arise from surface or thickness effects, and are not characteristic of the bulk glass structure. For the contrast seen in some electron micrographs, thickness variations of 50–100 Å would be required for amplitude contrast and steps of 20–40 Å for phase contrast. Such variations in thickness might well be observed with some frequency as various samples are prepared. The large area fraction represented by the observed heterogeneities pose, however, some significant problems in the areas of fracture mechanics and surface chemistry. Specifically, it is not clear why such an irregular topography should result from the fracture process used in preparing the samples. Certainly such fracture would result in a much greater total surface energy than if submicroscopically planar fracture occurred, neglecting the possibility of sizable variations in local cohesion on the scale of 50–100 Å (which would be reflected in large small angle scattering). It is possible, of course, that the heterogeneities might result in some cases from surface crystallization, or adsorption of impurities from the atmosphere or sample preparation materials, or from carbonaceous deposits due to the

* Small crystalline regions can show contrast effects as a result of Bragg scattering (diffraction) outside the objective aperture of the electron microscope. Such scattering from regions 30–200 Å in size would give rise to observable crystalline diffraction patterns in the electron microscope, and, if their volume fraction was sufficient, to observable large angle X-ray diffraction effects as well. Since such Bragg diffraction is in direct variance with experimental results for fused silica, this source of contrast can also be ignored in the discussion.

residual vapour of the electron microscope. Even in these cases, however, the detailed forms of the heterogeneities remain unexplained.

DENSITY FLUCTUATIONS: SMALL ANGLE X-RAY SCATTERING

We now consider the statistical fluctuations expected in a single phase fluid in order to determine whether the observed heterogeneities resemble in any way those we may expect on thermodynamic grounds.

According to thermodynamic fluctuation theory, the mean square density fluctuations expected in such a fluid may be represented (Callen, 1960):

$$\frac{\langle(\Delta\rho)^2\rangle}{\rho^2} = \frac{kTK_T}{V} \tag{3}$$

Here ρ is the average density; K_T is the isothermal compressibility; and V is the volume in which the fluctuations are taking place.

As Weinberg (1963) has discussed, a glass retains the configurational fluctuations, but not the vibrational fluctuations, characteristic of the liquid at the glass transition temperature, T_g. Consequently, Equation (3) can well be replaced by:

$$\frac{\langle(\Delta\rho)^2\rangle}{\rho^2} = \frac{kT_g K_T(T_g)}{V} \tag{4}$$

Taking $T_g = 1300\,°K$ and $K_T = 2{\cdot}1 \times 10^{-12}$ cm^2 dyne^{-1}, the rms density fluctuations become

$$\frac{\langle(\Delta\rho)^2\rangle^{\frac{1}{2}}}{\rho} \approx 6 \times 10^{-13} V^{-\frac{1}{2}} \tag{5}$$

To obtain some feeling for the magnitudes involved, it might be noted that for a cube 20 Å on a side, $\langle(\Delta\rho)^2\rangle^{\frac{1}{2}}/\rho \approx 0{\cdot}7\%$; while for a cube 100 Å on a side, $\langle(\Delta\rho)^2\rangle^{\frac{1}{2}}/\rho \approx 0{\cdot}06\%$.

The small angle X-ray scattering (SAXS) measurements of Weinberg (1963) and Levelut and Guinier (1967) indicate fluctuations somewhat larger than this, corresponding to $K_T = 6{\cdot}2 \times 10^{-12}$ and $7{\cdot}8 \times 10^{-12}$ cm^2 dyne^{-1}, respectively. These measurements indicate fluctuations over a (20 Å)^3 region of about 1% and over a (100 Å)^3 region of about 0·1%. But, it will be recalled that the variations in density needed for observable contrast in the electron microscope were as much as 50 to 100%. These are much larger than the values predicted by thermodynamic fluctuation theory for an isotropic material or the values measured experimentally by SAXS studies.

It has been argued (e.g. Warren, 1937) that the absence of significant small angle scattering indicates that silica glass is essentially continuous and is not composed of discrete ordered regions having definite boundaries and breaks. The force of this argument for the case of very small crystallites has recently been questioned by Bienenstock and Bagley (1966). These authors calculated an upper bound to the SAXS from an array of cristobalite crystallites with a packing density appropriate for fused silica; for crystallites about 8Å in size (about 1 unit cell), the calculated intensities were of the same order of magnitude as the measured intensities.

It was noted, however, that the calculated intensities depend strongly on the assumed crystallite size and on the relative densities of crystal and glass. These calculations have recently been extended by the present authors. The results indicate that for micells or paracrystals in the size range 50–100 Å (greater than about 5 unit cells across) with electron density differences between the paracrystals and the amorphous boundaries sufficient to provide contrast observable in the electron microscope, and occupying volume fractions corresponding to those of the heterogeneities in the published electron micrographs, the calculated SAXS intensities are 1–3 orders of magnitude larger than the measured intensities over the smallest range of k ($4\pi \sin \theta/\lambda$) measured.

These calculations represent something of an upper bound to the expected intensities, and one cannot say with confidence that intensities of the order of the calculated values would in fact be observed. The calculations do, however, remove the force of the Bienenstock and Bagley argument for the range of conditions corresponding to some of the electron microscope observations.

In this light, it seems preferable to estimate the SAXS intensity from fluctuation theory using relations such as equation (4). For relatively large volume fractions of heterogeneities, such as seen in the published electron micrographs in question, the root mean square fluctuation in density should correspond closely with the density variations estimated in discussing the electron microscopy if such variations are to be the effective source of contrast. As we have seen, however, the density fluctuations observed in SAXS studies are smaller by an order of magnitude or more than the variations needed to account for the electron microscope observations.

CONCLUSIONS

An extensive transmission electron microscope study of SiO_2 was performed in which thin sections were prepared by a number of techniques, including vacuum vapour deposition. A number of sections were prepared by each technique. The study indicates that heterogeneities on the scale of 30 to 100 Å are only sometimes seen, and—when seen—their scale depends on the technique employed in preparing the sections and is unchanged on extended heat treatment in the glass transition region.

The sources of contrast in transmission electron microscopy have been considered in some detail. For the case of silica glass, it has been shown that the variations in density and internal potential required to account for the observed heterogeneities far exceed those reasonably expected or those observed in SAXS studies. The heterogeneities have then been associated with surface effects rather than bulk structural features of the glass. Although some suggestions have been made, the origins of the surface features in particular cases remain to be elucidated satisfactorily. In any case, the heterogeneities sometimes seen in electron micrographs or fused silica cannot be taken as structural characteristics of the bulk glass.

ACKNOWLEDGEMENTS

The M.I.T. portion of this work was supported by the Advanced Research Projects Agency, under Contract ARPA SD-90. The authors thank Dr. Doris L. Evans for a critical review of the manuscript and valuable suggestions concerning the presentation of the material.

REFERENCES

Bienenstock, A., and B. G. Bagley (1966). *J. appl. Phys.*, **37**, 4840.
Callen, H. B. (1960). *Thermodynamics.* Wiley, New York.
Doherty, P. E., and R. R. Leombruno (1964). *J. Am. Ceram. Soc.*, **47**, 368.
Doherty, P. E., and R. R. Leombruno (1965). Private communication.
Levelut, A. M. and A. Guinier (1967). *Bull. Soc. fr. Minet. Crystallogr.* **90**, 445.
de Neufville, J. P. (1969). *A Study of Glass Formation in the* GeO_2–Ge *System*, PhD Thesis, Harvard University.
Oberlies, F. (1965). In *Proceedings of VII International Congress on Glass, Brussels, 1965*, Vol. II, Paper 316. Institut National de Verre, Charleroi, and Federation de L'Industrie du Verre, Brussels, Belgium.
Prebus, A. F., and J. W. Michener (1954). *Ind. Engng. Chem.*, **46**, 147.
Seward, III, T. P. (1967). *Structure and Phase Separation in Amorphous Silicates.* Ph.D. Thesis, Harvard University.
Seward, III, T. P., D. R. Uhlmann, D. Turnbull, and G. R. Pierce (1967). *J. Am. Ceram. Soc.*, **50**, 25.
Seward, III, T. P., and D. R. Uhlmann (1970). *Corning Glass Works Internal Report.*
Warren, B. E. (1937). *J. appl. Phys.*, **8**, 645.
Warshausky, M. and H. A. Robinson (1968). *J. appl. Phys.*, **39**, 156.
Weinberg, D. L. (1963). *Phys. Letters*, **7**, 324.
Zarzycki, J., and R. Mezard (1962). *Physics Chem. Glasses*, **3**, 163.

35

Study by Raman Spectroscopy of Crystalline and Glassy Diopside

J. ETCHEPARE

Laboratory of Optics and Molecular Physics, Ecole Polytechnique, Paris, France

Among materials so far studied by infrared or Raman spectroscopy in both crystalline and amorphous states, quartz and vitreous silica and germania are noteworthy. The observations suggest that silica has broadly the structure and even the symmetry found in α quartz. In an attempt to confirm this analogy measurements of the Raman spectra of a single crystal of diopside and of glass of the same stoechiometric composition have been made and are reported here. Diopside is a metasilicate and is characterized by equal numbers of bridging and non-bridging oxygens.

The main feature of the scattering spectra of α quartz and vitreous silica are shown in Figure 1. The figure shows the Raman spectrum of the crystal published by Scott and Porto (1967). Its most intense lines occur at 464 and 207 cm^{-1} and are of type A_1 (α quartz has D_3 symmetry) and correspond to completely symmetrical movements which may be assigned to deformations of the tetrahedral O–Si–O angles. The Raman spectrum of vitreous silica found by the author is added to the figure. It shows, in the same region, a broad, intense, and strongly polarized band which can be attributed to vibrations of the same type. The other much weaker and depolarized bands of the glass can similarly be said to correspond to the type E lines of the crystal; in particular the intense and probably depolarized band at about 60 cm^{-1} can be compared with the line at 128 cm^{-1} (type E) in α quartz. The latter assignment in the spectrum of the glass to a one phonon mode is confirmed by its behaviour as a function of temperature. Hass (1970) showed that the variations in intensity of the whole spectrum of vitreous silica in the temperature range 300–12 °K can be interpreted as a first order process giving Stokes lines of intensity proportional to $1 + n_0$;

$$\text{where } n_0 = 1/[\exp(\hbar w_0/kT) - 1]$$

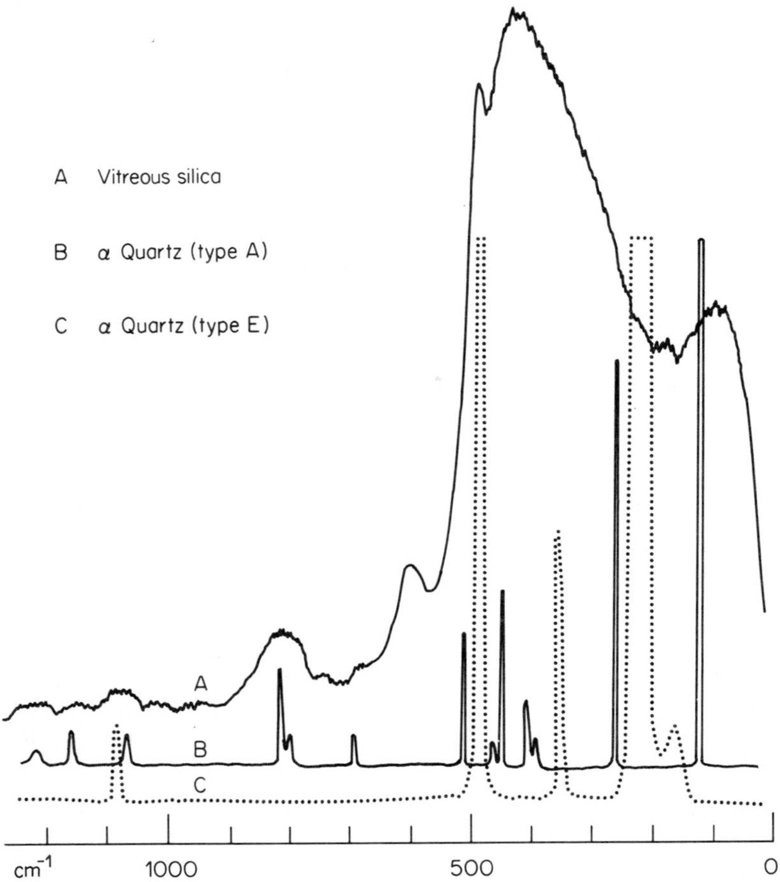

Figure 1. Raman spectra of vitreous silica and α quartz

The comparison of the scattering spectra of two forms of the same compound, where they exist, can thus be useful in interpreting the spectrum of the glass form.

SPECTROSCOPIC STUDY OF DIOPSIDE SINGLE CRYSTAL

A detailed study of the vibrational spectrum of diopside has been made with the hope of interpreting the Raman spectrum of the glass and helping to draw conclusions about its structure and local symmetry.

Diopside, a natural magnesium–calcium silicate, is characterized by the chain structure of a monoclinic pyroxene of formula $(Si_2O_6)_\infty$ (Figure 2).

Its unit cell contains $2[CaMg(Si_2O_6)]$ and its space group is C_{2h}^{6}. Among the thirty fundamental oscillations active in Raman scattering and predicted by group theory, twenty (10Ag + 10Bg) correspond to internal movements of the Si_2O_6 chain; the external motions are divided between translations (2Ag + Bg) and rotation (Bg) of the chains relative to each other and oscillation of the Ca and Mg cations (2Ag + 4Bg).

The spectra of diopside shown in Figures 3 and 4 correspond, according to the notation of Porto (1967), to the following orientations: $x(yy)z + x(yx)z$

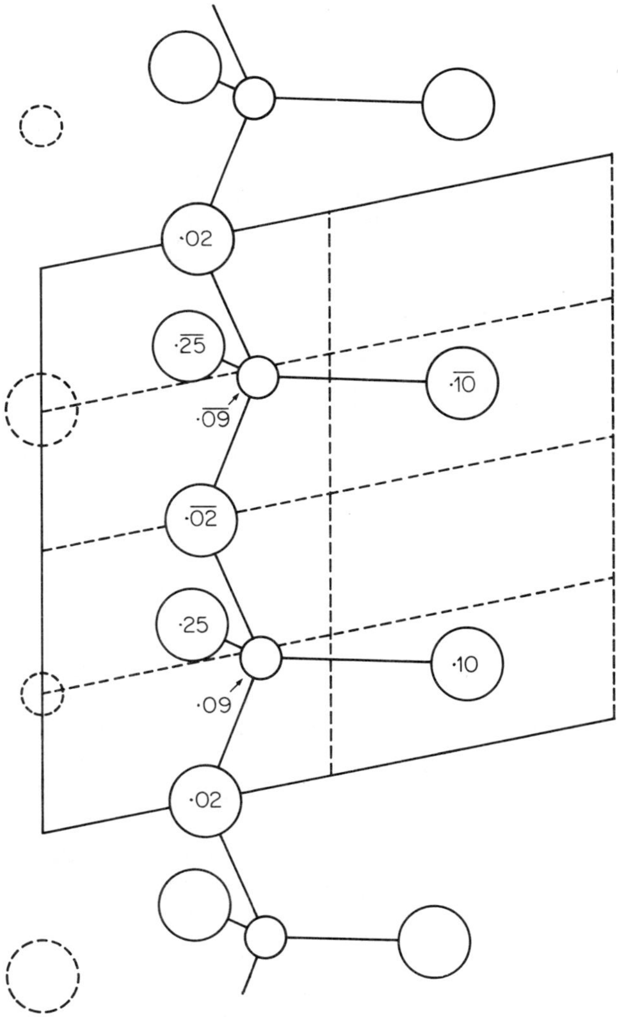

Figure 2. Configuration of the pyroxene chain

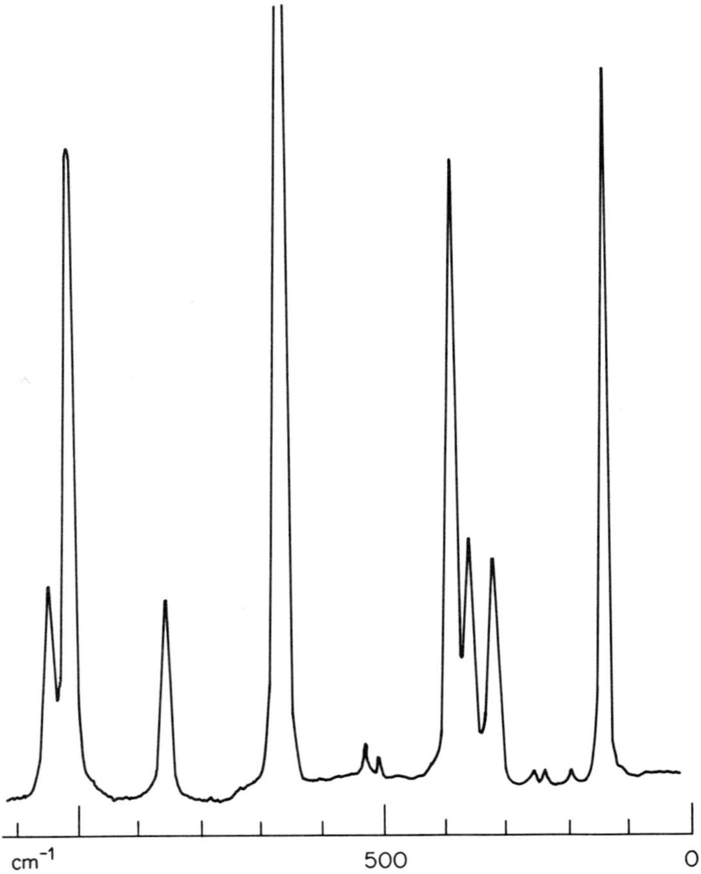

Figure 3. Raman spectrum of diopside crystal: type Ag vibrations

(type Ag movements) and $x(zy)z + x(zx)z$ (type Bg). The scattering spectra of single crystal diopside* show that certain lines correspond to Si–O vibrations (movements of the silicate chain) and others must be attributed to Ca–O or Mg–O bonds. So far as translations and rotations of the chains relative to each other are concerned, it is believed that the Raman lines which they produce are too weak to be identified with confidence. As the forces between the different chains are weak, coupling between them can be neglected in the first approximation and assignments of their internal motion can then be deduced from an elementary calculation of the normal modes of

* Natural crystals kindly supplied by Prof. Guillemin and Mr. Sainfeld (École Nationale Supérieure des Mines, Paris) and Mr. Fauquier (Museum of Natural History, Paris).

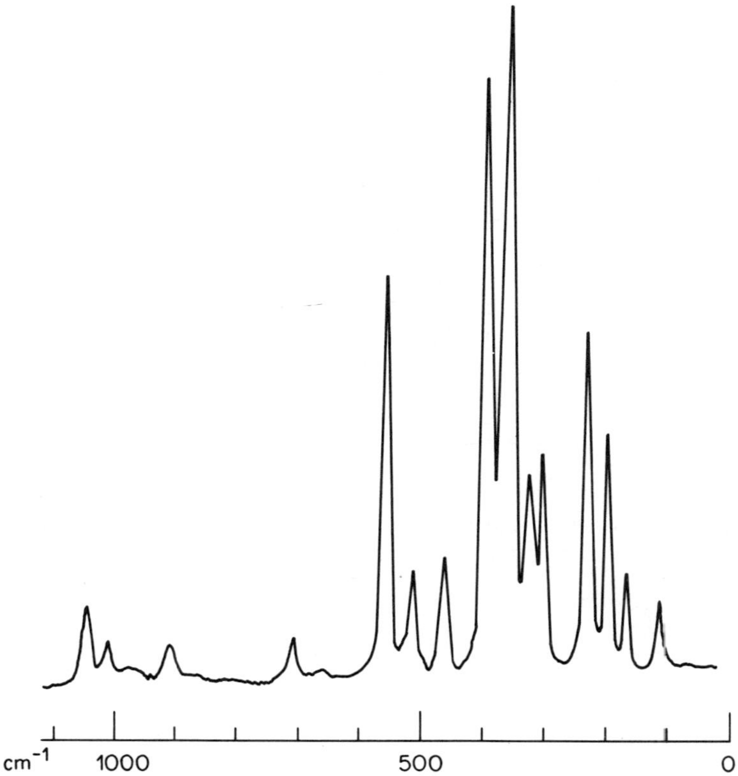

Figure 4. Raman spectrum of diopside crystal: type Bg vibrations

vibration of a silica chain structure (Etchepare, 1970). On the other hand the examination of the intensities of the lines at different orientations of the single crystal in relation to the plane of polarization of the incident laser beam allows certain lines to be assigned to well defined movements. The assignment of motion of cations takes account of the work of Tarte (1962, 1963) on orthosilicates using infrared spectroscopy. All these results are gathered together in Tables 1 and 2.

RAMAN SPECTRUM OF GLASSY DIOPSIDE

Figure 5 shows the Raman spectra of a glass* having the same composition as the crystal and its appears quite transparent and free from micro-crystallization. The features of the spectra are similar to those of other glasses of

* Made in a platinum crucible by Sovirel Société.

Table 1. Frequencies of vibration (cm^{-1}) of crystalline diopside, symmetry Ag.

Deformation vibrations and chain movements	140		
	182		
	235		
	256		
		326	M—O bonds vibrations
	360 ?		
		390	
	508 ?		
	530		
	668		
Si–O stretch	858		
	1016		
	1050		

Table 2. Frequencies of vibration (cm^{-1}) of crystalline diopside, symmetry Bg.

Deformation vibrations and chains movements	118		
	169		
	198		
	232		
	305		
		328	M–O bonds vibrations
	358		
		395	
	468		
	517		
	560		
Si–O stretch	710		
	920		
	1015 ?		
	1048?		

similar type: the bands are broad and the maxima poorly defined. However maxima are situated at 90, 340, 625 and 1000 cm^{-1}.

Theory tells us that, for the first order Raman spectrum of a crystal, only the optical branches of the phonon spectrum have non-zero frequencies for wave vector $\mathbf{k} = 0$ and this leads to a series of discrete lines. When going over to a disordered medium, the translational symmetry is destroyed and this invalidates the selection rule $\mathbf{k} = 0$. As a result the line spectrum of the crystal is replaced by a continuous spectrum which nevertheless reflects the main features of the frequency spectrum of the crystal.

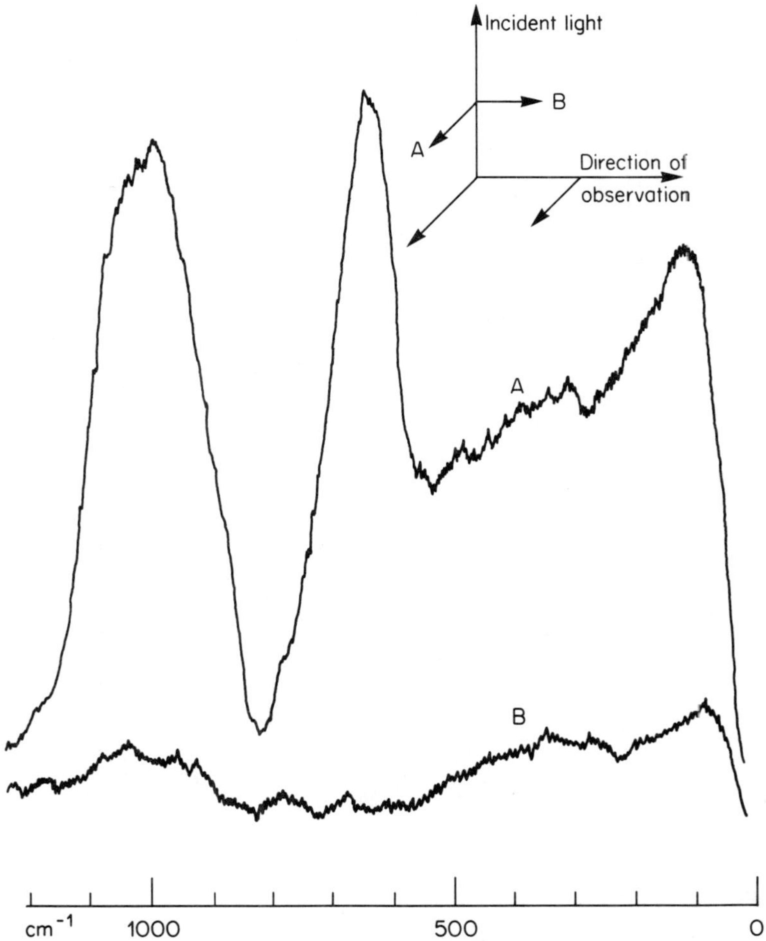

Figure 5. Raman spectra of glassy diopside

Thus one can relate (Figure 6) the bands at 625 and 1000 cm^{-1} to the lines of the diopside crystal at 668 and 860 + 1015 cm^{-1}, we attribute them by analogy to deformations of the O–Si–O angle and vibrations of the Si–O bonds, respectively. The broadening of the lines in this region can be interpreted as due to a statistical distribution of structural parameters such as bond lengths and bond angles, both Si–O–Si and O–Si–O. We indeed pointed out in an earlier work that the stretching vibrations and, above all, deformations of the SiO_4 tetrahedra are in many cases strongly coupled to movements of the Si–O–Si bridges and consequently they are sensitive to their geometrical arrangement.

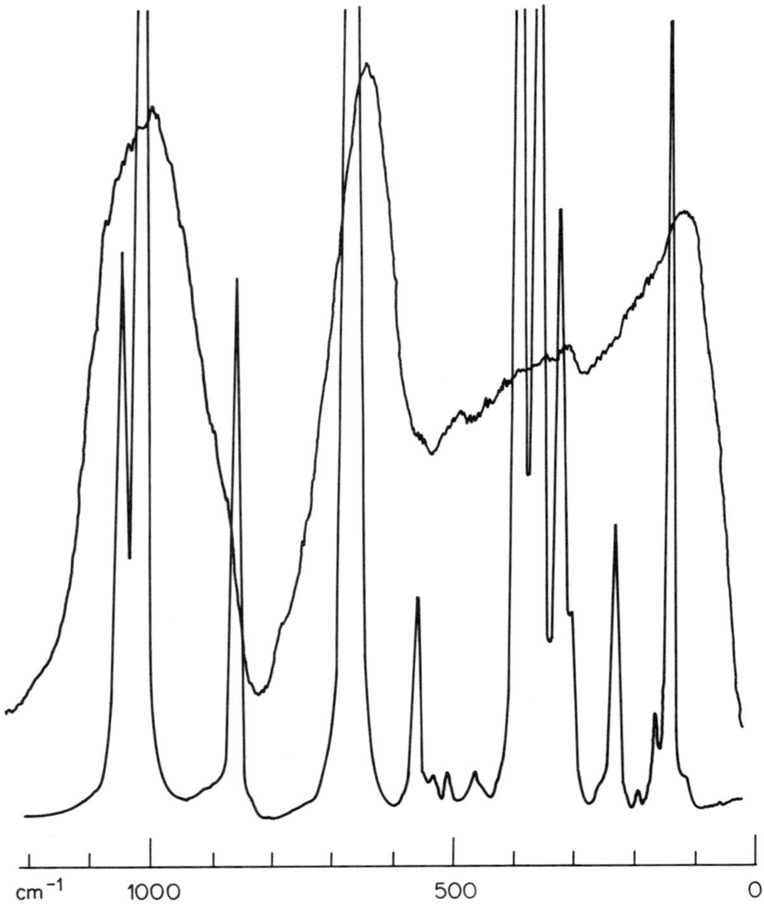

Figure 6. Comparison of Raman spectra of glassy and crystalline diopside

Interpretation of the region below 500 cm^{-1} is a distinctly more delicate matter. There is a continuous background from which emerge only two principal maxima. One of these, at about 340 cm^{-1}, may be attributed to movement of the modifier ions (Ca and Mg) in an oxygen cage of indefinite symmetry. This interpretation derives from a comparison between glass and crystal as well as the analogy with the spectra of other binary glasses with alkaline earth modifiers (Ba and Sr) where a maximum between 300 and 350 cm^{-1} can definitely be assigned to vibrations of the ion. The intense band at about 90 cm^{-1} is attributed to deformations of the –Si–O–Si– chain. Its temperature dependence is identical to that of the type Ag band in diopside crystal at 145 cm^{-1}; it is also similar to the low frequency band in vitreous silica and obeys the same law. However this region of the spectrum below 500

cm^{-1} differs from the corresponding region for the crystal in the presence of a continuous background of intense scattering. The background is partly due to deformations and it is a characteristic of the glassy structure. It thus depends on the great variety of the vibrating units and also on the proportion of bridging oxygens. The disordered structure of the glassy network on short or long range means that a large number of different deformations is possible. New types of vibrations, forbidden in the crystal by the selection rules, become possible and may contribute to scattering in this region, in particular the motion of different links relative to one another. It should however be noted that such movements can only cause weak changes in the polarizability and thus produce bands of only low intensity.

The state of polarization of the Raman spectrum is also of interest. As with all glasses the strongest bands (625 and 1000 cm^{-1} especially) are highly polarized. The most intense lines of the crystalline diopside spectrum, which correspond to those in the glass at 625 and 1000 cm^{-1}, are of type Ag and thus completely polarized. Because of the disordered structure of the glassy lattice, the orientations of all different oscillators are randomly distributed; so we must consider an average effect (similar to a liquid) over all orientations of the polarizability factors. Bands corresponding to 'symmetrical' vibrations will still be polarized; and the most intense bands of the crystal, especially in the high frequency region, being of type Ag, the glass Raman spectrum will appear to be highly polarized. This observation cannot therefore be used to support any particular theory of glass structure.

CONCLUSION

This study of diopside shows the usefulness of comparing the same material in both crystalline and glassy forms. The examination of the vibrations of the crystal allows an interpretation, by analogy, of the most intense bands in the spectrum of the glass and these can be attributed to well defined vibrations of the silica chain. The assignment of a band around 350 cm^{-1} to vibrations of cations is confirmed by earlier work on alkaline earth modifiers (Ba, Sr). Furthermore the polarization of the spectrum of the glass does not disagree with observations made on the single crystal and can be explained by a simple averaging process. Other studies are, however, necessary before definite conclusions can be drawn about the structural analogy between a glass and any specific crystal. For this purpose it would be useful to compare the scattering spectra of crystals having chain structures of different configurations because calculations have shown that the characteristic frequencies of the various vibrations could be significantly different. Infrared spectroscopy of the same materials as both crystal and glass would be useful as they could extend the information available about local symmetries.

ACKNOWLEDGEMENT

This paper has been translated by Dr. M. Cable of the Department of Glass Technology, University of Sheffield.

REFERENCES

Etchepare, J. (1970). *Spectrochim. Acta*, (in press).
Hass, M. (1970). *J. Physics Chem. Solids*, **31**, 415.
Scott, J. F., and S. P. S. Porto (1967). *Phys. Rev.*, **157**, 3.
Tarte, P. (1962). *Spectrochim. Acta*, **18**, 467.
Tarte, P. (1963). *Spectrochim. Acta*, **19**, 25.

36

Light Scattering Studies in Amorphous Media

G. ROSS*

Research Department, Physics Division, Imperial Chemical Industries Limited, Plastics Division, Welwyn Garden City, Great Britain

---

The phenomenon of scattering of electromagnetic radiation, together with other aspects of wave propagation and diffraction, has been, during the last twenty years, the object of an active interest and intensive investigation, largely prompted by the requirements of the fast-developing communication techniques and the study of the microstructure of matter. The aim of this paper is to discuss certain aspects of microstructural determinations, more exactly to try to answer a fundamental problem which arises when investigating the structure of amorphous media by scattering methods: that of performing meaningful measurements.

The study of the scattered radiation can provide substantial information about the morphology of a medium; in fact, one can distinguish two categories of applications. The first consists of the study of atomic, molecular and particulate motions, in other words of structural dynamics, about which one can learn from an examination of the *inelastic scattering*, by concentrating upon the shift in frequency between the incident and scattered radiation, which occurs in, for example, Raman, Brillouin or quasi-elastic Rayleigh scattering, or, indeed, any other Doppler effect. The second range of applications—and only this will be followed in this paper—involves a static approach, making abstraction of the very small proportion of energy absorbed and re-emitted with a different frequency (or making allowance for it), in other words assuming *elastic scattering*. The problem here consists of obtaining information on the spatial distribution of inhomogeneity and this is normally achieved by studying the angular distribution of the scattered radiation. The inhomogeneity may be regular or may present various degrees of 'randomness';

* Present address: Department of Physics, Queen Elizabeth College, University of London, Great Britain.

for the study of amorphous—and other disordered—media, one is usually interested in the latter aspect, even if concerned with the short-range order, since still then the regularity is only partial.

In this paper, only inhomogeneities on a scale larger than the characteristic length of the short-range order will be considered, for which a statistical representation is the most appropriate. The treatment, developed for electromagnetic radiation, will lead to conclusions valid for any wavelength, although only the visible range of the frequency spectrum will be considered experimentally. All deductions, in fact, could be easily extended to any undulatory perturbation, such as De Broglie waves.

ELASTIC SCATTERING IN RANDOM INHOMOGENEOUS MEDIA

For obtaining information about structural statistics, one must first describe the inhomogeneity of the medium; the most adequate representation, since propagation of electromagnetic waves is discussed, is probably to define the random field of the local deviation in permittivity from the average $\delta\varepsilon(\boldsymbol{r})$. Disregarding usually the conductivity and assuming a uniform magnetic permeability, the fundamental formula which governs the scattering phenomenon, customarily given in the literature (see, for example, Debye, 1960), expresses the scattered electric field as the Fourier transform of the inhomogeneity distribution function

$$E_s(\boldsymbol{s}) \propto \mathfrak{f}\ \{\delta\varepsilon(\boldsymbol{r})\} \tag{1}$$

where, if $\boldsymbol{m}$ is the unit vector of the direction of observation and $\boldsymbol{e}_x$ the unit vector of the direction of the incident field,

$$s = |\boldsymbol{s}| = |\boldsymbol{e}_x - \boldsymbol{m}| = 2 \sin \tfrac{1}{2}(\text{arc}\cos(\boldsymbol{e}_x \boldsymbol{m})) \tag{2}$$

Apply the convolution theorem to (1) and the relation normally used in practical determinations follows. This expresses the Fourier relationship between the intensity of the scattered field and the convolution square of the inhomogeneity distribution function over the irradiated volume V:

$$\delta I(\boldsymbol{s}) \propto \mathfrak{f}\ \{\overset{2}{\widetilde{\delta\varepsilon(\boldsymbol{r})}}\} = \mathfrak{f}\ \{\int_V \delta\varepsilon(\boldsymbol{r})\, \delta\varepsilon(\boldsymbol{r}+\boldsymbol{\rho})\, \mathrm{d}V\} \tag{3}$$

For convenience, the intensity is usually measured by the Rayleigh ratio (or scattering cross-section) defined by:

$$R(s) = \frac{1}{V}\frac{\delta I(s)}{\phi_0/\mathscr{A}_0} = \frac{1}{\Delta l}\frac{\delta I(s)}{\phi_0} \tag{4}$$

where ϕ_0 is the flux in the incident beam of cross-section $\mathscr{A}_0$ and Δl is the thickness of the sample.

Thus, for a *random* scattering medium of *infinite* extent,

$$R(s) \propto \mathfrak{f}\{F'(\boldsymbol{\rho})\} \tag{5}$$

where the autocovariance of the random field $F'(\boldsymbol{\rho})$ is given by:

$$F'(\boldsymbol{\rho}) = \lim_{V\to\infty} \frac{1}{V}\int_V \delta\varepsilon(\boldsymbol{r})\,\delta\varepsilon(\boldsymbol{r}+\boldsymbol{\rho})\,\mathrm{d}V \tag{6}$$

For simplicity, it is customary to normalize the autocovariance, and the autocorrelation function $F(\rho)$ is defined by

$$F'(\rho) = F(\rho)\,F'(0) = F(\rho)\,\langle\delta\varepsilon^2\rangle \tag{7}$$

where $\langle\delta\varepsilon^2\rangle$ is the variance of the random field.

This applies to any inhomogeneous medium. One ought to remark here that the inhomogeneity may be either continuous or discrete; the second type, although important in certain circumstances, is less typical for an amorphous medium that the first one (at least for the scale of inhomogeneity investigated) and will not be discussed here in more detail.

For a continuous inhomogeneous medium, the most common correlation functions have been examined by Ross (1968, 1969a); they represent monotonic functions, vanishing asymptotically, the correlation distance a representing roughly the width of the curve and describing the scale of the inhomogeneity.

Although, in principle, one could study the wavelength dependence of the scattered intensity at a given angle, it is easier to determine the angular distribution of intensity; the corresponding curve, monotonic decreasing, has a width roughly of the order $1/\langle k\rangle a$, $\langle k\rangle$ being the average wavenumber in the medium with average refractive index $\langle n\rangle$;

$$\langle k\rangle = \frac{2\pi}{\langle\lambda\rangle} = \frac{2\pi}{\lambda_0}\langle n\rangle$$

Obviously, not only for practical reasons, the scattering volume cannot have an infinite extent, but, indeed, the formula (1) has been obtained making certain simplifying assumptions which ought to be stated clearly, because all of them impose an upper limit on the extent of the scattering volume:
—Born approximation

$$\frac{\langle\delta\varepsilon^2\rangle}{\langle\varepsilon\rangle^2} \ll 1 \tag{8a}$$

$$\tfrac{1}{2}k_0{}^2[\delta\varepsilon]_{\max}\,r^2_{\max} \ll 1 \tag{8b}$$

where

$$|\delta\varepsilon(r)| \leqslant [\delta\varepsilon]_{\max}$$

and r_{max} is the maximum extent of the scattering volume

—Fraunhofer approximation

$$L \gg \frac{\langle k \rangle r_{max}^2}{2} \tag{9}$$

where L is the distance sample-detector

—Conditions for temporal and spatial coherence of the incident electromagnetic field:

—quasimonochromaticity, i.e. the frequency of the electromagnetic field restricted to a band width

$$\nu_0 - \Delta\nu \leqslant \nu \leqslant \nu_0 + \Delta\nu, \qquad \Delta\nu \ll \frac{c}{r_{max}} \tag{10}$$

— very high spatial coherence, i.e.

$$\eta_{coh} \gg \mathscr{D} \tag{11}$$

where $\mathscr{D}$ is the diameter of the irradiating beam and η_{coh} is the (average) distance over which the phases of the incident electromagnetic field are correlated (coherence distance).

For a finite extent of the scattering volume, the average, per unit volume, of the convolution square of the inhomogeneity distribution function ceases to be, and to have the properties of, an autocovariance.

Let us define the shape function of the scattering volume, $T(\boldsymbol{r})$:

$$T(\boldsymbol{r}) = \begin{cases} 1, & P(\boldsymbol{r}) \in V \\ 0, & P(\boldsymbol{r}) \notin V \end{cases} \tag{12}$$

where P is a generic point, characterized by the oriented distance $\boldsymbol{r}$ from an arbitrary origin.

Then,

$$\mathscr{T}(\boldsymbol{\rho}) = \frac{1}{V}\overset{2}{\widetilde{T(\boldsymbol{r})}} = \frac{1}{V}\int_V T(\boldsymbol{r})\, T(\boldsymbol{r} + \boldsymbol{\rho})\, \mathrm{d}V \tag{13}$$

describes (apart from the factor $1/V$) the convolution square of the shape function.

Obviously,

$$\int_V \delta\varepsilon(r)\, \delta\varepsilon(r + \rho)\, \mathrm{d}V = \int_{\overset{2}{\widetilde{T(r)}}} \delta\varepsilon(r)\, \delta\varepsilon(r + \rho)\, \mathrm{d}V = \overset{2}{\widetilde{\delta\varepsilon(r)}} \tag{14}$$

and hence, for a finite extent of the scattering volume,

$$R(\boldsymbol{s}) \propto \frac{1}{V}\mathfrak{f}\left\{\overset{2}{\widetilde{\delta\varepsilon(\boldsymbol{r})}}\right\} = \frac{1}{V}\mathfrak{f}\left\{\frac{\overset{2}{\widetilde{\delta\varepsilon(\boldsymbol{r})}}}{\overset{2}{\widetilde{T(\boldsymbol{r})}}}\,\overset{2}{\widetilde{T(\boldsymbol{r})}}\right\} = \mathfrak{f}\{F'_V(\boldsymbol{\rho})\ \mathscr{T}(\boldsymbol{\rho})\} \tag{15}$$

$F'_V(\rho)$ represents, therefore, the average value of the convolution square of the inhomogeneity distribution function over the volume common to the scattering volume and its 'ghost', for a displacement equal to ρ, i.e. the volume available for the average is expressed by $\overset{2}{\widetilde{T(r)}}$.

Nevertheless, if the maximum extent of the scattering volume is much larger than the average distance over which the local deviations in permittivity from the average are correlated, measured by the correlation distance a, i.e. if

$$r_{\max} \gg a \tag{16}$$

(and this is the typical case in practical determinations), it is easy to visualize that a representation of $F'_V(\rho)$ against ρ looks like Figure 1.

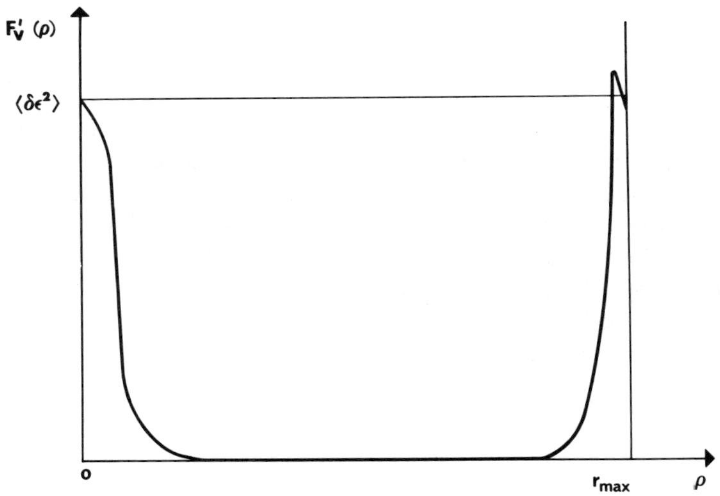

Figure 1. $F'_V(\rho)$ *vs.* ρ

In other words, with the condition (16), it is possible to express $F'_V(\rho)$ as a sum:

$$F'_V(\rho) = \frac{\left[\overset{2}{\widetilde{\delta\varepsilon(r)}}\right]_{\rho\simeq a}}{\overset{2}{\widetilde{T(r)}}} + \frac{\left[\overset{2}{\widetilde{\delta\varepsilon(r)}}\right]_{\rho\simeq r_{\max}}}{\overset{2}{\widetilde{T(r)}}} \simeq F'(\rho) + \langle\delta\varepsilon^2\rangle f_V(\rho) \tag{17}$$

because the first term is virtually identical to $F'(\rho)$.

f_V represents a correction term, accounting for the finite extent of the scattering volume. A quantitative expression for it is not obtainable; clearly, it depends strongly upon direction, its height being either larger or smaller than 1. Nevertheless, for a random medium, the average contribution from all directions corresponds to a maximum height roughly equal to 1. Obviously, the fact that, for ρ increasing, a smaller volume is available for the average, is accounted for by the weighting factor $\mathscr{T}(\rho)$. $f_V(\rho)$ is negligibly small, practically zero, for all values of ρ except those contained in a narrow interval, having a width of the order of a, situated at very large values of ρ, of the order of r_{max}.

Therefore, with the condition (16) and taking into account (17), (15) can be written (assuming, for simplicity, isotropy);

$$R(s) \propto \mathfrak{f}\{F(\rho)\,\mathscr{T}(\rho)\} + \mathfrak{f}\{f_V(\rho)\,\mathscr{T}(\rho)\} \tag{18}$$

Since, for $r_{max} \gg a$, the width of the curve $F(\rho)$ is much smaller than that of the curve $\mathscr{T}(\rho)$, one can assume

$$F(\rho)\,\mathscr{T}(\rho) \simeq F(\rho) \tag{19}$$

and hence

$$R(s) \propto \mathfrak{f}\{F(\rho)\} + \mathfrak{f}\{f_V(\rho)\,\mathscr{T}(\rho)\} \tag{20}$$

The first term, the Fourier transform of the autocorrelation function, represents the angular distribution of intensity for the ideal case of an infinite scattering volume, i.e. the power spectrum of the inhomogeneity distribution function, whilst the second describes the spurious oscillations superimposed over it, i.e. the 'speckle noise', observed whenever highly coherent radiation is scattered by a random inhomogeneous medium of finite extent.

Since the product $f_V(\rho)\,\mathscr{T}(\rho)$ is zero outside a (narrow) band of width roughly equal to a, centered at $\rho_0 \simeq r_{max}$ let

$$f_V(\rho)\,\mathscr{T}(\rho) = g_V(\rho - r_{max}) = g_V(\rho) * \delta(\rho - r_{max})$$

Hence,

$$\mathfrak{f}\{f_V(\rho)\,\mathscr{T}(\rho)\} \propto [\mathfrak{f}\{g_V(\rho)\}]\,\frac{\sin\langle k\rangle r_{max}s}{\langle k\rangle r_{max}s}, \tag{21}$$

where the sign $*$ denotes convolution, $\delta(\rho - r_{max})$ is the (shifted) Dirac distribution, and $\dfrac{\sin\langle k\rangle r_{max}s}{\langle k\rangle r_{max}s}$, its (three-dimensional) Fourier transform, is a highly oscillating function, which describes the speckle effect, indicating an extent of the speckles equal to the distance between two successive maxima,

$$\Delta s = \frac{\langle \lambda \rangle}{r_{max}} \tag{22}$$

In fact, of course, the pattern is more complicated, due to the fact that it consists of a superposition of the effects in all possible directions, in the three-dimensional scattering volume.

It is also possible to see, based on what was said above, that the relative amplitude of the noise, described by the signal/noise ratio, is, as a first approximation

$$\frac{S}{N} \propto \frac{r_{\text{max}}}{a} \tag{23}$$

The speckle noise represents a bar to exact measurements, imposing an upper bound to the resolution and ultimately limiting the accuracy in determining the correlation function.

One obvious way of increasing the signal/noise ratio is to increase the scattering volume. Nevertheless, the conditions (8–10), as well as practical reasons, prevent an indefinite increase in the extent of the sample.

There is, however, available, another means of reducing the speckle noise: by using partially coherent radiation, in other words by relaxing the condition (11).

Indeed, by taking into account the coherence characteristics of the incident electromagnetic field, one can show that, for an extended quasimonochromatic source, the angular distribution of scattered intensity is given, instead of (15), by (Ross 1969b, 1970):

$$R(s) \propto \mathfrak{f}\{F'_V(\rho)\,\mathscr{T}(\rho)\,|\mu|(\eta)\} \tag{24}$$

i.e., with the hypothesis (19)

$$R(s) \propto \mathfrak{f}\{F(\rho)\,|\mu|(\eta)\} + \mathfrak{f}\{f_V(\rho)\,\mathscr{T}(\rho)|\mu|(\eta)\} \tag{25}$$

where

$$\boldsymbol{\rho} = \boldsymbol{\xi} + \boldsymbol{\eta} = \boldsymbol{e}_x\xi + \boldsymbol{\eta}$$

$|\mu|$ is the coherence factor: its expression, for an experimental set-up using an extended incoherent quasimonochromatic (secondary) source, can be derived from the van Cittert–Zernike theorem:

$$|\mu| = 2\frac{J_1[k_0(\mathscr{D}_1/2\mathscr{F})\eta]}{k_0(\mathscr{D}_1/2\mathscr{F})\eta} \tag{26}$$

if $\mathscr{D}_1/2\mathscr{F} \ll 1$; $\mathscr{D}_1$ is the diameter of the (circular) secondary source, $\mathscr{F}$ the focal distance of the collimator and J_1 is the symbol for Bessel function of order 1.

The width of the curve $|\mu|(\eta)$ (i.e. the value of η corresponding to $|\mu| = 0{\cdot}5$) is the coherence distance:

$$\eta_{\text{coh}} = \frac{4}{k_0}\frac{\mathscr{F}}{\mathscr{D}_1}. \tag{27}$$

With a judicious choice of $\mathscr{F}$ and $\mathscr{D}_1$, so that

$$a \ll \eta_{\text{coh}} \ll r_{\text{max}} \tag{28}$$

the first term in (25) becomes practically $\mathfrak{f}\{F(\rho)\}$, whilst the second is drastically diminished; no speckle effect can be observed and the angular distribution of intensity is virtually identical to that corresponding to an infinite volume and complete coherence of the electromagnetic incident field. The partial coherence, therefore, effectively acts as a low-pass filter.

The second term in (25) can be written as

$$\mathfrak{f}\{f_V(\rho)\,\mathscr{T}(\rho)\,|\mu|\,(\eta)\} = [\mathfrak{f}\{f_V(\rho)\}\mathscr{T}(\rho)\}] * [\mathfrak{f}\{|\mu|(\eta)\}] \tag{29}$$

For a uniform distribution of intensity across the secondary source, obviously,

$$\mathfrak{f}\{|\mu|(\eta)\} = T(\sigma/2\mathscr{F}) \tag{30}$$

where the shape function $T(\sigma/2\mathscr{F})$ is defined by

$$T(\sigma/2\mathscr{F}) = \begin{cases} 1, \sigma \leqslant \mathscr{D}_1 \\ 0, \sigma > \mathscr{D}_1 \end{cases} \tag{31}$$

σ being the radial variable in the plane of the secondary source.

Hence, the effect of the low-pass partial-coherence filter may be expressed as

$$\mathfrak{f}\{f_V(\rho)\,\mathscr{T}(\rho)\,|\mu|\,(\eta)\} = [\mathfrak{f}\{f_V(\rho)\,\mathscr{T}(\rho)\}] * T(\sigma/2\mathscr{F}) \tag{32}$$

It ought to be observed that the same convolution may be obtained if, instead of using partially coherent radiation, i.e. an extended secondary source, the experiment is performed with (completely) coherent radiation, but one explores the Fraunhofer plane with an identical aperture, i.e. a circle with diameter $\mathscr{D}_1$ (for example the diaphragm of the microdensitometer which explores the photographic record or a pinhole in front of the measureing photomultiplier).

Indeed, it is easy to see that *both* the source plane $\boldsymbol{\sigma}$ *and* the Fraunhofer plane $\boldsymbol{s}$ are Fourier conjugates of the primary, i.e. sample, plane $\boldsymbol{n}$; the smearing out of the spurious oscillations (speckle noise) may therefore be performed by an extended aperture in *either* plane. In either situation, the result of the measurement may, in fact, be written:

$$R(s) \propto [(\mathfrak{f}\{F(\rho) + f_V(\rho)\}) * (\mathfrak{f}\{\mathscr{T}(\rho)\})] * T(\sigma/2\mathscr{F})$$

and with the assumption (28),

$$= \mathfrak{f}\{F(\rho)\} + [(\mathfrak{f}\{f_V(\rho)\}) * (\mathfrak{f}\{\mathscr{T}(\rho)\})] * T(\sigma/2\mathscr{F}) \simeq \mathfrak{f}\{F(\rho)\} \tag{33}$$

in other words the same result as in the ideal case.

Such smoothing or smearing out of the spurious oscillations is known in information theory as filtering while in optics it is traditionally called apodisation. It can be said, therefore, that the same effect can be achieved by the averaging, in the Fraunhofer plane, of the intensity distribution function over the area of the exploring diaphragm, or by a change in the phase distribution of the incident field in the primary plane, obtained by the aperture in the source plane; either procedure is equivalent to multiplying the convolution square of the inhomogeneity distribution function with a suitable 'apodising' function, namely the coherence factor (26).

As is well known, the suppression, or drastic reduction in the height of the 'feet' which occur in speckle noise (or diffraction patterns, which are produced, like the speckle noise, by the non-vanishing values of the function defined in the primary space in the available interval), in other words, an increase in signal/noise ratio, cannot be performed without a loss in resolution; it can be shown (Ross, 1970) that the maximum extent of the inhomogeneity still detectable is

$$r_{\max} = \frac{\langle \lambda \rangle}{\Delta s} \tag{34}$$

where Δs is the uncertainty in assessing the scattering angle, i.e.

$$r_{\max} = \frac{2\mathscr{F}}{\mathscr{D}_1} \langle \lambda \rangle \tag{35}$$

or, otherwise expressed, the maximum correlation distance ascertainable becomes

$$a_{\max} = \eta_{\mathrm{coh}} = \frac{4}{k_0} \frac{\mathscr{F}}{\mathscr{D}_1} = \frac{1}{\pi} \frac{2\mathscr{F}}{\mathscr{D}_1} \lambda_0 \tag{36}$$

It ought to be remarked that, although mathematically the smearing out *a priori* (in the source plane) and *a posteriori* (in the Fraunhofer plane) are identical, from a practical point of view the effects are different; the second alternative, offering the possibility of using lasers, is to be preferred whenever expeditiousness is the predominant consideration, whilst the first, presenting the advantage of allowing only partially coherent radiation throughout the optical system—whose defects' influence is thus diminished—is to be preferred when accuracy is the main concern. The Fraunhofer requirement, too, is, for partially coherent radiation

$$L \gg \frac{\langle k \rangle \eta^2_{\mathrm{coh}}}{2} \tag{37}$$

less stringent than for coherent radiation (9) and this points towards the advantage of the first alternative.

CONCLUSION

Once the optimum conditions for the experiment have been established and the power spectrum of the inhomogeneity distribution function has been recorded, the problem consists of determining the autocorrelation function. Obviously, it is not possible to compute directly the Fourier transform of the measured angular spectrum of intensity because, as can be easily seen from the Whittaker–Shannon sampling theorem, the resolution in determining $F(\rho)$ would be

$$\Delta\rho = \frac{\langle\lambda\rangle}{s_{\text{max}}}$$

therefore, at best,

$$\Delta\rho = \frac{\langle\lambda\rangle}{2}$$

This difficulty can be circumvented if a trial and error method is used, consisting of comparing the recorded intensity with the Fourier transforms of the most common types of autocorrelation function, precomputed and tabulated for a large spectrum of correlation distance. This is equivalent to—and may be interpreted as—an extrapolation of the values of $R(s)$ towards infinitely large values of s. Using this approach, one can assess, with visible light, inhomogeneities characterized by correlation distances which may be as small as 100 Å and—for very careful determinations—even less. Naturally, for such fine inhomogeneities, only the *dominant* scale is observable; smaller details are, in general, lost, especially if the dominant scale is lower than $\langle\lambda\rangle/2$.

The design of the experimental equipment, obviously, must follow the assumptions made in deducing the theory. For a mercury lamp (provided with a filter which selects the green line, thus realizing a quasimonochromatic radiation) as primary source, the diameter of the secondary source and the focal length of the collimator must be chosen so that relation (28) is fulfilled. For a beam diameter $\mathscr{D} = 1\text{mm}$, a good compromise is reached for $\mathscr{D}_1/2\mathscr{F} = 1{\cdot}25 \times 10^{-3}$. Surely, also the condition (37) must be obeyed. Details of the instrumentation, practical method used in determinations and some illustrative examples are given in Ross (1969a).

Here a few problems which occur in scattering determinations have been examined, trying to establish what do we really observe in an experiment and endeavouring to determine the conditions needed for a meaningful measurement. Based on this, the way in which an optimum compromise between various antagonistic requirements can be achieved, has been examined, as well as the accuracy of the results thus obtained.

REFERENCES

Debye, P. (1960). In *Non-crystalline Solids*. p. 1. (V. D. Fréchette, ed.), J. Wiley and Sons, New York.

Ross, G. (1968). *Optica Acta*, **15**, 451.

Ross, G. (1969a). *Optica Acta*, **16**, 95.

Ross, G. (1969b). *Optica Acta*, **16**, 611.

Ross, G. (1970). *Phil. Trans. R. Soc. (London)*, **A268**, 177.

NUCLEAR MAGNETIC RESONANCE AND ELECTRON SPIN RESONANCE STUDIES

37

N.M.R. Relaxations and Molecular Motion in Polymeric Glasses

G. ALLEN

Chemistry Department, University of Manchester, Great Britain

---

For a nucleus having spin angular momentum I placed in a powerful magnetic field H_0 the component of spin in the direction of H_0 (m_z) takes on the values

$$m_z = I, I-1, I-2, \ldots\ldots -I+2, -I+1, -I$$

Thus for a nucleus having $I = \frac{1}{2}$, two energy levels are obtained in the presence of the magnetic field, which are separated by

$$\Delta E = \frac{\gamma}{2\pi} \times H_0 = g\beta H_0$$

where $\gamma = 2\pi\, \mu/Ih$, μ is the magnetic moment of the nucleus, β is the nuclear magneton and g is the Landé factor. Transitions between the two levels are induced by irradiating the sample with radio frequency electromagnetic radiation of frequency ν such that

$$\Delta E = g\beta H_0 = h\nu$$

In an n.m.r. experiment the sample under investigation is placed in an r.f. coil, in a magnetic field H_0. A continuous r.f. current passes through the coil and is equivalent to an oscillating magnetic field H_1 orientated along the axis of the coils. In practice the magnitude of H_0 is about 10^4 gauss, whereas the magnitude of H_1 is of the order of 1 gauss. The resonance condition in continuous wave experiments is obtained by varying H_0 at a fixed radio frequency ν. The absorption signal is detected as an out of balance condition of the r.f. bridge of which the receiver coil forms part. Table 1 gives the resonance frequency for a variety of nuclei at a fixed value of H_0 (10^4 gauss). It also documents the resultant sensitivities at natural abundance for a variety of nuclei. It will be seen that the resonance frequency at a given field is specific for each nucleus. Furthermore, it is easy to see why ^1H is the most popular nucleus used in magnetic resonance, and also it is obvious why one would

Table 1.

	^{16}O,^{12}C.	^1H	^{31}P	^{29}Si	^{13}C	^2H	^{14}N	^7Li	^{11}B
I	0	$\frac{1}{2}$	$\frac{1}{2}$	$\frac{1}{2}$	$\frac{1}{2}$	1	1	$\frac{3}{2}$	$\frac{3}{2}$
Rel. sensitivity at natural abundance	0	1	0·06	3.10^{-4}	2.10^{-4}	$1{\cdot}5.10^{-4}$	10^{-3}	0·3	0·16
ν_{res} at 10^4 gauss	—	42·6	17·2	8·8	10·6	6·5	3·1	16·5	13·7

almost certainly talk about n.m.r. studies of ^7Li and ^{11}B in discussing the properties of inorganic glasses. As I shall be concerned entirely with synthetic organic polymers, my talk will be confined to proton magnetic resonance.

N.M.R. SPECTRA OF LIQUIDS AND SOLIDS

The line shapes observed in n.m.r. spectra give important clues to the physical state within the sample under investigation. Mobile liquids such as benzene give a single very sharp resonance signal with a width at half peak height (half band width) of about 0·3 to 0·5 Hz. Solid naphthalene on the other hand, and for that matter polystyrene or polyethylene or any other polymer, give extremely broad signals with half band widths of the order of 10^4 Hz. The shape of the broad lines may be smooth Lorentzian or Gaussian contours, or they may have some additional structure, but for a simple liquid the resonance line is always Lorentzian and sharp.

The structural line width is determined by the lifetime (τ) of the upper state involved in the transition. In a liquid, the magnetic dipolar coupling between nuclei is weak because of the constant tumbling motion of the molecules, consequently the transfer of the excess energy in the upper state to another spin or to the thermal motion of the lattice is inefficient, hence τ is long. Since, from Heisenberg's Uncertainty Principle

$$\Delta E \times \Delta\tau \sim \text{constant}$$

the uncertainty in E is small, when the lifetime of the upper state, and hence uncertainty in $\Delta\tau$, is large, and under these conditions the spectral line is very narrow. In a solid, the dipolar coupling between nuclear spins is much more effective since molecules are fixed in their relative orientations. Consequently the lifetime of the upper state is short and $\Delta\tau$ is small. Thus broad lines are observed in solids. However, if some groups in the molecule are rotating even when the main part of the molecule has solidified into a rigid lattice structure, the contribution from these mobile groups to the line width will be small. When the frequency of rotation of these groups is less than their contribution to the line width, i.e. about 10^4 Hz, then the additional broadening due to this motion will reappear. Van Vleck has made a major contribution to the calculation of n.m.r. line profiles for molecules, in rigid lattices of various geometries, and also for the contributions which disappear when the onset of rotation of different parts of the molecule occurs. Hence, by studying the line widths in solids as a function of temperature the onset of molecular motion of specific groups can be detected by the occurrence of line-narrowing. Ultimately the line width will collapse totally as the substance melts and enters the liquid phase. The melting of a glassy organic polymer produces a rubber rather than a mobile liquid and consequently the collapsed line-width is likely to be 100–500 Hz rather than ~0·5 Hz as for a mobile liquid.

RELAXATION PROCESSES

There are two main relaxation processes which control τ in a solid. The first-order time constants (T_1 and T_2) which characterize these processes can be measured by pulsed n.m.r. spectroscopy.

(a) The spin-lattice relaxation mechanism (relaxation time T_1) corresponds to the relaxation of the nuclear spin vector in the direction of H_0. Nuclear spins in the upper state relax to the lower energy state by transfer of their excess energy to the thermal motion of the molecular lattice. The nett amount of energy transferred is minute (about 10^{-3} calorie per gram atom of H) and the change in temperature of the lattice is undetectably small.

(b) The spin–spin relaxation mechanism (relaxation time T_2) corresponds to the relaxation of the phase-coherence of the precessing nuclei in the plane perpendicular to H_0. The excess energy of, for example, spin A in an upper energy state is passed on to spin B in the lower energy state. Thus spin B is promoted as spin A returns to the lower state. The relative population of the two states is not affected, but τ and hence $\Delta\tau$ are reduced and the line width is increased as T_2 decreases.

The variation of T_1 and T_2 is shown schematically in Figure 1 for a typical simple organic material such as glycerol; the curves are plotted as a function of molecular motional frequency in the sample. T_1 passes through a minimum value when the average correlation frequency of molecular motion is of the same magnitude as the n.m.r. frequency because under that condition coupling with the lattice is most efficient. T_2 declines to a limiting value as the molecular motion slows down and this corresponds to the appearance of a broad line spectrum. Temperature could have been plotted as an alternative to frequency along the abscissae in Figure 1, since an increase in temperature would correspond to an increase in molecular motional frequencies.

Clearly the measurement of the spin-lattice relaxation time T_1 as a function of temperature is important because the temperature at which $T_1^{\min}$ is observed establishes the temperature at which the average correlation frequency

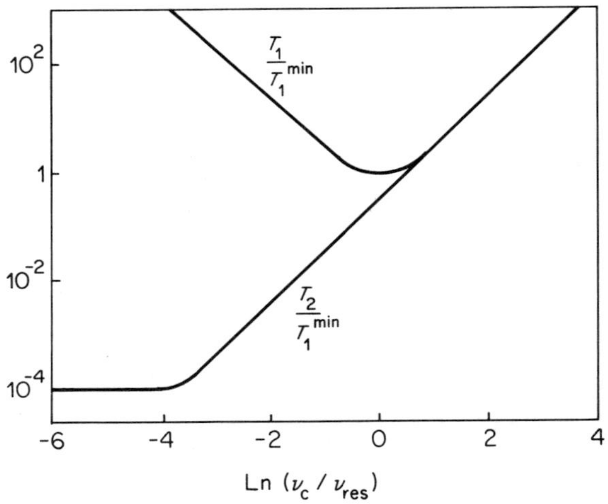

Figure 1. The variation of T_1 and T_2 with molecular motional correlation frequency (ν_c) or temperature

of molecular motion is directly related to the resonance frequency of the experiment. It is a more precise way of establishing the correlation frequency for molecular motion than is the observation of line-narrowing noted in the previous section. The use of T_1–temperature curves is complicated by the existence of distributions of (rather than single) relaxation times and also by the appearance of several $T_1^{\min}$ in the T_1-temperature curves for polymers. Nevertheless, as we shall see, the method provides extremely important results.

To summarize, there are two main ways of studying molecular motion in glasses by n.m.r. spectroscopy:

(a) from the measurement of line widths over a temperature range

(b) from the measurement of spin lattice times over a wide temperature range.

BROAD LINE STUDIES OF GLASSY POLYMERS

Broad-line n.m.r. spectrometers usually record the first derivation of the absorption line with respect to field strength, rather than the absorption curve itself. A typical absorption curve and its first derivative is shown in Figure 2. The absorption curve for a polymer is often not a single smooth curve; it may have multiple components which change in magnitude with temperature corresponding to the onset of motion in different parts of the lattice.

At room temperature high molecular weight polystyrene shows a simple single broad line of half band width 8·5 gauss (35·7 kHz) (Figure 3). In the same figure is shown the relatively sharp line superimposed on the broad

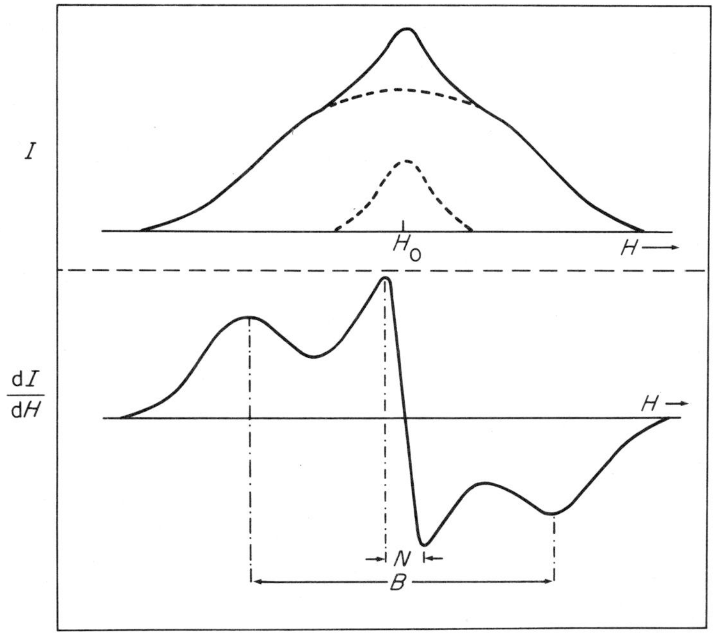

Figure 2. A typical broad line n.m.r. observation curve showing two components of different widths. The half band widths of the two lines are N and B respectively. [From R. L. Miller in *Encyclopedia of Polymer Science and Technology*, Vol. 4, Ed. H. F. Mark, N. G. Gaylord and N. M. Bikales, p. 450. Copyright © 1966 John Wiley & Sons Inc. by permission of John Wiley & Sons Inc.]

line of the polymer due to a small amount of imbibed benzene. Figure 4 shows the high-resolution n.m.r. spectrum of a 10% solution of polystyrene in CCl_4. The width of the whole trace is only 400 Hz and the line widths are very much smaller. The different chemical shifts, τ, of the aromatic ($\tau \sim 3$) and aliphatic protons (–>CH at $\tau \sim 7$ and >CH_2 at $\tau \sim 8$) can readily be seen.

The variation in the line-widths with temperature for three glassy polymers, polystyrene, poly(vinyl chloride) and poly(methyl methacrylate) is shown

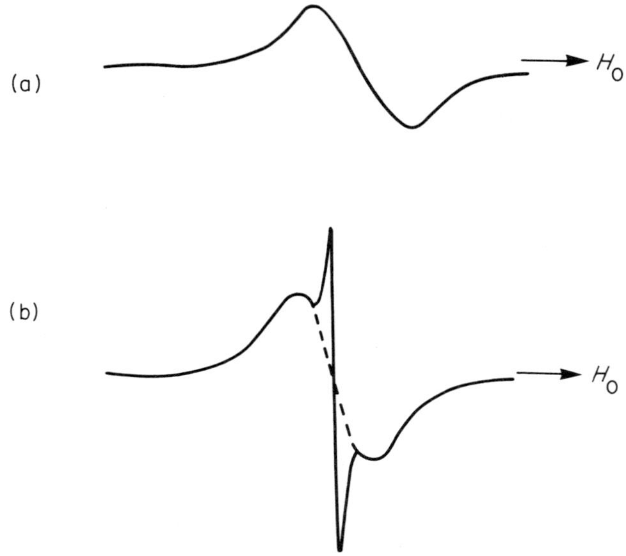

Figure 3. (a) A broad line n.m.r. spectrum of polystyrene glass. (b) containing 10% benzene

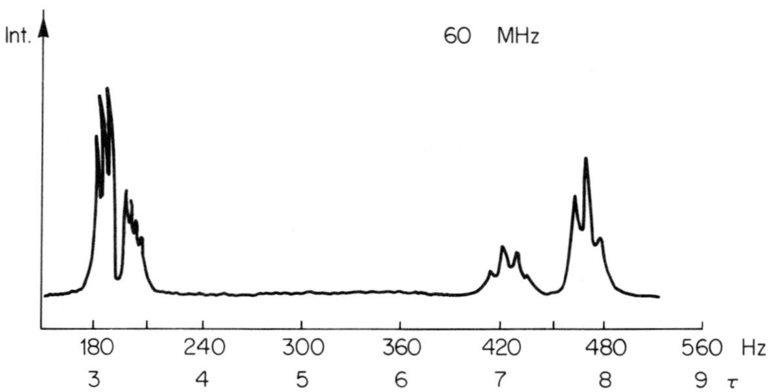

Figure 4. High resolution n.m.r. spectrum of a solution of carbon tetrachloride containing 10% of polystyrene at 60 MHz

in Figure 5. For polystyrene and poly(methyl methacrylate) at 120 °C and p.v.c. at 100 °C the sharp decrease corresponds to the glass/rubber transition region, i.e. the onset of long range molecular motion in the polymer chains. The regions are displaced some 25 °C above the corresponding dilatometric values because the n.m.r. measurement corresponds to a time scale equivalent to the linewidth (i.e. ~ 50 kHz). The curve for polystyrene also shows an inflection at 50 °C, in the region where a secondary loss peak is observed in mechanical relaxation studies. This is attributable to local motion involving the phenyl side groups. For the n.m.r. spectrum the assignment can be confirmed by calculation of the expected contribution to the overall line width from rigid phenyl side groups. Thus we might infer that this particular motion reaches a frequency of about 10 kHz at 50 °C and that below this temperature the group is effectively rigid.

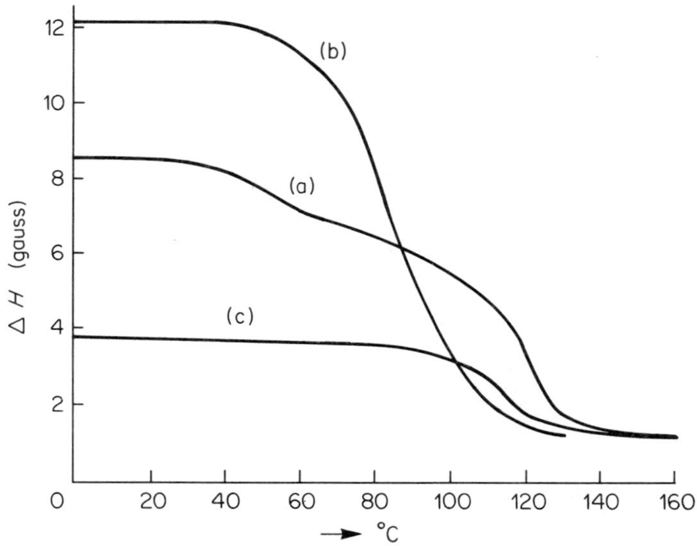

Figure 5. The temperature dependence of the broad line-widths for (a) polystyrene, (b) poly(vinyl chloride), (c) poly(methyl methacrylate)

It is noteworthy that even at room temperature, poly(methyl methacrylate) has a line width of only 4 gauss (~16 kHz). The line width calculated for the rigid molecule is about 10–12 gauss. Thus there must be considerable molecular motion in the glassy state at 80 to 100 °C below T_g. An investigation of the line profile at low temperatures shows that rotation of the side group

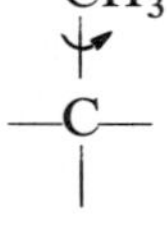

sets in at about $-140\ ^\circ C$ and is not completely averaged out until $-20\ ^\circ C$. The rotation of the ester-methyl group

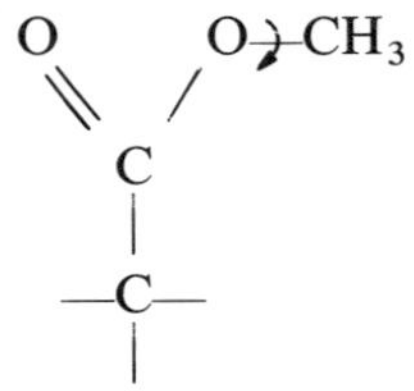

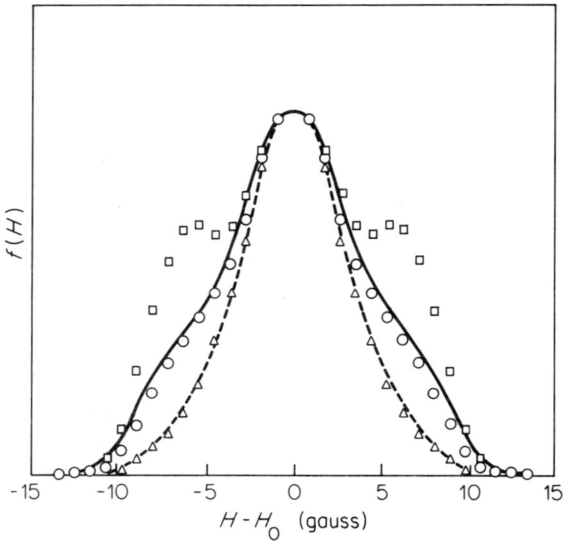

Figure 6. Comparison of the experimental line shape in polymethyl methacrylate at 77 °K (solid line) and 300 °K (broken line) with the shapes calculated for the rigid structure □, and the cases of rotation of one ○ and both △ methyl groups. [From K. M. Sinnott, *J. Polym. Sci.*, **42**, 3 (1960). Copyright © 1960 John Wiley & Sons Inc. by permission of John Wiley & Sons Inc.]

is free even at $-200\ ^\circ C$. Comparison between the experimental and calculated curves shown in Figure 6 confirms that both methyl groups are undergoing full motional line narrowing at room temperature.

SPIN LATTICE RELAXATION STUDIES OF GLASSY POLYMERS

The study of molecular motion by broadline measurements is confined to time scales dictated by the n.m.r. line width. The importance of relaxation time measurements is that the time scale at which the observation is made is defined by the n.m.r. frequency. Depending on the value of H_0 this will range from about 10 to 200 MHz. The method used to measure T_1 for a polymer uses a pulsed, rather than a continuous wave spectrometer.

Recently a new pulse technique has been introduced, the so-called rotating frame method, which allows the relaxation time $T_{1\rho}$ to be measured for relaxation along the direction of H_1. From the discussion of the basic n.m.r. experiment it will be recalled that H_1 is about 10^3 times smaller than H_0, so that $T_{1\rho}$ is *measured at an effective frequency* in the range 10–100 kHz. Thus as in mechanical and dielectric relaxation, a range of frequencies extending over decades now can be covered.

Figure 7 shows T_1-temperature plots for amorphous poly(propylene oxide) $-[-CH(CH_3)CH_2O-]_n-$ and poly(propylene oxide) d_3 $-[-CH(CD_3)CH_2O-]_n-$.

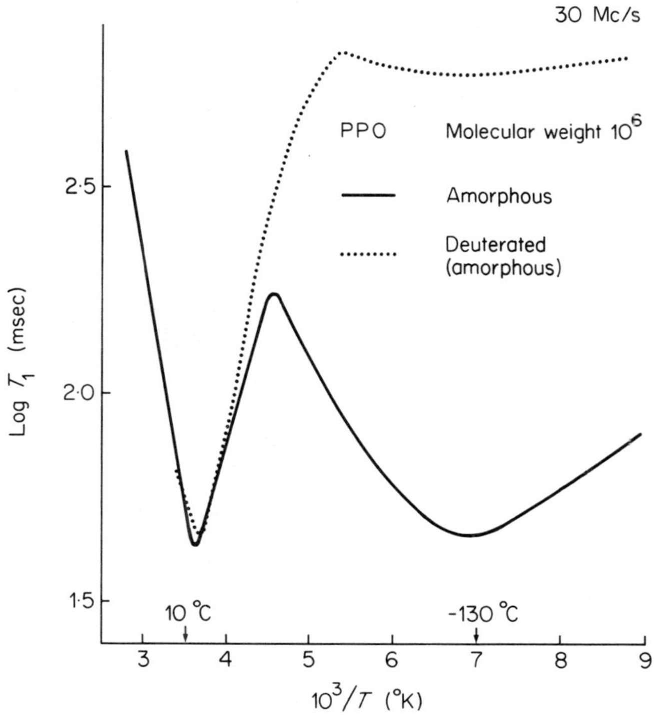

Figure 7. Spin lattice relaxation time curve for poly(propylene oxide) and poly(propylene oxide) d_3 at 30 MHz [From G. Allen, *Mol. Spectr., Rep. Conf.*, **1968**, 408]

The high temperature minimum in poly(propylene oxide) is the glass transition relaxation displaced from its dilatometric value, -70 °C, to 10 °C because the n.m.r. frequency is 30 MHz. The origin of the second minimum at -130

°C is the motion of the –CH_3 side group. The deuterated sample proves this assignment since the corresponding minimum is very much weaker, whereas the T_g minimum corresponding to the backbone motion of the polymer is virtually unchanged. Thus for C–CH_3 in poly(propylene oxide) $\nu_{C(CH_3)} \sim 30$ MHz at -130 °C.

T_1 measurements on a solution of the polymer in a magnetically inert solvent (CS_2) shows that the glass transitional motion is strongly influenced by intermolecular interactions (Figure 8). The T_g minimum moves to lower

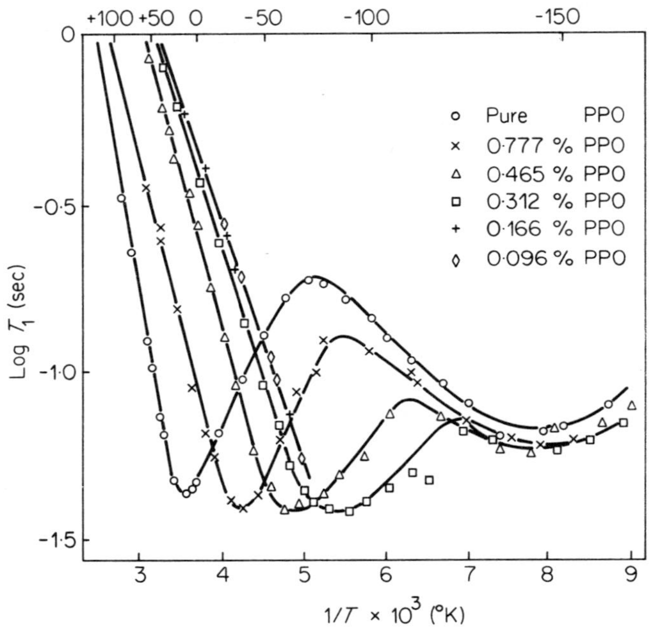

Figure 8. Spin lattice relaxation times for solutions of poly(propylene oxide) in CS_2 at 30 MHz. [From T. M. Connor, D. J. Blears and G. Allen, *Trans. Faraday Soc.*, **61**, 1100 (1965)]

temperatures as the solution becomes more dilute. The side group motion is dominated apparently by intramolecular effects since the location and breadth of its minimum is not seriously affected by dilution.

Figure 9 shows that the minima in $T_{1\rho}$ lie at lower temperatures than the corresponding minima in T_1, due to the lower effective frequency in the $T_{1\rho}$ measurements. Figure 10 shows the overall agreement for the average molecular relaxation times ($\tau \sim \nu_c^{-1}$) obtained from mechanical, dielectric and n.m.r. results.

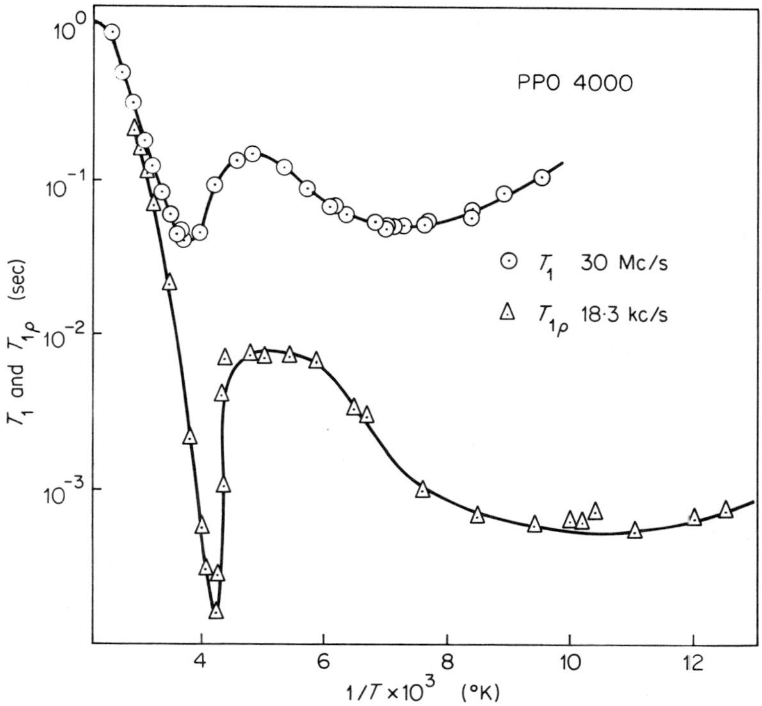

Figure 9. T_1 and $T_{1\rho}$ curves for poly(propylene oxide). [From T. M. Connor and A. Hartland, *Polymer*, **9**, 591 (1968)]

From the T_1 and $T_{1\rho}$ measurements the overall rate of rotation of the $-CH_3$ group at room temperature is estimated to be about 5×10^9 Hz. The activation energy is 4·0 kcal $mole^{-1}$. From neutron inelastic incoherent scattering spectroscopy we have measured the fundamental frequency of torsional oscillation of the $-CH_3$ group in poly(propylene oxide) to be 6×10^{12} Hz and the threefold barrier to internal rotation is calculated to be 3·5 kcal $mole^{-1}$. Thus we have a fairly detailed picture of the $-CH_3$ group motion in this polymer together with the conclusion that the barrier to internal rotation is largely responsible for the activation energy of the rate process. A more detailed study of the methyl group motion would need to consider the effects of quantum mechanical tunnelling, for which there is not time.

In conclusion, consider the T_1–temperature plot for poly(methyl methacrylate) shown in Figure 11; the curve is more complicated than that for poly(propylene oxide) because of the possibility of several side group motions. Comparison with the corresponding curve for poly(methylacrylate)

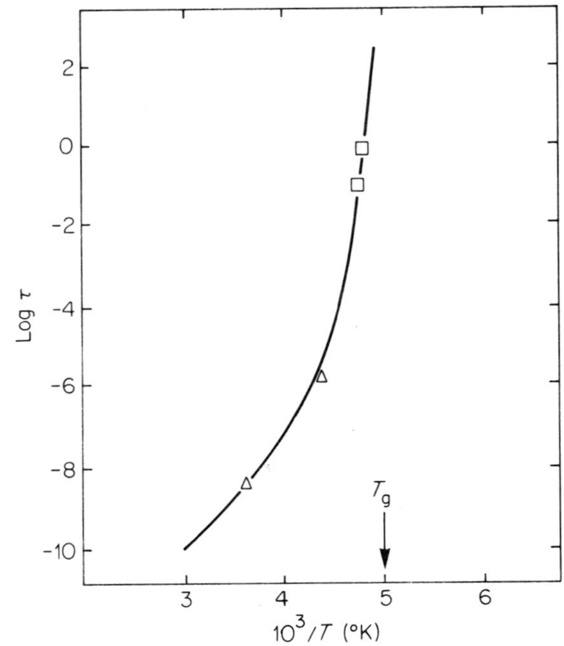

Figure 10. Relaxation times (τ) obtained from mechanical □, dielectric △, and T_1, $T_{1\rho}$ measurements for poly(propylene oxide) in the region of the glass transition temperatures. [From G. Allen, *Mol. Spectr., Rep. Conf.*, **1968**, 408]

(Figure 12) shows that the α-CH_3 group motion must be responsible for the T_1 loss at −73 °C at 30 MHz.

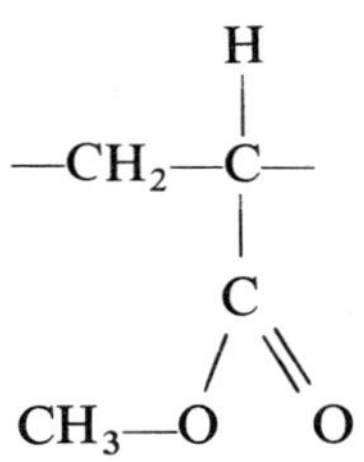

PMA

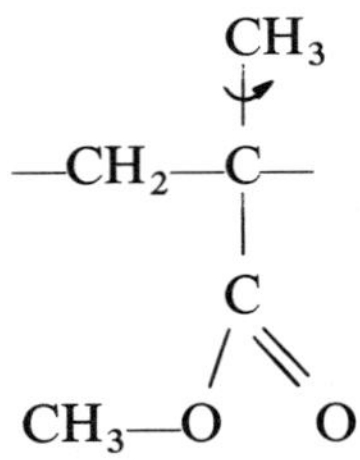

PMMA

In the poly(methyl methacrylate) T_1 curve, the minima for the –C–CH_3 and the –O–CH_3 motions are observed, corresponding to the results discussed in the preceeding section on broadline studies. Since the α-CH_3 group minimum lies at a much higher temperature to the O–CH_3 group minimum, the conclusion to be made is that the energy barrier to rotational motion is higher for the α-CH_3 group. From neutron scattering studies these barriers

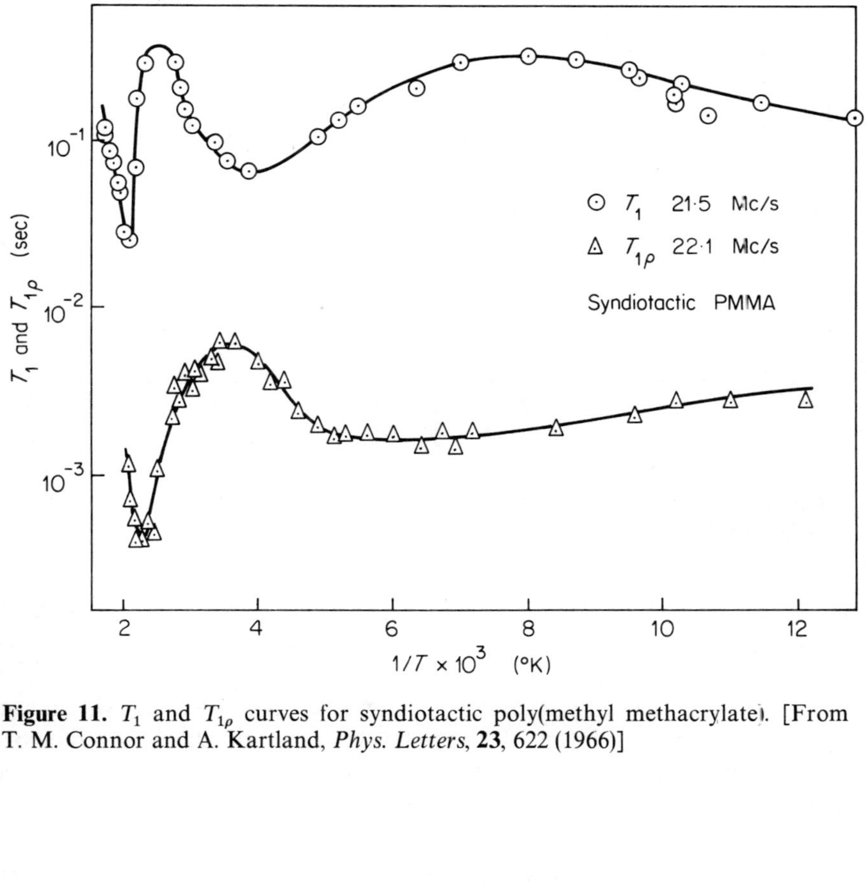

Figure 11. T_1 and $T_{1\rho}$ curves for syndiotactic poly(methyl methacrylate). [From T. M. Connor and A. Kartland, *Phys. Letters*, **23**, 622 (1966)]

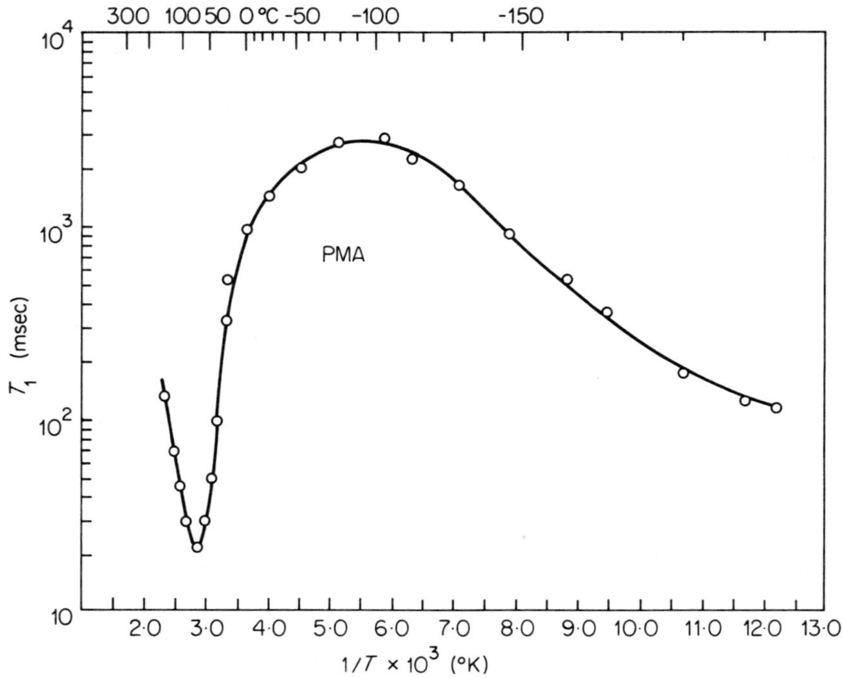

Figure 12. T_1 for poly(methyl acrylate) at 30 MHz. [From T. Kawai, *J. Phys. Soc. Japan*, **16**, 1221 (1961).]

have been shown recently to be respectively ~7 and ~1 kcal mole^{-1}.

In this paper the author has tried to show that n.m.r. spectroscopy is a powerful tool for studying molecular motions in glassy polymers. Although I have concentrated on proton m.r. the method can be used for other nuclei which also possess spin angular momentum. It is important to remember however that often several relaxation techniques in combination have to be used to give a comprehensive picture.

ACKNOWLEDGEMENT

I wish to thank Dr. R. M. John, Mr. R. F. Warren of this Chemistry Department for permission to use unpublished results, and also Dr. T. M. Connor (N.P.L.).

38

Broad-Line N.M.R. of Epoxy Resins

B. ELLIS

Department of Glass Technology, University of Sheffield, Great Britain

The epoxy resins discussed in this paper have the structure I, with a molecular weight range of 350 to 4,000, i.e. $n = 0$ to $n \sim 10$.

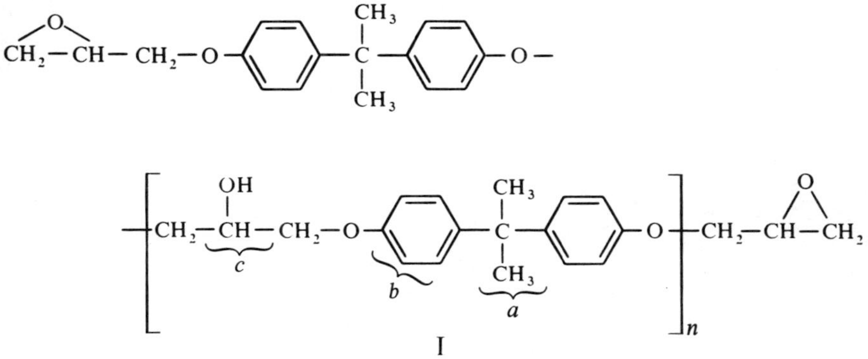

I

In broad-line n.m.r. measurements it is usual to record the first derivation of the absorption signal and typical spectra are given in Figure 1 for a resin with a molecular weight of 945. At low temperatures (180 °K) there is a broad line (Figure 1a), with an approximately Gaussian line shape; as shown by a simple test (Alger, 1968) where the ratio between the slopes of the derivative curve at the centre of resonance and at the point of maximum slope at higher and lower magnetic field is measured, for a Gaussian curve this ratio is 2·2; for the spectrum in Figure 1a it is $2{\cdot}3 \pm 0{\cdot}1$. In the glass transition region ($T = 313$ °K in Figure 1b) two lines are partially resolved, a narrow line and also a broad line which is narrower than the broad line at lower temperatures.

Above the glass transition region only a narrow line is obtained (Figure 1c).

The line width, δH, (Slichter, 1958) defined as the distance in gauss between points of maximum and minimum slope is related to the spin–spin relaxation time T_2:

$$\delta H = \frac{\text{const.}}{\gamma T_2} \tag{1}$$

where the constant depends on the line shape, γ is the gyromagnetic ratio.

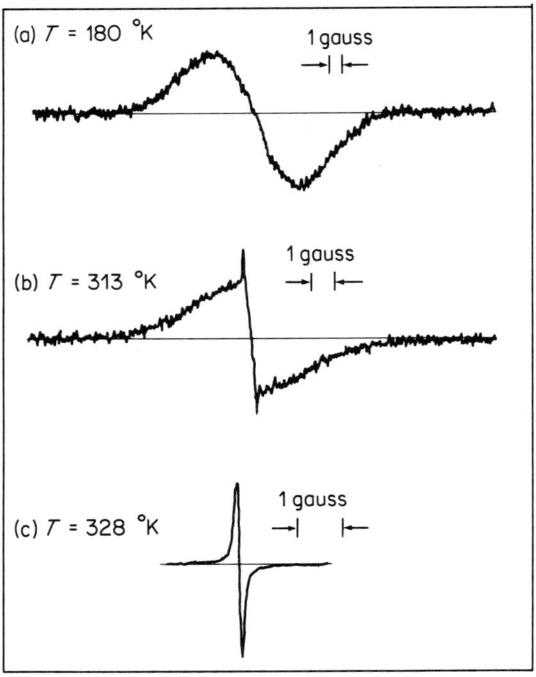

Figure 1. Typical n.m.r. spectra. Resin No. 3. (a) T = 180 °K, (b) T = 313 °K, (c) T = 328 °K

It is T_2 which is sensitive to the extent of molecular motion. Hence, it is possible to say that Figure 1a is typical of little or only limited molecular motion, Figure 1b indicates fairly extensive motion and Figure 1c indicates that there is isotropic rotation of molecular segments. It is the purpose of this paper to analyse in more detail the extent of such molecular motion and the effect of molecular weight on the temperature at which rapid isotropic rotation occurs.

The other parameter of the n.m.r. spectra that is useful for discussion of molecular motion is the second moment which can be calculated for a rigid lattice (not necessarily a regular crystalline lattice) when all the inter-proton distances are known from the Van Vleck formula (Smith, 1965).

$$S = \frac{6}{5}\frac{I(I+1)}{N} g^2\beta^2 \sum_{ij} r_{ij}^{-6} \tag{2}$$

which applies for an unorientated solid in the absence of any other nuclear species. I is the nuclear spin ($=\frac{1}{2}$ for protons), N is the number of nuclei at resonance, g is the nuclear equivalent of the Landé constant and β is the nuclear magneton. r_{ij} is the inter nuclear distance and because it occurs in equation (2) as the inverse sixth power it is found that for $r > 3$ Å inter-proton interaction is small and can be neglected for $r > 5$ Å without great error.

Although there have been other studies (Clark-Monks and Ellis, 1970; Slonim and Lyubimov, 1970) of broad-line n.m.r. of epoxy resins the effect of molecular weight on the transition behaviour has not been studied previously.

EXPERIMENTAL

Measurements were made on a series of uncured commercial bisphenol-A-epichlorhydrin epoxy resins I, $n = 0$ to ~ 10; details of these resins are given in Table 1. The cured resin was prepared from resin 3 with a stoichio-

Table 1.

Resin	Molecular[a] Weight	Epoxy[b] Content	Density of[c] resin at 25 °C
1	355	5·15	(1·06)
2	755	2·43	1·11
3	945	2·04	1·16
4	1460	1·17	1·16
5	2625	0·52	1·18
6	4935	0·31	1·19

[a] M_n determined at Ciba (ARL) Ltd., using a Perkin–Elmer–Hitachi Model 115 Vapour Pressure osmometer—accuracy is claimed to be ±5%.

[b] Epoxy equivalents determined at Ciba (ARL) Ltd. by titration with perchloric acid in acetic acid in the presence of tetraethylammonium bromide (Dobinson *et al.*, 1969a), except resin 1 which was determined by Mrs. L. Hemshall using the HCl/acetone method (Dobinson *et al.*, 1969b).

[c] Determined by Dr. C. Clark-Monks (1970) using a pyknometer with distilled water as the displacement fluid. The average of 3 repeat determinations is given and the accuracy is estimated as ±0·01(5) gm/ml. Resin I is liquid at 25 °C and the value given is that of the solid obtained by curing with about $\frac{1}{3}$ of the stoichiometric amount of 4,4′-diaminodiphenyl methane for 1 hour at 110 °C and 1 hour at 140 °C.

metric amount (10% by weight) of 4,4′-diaminodiphenyl methane, and heated for 2 hours at 110 °C and 1 hour at 140 °C.

N.m.r. measurements were made on a JEOL Spectrometer operating at 60 Mc/s (Model JNM-3H.60) fitted with variable temperature probe. The experimental second moments S_e were calculated by numerical integration using equation (3):

$$S_e = \tfrac{1}{3} C \frac{\Sigma x^3 y}{\Sigma xy} \tag{3}$$

where C is an instrumental calibration constant relating x on the derivative curve to the field sweep in gauss and y is the modulus of the amplitude of the derivative curve at x (both x and y are measured in suitable scale units). Second moments are reported as the average of six repeat determinations.

RESULTS AND DISCUSSION

Second moments

The change in experimental second moments with temperature are shown in Figure 2. The second moment $S = S_1 + S_2$ where S_1 is the rigid lattice intramolecular contribution and S_2 is the intermolecular contribution. has been calculated previously to be 14·6 G^2 (Clark-Monks and Ellis, 1970). In this calculation it was estimated that the end groups made a contribution equal to that for sub-unit *c*. The present results confirm this estimate because the second moments at the lowest temperatures are essentially the same for all the six resins with a change in end group concentration by a factor of about × 17.

The measured second moment at the lowest temperature attainable was ~ 16 G^2 and is in good agreement with the calculated value for S_1, with the intermolecular contribution S_2 = 1·4 G^2. This is reasonable since it would not be expected that there should be any very close intermolecular interactions.

From the present second moment results, which show that a rigid lattice value is attained at *ca* 100 °K, it would appear that methyl group rotation is essentially stopped ($f < 10^4$ c/s). Thus, quantum tunnelling of methyl groups (McCall *et al.* 1969) which has had to be introduced to account for the low experimental second moments at liquid helium temperatures for some substances does not occur in the case of these resins. It is difficult at present to be sure *a priori* when quantum tunnelling will occur but it may be that it is inhibited in this case since either coordinated tunnelling by both methyl groups or the tunnelling of one methyl group must be coupled with oscillations of the phenylene group. Allen (1968) has recently examined methyl group tunnelling theoretically and compared predictions with experiment for ‘simple’ organic substances (e.g. toluene) and Eisenberg and Reich

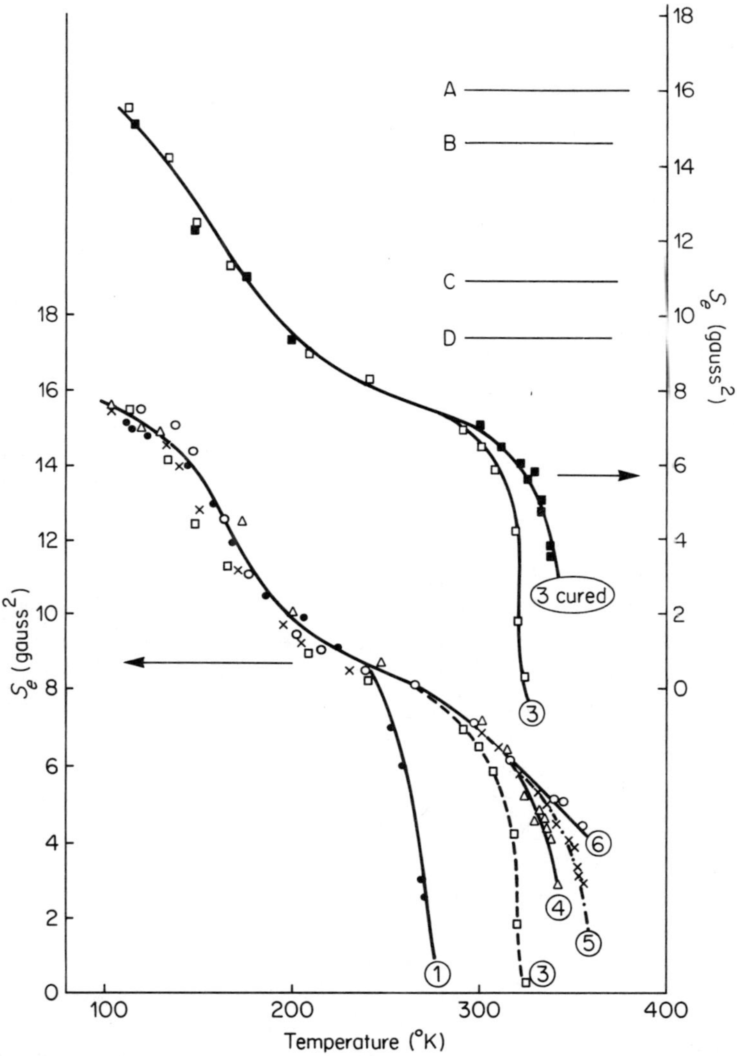

Figure 2. Second moment *vs.* temperature. Resin: 1 ●; 2 left out for clarity; 3 □; 3 cured ■; 4 △; 5 ×; 6 ○. A denotes S_e (110 °K) = 16·0 G^2; B, S_1 = 14·6 G^2 (rigid lattice); C. S_e (190°K) = 10·8 G^2; D, S_1 = 9·4 G^2 (methyl group rotation); S_2 = constant = 1·4 G^2

(1969) have discussed methyl group tunnelling for polymers, but none of their examples has two methyl groups attached to one carbon atom. Thus, even in the absence of measurements at liquid helium temperatures it is possible to be certain that the experimental second moment is not less than that predicted by using equation (2) with 'rigid' methyl groups and hence conclude that tunnelling does not take place. In a polycarbonate (McCall and Falcone, 1970) which contains sub-units *a* and *b* (*c* is replaced by –O–C:O–O–) it has been found from T_2 measurements that the methyl groups are 'rigid' at low temperatures below 150 °K and attention is drawn to the steric effect on the rotation of two methyl groups on the same carbon atom. Thus the behaviour of sub-units *ab* is similar in both these epoxy resins and a polycarbonate.

Previously, the decrease in second moment between 110 °K and 200 °K was attributed to increasing extents of methyl group rotation; the intramolecular contribution for a rigid resin but with rotating methyl groups would be $S_1 = 9{\cdot}4\ G^2$, so that with a constant $S_2 = 1{\cdot}4\ G^2$, $S_1 + S_2 = 10{\cdot}8\ G^2$ which agrees with the experimental second moment $S_e = 10{\cdot}8\ G^2$ at *ca.* 180°K for the present range of resins. Thus, it is established that change of end group concentration (by × 17) does not influence the change in extent of methyl group rotation, as is to be expected. Further support for this conclusion is provided by the observation that this motion is unaffected by cure, that is cross-linking, of the resin by reaction of the epoxy end-groups and the hydroxyl groups in sub unit *c*.

With increase in temperature from 200 to 300 °K, Figure 2 the experimental second moment S_e decreases from 10 to 7 G^2 without any significant differences due to the molecular weights of the resins being observed. An averaging of the dipolar broadening due to rotation of the phenylene groups would not account for a decrease in second moment of as large as 3 G^2 (only about 1 G^2), therefore it would appear that more extensive motions than just rotation of phenylene groups are taking place in this temperature range. This observation could be interpreted along the lines of Goldstein (1969) whose theory predicts that there is some molecular mobility even though the glass transition has occurred. The effect of cross-linking the resin is to raise the temperature at which this relaxation occurs.

In the glass transition region the experimental second moment decreases very rapidly with rise in temperature, thus with a 2–5 °K increase in temperature all groups start rotating around one axis and with a temperature rise of about 10 °K isotropic rotation is shown to occur because the measured second moments decrease to less than 1 G^2. The temperature at which this very rapid decrease in S_e occurs is sensitive to the molecular weight of the resins and also to cross-linking.

Line widths

The line width changes with temperature (Figures 1 and 3, where in the transformation range the wider line width of a composite spectrum is plotted).

At temperatures above the glass transition region a narrow line ($\delta H \sim 0{\cdot}2$

gauss) is obtained; through the glass transition range the line width increases rapidly, 5–6 gauss for a decrease in temperature of about 20 °K. The change in line width is larger and the temperature range of the transition is smaller the lower the molecular weight of the resin, compare resins 1 and 6, Figure 3.

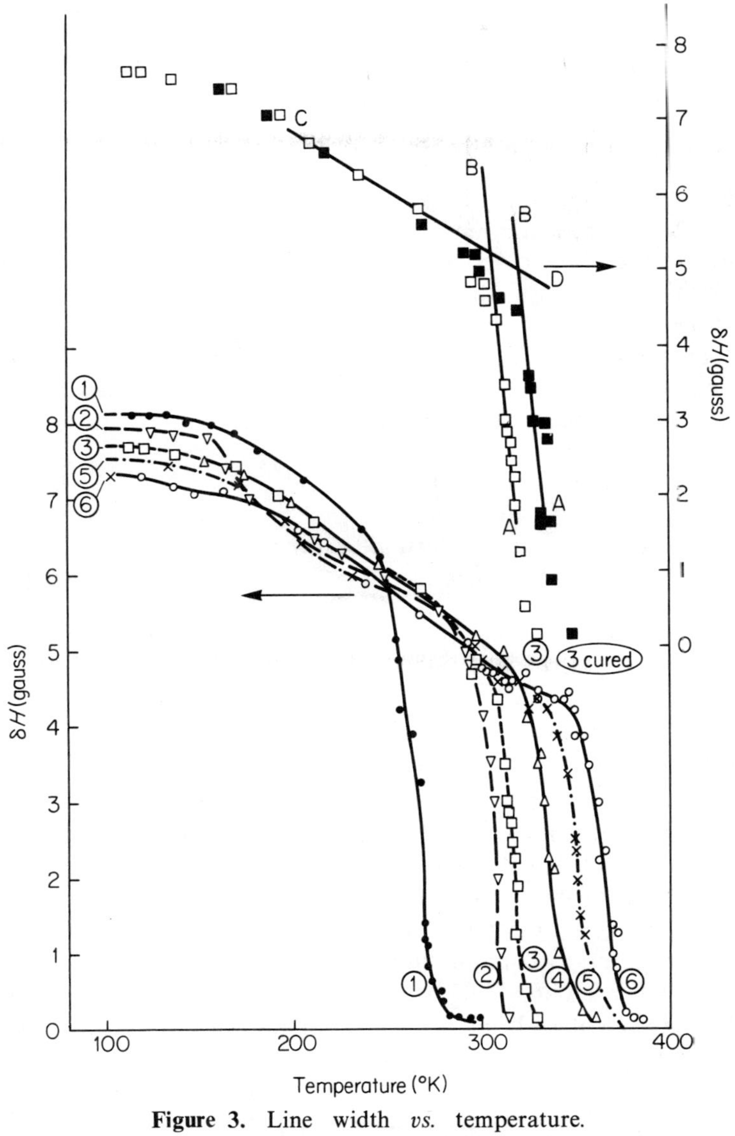

Figure 3. Line width *vs.* temperature.
Resin: 1 ●; 2 ▽; 3 □; 3 cured ■; 4 △; 5 ×; 6 ○

For resin No. 3 (Table 1) it has been shown (Clark-Monks and Ellis, 1970) that the transition region (300–320 °K) for n.m.r. line width is also observed as a broad transition in a D.T.A. curve. The inferences that can be drawn from the change in line width with temperature are consistent with those drawn from the changes in second moment with temperature and will not be discussed again. There is one additional interesting feature for which we do not have an explanation; the line width at the lowest temperatures is inversely proportional to the molecular weight, see Figure 3. This effect is small but would seem to be established by the data given in Figure 3. Thus the observations of Slichter and Davis (1964) for polyisobutylene and McCall *et al.* (1962) for polyethylene at $T > T_m > T_g$ that the spin–spin relaxation time [$T_2 \propto 1/\delta H$—equation (1)] is inversely proportional to the molecular weight are not applicable for the case of epoxy resins at $T < (T_g - 100)$ °K. Within the experimental limits of broad-line n.m.r. it was not possible to investigate the effect noted by Slichter *et al.* but would be worth study because they were not able to offer a satisfactory explanation and the fortuitous apparent correlation with the chain entanglement length obtained from melt viscosity measurements requires confirmation.

Activation energies can be calculated if there is an Arrhenius type temperature dependence for the correlation frequency ν_c

$$\nu_c = \nu_{co}\, e^{-E/RT} \tag{4}$$

the correlation frequency ν_c can be estimated from the changes in line width with temperature using:

$$\nu_c = \frac{\gamma \delta H}{2}\left\{\tan\left[\frac{\pi}{2}\cdot\frac{(\delta H)^2 - B^2}{C^2 - B^2}\right]\right\}^{-1} \tag{5}$$

where C and B are the experimental line widths at temperatures below and above the transition respectively.

For resin No. 3 ν_c was previously (Clark-Monks and Ellis, 1970) calculated using equation (5) and from a plot of log ν_c *vs.* $1/T$ the activation energies were found to be $1 \cdot 8 \pm 0{\cdot}3$ kcal/mole for the low temperature region 110 °K → 200 °K and this was attributed to methyl group rotation. The apparent activation energy for the glass transition was found to be 30 ± 3 kcal/mole. For methyl group rotation preliminary calculations from T_1 measurements on resin No. 3 (John, 1970) yield a value for $\eta E = 2{\cdot}0 \pm 0{\cdot}1$ kcal/mole in good agreement with the previous estimate. However, since $0 < \eta < 1$, where η is a parameter depending on the width of the distribution of correlation times, the implication is that the activation energy for methyl group rotation obtained by the use of equation (4) and (5) is low by the factor η which still requires determination.

Effect of molecular weight on glass transition temperature T_g

Whilst it is clear from Figure 3 that there is a *range of temperature* over which the line width changes rapidly, it is convenient to define a glass transition temperature, T_g. Gee (1966) in a discussion of the nature of the glass transition in polymers has pointed out that the experimental procedure used must be taken into account in any operational definition of T_g. For the present purposes T_g is defined as the temperature of intersection of lines AB and CD, Figure 3, drawn linearly through the curve of δH *vs.* temperature in the transition region and at temperatures immediately below this region. This procedure is preferred to that of Faucher *et al.* (1966), who assigned T_g to the lowest temperature at which the narrow component (Figure 1b) could be clearly seen, because this depends on the instrumental settings of H_1 and modulation, but either method will yield essentially the same value of T_g. For resin No. 3 it was found previously (Clark-Monks and Ellis, 1970), that the T_g determined by the present method agreed with that found from rapid decrease in S_e, the D.T.A. peak temperature and the 'break' in the Arrhenius plots of correlation frequency.

The effect of cure on the glass transition is shown in Figure 3 and it is clear that T_g has increased with cure. It has been suggested that the increase of T_g with degree of cross-linking can be represented by

$$T_g = T_g^0 + \frac{D}{\overline{M}_c} \tag{6}$$

where D is a constant and $\overline{M}_c$ is the average molecular weight between cross-links. Unfortunately it is not at present possible to confirm such a relationship for cured epoxy resins because there was not a reliable method for determination of $\overline{M}_c$ although Bell (1970) has recently suggested a method. It may in the future be possible to use such a relationship to determine $\overline{M}_c$.

The effect of molecular weight on the T_g of polymers has been reviewed by Shen and Eisenberg (1966, 1970), and it is generally observed that T_g increases monotonically with polymer molecular weight. An exception (Faucher *et al.*, 1966) is the case of poly(ethylene oxide) where there is agreement between dynamical mechanical (torsion pendulum) and broad-line n.m.r. determinations of T_g but it was found that there is a maximum in the T_g *vs.* $\overline{M}$ curve. In the present experiments only the low molecular weight end of the scale was studied, and the possibility of such a maximum for the epoxy resins is not specifically excluded but neither is it expected.

The simplest theory predicting the effect of molecular weight on T_g is a free volume theory of Fox and Flory which leads to the relationship:

$$T_g(M) = T_g^0 - \frac{K}{\overline{M}_n} \tag{7}$$

where
$$K = \frac{2\rho N_A \theta}{\alpha}$$

and $\overline{M}_n$ is the number average molecular weight.

ρ is the polymer density, N_A is Avogadro's number, α is the thermal expansion coefficient of the glass and θ is the free volume contributed by a chain end. This relationship whilst apparently satisfactory for polystyrene has not proved to be general, it does not represent the present data, Figure 4; and similar deviations from equation (7) were found for polymethylmethacrylate (Beevers and White, 1960) and poly-α-methylstyrene (Cowie

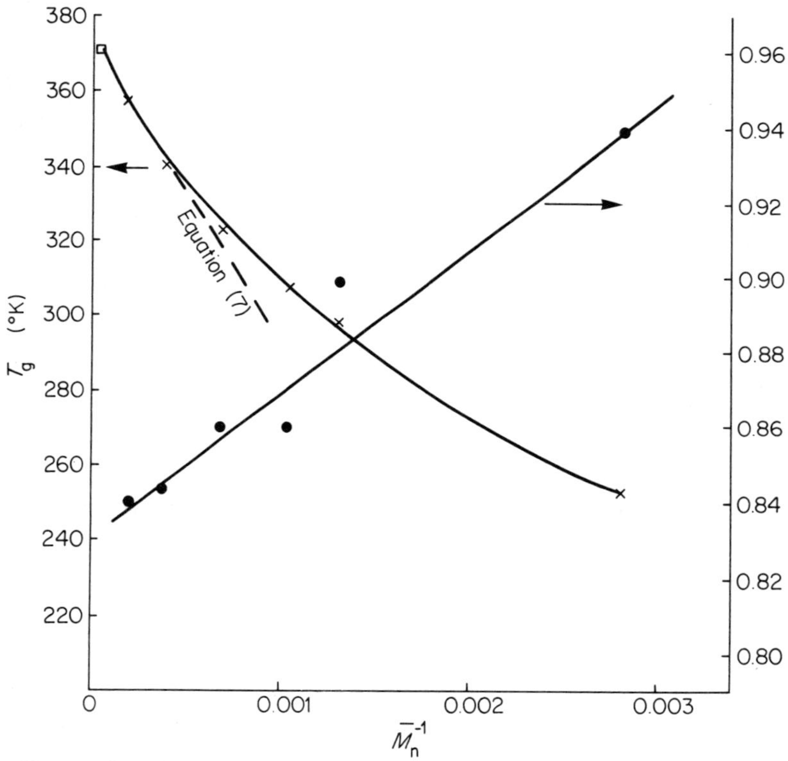

Figure 4. T_g and specific volume *vs.* reciprocal molecular weight □ T_g for phenoxy resin [Reinking, N. H., A. E. Barnabeo, and W. F. Hale, (1963). *J. appl. Polym. Sci.*, 7, 2135] which has the same repeat unit as the epoxy resin I

and Toporowski, 1968). This is despite the fact that the specific volume of polymethylmethacrylate and also of the epoxy resins, Figure 4, are approximately linear functions of the inverse molecular weight which is a consequence of the free volume theory in which there is an allowance for additional free volume for each chain end.

The more sophisticated statistical mechanical theory of Gibbs (1956) and Gibbs and Di Marzio (1958) has been used to interpret the experimental data on polymethylmethacrylate (Beevers and White, 1960) and poly-α-methylstyrene (Cowie and Toporowski, 1968). It would be possible to use this theory with the present data, but this has not been attempted because of the unreality of the model when applied to an epoxy resin.

An alternative approach to the relationship of T_g to molecular weight of a polymer is given in equation (8):

$$\frac{1}{T_g} = \mathscr{E}' \cdot \frac{1}{P} - \Gamma' \tag{8}$$

where $\mathscr{E}'$ and Γ' are constants and P is the degree of polymerization. In this equation the relationship between T_g and molecular weight is the inverse of that in equation (7). An equation of this form, equation (8), was originally suggested on an empirical basis by Ueberreiter and Kanig (1952) and subsequently Kanig (1963) has given a thermodynamic analysis which leads to an equation of this form, equation (8), in which (using Kanig's symbols):

$$\mathscr{E}' = \frac{V^*_{f_1}}{V^*_{fM}} \cdot \frac{R}{\phi^*_{f_2} A^*_{MM}}$$

and

$$\Gamma' = \frac{R(\ln \phi^*_{f_2} + \phi^*_{f_2})}{(\phi^*_{f_2})^2 \cdot A^*_{MM}} \tag{9}$$

where R is the gas constant and $V^*_{f_1}/V^*_{fM}$ is the ratio of the hole volume to the vibrational expansion volume, A^*_{MM} is a work parameter related to the formation of holes which can be identified with intermolecular or cohesive forces in the system, ϕ^*_{f1} is the ratio $(\alpha_1 - \alpha_g)/\alpha_g$, $\phi^*_{f2} = (1 - \phi^*_{f1})$. From equations (8) and (9) it can be seen that for a polymer with infinite molecular weight the glass temperature predicted is $T_g = 1/\Gamma'$ and because Kanig argues that $\phi^*_{f_1}$ is approximately constant (~0·6) (Kanig, 1963) it follows that the limiting glass temperature is determined by the intermolecular forces, that is, A^*_{MM}, which is partially confirmed by the general trend of T_g increasing with the melting point T_m of both polymeric and low molecular weight organic glasses (Ellis and McDonald, 1969). However, in general such a single parameter must be insufficient to determine T_g (Gee 1966, 1970). Some progress has been made in deriving a relationship between T_g and $\overline{M}_n$ on the basis that the average spin–spin relaxation time T_2 is a constant at T_g and that the n.m.r. line width is determined by an average of $T_{2 \cdot c}$ for chain segments and $T_{2 \cdot e}$ for chain ends. However, to obtain a relationship similar to equation

(8) several approximations have to be made for which justification is required, but it would appear that it will be possible to obtain a relationship in terms of the correlation frequencies of the molecular motion in a polymer in the glass transition region.

For a test of equation (8) it is most convenient to rearrange it to equation (10)

$$\frac{\overline{M}_n}{T_g} = \mathscr{E} - \Gamma \overline{M}_n \tag{10}$$

where $\mathscr{E}$ and Γ are constants adjusted from $\mathscr{E}'$ and Γ' to take account of the change of variable from P to $\overline{M}_n$. That such a relationship fits the present data is shown clearly in Figure 5. It is also of interest to note that both the

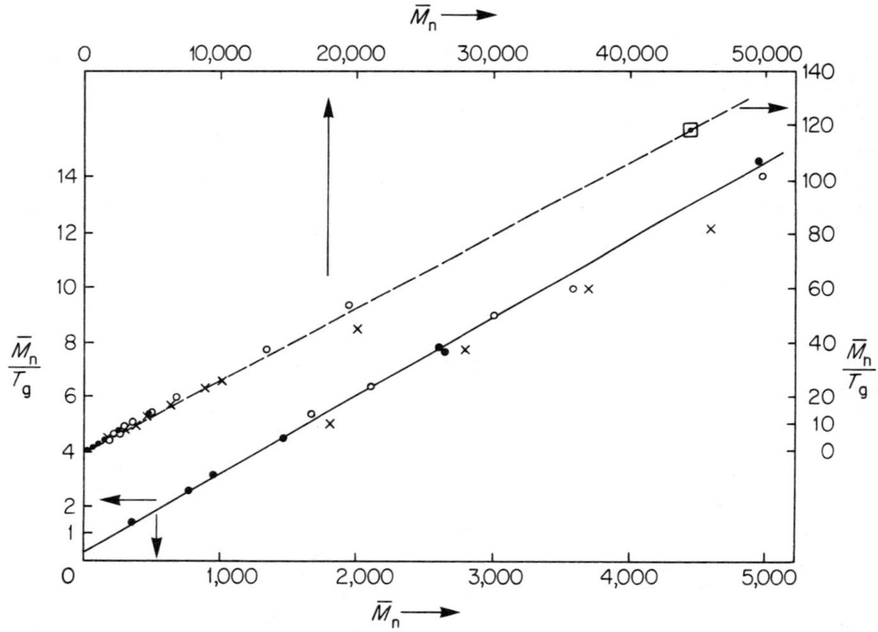

Figure 5. $\overline{M}_n/T_g$ *vs.* $\overline{M}_n$

●Present data

□T_g for phenoxy resin [Reinking, N. H., A. E. Barnabeo, and W. F. Hale (1963). *J. appl. Polym. Sci.*, **7,** 2135]

×Data for polystyrene from Fox, T. G., and P. J. Flory (1954). *J. Polym. Sci.*, **14**, 315.

○Data for poly-α-methylstyrene [Cowie and Toporowski (1968)]

data for polystyrene and poly-α-methylstyrene agree with only a 5–10% change in Γ. This is as would be expected from Kanig's (1963) observation that $\phi^*_{f_1}$ is approximately constant and then Γ would be determined mainly by the intermolecular forces i.e. $A^*_{MM'}$ which could well be of similar magnitude for both polystyrene and poly-α-methylstyrene. Thus, if Kanig's theory is correct the present results indicate that the intermolecular forces in these epoxy resins are of similar magnitude to those in polystyrene. When a plot of the data on polystyrene and poly-α-methylstyrene according to equation (10) is extended to higher molecular weights it is still possible to fit a straight line to the data but the slope is less. This suggests that the true relationship between $\overline{M}_n/T_g$ and $\overline{M}_n$ is a curve of slowly changing slope. However, for a limited range of molecular weight it is established that equation (10) is a good representation of the data.

ACKNOWLEDGEMENTS

Thanks are due to the Science Research Council for financial support, to Mrs. L. Hemshall for her assistance with the experimental work, to L. Tye for servicing of the n.m.r. spectrometer, to Dr. B. P. Stark, Ciba (ARL) Ltd., for the determination of molecular weights and epoxy contents and to Ciba (ARL) Ltd. for samples of epoxy resin.

REFERENCES

Alger, R. S. (1968). *Electron Paramagnetic Resonance-Techniques and Applications.* Interscience. pp. 42–43.

Allen, P. S. (1968). *J. chem. Phys.*, **48**, 3031.

Beevers, R. B., and E. F. T. White (1960). *Trans. Faraday Soc.*, **56**, 744.

Bell, J. P. (1970). *J. appl. Polym. Sci.*, **14**, 1901.

Clark-Monks, C. (1970). *Ph.D. Thesis*, Sheffield.

Clark-Monks, C., and B. Ellis (1970). *J. Polym. Sci. A2*, **8**, 2203.

Cowie, J. M. G., and P. M. Toporowski (1968). *Eur. Polym. J.*, **4**, 621.

Dobinson, D., W. Hoffmann, and B. P. Stark (1969). *The Determination oi Epoxide Groups.* Pergamon. (a) p. 40; (b) p. 29.

Ellis, B., and M. P. McDonald (1969). *J. non-cryst. Solids*, **1**, 186.

Eisenberg, A., and S. Reich (1969). *A.C.S. Polymer Preprints*, **10**(2), 655.

Faucher, J. A., J. V. Koleske, E. R. Santee, J. J. Stratta, and C. W. Wilson (1966). *J. appl. Phys.*, **37**, 3962.

Gee, G. (1966). *Polymer*, **7**, 177.

Gee, G. (1970). *Contemp. Phys.*, **11**, 313.

Gibbs, J. H. (1956). *J. chem. Phys.*, **25**, 185. See also (1960). In J. D. Mackenzie (Ed.), *Modern Aspects of The Vitreous State.* Vol. I. Butterworths. p. 152.

Gibbs, J. H., and E. A. Di Marzio (1958). *J. chem. Phys.*, **28**, 373, 807.

Goldstein, M. (1969). *J. chem. Phys.*, **51**, 3728. (See also this volume p. 23).

John, R. M. (1970). Private communication.

Kanig, G. (1963). *Kolloidzeitschrift. Z.f. Polymere*, **190**, 1.

McCall, D. W., D. C. Douglas, and E. W. Anderson (1962). *J. Polym. Sci.*, **59**, 301.

McCall, D. W., D. C. Douglas, V. J. McBrierty, and M. J. R. Hoch (1969). *Discuss. Faraday Soc.*, No. **48**, 205.

McCall, D. W., and D. R. Falcone (1970). *Trans. Faraday Soc.*, **66**, 262.

Shen, M. C., and A. Eisenberg (1966). *Prog. Solid St. Chem.*, **3**, 407. (1970) *Rubb. Rev.*, **43**, 95. This is reprint of above with an addendum.

Slichter, W. P. (1958). *Adv. Polym. Sci.*, **1**, 35.

Slichter, W. P., and D. D. Davis (1964). *J. appl. Phys.*, **34**, 98.

Slonim, I. Ya., and A. N. Lyubimov (1970). *The N.M.R. of Polymers.* English Edition translated by C. N. Turton and T. I. Turton. Plenum Press.

Smith, G. W. (1965). *J. chem. Phys.*, **42**, 4229.

Ueberreiter, K., and G. Kanig (1952). *J. Colloid Sci.*, **7**, 569.

39

Relaxation Studies Below the Glass Temperature in Polymers Containing *para*-Phenylene Linkages

G. ALLEN, R. M. JOHN

Department of Chemistry, The University, Manchester, Great Britain

G. M. JEFFS and J. A. McAINSH

I.C.I. Ltd., Petrochemicals and Polymers Laboratory, The Heath, Runcorn, Cheshire, Great Britain

---

Polymers containing *para*phenylene linkages are of interest because they include many of the amorphous organic polymers which have glass temperatures greater than 140 °C. Four typical examples are listed in Table 1.

Each polymer shows the expected strong mechanical and dielectric energy loss process in the glass transition region. Below T_g, however, the polymers have surprisingly few subsidiary loss features. McCall (1970) has reported n.m.r. spin lattice relaxation times (T_1) in the main magnetic field H_0, and also spin lattice relaxation times ($T_{1\rho}$) measured in the rotating frame for bisphenol 'A' polycarbonate. A low temperature minimum was assigned, without confirmation, to methyl group motion, and the possibility of contributions to T_1 from local motions involving the phenylene group was mentioned but not detected in the pulsed n.m.r. measurements, though it was tentatively invoked as the origin of a broad, weak dielectric loss.

In the present paper similar mechanical, dielectrical and pulsed n.m.r. studies are described on Union Carbide bisphenol 'A' polysulphone (McAinsh, 1969) and some related bisphenol 'A' cyanurates. (John, 1970). All of the polymers used have molecular weights in excess of 10^5 and have been fractionated to remove low molecular weight material.

Table 1. Glass transition temperatures of *para*phenylene linked polymers

Polymer	T_g °C (dilatometer)	T_g °C (DTA)
Bisphenol 'A' polysulphone	176	179
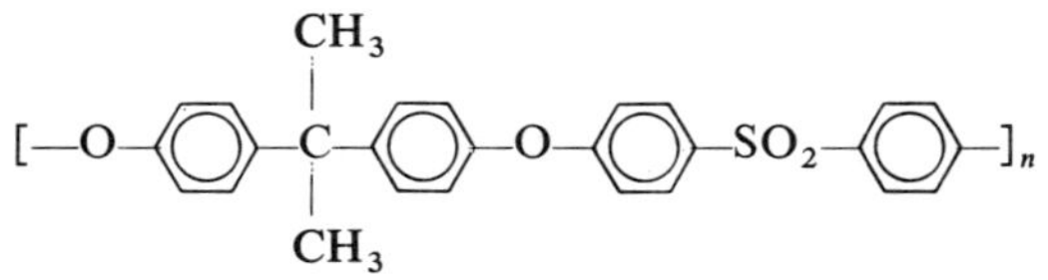		
Bisphenol 'A' polycarbonate	145	—
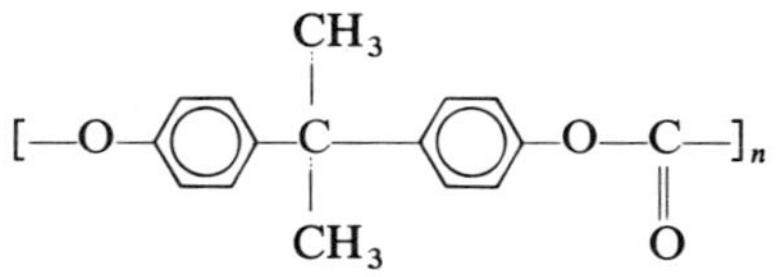		
poly (2,6 dimethyl phenylene oxide)	209	214
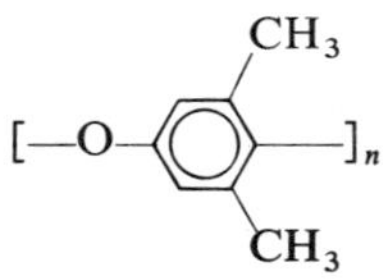		
polysulphone	–	214
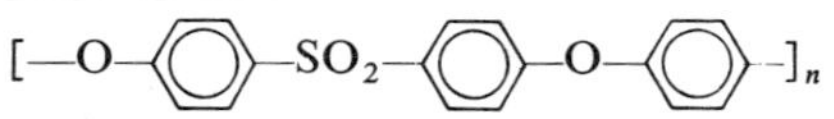		

MECHANICAL LOSS MEASUREMENTS

(Allen, Jeffs and McAinsh, 1971)

A conventional torsion pendulum was used at 1 Hz to obtain results in the temperature range -150 °C to $+150$ °C. The design of the environmental chamber did not allow the sample to be kept completely dry and it was found that during a run the water content of the sample could not be maintained below 0·07% by weight. Figure 1 shows results for two samples, containing respectively 0·073 and 0·31% of absorbed water. The tan δ plot shows a β relaxation peak at -100 °C at 1 Hz and the strength of the loss is proportional to the amount of water in the sample. The α-relaxation (glass–rubber transition) lies above 200 °C and could not be investigated with this apparatus, a feature common to all subsequent measurements.

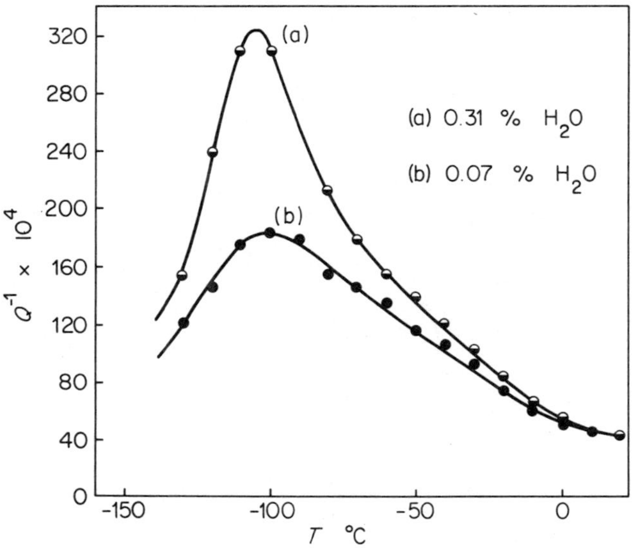

Figure 1. Effect of water on the low temperature mechanical loss in polysulphone. [From *Polymer*, **12**, 85 (1971)]

DIELECTRIC LOSS MEASUREMENTS

(Allen, Jeffs and McAinsh, 1971)

A General Radio transformer ratio arm bridge (type 1615A) was used at frequencies of 1 kHz and 10 kHz over a temperature range of −120 to +100 °C. The results shown in Figure 2 are similar in pattern to the mechanical loss data. Again the strength of the β peak is related to the water content of the sample. At these higher frequencies however the loss peak has moved up to −70 °C (1 kHz) and −45°C (10 kHz).

Despite careful sample preparation it is not clear whether a loss peak would be observed in this region in a truly dry sample or whether the relaxation process is intimately associated with water bound to the main polymer chain.

PULSED N.M.R. RELAXATION MEASUREMENTS (John, 1970)

In an attempt to obtain more precise information concerning local motions in bisphenol 'A' polysulphone, T_1 measurements at 30 MHz and $T_{1\rho}$ measurements at 105 kHz have been made using a Polaron pulsed n.m.r. spectrometer over the temperature range −150 to 200 °C.

The T_1 and $T_{1\rho}$ measurements on bisphenol 'A' polysulphone are summarized in Figure 3. A pronounced minimum at −30 °C is observed in the

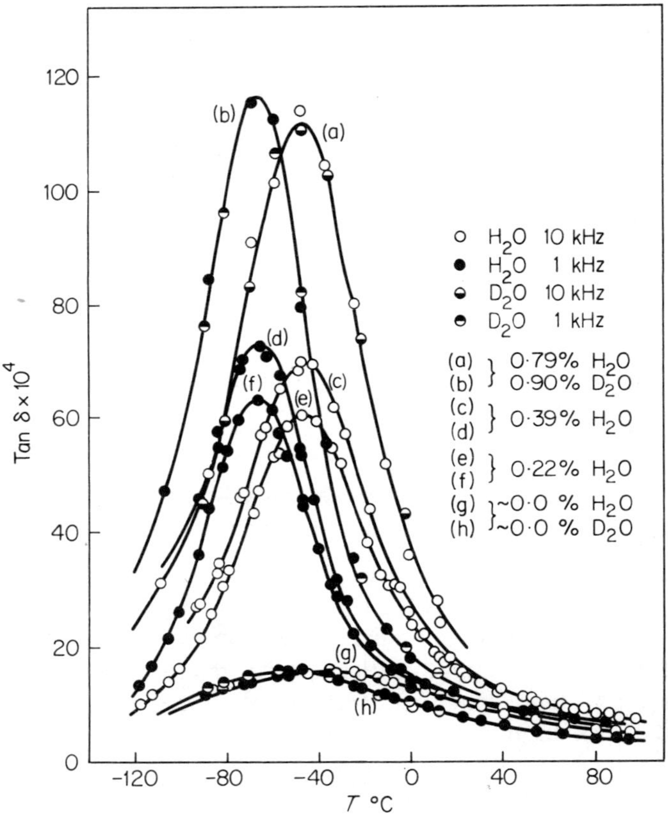

Figure 2. Effect of water (H_2O and D_2O) on the low-temperature dielectric loss in polysulphone. [From *Polymer*, **12**, 85 (1971)]

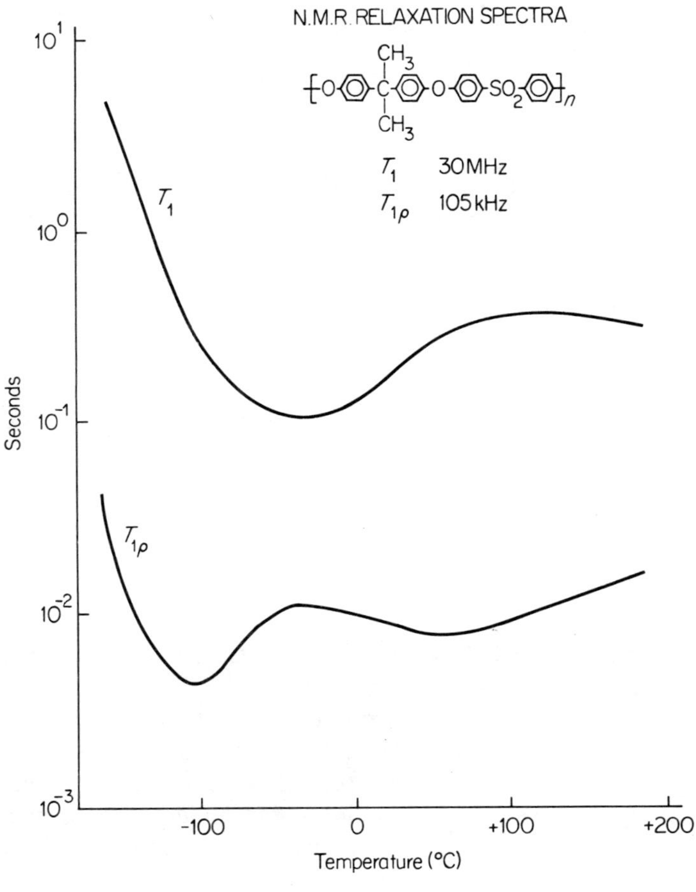

Figure 3. T_1 and $T_{1\rho}$ for bisphenol 'A' polysulphone

ln T_1–temperature plot and the corresponding minimum in ln $T_{1\rho}$ lies at -100 °C. The ln $T_{1\rho}$ plot also shows a broader, more shallow, minimum at $+50$ °C. Neither of these minima appear to be related to the β-loss peak observed in the dielectric or mechanical measurements. Minima corresponding to this β process would be expected at about -30 °C in $T_{1\rho}$ and about 0 °C in T_1. Furthermore the pulsed n.m.r. results are independent of the water content (up to 0·7%) of the sample.

A more complete understanding of the n.m.r. relaxation processes has been obtained from a study of two related cyanurate polymers (a) and (b)

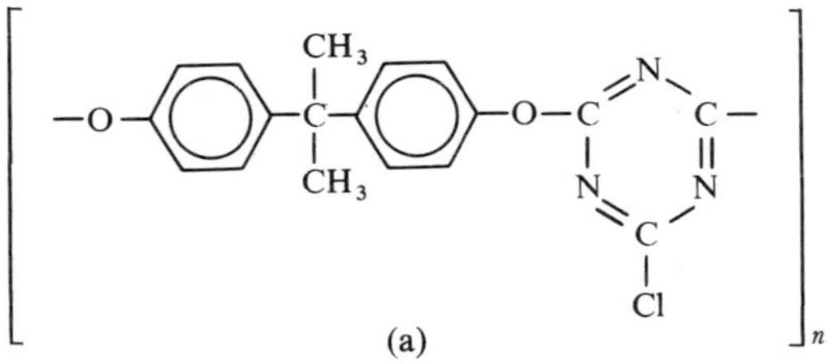

(a)

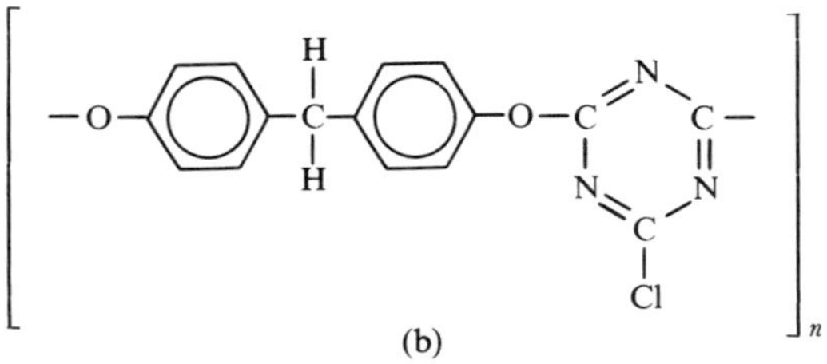

(b)

Inspection of Figure 4 shows that the ln T_1–temperature plot for polymer (a) displays the same pronounced minimum at -30 °C as does the bisphenol 'A' polysulphone, whereas the relaxation process is entirely absent from the corresponding plot for polymer (b). Thus the origin of this relaxation process in both of the bisphenol 'A' polymers is associated with the motion of the methyl group. In turn this observation confirms the assignment made by McCall for the corresponding n.m.r. relaxation process in bisphenol 'A' polycarbonate.

The ln $T_{1\rho}$–temperature plot for polycyanurate (b) shown in Figure 5 does have a broad, shallow minimum centred at approximately $+30$ °C and this relaxation process *is* obviously similar to the second minimum at $+50$ °C observed in the corresponding $T_{1\rho}$ plot for the bisphenol 'A' polysulphone. Since it is known that n.m.r. loss processes associated with phenyl motion tend to give rather shallow minima at relatively large values of T_1 [compare for example the side-group relaxation processes in poly(t-butyl ethylene oxide) and poly(styrene oxide) (Allen, Blears and Connor, 1967)], it

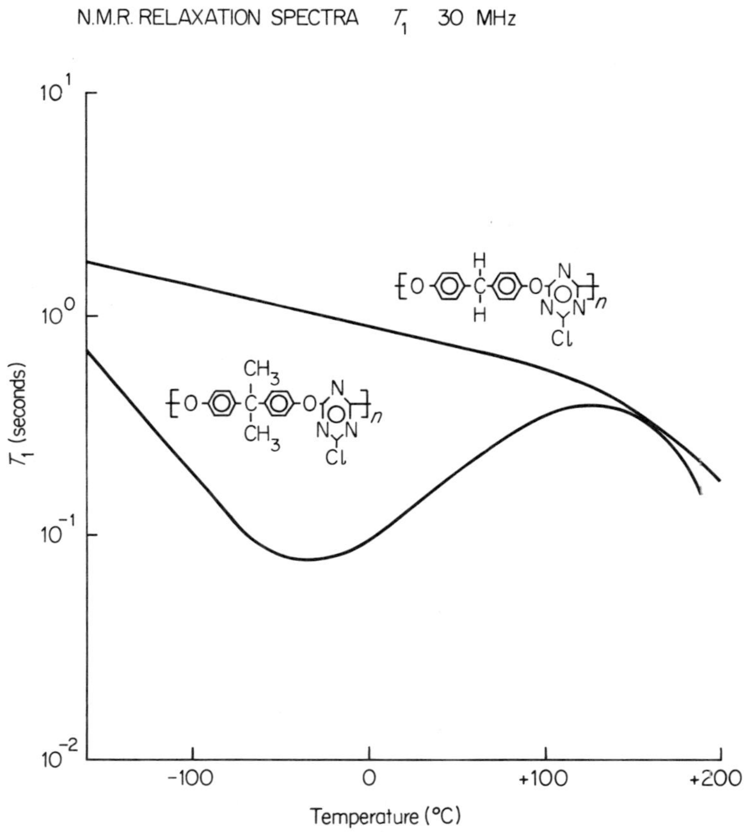

Figure 4. T_1 for bisphenol 'A' polycyanurate and its methylene analogue

is reasonable to associate this loss process with the motion of the phenylene groups about their *para*-linkages. McCall (1970) speculated that there should be such a local contribution to n.m.r. relaxation in bisphenol 'A' polycarbonate but none was observed. Certainly the relaxation process would not be expected to be pronounced since it involves interaction with four protons placed on the periphery of a benzene ring. Further support for this assignment has been obtained by crosslinking the bisphenol 'A' polycyanurate through the Cl atom on the cyanurate ring with more bisphenol 'A' to produce a three dimensional network. Figure 6 shows that even in the cross-linked material, where only local phenylene motion will be possible, the relaxation process is still apparent.

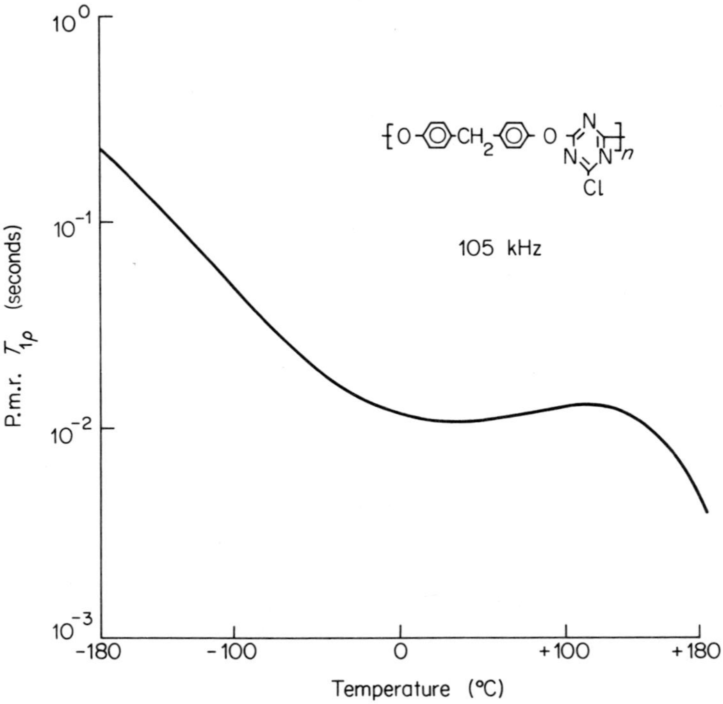

Figure 5. $T_{1\rho}$ for the methylene analogue of bisphenol 'A' polycarbonate

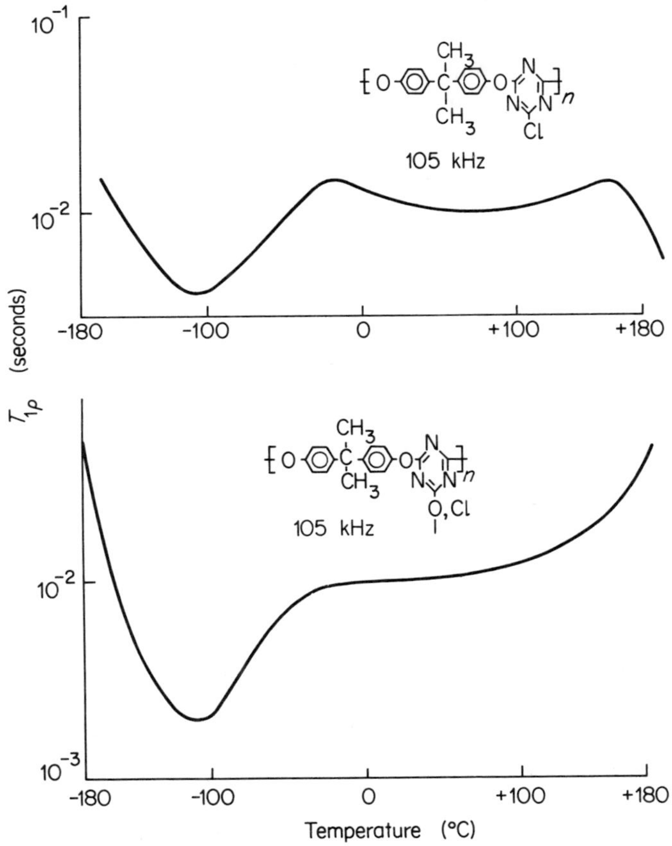

Figure 6. Comparison of $T_{1\rho}$ for linear bisphenol 'A' polycyanurate and the corresponding polymer crosslinked with additional bisphenol 'A'

CONCLUSION

The local methyl and phenylene group motions in the bisphenol 'A' family of polymers have been located by pulsed n.m.r. relaxation spectroscopy. The β-process observed in dielectrical loss and mechanical loss is related to the water content of the material and is possibly due to hydrogen bonded water–polymer interactions. Although the exact location of this β-process in a dry polymer has not been definitely identified, it seems possible that it may lie some 20–30 °C higher than in the wet samples, and as such will be associated with local motion of the paraphenylene structural units. It is clearly very important to assay the water content of polymers containing paraphenoxy linkages when studies are made of dielectrical or mechanical loss.

ACKNOWLEDGEMENT

We thank Mr. M. Coville for preparation of the cyanurate used in this work and Mr. R. F. Warren for advice and help with the n.m.r. measurements.

REFERENCES

Allen, G., D. J. Blears, and T. M. Connor (1967). *Polymer* (London), **8**, 401.
Allen, G., G. M. Jeffs, and J. A. McAinsh (1971). *Polymer* (London), **12**, 85.
John, R. M. (1970). *Ph.D. Thesis*, University of Manchester.
McAinsh, J. A. (1969). *Ph.D. Thesis*, University of Manchester.
McCall, D. W., and D. R. Falcone (1970). *Trans. Faraday Soc.*, 262.

40

E.S.R. and Optical Studies of Mo^{5+} Ions in Phosphate Glasses

J. BAUGHER

James Franck Institute, University of Chicago, U.S.A.

and S. PARKE

Department of Glass Technology, University of Sheffield, Great Britain

---

The purpose of the present paper is to establish that the ligand field around the pentavalent molybdenum ion in glasses is so highly distorted from octahedral symmetry that it is better to describe it as tetragonal symmetry. It is conventional to indicate that one of the oxygen atoms may be closer to the metal ion than with octahedral symmetry by writing MoO^{3+}. The latter is called a molybdenyl ion and the low symmetry associated with it results in an e.s.r. resonance at g values smaller than 2. In addition, as the metal ion possesses a nuclear spin, hyperfine structure is readily seen and is a sensitive indicator of the presence of covalent bonding. There has been little previous work on the e.s.r. spectra of molybdenum in oxide glasses although Landry (1968) published an X-band spectrum of Mo^{3+} in a zinc-aluminium-phosphate glass but did not discuss the $4d^1$ ion Mo^{5+}. There have been some studies of Mo^{5+} in supercooled solutions (Garifyanov and Fedotov, 1963; Abraham *et al.*, 1966), which establish the presence of the molybdenyl ion.

Mo^{5+} e.s.r. spectra observed at room temperature in a series of phosphate glasses are described in this paper. Optical spectra have also been studied in order to understand more clearly the e.s.r. spectra.

EXPERIMENTAL

The glasses were melted from the orthophosphates of zinc, aluminium, calcium and sodium. Additional phosphate was introduced as ammonium phosphate. The molybdenum was added as the metal. The melts were prepared under reducing conditions in a gas furnace.

The e.s.r. spectra were taken at room temperature with a Hilger and Watts Microspin spectrometer. The field was calibrated using a proton resonance frequency meter. Absorption spectra were measured on a Cary 14 spectrophotometer.

RESULTS

Optical spectra

In all phosphate glasses which stabilize the trivalent molybdenum ion, the spectrum is as shown by the dotted curve in Figure 1 (Parke and Watson,

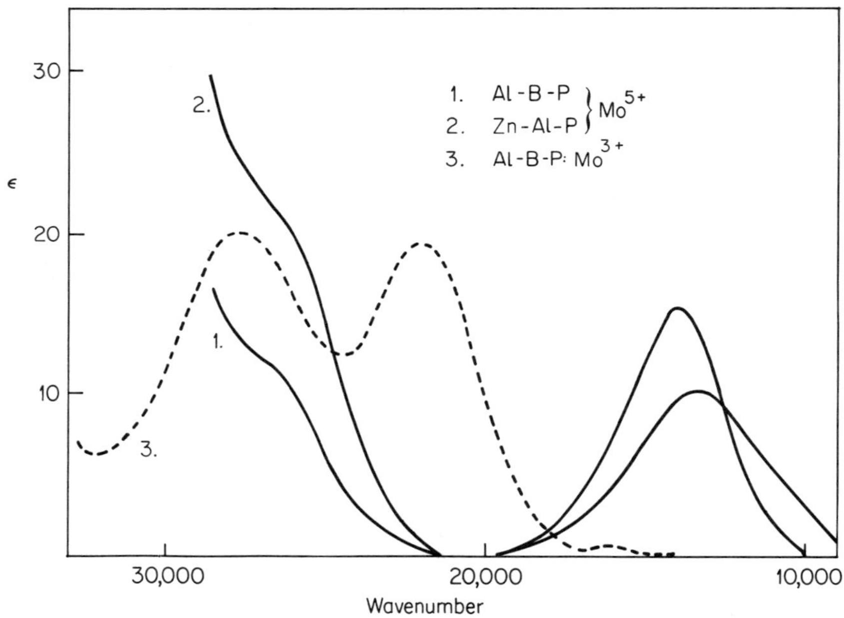

Figure 1. Optical spectra of Mo^{3+} and Mo^{5+} ions in phosphate glasses

1969). The positions of the bands at 27,800 cm^{-1} and 22,000 cm^{-1} are almost independent of the nature of the cation. Under less severe reducing conditions there are bands at 26,500 cm^{-1} and 13,000–14,000 cm^{-1} but no band at 22.000 cm^{-1} (Figure 1). It is assumed, therefore, that the bands shown in curves 1 and 2 may be assigned to Mo^{5+}. In some glasses where there is evidence of both Mo^{3+} and Mo^{5+} there is a band at 18200 cm^{-1} which cannot be associated with either ion. It could be assigned to Mo^{4+}, the other bands belonging to this ion being overlapped by Mo^{3+} and Mo^{5+} bands.

The band positions for Mo^{5+} in phosphate glasses are given in Table 1.

Table 1. Optical absorption bands of Mo^{5+} in glasses

Composition	Band position (cm^{-1})	
ZnO-Al_2O_3-P_2O_5	26400	14100
B_2O_3-Al_2O_3-P_2O_5	26600	13400
CaO-P_2O_5	28400	12900
Na_2O-P_2O_5	28100	12900

E.s.r. spectra

An example of a typical spectrum observed is shown in Figure 2. All of the Mo^{5+} e.s.r. spectra obtained were similar except for slight changes in the relative intensities of some components. The strong line in the centre arises from Mo^{5+} ions containing the non-magnetic 'even' nuclear isotopes of molybdenum, and can be described by the axial spin Hamiltonian

$$\mathscr{H}_e = \beta g_{\parallel} H_z S_z + g_{\perp}(H_x S_x + H_y S_y) \tag{1}$$

where β is the Bohr magnetron, $g_{\parallel}$ and $g_{\perp}$ are the parallel and perpendicular components of the g-tensor, H_x, H_y, H_z are the components of the magnetic field, and S_x, S_y, S_z are the components of the electron spin.

The Hamiltonian (1) leads to the resonance condition

$$h\nu_0 = g\beta H \tag{2}$$

where ν_0 is the applied frequency, H is the magnitude of the applied field, and

$$g = (g_{\perp}^2 \sin^2 \theta + g_{\parallel}^2 \cos^2 \theta)^{\frac{1}{2}} \tag{3}$$

Here θ is the angle between the magnetic field and the symmetry axis of the site. In a glass sample, one must average over all values of θ. An envelope of resonances, known as a powder pattern, is obtained. The powder pattern for the resonance condition of equation (2) is shown as the solid line in Figure 2 (see, for example, Cohen and Reif, 1957). A divergence appears at a field of $H_{\perp} = h\nu_0/g_{\perp}\beta$ and a shoulder appears at $H_{\parallel} = h\nu_0/g_{\parallel}\beta$. There are two molybdenum isotopes which have a magnetic moment. The magnetic moments of the isotopes Mo^{95} and Mo^{97} differ by only two per cent and the nuclear spin of each is $\frac{5}{2}$. The abundance ratio of even to odd isotopes is approximately 3:1. The hyperfine structure, due to the odd isotopes can be

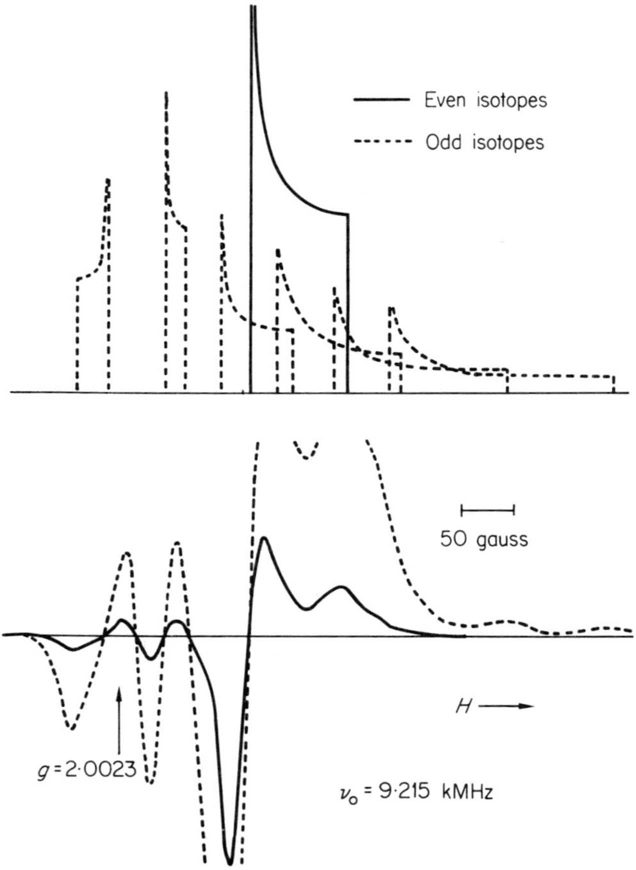

Figure 2. E.s.r. spectrum of Mo^{5+} in alumino-boro-phosphate glass

seen in the 'wings' of the experimental spectrum of Figure 2. In this case, the spectra can be described by the spin Hamiltonian

$$\begin{aligned}\mathscr{H}_0 = \beta g_{\parallel} H_z S_z + g_{\perp}(H_x S_x + H_y S_y) + A_{\parallel}^{95} I_z S_z \\ + A_{\perp}^{95}(I_x S_x + I_y S_y) + A_{\parallel}^{97} I_z S_z \\ + A_{\perp}^{97}(I_x S_x + I_y S_y) \qquad (4)\end{aligned}$$

where $A_{\parallel}$, $A_{\perp}$ are the parallel and perpendicular principal components of the hyperfine coupling tensor, and I_x, I_y, I_z are the nuclear spin components. Since the magnetic moments of the two odd molybdenum isotopes are almost identical, their separate contribution to the hyperfine structure cannot be resolved. The Hamiltonian of equation (4) leads to the resonance condition

$$h\nu_0 = g\beta H - Km - \frac{A_\perp^2}{4\beta g H_0}\left(\frac{A_\parallel^2 + K^2}{K^2}\right)[I(I+1) - m^2] + (2\beta g H_0)^{-1} \times \left(\frac{A_\parallel^2 - A_\perp^2}{K}\right)^2 \left(\frac{g_\parallel g_\perp}{g^2}\right)^2 m^2 \sin^2\theta \cos^2\theta \qquad (5)$$

correct to second order in the hyperfine energy. In equation (5), m is the nuclear magnetic quantum number and

$$K = \frac{1}{g}(g_\perp^2 A_\perp^2 \sin^2\theta + g_\parallel^2 A_\parallel^2 \cos^2\theta)^{\frac{1}{2}} \qquad (6)$$

H_0 is defined such that $h\nu_0 = g\beta H_0$ and the separate contribution of the two isotopes is ignored. The powder patterns resulting from equation (5) have been discussed by Sands (1955). In the case of molybdenum there will be six envelopes whose divergences occur at

$$\frac{h\nu_0}{g_\perp\beta} - \frac{A_\perp m}{g_\perp\beta} - \frac{A_\parallel^2 + A_\perp^2}{4H_0^\perp(g_\perp\beta)^2}[I(I+1) - m^2]$$

and shoulders at

$$\frac{h\nu_0}{g_\parallel\beta} - \frac{A_\parallel m}{g_\parallel\beta} - \frac{A_\perp^2}{2H_0^\parallel(g_\parallel\beta)^2}[I(I+1) - m^2]$$

These six envelopes are shown in Figure 2 by the dotted lines.

The spin Hamiltonian parameters may be fitted by two different methods. The shoulder belonging to the central transition is taken as $g_\parallel$ and on the high field side two parallel components of the hyperfine spectrum are well resolved. The separation of these, ignoring the second order terms, gives $A_\parallel$. Having determined $g_\parallel$ and $A_\parallel$ the parallel line spectrum can be drawn in. The determination of $g_\perp$ is more difficult but a good estimate may be obtained by comparison with computer-evaluated spectra for axial symmetry. On the low field side of the central line two perpendicular components may be found, and from them $A_\perp$ may be determined. In the second method, the high field parallel components are assumed to correspond to $m = -\frac{5}{2}$ and $m = -\frac{3}{2}$. On the low field side the components (in order of increasing field) $\frac{5}{2}$ (parallel), $\frac{5}{2}$ and $\frac{3}{2}$ (perpendicular) and $\frac{3}{2}$ (parallel) are identified. These six equations are more than sufficient to fix $g_\parallel, g_\perp, A_\parallel$ and $A_\perp$ by means of an iteration procedure.

The results are shown in Table 2. It must be noted that only the magnitudes of $A_\parallel$ and $A_\perp$, and not their signs, can be obtained.

A Q-band spectrum for the zinc-aluminium-phosphate glass has hyperfine lines that are less well resolved but values of $g_\parallel$ and $g_\perp$ obtained from the even isotopes line are in agreement with those found at X band.

Table 2. g-values and hyperfine constants of Mo^{5+} in glasses

Composition	$g_{\parallel}$	$g_{\perp}$	$\lvert A_{\parallel}\rvert$ $\times 10^{-4}\ cm^{-1}$	$\lvert A_{\perp}\rvert$
ZnO-Al_2O_3-P_2O_5	1·881	1·930	94·8	54·6
B_2O_3-Al_2O_3-P_2O_5	1·884	1·935	91·3	50·4
CaO-P_2O_5	1·881	1·928	95·9	53·6
Na_2O-P_2O_5	1·884	1·923	97·2	59·8

DISCUSSION

Analysis of the g-tensor

The Mo^{5+} ion is a $4d^1$ ion, having a single d electron outside a closed shell. The values of $g_{\parallel}$ and $g_{\perp}$ which are obtained depend critically on the nature of the crystal field which acts upon the ion. Tetrahedral complexes usually have relatively small crystal field splittings and optical absorption bands in the infrared. An octahedral coordination scheme is much more likely to yield bands in the optical region and such a scheme will be assumed for the Mo^{5+} ions in this study.

If a complex has perfect octahedral symmetry then the ground state is an orbital triplet, the g-tensor vanishes, and there is no observable e.s.r. For a resonance to be detected some sort of distortion must take place. The e.s.r. results indicate that axial symmetry is present. Generally, the axial symmetry can be either threefold or fourfold.

Gladney and Swalen (1965) have given a detailed analysis of the g-tensor in the presence of trigonal symmetry. Either a singlet or a doublet can be the ground state, depending on whether the distortion is an elongation or compression of the octahedron along the polar axis. When the doublet is the ground state both $g_{\parallel}$ and $g_{\perp}$ are near zero. When the singlet is lowest, $g_{\parallel} = 2$ and $g_{\perp}$ is less than, but close to, 2. Since it is found that $g_{\perp} > g_{\parallel}$ and that both are significantly different from 2, any octahedral site with a distortion along a threefold axis can be ruled out.

The energy level for fourfold symmetry, is given in Figure 3. If the doublet (d_{xz}, d_{yz}) is the ground state, both g-values are near zero (Griffith, 1964). When the orbital singlet d_{xy} is lowest one can show, using the formalism of Pryce (1950), that

$$\begin{aligned} g_{\parallel} &= 2{\cdot}0023\left(1 - \frac{4\lambda}{E(x^2 - y^2) - E(xy)}\right) \\ g_{\perp} &= 2{\cdot}0023\left(1 - \frac{\lambda}{E(xz, yz) - E(xy)}\right) \end{aligned} \tag{7}$$

where λ is the spin–orbit coupling constant. Using the experimental values of $g_{\parallel}$ and $g_{\perp}$ given in Table 2, and taking λ to be equal to its free ion value of

1030 cm^{-1}, one obtains optical bands which are much higher in energy than those observed experimentally. This discrepancy is often noted in a comparison of optical and e.s.r. data and is due to the presence of covalency in the metal–ligand bonds. Covalency has the effect of reducing the spin–orbit

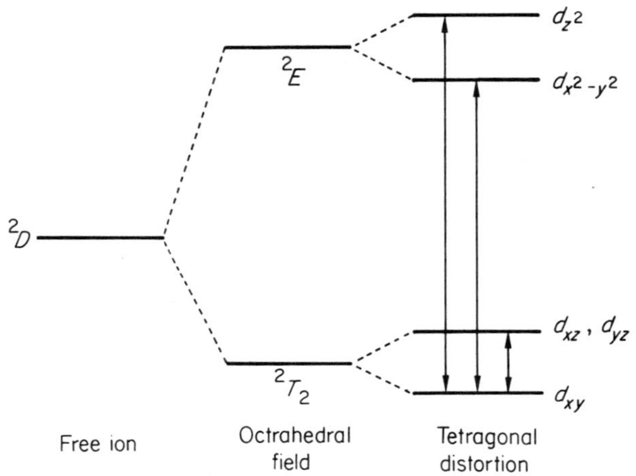

Figure 3. Energy level scheme for tetragonal symmetry

coupling constant λ. In order to give reasonable agreement with the optical data λ would have to be reduced to a value near 450 cm^{-1}. However, λ may be eliminated from equation (7) to give the relative positions of the levels $x^2 - y^2$ and (xz, yz) and these are consistent with the optical results as shown in Table 3. The e.s.r. results on the glasses are in agreement with a tetragonally

Table 3.

Composition	$E(x^2 - y^2) - E(xy)/E(xz, yz) - E(xy)$	
	optical	e.s.r.
$ZnO\text{-}Al_2O_3\text{-}P_2O_5$	1·87	2·38
$B_2O_3\text{-}Al_2O_3\text{-}P_2O_5$	1·99	2·27
$CaO\text{-}P_2O_5$	2·20	2·45
$Na_2O\text{-}P_2O_5$	2·18	2·68

distorted octahedron having an orbital singlet as the ground state. This ground state can arise when the octahedron is compressed along the fourfold axis. The compression is quite large since $E(x^2 - y^2) - E(xy)$ is about twice $E(xz, yz) - E(xy)$. This conclusion is in agreement with the concept of the molybdenyl ion as discussed in the introduction.

Analysis of the hyperfine tensor

When the ground state of the Mo^{5+} ion is d_{xy}, the components $A_{\|}$ and $A_{\perp}$ of the hyperfine tensor may be written as

$$A_{\|} = -P(\tfrac{4}{7} + \Delta g_{\|} + \tfrac{3}{7}\Delta g_{\perp} + \kappa)$$
$$A_{\perp} = -P(-\tfrac{2}{7} + \tfrac{11}{14}\Delta g_{\perp} + \kappa) \qquad (8)$$

where $P = 2\gamma\beta\beta_N\langle r^{-3}\rangle$ and $\Delta g = 2{\cdot}0023 - g.\ g_{\|}$ and $g_{\perp}$ are given by equation (7), β_N is the nuclear magneton and γ is the gyromagnetic ratio. For the isotopes Mo^{95} and Mo^{97} the latter is equal to $-0{\cdot}3716$ and $-0{\cdot}3755$ respectively. κ is a pure number which expresses the magnitude of the isotropic Fermi contact interaction.

Since κ is positive for transition ions and P is negative, a consistent scheme is derived only if both $A_{\|}$ and $A_{\perp}$ are assumed to be positive. P can be determined from the difference

$$A_{\|} - A_{\perp} = -P(\tfrac{6}{7} + \Delta g_{\|} - \tfrac{5}{14}\Delta g_{\perp}) \qquad (9)$$

Inserting values of $g_{\|}$, $g_{\perp}$, $A_{\|}$ and $A_{\perp}$ from Table 2 we obtain the values of P and κ given in Table 4. For the free Mo^{5+} ion $P = -66{\cdot}7 \times 10^{-4}\ cm^{-1}$

Table 4. Values of P and κ for molybdenum in glasses

Composition	P $\times 10^{-4}\ cm^{-1}$	κ	$P\kappa/\gamma$ cm^{-1}
ZnO-Al_2O_3-P_2O_5	−42	1·5	0·017
B_2O_3-Al_2O_3-P_2O_5	−43	1·4	0·016
CaO-P_2O_5	−44	1·4	0·017
Na_2O-P_2O_5	−40	1·7	0·018

(McGarvey, 1969). The fact that in all cases the magnitude of P is smaller than the free ion value is indicative of the fact that $\langle r^{-3}\rangle$ is smaller for the bound ion. Hence, just as in the case of the g-tensor, the hyperfine tensor reflects a considerable degree of covalency in the bonding. The quantity κ should ideally be zero for a $4d^1$ ion since the Fermi contact interaction vanishes for d-orbitals. A slight admixture of the $5s$ excited state into the $4d$ ground state could explain the existence of a non-zero Fermi contract term.

For $V^{4+}(3d^1)$ the hyperfine constants are both negative and κ and P are positive. From the data given by Hochstrasser (1966) one can calculate

$\kappa = 0{\cdot}76$, $P = 123 \times 10^{-4}\ \mathrm{cm}^{-1}$. The ratio $P\kappa/\gamma$ is found to be $0{\cdot}0063\ \mathrm{cm}^{-1}$ in agreement with Bleaney and Stevens (1953) who give $0{\cdot}0060\ \mathrm{cm}^{-1}$, $0{\cdot}0054\ \mathrm{cm}^{-1}$, $0{\cdot}0067\ \mathrm{cm}^{-1}$ and $0{\cdot}0060\ \mathrm{cm}^{-1}$ for the divalent ions of copper, cobalt, manganese and vanadium respectively. It would be interesting to know if the ratio $P\kappa/\gamma$ also has a nearly constant value throughout the second row of transition ions. The results of Kim *et al.* (1968) on Nb^{4+} lead to $P\kappa/\gamma = 0{\cdot}019\ \mathrm{cm}^{-1}$ which is of the same order as the values listed in the final column of Table 4.

Effects of covalency

It is not the intention here to give a full molecular orbital analysis of our results but it is worthwhile to indicate the direction in which a future analysis must proceed. Kivelson and Lee (1964) have shown that equations (7) and (8) become

$$\begin{aligned}
g_{\parallel} &= 2{\cdot}0023\left(1 - \frac{4\alpha^2\beta^2\lambda'}{E(x^2 - y^2) - E(xy)}\right)\\
g_{\perp} &= 2{\cdot}0023\left(1 - \frac{\delta^2\beta^2\lambda'}{E(xz, yz) - E(xy)}\right)\\
A_{\parallel} &= -\ P[\beta^2(\tfrac{4}{7} + \kappa) + \Delta g_{\parallel} + \tfrac{3}{7}\Delta g_{\perp}]\\
A_{\perp} &= +\ P[\beta^2(\tfrac{2}{7} - \kappa) - \tfrac{11}{14}\Delta g_{\perp}]
\end{aligned}$$

where λ' is now the spin–orbit coupling constant of the bound ion and α^2, β^2, δ^2 are measures of in-plane σ bonding, in-plane π bonding and out-of-plane π bonding respectively. These coefficients have the value 0·5 when the bonding is purely covalent and approach unity when the bonding is ionic.

McGarvey (1969) has given an MO analysis for +5 and +4 ions of configuration d^1. From his equations it is possible to deduce that the effective charge on the molybdenum ion in the present glasses is +3 and that $\beta^2 \sim 0{\cdot}77$. Hence, assuming the spin–orbit constant for Mo^{3+} ($817\ \mathrm{cm}^{-1}$) we find $\alpha^2 \sim 0{\cdot}63$ and $\delta^2 \sim 0{\cdot}81$. Although these numbers must not be regarded as quantitative until a more exact treatment is attempted, they are self-consistent and show that substantial π bonding is present in the glasses.

REFERENCES

Abraham, M. M., J. P. Abriata, M. E. Foglio, and E. Pasquini (1966). *J. chem. Phys.*, **45**, 2069.

Bleaney, B., and K. W. H. Stevens (1953). *Rep. Prog. Phys.*, **16**, 108.

Cohen, M. H., and F. Reif (1957). *Solid St. Physics*, Vol. 5. Academic Press.

Garif'yanov, N. S., and V. N. Fedotov (1963). *Soviet Phys. JETP*, **16**, 269.

Gladney, H. M., and J. D. Swalen (1965). *J. chem. Phys.*, **42**, 1999.

Griffith, J. S. (1964). *The Theory of Transition Metal Ions.* Cambridge University Press.

Hochstrasser, G. (1966). *Physics Chem. Glasses*, **5**, 178.
Kim, Y. M., D. E. Reardon, and P. J. Bray (1968). *J. chem. Phys.*, **48**, 3396.
Kivelson, D., and S. K. Lee (1964). *J. chem. Phys.*, **41**, 1892.
Landry, R. J. (1968). *J. chem. Phys.*, **48**, 1422.
Parke, S., and A. I. Watson (1969). *Physics Chem. Glasses*, **10**, 37.
Pryce, M. H. L. (1950). *Proc. phys. Soc. Lond.*, **A63**, 25.
Sands, R. H. (1955). *Phys. Rev.*, **99**, 1222.

41

E.S.R. and Optical Absorption Spectra of Irradiated Lead Borate Glasses

S. ARAFA and A. BISHAY

Solid State and Materials Research Center, The American University in Cairo, Cairo, Egypt, U.A.R.

Observations made on some lead borate and lead-alumino-borate glasses (Bishay, 1960) have shown that a new optical absorption band is induced at about 1·5 eV (825 mμ) by gamma irradiation. A similar band was also observed in gamma irradiated alkali alumino lead phosphate glasses (Bishay, 1961) and lead silicate glasses (Bishay and Ferguson, 1962). Experimental evidence proved that this band is associated with Pb^{2+} ions in the glass structure. This band was found to be rather unstable and saturates at a dose of about 10^6 rads (Bishay, 1960).

Bishay and Maklad (1966) studied the induced optical absorption in binary lead borate glasses in relation to structural changes. They suggested that the band induced in these glasses at about 1·6 eV is associated with an electron trapped in the vicinity of Pb^{2+} ions. The intensity of this band was reported to increase with increasing PbO content (from 18–25 mole % PbO), reaching a maximum at about 25 mole % PbO.

Bishay and Maklad (1967) reported later that gamma irradiation of potassium borate glasses containing about 1·8 mole % Tl^{+}, Pb^{2+} or Bi^{3+} resulted in two induced optical absorption bands which they called T and L bands. The T and L bands for the glass containing Pb^{2+} ions were observed at about 1·55 eV and 2·6 eV, respectively. The same authors also reported that these two bands (T and L) are associated with electrons trapped near the Pb^{2+} ions.

No electron spin resonance results of irradiated lead borate glasses could be found in the open literature. The e.s.r. of some irradiated borate glasses containing lead oxide is discussed in this paper in relation to the induced optical absorption. A comparison is made between the spectra induced in some of these glasses and alkali borate glasses.

EXPERIMENTAL

Sample preparation

Analytical reagent chemicals were used in the preparation of glasses of molar composition $(100 - x)\, B_2O_3 \,.\, xPbO$. Boric oxide was introduced as boric acid, and lead as PbO.

All batches were sintered for about one hour at 600 °C, and melted in platinum crucibles for four hours at about 1000 °C under normal conditions. The poured discs were annealed from 450 °C at a rate of about 0·6 °C/min. Discs of different thicknesses were ground and polished for optical measurements; powdered samples were used for e.s.r. measurements.

Optical, e.s.r. and irradiation studies

Optical absorption measurements were made on a recording 700 SP Unicam Spectrophotometer before and after irradiation. The range of wavelengths studied was usually between 200 mμ and 1μ (about 6–1 eV). The optical spectra were measured directly after irradiation.

The e.s.r. spectra of irradiated powdered glass samples were obtained at room temperature as well as at −160 °C by the use of a V–4502 X-band spectrometer employing 100 kc/sec modulation.

All samples were irradiated at room temperature in the dark at an intensity of $3{\cdot}7 \times 10^5$ rads/hr using a ^{60}Co gamma cell-220 ($\sim$6000 curie). On the other hand, some samples were irradiated with X-rays from a tungsten tube (General Electric XRD-6) while others were exposed to a mixed flux from the reactor.

RESULTS AND DISCUSSION

Figure 1(a) shows the induced optical absorption of a lead borate glass of the composition $70\, B_2O_3 \cdot 30\, PbO$ after γ-irradiation. The gaussian resolution of the observed spectrum indicates the presence of three bands induced at about 1·6, 2·4 and 3·5 eV, respectively.

Figure 1(b) shows the e.s.r. spectrum induced in a γ-irradiated lead borate glass of the same composition, $70\, B_2O_3 \cdot 30\, PbO$, measured at room temperature. The spectrum displays a weak signal with no observed hyperfine structure. The shape of this line indicates anisotropy in the g-value. Its width, measured as the separation between the two extremum points, is about 32 gauss, and the g-value, measured at the mid-point between the two extremum points, is 1·996. This line is very unstable at high temperatures; it nearly disappears after heating the irradiated sample to 150 °C. The bleaching behaviour of this line is very similar to that observed for the rather unstable optical absorption band induced at 1·6 eV in glasses of the same composition. Figure 2 shows the percentage fading for both the 1·6 eV and the $g = 1{\cdot}996$ e.s.r. line induced in the irradiated lead borate glass of the composition

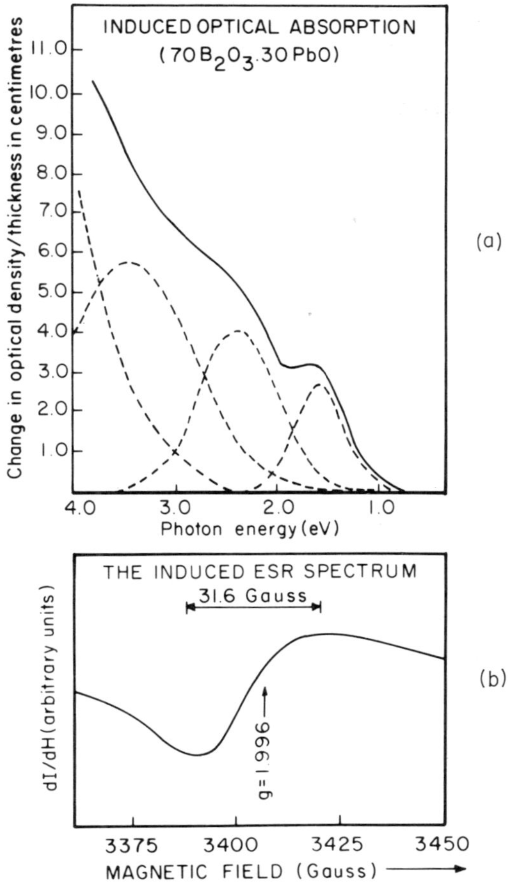

Figure 1(a). Resolution of induced optical absorption for the (70 $B_2O_3 \cdot 30$ PbO) glass. (b) Induced electron spin resonance for the (70 $B_2O_3 \cdot 30$ PbO) glass

70 B_2O_3 . 30 PbO. Figure 3 shows the growth curve for the 1·6 eV optical band and the $g = 1{\cdot}996$ e.s.r. line induced in the same lead borate glass. The intensity of the e.s.r. line was measured as the height between the two extremum points. The intensity of the optical band and e.s.r. line reaches a saturation value after a gamma irradiation does of about 10^6 rads. The results shown in Figures 2 and 3 indicate that both the induced band and e.s.r. line may have the same origin.

The studies made by Bishay and Maklad (1966) indicate strongly that the optical absorption band induced at about 1·6 eV in lead borate glasses is associated with an electron trap centre in the vicinity of Pb^{2+} ions. It is

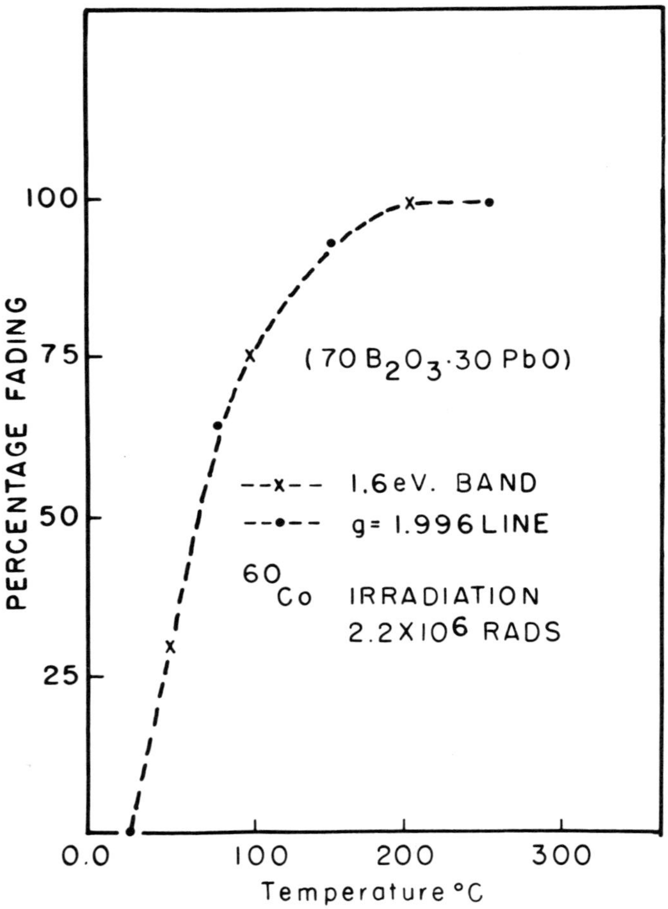

Figure 2. Thermal bleaching of induced optical absorption and e.s.r. in a lead borate glass (70 $B_2O_3 \cdot 30$ PbO)

now postulated that the $g = 1{\cdot}996$ e.s.r. line induced in these glasses is also associated with the same centre.

It should be stressed here that we cannot neglect the possibility that the electron trapped in the vicinity of the Pb^{2+} ions is actually orbiting around the lead (i.e. $Pb^{2+} + e$, or Pb^{+}) instead of being trapped in an oxygen vacancy in the vicinity of the Pb^{2+} ion as postulated for Centre V in potassium borate glasses (Arafa and Bishay, 1969). Hyperfine interaction with the lead nucleus could have helped in solving this problem. However, about 88% of natural lead has a nuclear spin of zero and 22% has a nuclear spin of $\frac{1}{2}$. The available results cannot clearly prove the definite presence of a two line resonance associated with ^{207}Pb nucleus (nuclear spin $\frac{1}{2}$). Further systematic work will

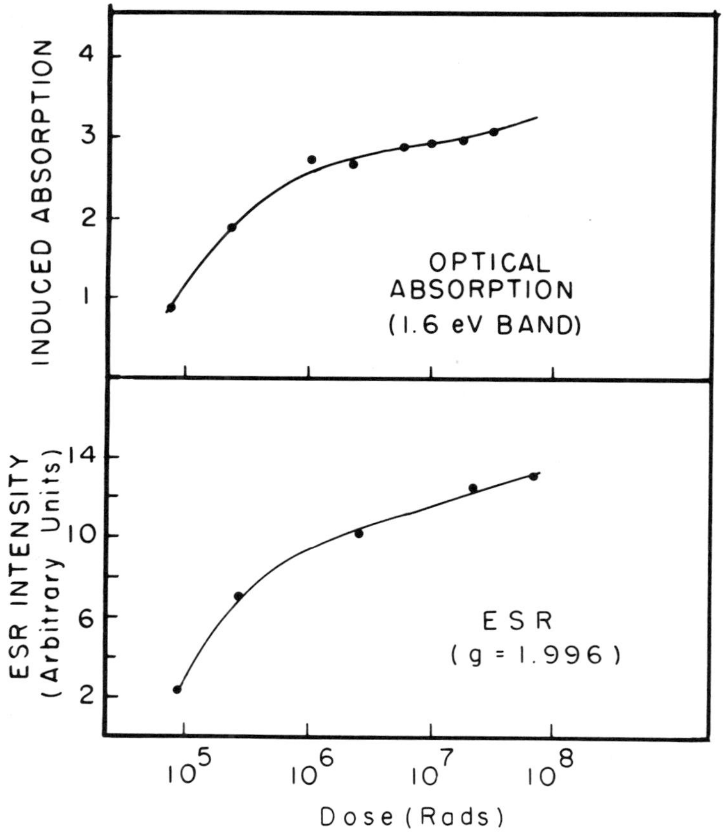

Figure 3. Growth curves for the 1·6 eV optical band and the $g = 1{\cdot}996$ e.s.r. line induced in a lead borate glass (70 $B_2O_3 \cdot 30$ PbO)

thus be needed before a definite conclusion can be reached regarding the similarities between Centre V observed in high alkali borate glasses and the centre observed in glasses containing Pb^{2+} ions.

An interesting feature of the induced e.s.r. spectrum of the lead borate glass containing 30 mole % PbO [Figure 1(b)] is that the complex signal characteristic of irradiated alkali borate glasses is not observed when the e.s.r. for this glass is measured at room temperature. Figure 4 shows the induced e.s.r. spectra for four X-irradiated lead borate glasses. The complex resonance characteristic of irradiated borate glasses is clearly observed in the spectra of the glasses containing 18 and 20 mole % PbO. This complex resonance nearly disappears in the spectrum of the glass containing 22 mole % PbO and was not observed for glasses containing more than 25 mole % PbO when these glasses were measured at room temperature. This may support the structural change suggested earlier at about 25 mole % PbO (Bishay and Maklad, 1967).

The intensity of the structureless line induced in borate glasses containing different lead oxide concentrations and measured at room temperature reaches a maximum at about 25 mole % PbO after which it decreases with

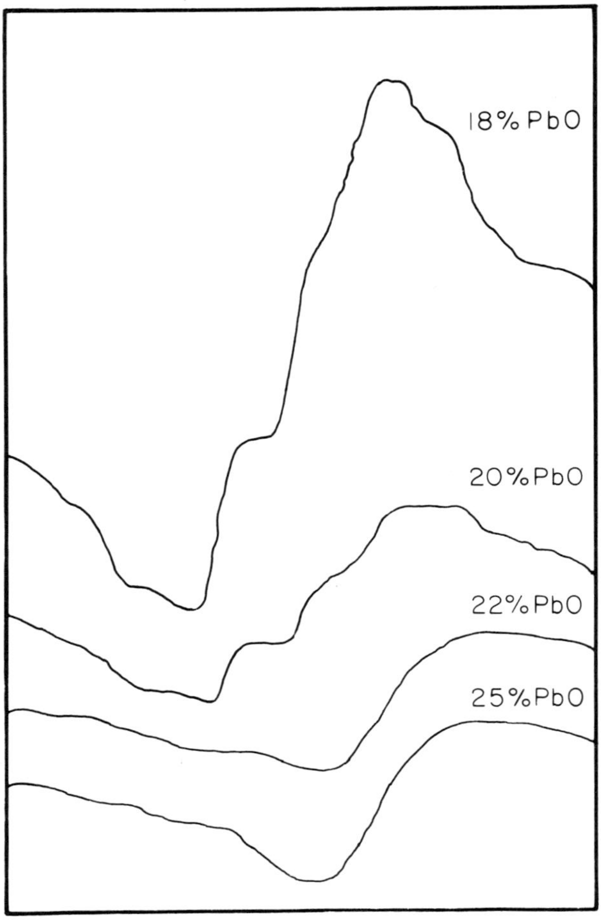

Figure 4. Effect of increasing PbO on e.s.r. spectra (18, 20, 22 and 25 mole % PbO)

increasing lead content beyond 25 mole %. This is illustrated in Figure 5 in which the intensity of the structureless broad line at $g = 1{\cdot}996$ (measured as the height between the two extremum points) is plotted versus mole % PbO. The same figure shows the plot of (α_mU) for the optical absorption band induced at about 1·6 eV *vs.* the PbO content in the same series of glasses (after Bishay and Maklad, 1966). Both the optical band and the e.s.r. line change in the same way as a result of increasing the PbO content, with a maximum intensity at about 25 mole %.

The above observations (Figure 5) support strongly the previous assumption, namely that the e.s.r. line induced at $g = 1{\cdot}996$ and the optical absorption band induced at about 1·6 eV in irradiated lead borate glasses are due to the same colour centre.

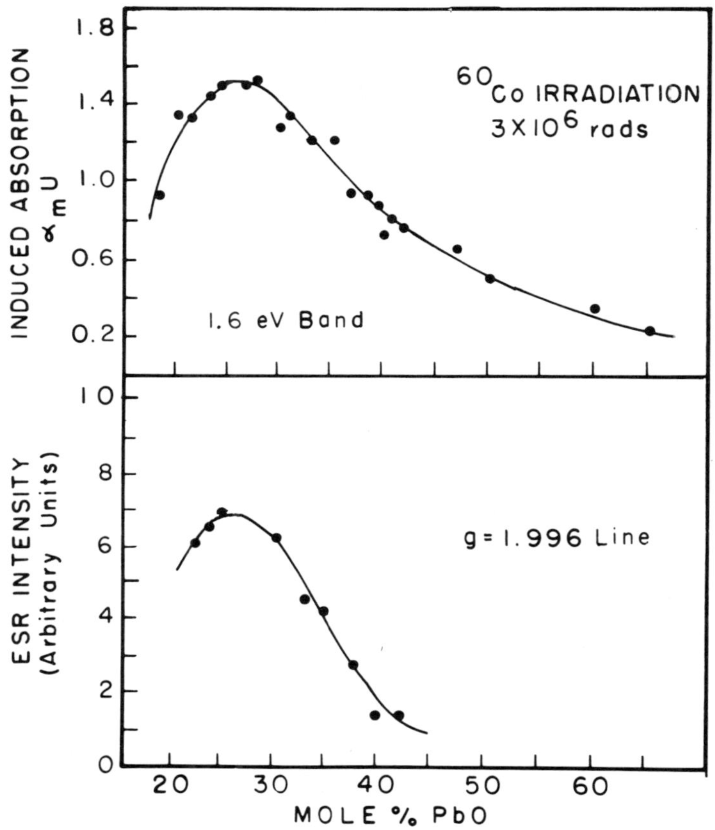

Figure 5. Effect of increasing PbO on induced absorption ($\alpha_m U$) for the 1·6 eV band and e.s.r. intensity for the $g = 1{\cdot}996$ line.

Figure 6 shows the induced e.s.r. spectrum for a lead borate glass containing 30 mole % PbO and measured at both −160 °C and room temperature. At −160 °C the complex hyperfine resonance is observed. While the complex resonance cannot be detected at room temperature in this glass, the structureless broad line at $g = 1{\cdot}996$ is clearly observed. Similar results were obtained for all glasses containing higher PbO concentrations. The double integration method was used for calculating the area under each spectrum. In each case calculations were made relative to the spectrum of a standard sample (pitch in KCl) measured under exactly the same conditions; at room temperature and at − 160 °C, respectively. The results obtained showed clearly that the relative area at −160 °C was double the relative area at room temperature. It is, therefore, very clear that the structureless resonance and the hyperfine resonance are not due to the same centre.

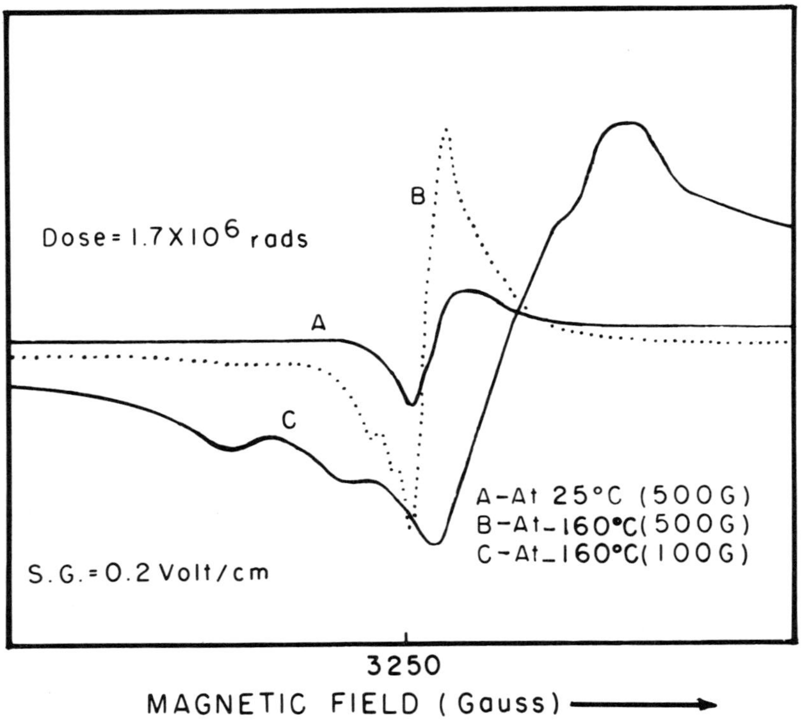

Figure 6. The e.s.r. induced in a lead borate glass containing 30 mole % PbO at −160 °C and room temperature

The fact that the hyperfine resonance in the irradiated lead borate glasses containing high lead oxide concentrations (25 mole %) is observed at −160 °C and not at room temperature, may be explained by one or both of the following assumptions:

(a) the structure of these high lead borate glasses changes the spin-lattice relaxation time of the complex borate centre in such a way that its resonance can only be observed at liquid nitrogen temperature.

(b) at room temperature, in glasses containing high lead concentrations, the trapped holes or electrons are jumping with relatively high frequency over many sites thus resulting in a smearing of the signal over a wide range of the magnetic field so that it is very difficult to observe. At $-160\ °C$ or lower, on the other hand, this motion will be frozen and the trapped hole or electron is now localized on one site resulting in the observed spectrum.

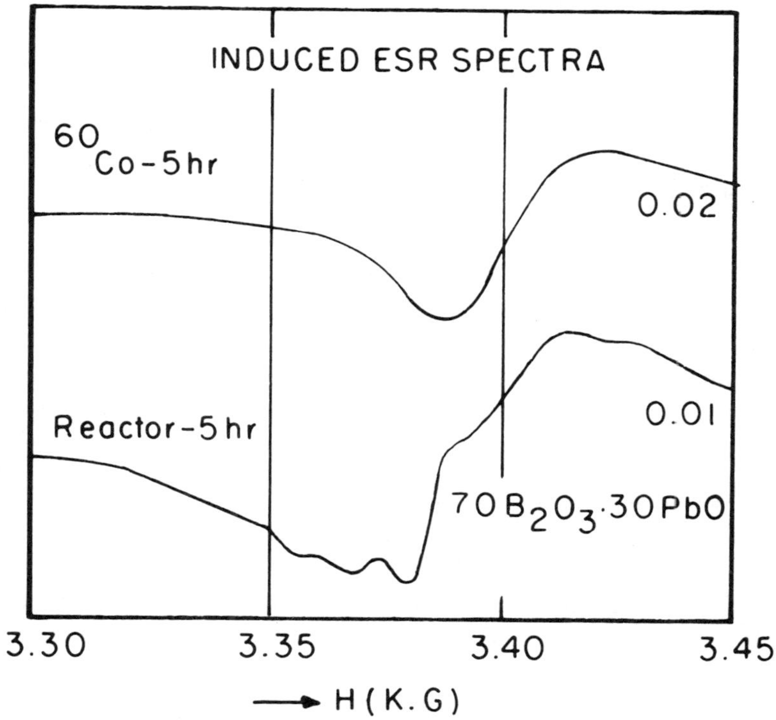

Figure 7. The e.s.r. spectra induced in the (70 $B_2O_3 \cdot 30$ PbO) glass irradiated for five hours by ^{60}Co or in the reactor (mixed flux). Measurements were made at room temperature

The above two assumptions are possible since the Pb^{2+} ions are known to take a different structural role, when present in high concentrations, as compared with alkali or alkaline earth ions.

Figure 7 shows the effect of reactor irradiation on the e.s.r. spectrum of a lead borate glass of the composition 70 $B_2O_3 \cdot 30$ PbO compared with the e.s.r. spectrum induced in a similar sample as a result of ^{60}Co-irradiation. Generally the whole spectrum decreased in intensity by a factor of about 50% in the case of reactor irradiation as compared with that of ^{60}Co-irradiation. This is not surprising, as the broad e.s.r. line associated with Pb^{2+} ions was found to be rather unstable. The substantial increase in temperature of the borate samples during reactor irradiation as a result of the reaction $^{10}B + n = {}^{7}Li + {}^{4}He$ may be enough to cause annealing of most of the unstable centres. Furthermore, it should be noted that the sample irradiated in the reactor was kept in the hot cell for one day before e.s.r. measurements were taken.

The complex borate hyperfine structure was observed only in the reactor irradiated sample and was superimposed on the broad anisotropic line associated with Pb^{2+} ions. The defects introduced by reactor irradiation may affect the glass structure in a way that makes it possible to observe the complex hyperfine resonance at room temperature.

A detailed analysis of the observed e.s.r. spectra as well as systematic measurements of spin–lattice relaxation times of the irradiated lead borate glasses are still needed.

CONCLUSIONS

The foregoing results indicate clearly that the optical band (at about 1·6 eV) and the e.s.r. line (at $g \simeq 1{\cdot}996$) induced in irradiated lead borate glasses are due to the same centre. This centre is an electron trap centre in the vicinity of lead ions and the associated band and e.s.r. line reach their maximum intensity in glasses containing about 25 mole % PbO.

The hyperfine structure characteristic of the e.s.r. of irradiated alkali borate glasses was observed at room temperature in glasses containing less than 25 mole % PbO but was impossible to observe at that temperature for glasses containing higher concentrations of PbO. However, measurements at −160 °C or irradiation in the reactor made it possible to observe the hyperfine complex resonance.

The results indicate that there must be a change in the role of PbO in the glass structure as the PbO content exceeds 25 mole %. It is postulated that lead has a much higher ionic character in glasses containing less than 25 mole % PbO as compared with those containing higher concentrations of PbO.

REFERENCES

Arafa, S., and A. Bishay (1969). *Cryst. Latt. Defects*, **1**, 91.
Arafa, S., and A. Bishay (1970). *J. Am. Ceram. Soc.*, **53**, 390.
Bishay, A. (1960). *J. Am. Ceram. Soc.*, **43**, 417.
Bishay, A. (1961). *J. Am. Ceram. Soc.*, **44**, 545.

Bishay, A., and K. Ferguson (1962). *Advances in Glass Technology.* Plenum Press, New York. p. 133.
Bishay, A., and M. Maklad (1966). *Physics Chem. Glasses*, **7**, 149.
Bishay, A., and M. Maklad (1967). *J. Am. Ceram. Soc.*, **50**, 503.

VIBRATIONAL PROPERTIES

42

Neutron Scattering in Amorphous Solids

A. J. LEADBETTER, A. C. WRIGHT* and A. J. APLING

School of Chemistry, University of Bristol, Bristol, Great Britain

---

General discussions of neutron scattering from solids and liquids have been given by Lomer and Low and others in *Thermal Neutron Scattering* (1965), by Turchin (1965) and a general discussion of neutron diffraction by Bacon (1962). Egelstaff (1965) has previously discussed applications of neutron scattering to glasses but there were then few experimental data available.

The scattering of neutrons, particles and radiation is characterized by the change of energy $\hbar\omega$ and momentum $\hbar Q$ on scattering. Q is the scattering vector defined by

$$\boldsymbol{Q} = \boldsymbol{k} - \boldsymbol{k}_0 \tag{1}$$

$$= 4\pi \sin\theta/\lambda \text{ if } k = k_0 \text{ (elastic scattering)}$$

where

$$k = 2\pi/\lambda \text{ and } 2\theta \text{ is the scattering angle.}$$

Neutrons are unique in the wide range of energy and momentum change readily available to experimental study. The results of experimental observations of scattered neutron spectra are conveniently expressed as the double differential cross section per unit solid angle, unit energy interval $d^2\sigma/d\Omega dE$. Following Van Hove (1954) this may be written, as the sum of coherent and incoherent components as follows (for a monatomic material)

$$\frac{d^2\sigma}{d\Omega dE} = \frac{d^2\sigma^{\text{inc}}}{d\Omega dE} + \frac{d^2\sigma^{\text{coh}}}{d\Omega dE}$$

$$= \frac{N}{h}\frac{k}{k_0}\Big[(\overline{b^2} - \overline{b}^2)\int G_s(\boldsymbol{r}, t)\exp - i(\boldsymbol{Q}\cdot\boldsymbol{r} + \omega t)\, d\boldsymbol{r}dt$$

$$+ \overline{b}^2 \int G(\boldsymbol{r}, t)\exp - i(\boldsymbol{Q}\cdot\boldsymbol{r} + \omega t)\, d\boldsymbol{r}dt\Big] \tag{2}$$

* Now at: J. J. Thomson Physical Laboratory, Whiteknights, Reading, Great Britain.

Here, b is a neutron scattering length, $G_s(\boldsymbol{r}, t)$ is the self-correlation function in space and time and represents classically the probability of finding an atom at $\boldsymbol{r}$ at time t if it was at the origin at time zero.

$G(\boldsymbol{r}, t)$ is the total correlation function, or time dependent pair distribution function. Classically this is the average number density of atoms at $\boldsymbol{r}$ at t if some atom was at $\boldsymbol{r} = 0$ at $t = 0$.

The simple neutron diffraction experiment analogous to X-ray diffraction measures *total neutron scattering*. This means that both elastic and (phonon) inelastic events of all energy transfer are detected and the observed cross section is then given by integrating equation (2) over all ω. In order to carry out the integration it is assumed that the energy transfer is negligible compared with the incident energy i.e. $k = k_0$, this is called the static approximation. The incoherent cross section is of little interest, since $d\sigma^{\text{inc}}/d\Omega = N(\overline{b^2} - \bar{b}^2)$ but the coherent cross-section is then related to $G(\boldsymbol{r}, 0)$, the instantaneous pair correlation function, which may be split into self and distinct parts as follows

$$G(\boldsymbol{r}, 0) = G_s(\boldsymbol{r}, 0) + G_d(\boldsymbol{r}, 0) = \delta(\boldsymbol{r}) + \rho(\boldsymbol{r}) \tag{3}$$

where $G_d(\boldsymbol{r}, 0) = \rho(\boldsymbol{r})$ is the pair distribution function describing the number density of atoms at a distance $\boldsymbol{r}$ from the origin atom at the same instant in time. Hence,

$$\frac{d\sigma^{\text{coh}}}{d\Omega} = N\bar{b}^2(1 + \rho(\boldsymbol{r}) \exp(-i\boldsymbol{Q} \cdot \boldsymbol{r}))\, d\boldsymbol{r} \tag{4}$$

In the static approximation the X-ray and neutron diffraction experiments are thus entirely analogous, but while this approximation is an excellent one for X-rays, it is less good for neutrons, where $\hbar\omega \nll \hbar^2 k_0^2/2m$.

A method of correcting for the inadequacy of the static approximation in the neutron case has been proposed by Placzek (1952). The correction is very small for heavy atoms and appears in general to be of sufficient accuracy to give no serious error in the derived distribution functions. (Enderby 1968, Wright 1969, Lorch 1970).

The generalization of equation (4) and its application to polyatomic glasses will be discussed in the third section below.

For solids, where the atoms are in general localized in small regions of space, the incoherent and coherent scattering may each be separated into elastic and inelastic components. The elastic scattering ($\hbar\omega = 0$) is related to the asymptotic (time-independent) limits of the space–time correlation functions at $t \to \infty$. This corresponds to a neutron–solid interaction extending over a time very long compared with vibrational periods, and hence to the average (time-independent) positions of the atoms. The *incoherent elastic* scattering cross-section is related to $G_s(r, \infty)$ and noting that $\int G_s(\boldsymbol{r}, \infty)\, d\boldsymbol{r} = 1$, then using the harmonic approximation to describe the displacements of the atoms about their equilibrium positions, it becomes

$$d\sigma_{el}^{inc}/d\Omega = N(\overline{b^2} - \bar{b}^2)\,e^{-2W} \tag{5}$$

It is thus simply related to the Debye–Waller factor $W = \frac{1}{2}\overline{(\boldsymbol{Q}\cdot\boldsymbol{u})^2}$. The *coherent elastic* cross section describes, for a crystal, the intensity of Bragg scattering which may be determined with both X-rays and neutrons. For a disordered solid, however, this cross-section is best described using the total correlation function $G(\boldsymbol{r}, \infty)$ which may be split into a self part, as above, and a distinct part $G_d(\boldsymbol{r}, \infty)$ which is simply the pair distribution function describing the time-averaged atomic positions $(\rho_e(r))$. These may again be described in terms of equilibrium positions and gaussian displacements to give

$$\frac{d\sigma_{el}^{coh}}{d\Omega} = N\bar{b}^2[1 + \int \rho_e(r)\exp(-i\boldsymbol{Q}\cdot\boldsymbol{r})\,d\boldsymbol{r}]\,e^{-2W} \tag{6}$$

where $\rho_e(r)$ is the pair distribution function for the equilibrium atomic positions. This cross-section may be experimentally measured using neutrons, but not X-rays, because only in the former case is adequate energy analysis possible.

For normal liquids, the atoms are not localized in space on the time scale measureable in neutron scattering experiments ($t \sim 10^{-11}$ to 10^{-14}s). The local structure will change significantly in a time of less than 10^{-11}s and hence the scattering may not rigorously be divided into elastic and inelastic regions, although a division into quasi-elastic (including diffusive motion components) and inelastic regions is often a useful approximation. Because of the relatively long relaxation times in the viscous melts of typical glass-forming liquids these will behave as solids on the time scales measurable in neutron scattering experiments, up to temperatures considerably in excess of T_g.

The *incoherent inelastic* cross section, which is related to $G_s(\boldsymbol{r}, t)$ and is dependent on the motion of individual atoms, is simpler to interpret than the coherent inelastic cross section, which depends also on the relative motion of different atoms. In general $G_s(\boldsymbol{r}, t)$ may readily be related to motions on different time-scales, from self-diffusion to high frequency vibrations, but for simple glasses at $T \lesssim T_g$ we need only consider the more-or-less harmonic vibrational motion about the quasi-equilibrium sites. In this approximation a glass has proper normal modes of vibration exactly as does a crystal, although for the glass these will not in general be plane waves. The incoherent inelastic scattering is then related to the frequency distribution $g(\omega)$ of the normal modes, although the relation is not simple and direct except in certain special cases. The scattering from a glass will be mainly incoherent only if it contains enough hydrogen or vanadium, which have large incoherent cross sections.

However, coherence effects become less important at high Q because the scattering experiment is then effectively examining a smaller region of space so that the self part of the space-time correlation function tends to dominate the scattering. Hence the coherent cross section tends towards the form of the incoherent as Q increases, so that in all cases it is possible to learn something of $g(\omega)$.

For disordered solids and liquids there is as yet no proper general theory of *coherent inelastic* scattering because this would involve a general description of the collective atomic motions which does not exist for these materials. However, it is reasonable to assume that for a glass these collective excitations are approximately harmonic vibrations and in the long wave-low frequency limit these will be the elastic waves, precisely as for a crystal. Furthermore, the short range order in many simple glasses is closely similar to that in the crystal forms of the same substances, so it may be expected that the correlated motions at short distances might be similar to those in the crystal. It should therefore be a reasonable approximation to describe the collective motions in terms of phonons as for the crystal. The disorder in the glass will result in an orientational averaging of the phonon branches similar to that for a polycrystalline specimen and also a reduction in the phonon lifetimes. The collective excitations in a glass might therefore be crudely likened to those of a very anharmonic polycrystal. If this is correct it should be possible to use the energy and wave vector conservation conditions of the coherent scattering process to determine one or more dispersion curves [$\omega(Q)$ relations] for the glass (see second section below).

The various types of neutron scattering experiments possible and the information they give are summarized in a simplified form in Table 1.

Table 1. Summary of neutron scattering experiments

	Coherent	Incoherent
Total	$\rho(r)$	$\overline{b^2} - \bar{b}^2$
Elastic	$\rho_e(r)$	$\overline{u^2}$
Inelastic	$\omega(Q)$	$g(\omega)$

Only one coherent elastic experiment has so far been reported for a glass (Lorch 1970) and this technique will not be discussed further here. Several total neutron scattering experiments have been reported for vitreous SiO_2 (see e.g. Henninger *et al.*, 1967 and Lorch, 1969) and two for vitreous GeO_2 (Lorch, 1969 and Ferguson and Hass, 1970) and in the third section we shall discuss GeO_2 and new results on BeF_2 and As_2Se_3.

Inelastic scattering experiments have been reported for several simple glasses, all of which are largely coherent scatterers: SiO_2 (Egelstaff, 1965; Leadbetter, 1969) GeO_2 (Leadbetter and Litchinsky, 1970), BeF_2 (Leadbetter and Wright, 1970), Se (Gissler *et al.*, 1968). Some of the main features of this work will be summarized briefly in the second section.

In the following sections experimental results will be discussed but no experimental details will be included. These have either been described already or will be given when full details of the results are published later.

INELASTIC SCATTERING

Almost all simple inorganic glasses are composed of atoms which are largely coherent scatterers, although a number of metals (e.g. Li and Na) which may be important constituents of polycomponent glasses have larger incoherent than coherent cross sections. We therefore consider primarily the coherent scattering from polyatomic glasses on the quasi-crystalline basis outlined in section 1.

Coherent inelastic scattering is allowed only when energy and wave vector conservation conditions are satisfied. These may be written, for a crystal,

$$\hbar\omega(\text{neutron}) = \hbar\omega(\text{crystal}) \tag{7}$$

and

$$\boldsymbol{Q} \pm \boldsymbol{q} = 2\pi\boldsymbol{\tau} \tag{8}$$

where $\boldsymbol{\tau}$ is a reciprocal lattice vector and $\boldsymbol{q}$ is a phonon wave vector.

For a polycrystalline specimen, because $\boldsymbol{Q}$ can take all orientations with respect to the lattice, (8) becomes

$$Q + q \geqslant 2\pi\tau \geqslant Q - q \tag{9}$$

For the glass, the reciprocal lattice vectors in (9) are replaced by the broad peaks of the structure factor $S(Q)$ which is just the diffraction pattern and is defined as

$$S(Q) = \frac{1}{N_{\text{u}}}\frac{\mathrm{d}\sigma^{\text{coh}}}{\mathrm{d}\Omega}\frac{1}{\sum \bar{b}_i^2} \tag{10}$$

where N_{u} is the number of units of composition (e.g. SiO_2 units) in the sample and the summation is over the atoms in this unit.

Typical inelastic neutron spectra are shown in Figure 1 where scattering from polycrystalline and vitreous forms of BeF_2 and GeO_2 are shown. Incident neutrons were of energy ~ 4 meV, scattering angles are indicated in the figure and energy discrimination was by time- of-flight analysis.

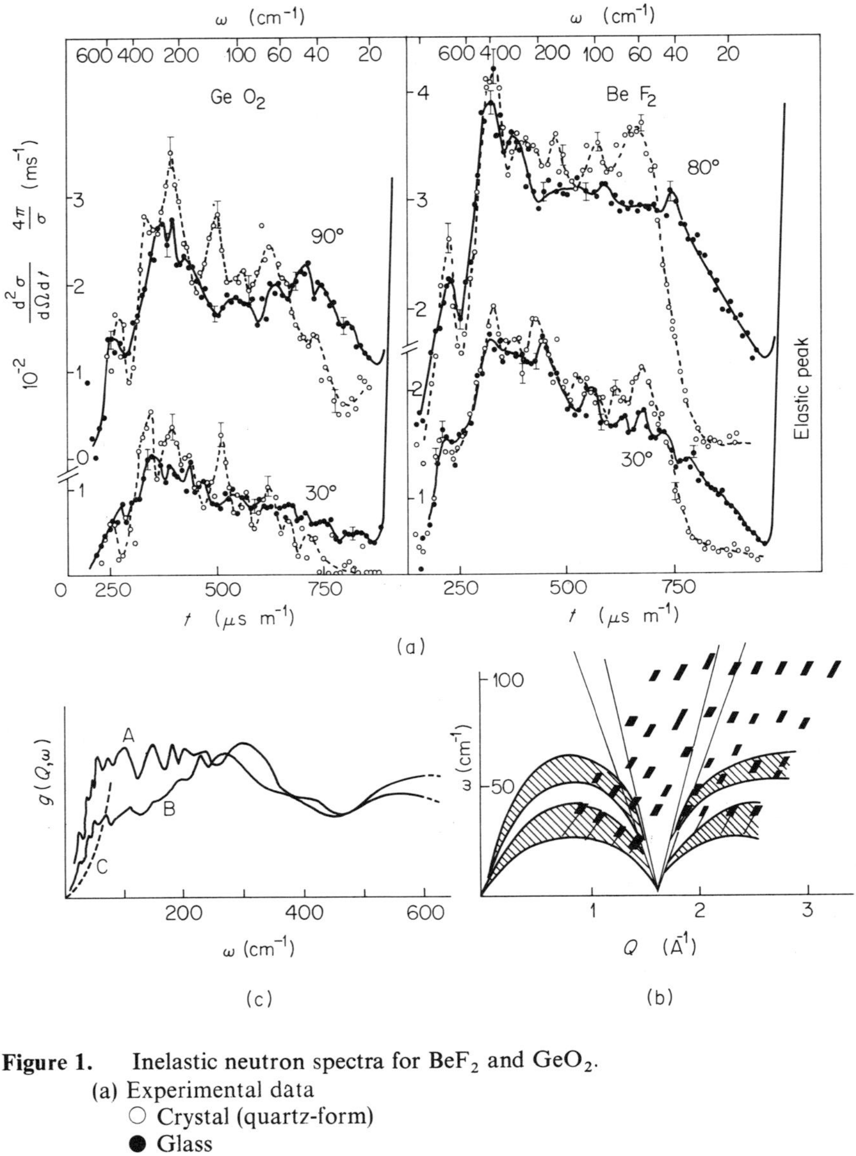

Figure 1. Inelastic neutron spectra for BeF_2 and GeO_2.

(a) Experimental data

○ Crystal (quartz-form)

● Glass

(b) $\omega(Q)$ relations for vitreous BeF_2.

Squares represent peak positions and bars extend to peak edges at half-height. Estimates are shown of two average acoustic branches, plus transverse and longitudinal sound velocities, based on the structure factor maximum at $Q \sim 1{\cdot}6\ \text{A}^{-1}$.

(c) A plot of $g(Q, \omega)$ against ω for vitreous GeO_2 at two scattering angles: A, 30°; B, 90°. Curve C represents the Debye spectrum based on the sound velocities and normalized to B

The formula for the intensity of coherent inelastic scattering from a polycrystalline polyatomic specimen is rather complex and will not be discussed here. However, by using (7) and (9), together with a knowledge of the general form of the cross-section, it is possible to estimate the shape of one or more low frequency dispersion ($\omega(Q)$) curves by examination of the Q-dependence of the positions of peaks and edges in spectra like those of Figure 1(a). The results of such an analysis are shown for vitreous BeF_2 in Figure 1(b), which indicates that at least two acoustic branches of the spectrum exists for the glass into the equivalent of the second Brillouin zone and that the first peak in $S(Q)$ at $Q \geqslant 1{\cdot}6\ \text{A}^{-1}$ is acting like a spherically averaged reciprocal lattice point. It is therefore meaningful to discuss the collective excitations in such glasses in quasi-crystalline (phonon) terms, at least for the acoustic branches of the spectrum.

At sufficiently high Q, as indicated in the first section, the coherent cross section approaches that for incoherent scattering. For a polyatomic glass containing p atoms per unit of composition (e.g. $p = 3$ for GeO_2) the double differential cross section may be written

$$\frac{1}{N_u}\frac{d^2\sigma}{d\Omega dE} = \frac{k}{k_0}\frac{nQ^2}{2\omega}g(Q,\omega) \tag{11}$$

and in the incoherent approximation

$$g(Q,\omega) = \langle \sum_j \sum_l \bar{b}_l^2(\hat{Q}\cdot\boldsymbol{\alpha}_l^j)^2 M_l^{-1} e^{-2W_l}\ N_u^{-1}\,\delta(\omega-\omega_j)\rangle \tag{12}$$

$\hat{Q}$ is a unit vector along $\boldsymbol{Q}$, $n = (\exp \hbar\omega/kT - 1)^{-1}$ is the phonon occupation number, the atoms are labelled l, the normal modes j, α_l^j is the displacement of the lth atom in the jth mode and the average is over all orientations of the unit of composition in the structure with respect to $\boldsymbol{Q}$

$$\sum_j \delta(\omega - \omega_j) = 3pNg(\omega) \tag{13}$$

where $g(\omega)$ is the frequency distribution normalized to unit area.

Equation (12) may be simplified by separating the contributions from the different atoms, averaging over all orientations and assuming chemically identical atoms to be dynamically equivalent so that there is only one term per atom type. The result may be written (Leadbetter and Litchinsky 1970)

$$g(Q,\omega) = f_N(\omega)\,g(\omega) \tag{14}$$

and for an A_xB_y glass $f_N(\omega)$ may be written

$$f_N(\omega) = [\overline{\alpha_A^2}\,e^{-2W_A} + K\overline{\alpha_B^2}\,e^{-2W_B}]\frac{\bar{b}_A^2}{M_A} \tag{15}$$

where K is a constant. This has been applied to SiO_2, GeO_2 and BeF_2 (Leadbetter and Litchinsky 1970) and the results for GeO_2 ($K = 4{\cdot}2$) are shown in Figure 1(c), where the values of $g(Q, \omega)$ derived from the observed cross section using equation (11) are shown as a function of ω for two angles of scatter.

Coherence effects manifest themselves as a dependence of $g(Q, \omega)$ on Q, and hence on scattering angle, so the differences between the two curves of Figure 1(c) are a measure of their importance. The incoherent approximation obviously improves with increasing ω (and hence Q) so that (14) and (15) should be a fair description of the higher frequency data.

It should be emphasized, however, that without some information about the frequency-dependent relative atom displacements the experiments do not give $g(\omega)$ directly, even if the incoherent approximation was valid.

At low Q (and ω) coherence effects add further complication but even here Q-independent peaks in $g(Q, \omega)$ undoubtedly correspond to peaks in $g(\omega)$, although the detailed shape of $g(\omega)$ is not given. Coherent scattering experiments can thus give considerable information about the main features of $g(\omega)$ at all frequencies. For experiments of the type described here, namely time-of-flight analysis of scattered cold neutrons, there is in fact a practical upper limit of about 80 meV ($\sim$ 600 cm^{-1}) on the energy transfers detectable so that the highest frequency part of $g(\omega)$ may not be observed.

TOTAL SCATTERING

Theoretical outline

Equation (4) is readily generalized to a polyatomic system and averaging over all orientations of $\boldsymbol{Q}$ with respect to $\boldsymbol{r}$ gives

$$\frac{1}{N_u}\frac{d\sigma^{coh}}{d\Omega} = I(Q) = \sum_i \bar{b}_i^2 + \sum_i\sum_j \bar{b}_i\bar{b}_j \int_0^\infty 4\pi r^2 \rho_{ij}(r)\frac{\sin Qr}{Qr}\,dr \tag{16}$$

i labels the atoms in a unit of composition (e.g. Ge and 2 O for GeO_2) and j labels the chemically distinct atom types (e.g. Ge and O). $\rho_{ij}(r)$ is the number density of atoms j at a distance r from atom i. The formula for X-ray diffraction is identical to equation (16) except that $f_j(Q)$ replaces $\bar{b}_j$.

We define the correlation function $d_{ij}(r)$ by

$$d_{ij}(r) = 4\pi r[\rho_{ij}(r) - \rho_j^0] \tag{17}$$

and $d_j^0 = 4\pi r\rho_j^0$ where the superscript zero indicates the average value.

The (unobserved) zero scattering $I_0(Q)$ from the homogeneous bulk solid of average atomic density ρ^0 (number of units of composition per unit volume) may be separated in the usual way and

$$Qi(Q) = \sum_i\sum_j Qi_{ij}(Q) = \sum_i\sum_j \bar{b}_i\bar{b}_j \int_0^\infty d_{ij}(r)\sin Qr\,dr \tag{18}$$

where

$$i(Q) = I(Q) - \sum_i \bar{b}_i^2 - I_0(Q) \tag{19}$$

and

$$Qi_{ij}(Q) = \bar{b}_i\bar{b}_j \int_0^\infty \mathrm{d}_{ij}(r) \sin Qr \,\mathrm{d}r \tag{20}$$

Equation (20) may be Fourier transformed to give

$$d_{ij}(r) = \frac{2}{\pi\bar{b}_i\bar{b}_j} \int_0^\infty Qi_{ij}(Q) \sin Qr \,\mathrm{d}Q \tag{21}$$

and if $D^{\mathrm{N}}(r)$ is defined as the Fourier transform of $Qi(Q)$ then

$$D^{\mathrm{N}}(r) = \sum_i \sum_j \bar{b}_i\bar{b}_j \mathrm{d}_{ij}(r) \tag{22}$$

The 'total radial distribution function' is defined as

$$\sum_i \sum_j \bar{b}_i\bar{b}_j 4\pi r^2 \rho_{ij}(r) = r[\mathrm{D}^N(r) + D_0^{\mathrm{N}}(r)] \tag{23}$$

where

$$D_0^{\mathrm{N}}(r) = \sum_i \sum_j \bar{b}_i\bar{b}_j d_j^0(r) = (\sum_i \bar{b}_i)^2 4\pi r \rho^0 \tag{24}$$

For X-rays the corresponding results are (Waser and Schomaker, 1953)

$$D^{\mathrm{X}}(r) = \frac{2}{\pi} \int_0^\infty \frac{Qi(Q)}{f_0^2(Q)} \sin Qr\mathrm{d}Q \tag{25}$$

$$d_{ij}^{\mathrm{X}}(r) = \frac{2}{\pi} \int_0^\infty \frac{Qi_{ij}(Q)}{f_0^2(Q)} \sin Qr\mathrm{d}Q \tag{26}$$

$$D^{\mathrm{X}}(r) = \sum_i \sum_j \mathrm{d}_{ij}^{\mathrm{X}}(r) \tag{27}$$

where

$$f_0(Q) = \frac{\sum_i f_i(Q)}{\sum_i Z_i}$$

is the average form factor per electron, and $f_0^2(Q)$ is used as a sharpening function.

Because of the Q dependence of the X-ray form factor f_i the relation between $\mathrm{d}_{ij}^{\mathrm{X}}$ and d_{ij} is a convolution of the form

$$d_{ij}^{\mathrm{X}}(r) = \int_{-\infty}^{\infty} d_{ij}(u)\, P_{ij}(r - u)\,\mathrm{d}u \tag{28}$$

$P_{ij}(r)$ is the peak function for the given atom pair ij and is defined by the cosine transform

$$P_{ij}(r) = \frac{1}{\pi} \int_0^\infty \frac{f_i(Q) f_j(Q)}{f_0^2(Q)} \cos rQ \mathrm{d}Q \tag{29}$$

Also

$$D_0^X (r) = (\sum_i Z_i)^2 d^0(r) \tag{30}$$

Traditionally the approximation has been made that

$$f_i(Q) = \overline{K}_i f_0(Q) \tag{31}$$

where $\overline{K}_i$ is a constant representing an effective average atomic number whence

$$D^X(r) = \sum_i \sum_j \overline{K}_i \overline{K}_j d_{ij}(r) \tag{32}$$

in complete analogy with equation (22) for neutrons.

In practice, for both neutrons and X-rays, experimental data are limited to a finite range of Q, below Q_m. This means that $i(Q)$ is necessarily multiplied by some modification function $M(Q)$ before Fourier transformation. In order to minimize the so-called termination ripples a function of the form $M(Q) = e^{-\beta Q^2}$ has usually been used (e.g. Mozzi and Warren, 1969), but we prefer to use the resolution function proposed by Lorch (1969) which has a clear physical basis and a transform which is slightly better behaved (Wright, 1969). This is equivalent to using a modification function of the form

$$M(Q) = \frac{\sin Q\Delta/2}{Q\Delta/2} \tag{33}$$

where

$$\Delta = 2\pi/Q_m$$

For X-rays and neutrons we then obtain

$$D'(r) = \sum_i \sum_j \mathrm{d}'_{ij}(r) \tag{34}$$

where

$$d'_{ij}(r) = \int_{-\infty}^{\infty} d_{ij}(u)\, P'_{ij}(r - u)\, \mathrm{d}u \tag{35}$$

with

$$P_{ij}^N(r) = \frac{b_i b_j}{\pi} \int_0^{Q_m} \frac{\sin Q\Delta/2}{Q\Delta/2} \cos Qr \mathrm{d}Q \tag{36}$$

for neutrons
and

$$P_{ij}^{X}(r) = \frac{1}{\pi}\int_{0}^{Q_m} \frac{f_i(Q)f_j(Q)}{f_0^2(Q)}\frac{\sin Q\Delta/2}{Q\Delta/2}\cos Qr\,dQ \qquad (37)$$

for X-rays.

The above discussion shows that for a polyatomic system a full description of the average structure requires several pair distribution functions, while a single scattering experiment only gives a weighted sum of these several functions, convoluted with the peak functions P_{ij}. Because of the Q-dependence of the X-ray form factors the neutron peak function is simpler than the X-ray, particularly if Q_m has a high value, when $P^N(r)$ only produces slight broadening of the peaks of $d_{ij}(r)$. The Q dependent form factors also result in the X-ray intensity decreasing rapidly with Q.

It is preferable to discuss the structure in terms of the correlation function $D(r)$ rather than the total radial distribution $r(D(r) + D^0(r))$ which has often been used. The former function is composed additively of symmetrical peaks whose position is not influenced by the multiplicative factors affecting the intensity and resulting in the convolutions (35) to (37) (Waser and Schomaker, 1953).

For a polyatomic glass several pair distribution functions are required to describe fully the average structure. We shall consider only diatomic systems for which three distribution functions are required. For an AB_2 glass, for example, we need d_{AA}, d_{BB} and d_{AB} ($d_{AB} = 2d_{BA}$). These may in principle be obtained if three separate scattering experiments can be performed (e.g. see Enderby *et al.* 1966) but this is not easily possible. In practice, therefore, one has to attempt to determine the structure using inadequate data. In order to make any progress some other information is essential and this is generally a knowledge of the crystal chemistry of the material e.g. the importance of SiO_4 tetrahedra in the structure of the silica polymorphs.

This information may be utilized in a variety of possible approaches. First the assumed basic structural unit (e.g. SiO_4 tetrahedra) can be confirmed as being consistent with the data by measuring the areas of peaks in the distribution function. Thereafter one may try to establish the short range correlation of these units by a trial and error procedure computing the $D(r)$ using equations (34) to (37) (Mozzi and Warren, 1969). Alternatively, one may compare either the scattered intensity or the correlation function with that calculated from some random model (Bell and Dean, 1966; King, 1967). We have adopted a third approach—using a quasi-crystalline model to calculate $i(Q)$ and thence $D(r)$ by inversion as for the experimental data. No method will give a unique result for the structure from inadequate data and it is therefore important to have both X-ray and neutron data where possible in order to narrow down the range of possible structures.

Quasi-Crystalline Model

This model has been described in detail elsewhere (Wright, 1969; Leadbetter and Wright, 1971) so only a brief outline will be given here. It is very similar to the model used by Kaplow *et al.* (1965) in their work on the structure of liquid metals. The essence of the model is to describe the short range order in the glass in terms of a crystal structure and to reduce the correlation in atom positions $d_{ij}(r)$ gradually so as to reach zero at a critical correlation length L. This is done using only the parameter L by calculating the intensity using a formula of the form

$$Qi(Q) = \sum_i \sum_j \bar{b}_i \bar{b}_j \int_0^L d_{ij}(r) F(r) \sin Qr dr$$

with $F(r) = (r/L - 1)^2 (r/L + 2)/2$ which is unity at $r = 0$ and zero at $r = L$.

This is equivalent to calculating the scattering intensity from a sphere of crystal of diameter L, averaged over all orientations, and set in a matrix of average atomic density ρ^0 with zero correlation in atom positions ($d_{ij}(r) = 0$), which removes small angle scattering.

It is not expected that this simple model will be a complete representation of the structure for any glass but should serve as a basis from which to discuss the structure of particular glasses, the model being modified as required in each individual case. The model enables the individual components $Q_{ij}(Q)$ to be calculated and Fourier transformation of these with the same Q_m and f_i as for the experiments gives the individual $d_{ij}^{X}(r)$ and $d_{ij}^{N}(r)$ and the convolution with the P_{ij} is automatically included. Thus the calculated $D^X(r)$ and $D^N(r)$ are directly comparable with the experimental functions and the availability of the individual pair correlation functions should show where the model might be improved.

As_2Se_3

As_2Se_3 is a particularly interesting case because the atomic numbers of As and Se are respectively 33 and 34. This means that equation (31) will be an excellent approximation for the X-ray form factors but it will not be possible to distinguish between the two types of atoms in the structure. In other words As_2Se_3 will behave effectively as a monatomic substance with respect to X-ray diffraction. On the other hand the neutron scattering lengths for As and Se are respectively 0·63 and 0·89 $\times$ 10^{-12}cm so the various possible pair contributions $\bar{b}_i \bar{b}_j d_{ij}$ to the total correlation function $D(r)$ will be distinguishable.

In Figure 2(a) are shown X-ray and neutron intensity data on vitreous As_2Se_3 obtained by the authors. These show very clearly the difference between the two experiments due to the rapid decrease of the X-ray form factors with increasing Q. The X-ray data were obtained in the usual way using a LiF monochromator in the diffracted beam (Wright, 1969). The

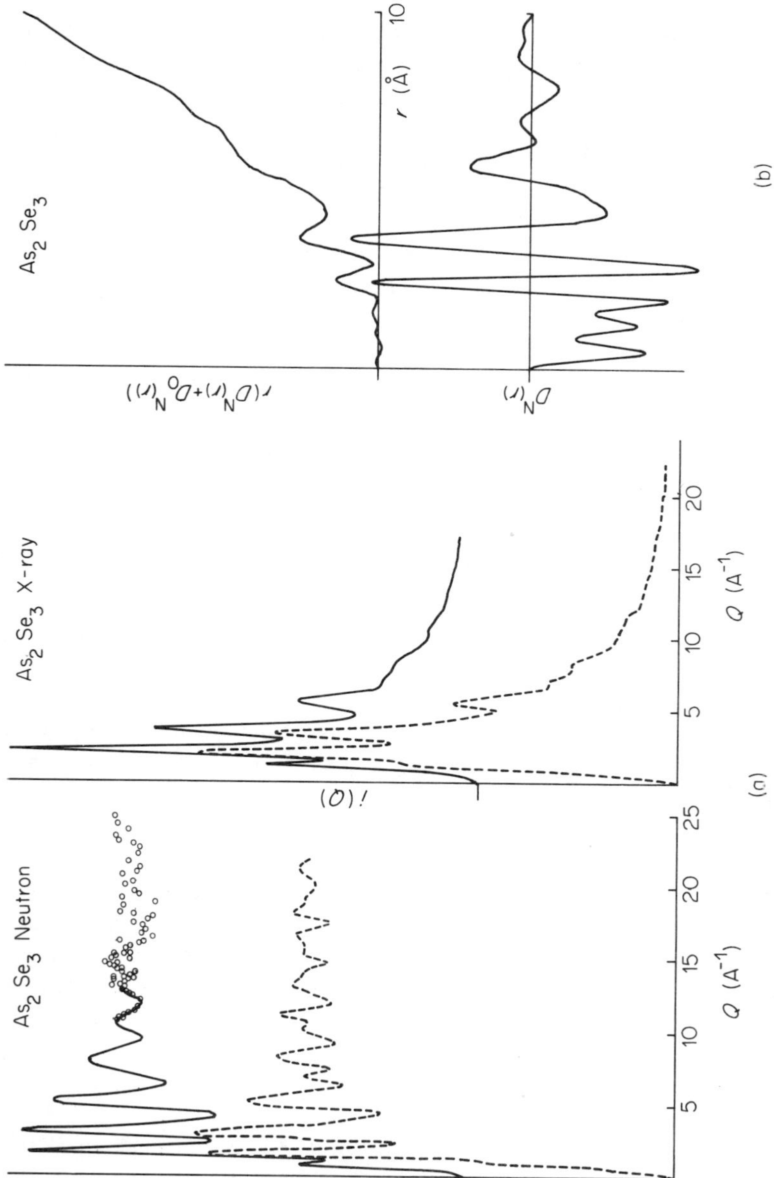

Figure 2. Intensities of total neutron and X-ray scattering from vitreous As_2Se_3

(a) Intensities

— conventional experiment

○ pulsed, time-of-flight, neutron diffraction

- - - quasi-crystalline model with critical correlation length, $L = 10$ Å

(b) Neutron correlation function $D^N(r)$ and total radial distribution function $r(D^N(r) + D_0^N(r))$ for vitreous As_2Se_3

neutron data were obtained at AWRE Aldermaston in three separate experiments. For $Q < 13{\cdot}5\ \text{A}^{-1}$ the results were obtained with crystal monochromated incident radiation of 1·14 Å and 0·76Å. These data are in excellent agreement in the large overlap region and are not shown separately. The third set of data shown as individual points are preliminary results obtained by a pulsed diffraction technique in which neutrons of different wavelength are scattered through a fixed angle and separated by time of flight analysis. The importance of this technique is that it may provide a means of obtaining data out to high $Q(Q \sim 25\ \text{A}^{-1})$ in reasonable counting times and as the resolution in $D(r)$ is directly related to the value of Q_m, such data are obviously valuable (cf. Mozzi and Warren, 1969).

Detailed analysis of these data is not complete but the neutron data up to $Q_m = 13{\cdot}5\ \text{A}^{-1}$ have been transformed and the correlation function $D^N(r)$ together with the total radial distribution function are shown in Figure 2(b). The ripples at low r in $D^N(r)$ are due to errors in the data at higher Q. It is worth noting that these ripples are much less pronounced in $r(D^N(r) + D_0^N(r))$. The area under the first peak in the distribution function is consistent with a coordination of 3 Se atoms about each As, plus a small additional Se–Se pair contribution arising from a slight excess of Se in the composition.

Preliminary calculations have been made of neutron and X-ray intensities for a quasi-crystalline model based on an orpiment structure. No detailed crystal structure determinations have been reported for As_2Se_3 so the parameters of orpiment itself (As_2S_3) were used, with the lattice expanded uniformly to give the same density as for the selenide glass. The results of these calculations for a critical correlation length L of 10 Å are shown in Figure 2(a). Although there are differences of detail the agreement is sufficiently good to indicate that this model is a promising approach to a description of the short range structure of As_2Se_3 glass.

GeO_2

In Figure 3 are shown $D^N(r)$ for vitreous GeO_2 determined by Lorch (1969) and by Ferguson and Hass (1970) together with $D^X(r)$ determined by us, restricted for the present discussion to $r < 5$Å. The two neutron experiments agree well as to peak positions but less well for relative intensities. The X-ray correlation function agrees well with that determined by Zarzycki (1957) but is of higher resolution.

All the data are consistent with tetrahedral coordination of O about Ge with a Ge–0 distance of 1·74 Å.

Both Lorch and Ferguson and Hass interpret the second peak in $D^N(r)$ at ~2·85 Å as arising from the O–O pairs of the GeO_4 tetrahedron and the third peak at ~3·45 Å as due to the four first-neighbour Ge–Ge pairs, and compare this result with the second peak in $D^X(r)$ at 3·18 Å attributed to the same pairs. It is clear that this interpretation cannot be correct and that the differences in the neutron and X-ray correlation functions must arise from the different relative scattering lengths of Ge and O for neutrons and X-rays. (For Ge, Z is 32 and $\bar{b}$ is 0·84 while for O the values are respectively 8 and 0·58.)

The results of calculations using the quasi-crystalline model based on the quartz form of GeO_2 with a correlation length $L = 10{\cdot}5$ Å are also shown in Figure 3. The individual pair correlation functions $d_{ij}^{X}(r)$ and $d_{ij}^{N}(r)$ for the simple model are shown and when summed these give fairly good agreement with experiment. A slight modification of this model, however, makes the agreement excellent: in the quartz structure there is a wide distribution of

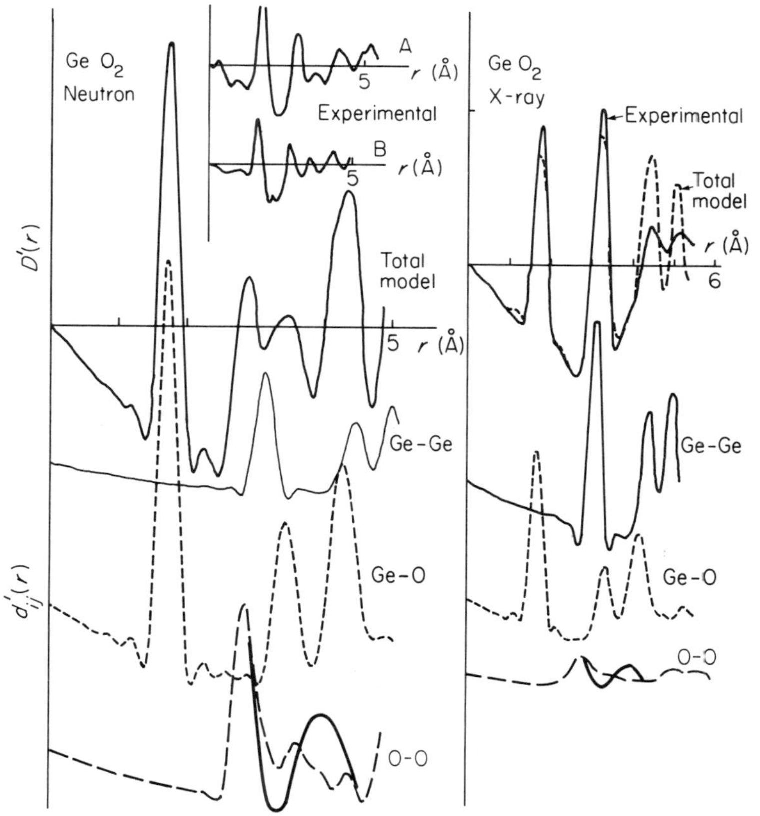

Figure 3. Correlation functions for vitreous GeO_2. Experimental neutron data: A, Lorch (1969): B. Ferguson and Hass (1970). The bold line on the O–O pair correlation curves show the modification from the simple quartz structure (see text). The individual $d'_{ij}(r)$ functions have been internormalized according to the summations of equation (34) (i.e., $d'_{Ge-Ge}(r)$: $2d'_{Ge-O}(r)$: $2d'_{O-O}(r)$. N.B. $d'_{Ge-O}(r) = 2d'_{O-Ge}(r)$)

O–O spacings involving 8 pairs between 3 and 4 Å and the modification involves an increase in the average separation of these pairs amounting to a net increase of about 0·7 Å for three pairs. This is equivalent to a slight regorganization of the GeO_4 tetrahedra towards the arrangement they would have in a cristobalite-like structure and to the lower density of the glass compared with the quartz structure. This modified $d_{O-O}(r)$ correlation is shown in Figure 3 and the total correlation functions shown were computed using this modified form. The peaks in $D^N(r)$ at 1·74 and 2·85 may thus be identified with Ge–O and O–O pairs respectively and that in $D^X(r)$ at 3·18 Å mainly with the Ge–Ge pair but this peak also contains contributions from Ge–O and O–O pairs. The broad peak at 3·45 Å in $D^N(r)$ contains contributions from Ge–O and O–O pairs and also from the first Ge–Ge pair which is not itself observed as a distinct peak.

Sufficiently accurate neutron data and analysis could in principle have shown that the suggested interpretation by the previous workers was erroneous but the above discussion shows the value of having both neutron and X-ray data. Thus the O–O pair at 2·85 Å spacing is only seen clearly in the neutron experiment and the Ge–Ge pair at 3·18 Å in the X-ray experiment. It also shows the danger of a too naive identification of peaks in $D(r)$ with single pair contributions even at relatively low r. For $r \gtrsim 5$ Å the agreement of the modified quasi-crystalline quartz model with both X-ray and neutron experiments is excellent, indicating somewhat distorted tetrahedra in the glass as in the crystal and an average Ge–O–Ge angle of about 133°. Note that the coincidence of the O–O and Ge–Ge peaks make an estimate of the distribution of Ge–O–Ge angles very difficult. The good agreement between model and experiment emphasizes the insensitivity of the density to the very short range order. At higher r the agreement worsens because the model is too ordered, but this will be discussed elsewhere.

BeF_2

Because of the low atomic numbers of both atoms, BeF_2 (like B_2O_3) is a difficult case for X-ray study because of low coherent intensities and high Compton modified intensities. The light atoms also mean that static approximation errors in total neutron scattering experiments will be appreciable, but these may be adequately corrected for by Placzek's method (see first section, Enderby, 1968 and Wright, 1969).

We have done both X-ray and neutron diffraction experiments but a brief report only of the latter will be given here.

The intensity function $Qi(Q)$ and its transform $D^N(r)$ are shown in Figure 4. Also shown are the results of calculations using the quasi-crystalline model based on the α-cristobalite structure with a critical correlation length L of 11 Å which gave the best fit of the various crystal structures and L values tried. No full crystal structure determination of the α-cristobalite form of BeF_2

has been done so the positional parameters from the silica structure were used. The values of $Qi(Q)$ and $D^N(r)$ computed for the model are also shown in Figure 4. The general agreement is very good.

The above calculations show that vitreous BeF_2 is built of BeF_4 tetrahedra (experimental coordination number 3·99) with a Be–F distance of 1·54 Å and an F–F distance of 2·56 Å, the Be–Be pair contribution is not detected as a separate peak. The data are consistent with an overall structure out to $r \sim 10$ Å closely similar to the assumed α-cristobalite model. Figure 4 shows the way in which the simple model still differs from the glass structure. First the Be–F distances within a tetrahedra are slightly too sharp, the experimental

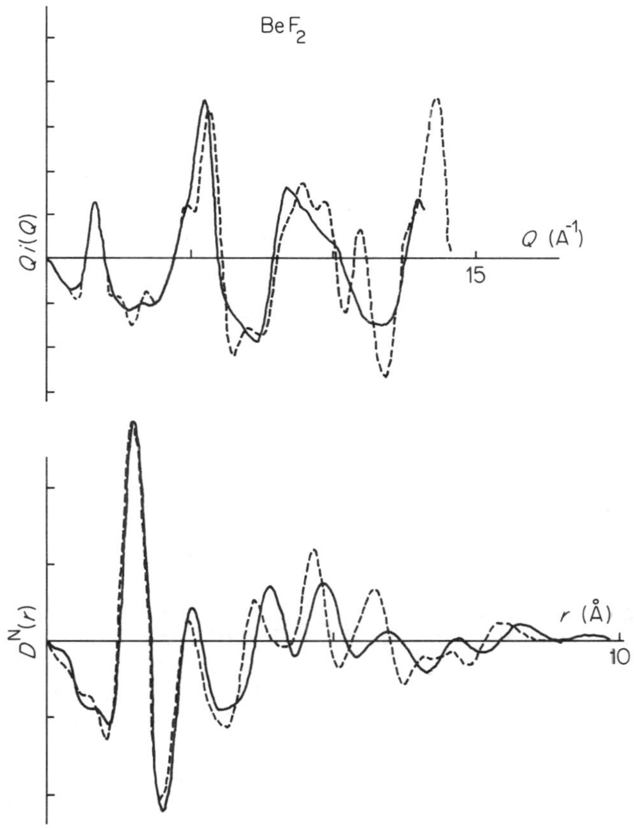

Figure 4. Neutron intensity and correlation functions for vitreous BeF_2.
—— experimental
--- quasi-crystalline model with $L = 11$ Å

broadening could be due to thermal motion or a distribution of equilibrium bond lengths. Secondly the model is too dense: much better agreement would be obtained by a uniform expansion of about 5 % and thirdly the real structure is slightly more disordered (broader lower peaks in $D^N(r)$) than the model provides for at $r \gtrsim 5$ Å.

Our X-ray data show the first Be–F distance to be at 1·54 Å and the intra-tetrahedral F–F distance to be 2·53 Å, in good agreement with the neutron work. The X-ray work of Batsanova *et al.* (1968) gave a Be–F distance of only 1·43 Å although the F–F distance of 2·54 Å is in good agreement with our work. The reason for the poor result for the Be–F distance probably arises because this peak is very small in X-ray work and is therefore particularly prone to distortion by error ripples and also by convolution effects in the total radial distribution function which was in fact used.

ACKNOWLEDGEMENTS

We are grateful to the Science Research Council for financial support and for research studentships to ACW and AJA. The help and cooperation of the University Support Group at AWRE Aldermaston is also gratefully acknowledged.

REFERENCES

Bacon, G. E. (1955). *Neutron Diffraction.* O.U.P., London.
Batsanova, L. R., G. S. Yur'ev, and V. P. Doronina (1968). *J. struct. Chem.*, **9**, 63.
Egelstaff, P. (1965). In J. A. Prins (Ed.), *Proc. Int. Conf. on Physics of Non-crystalline Solids*, Delft 1964. North-Holland, Amsterdam. p. 127.
Enderby, J. E. (1968). In H. N. V. Temperley, J. S. Rowlinson, and G. S. Rushbrooke, (Eds.), *Physics of Simple Liquids.* North Holland, Amsterdam.
Enderby, J. E. D. M. North, and P. A. Egelstaff (1966). *Phil. Mag.*, **14**, 961.
Ferguson, G. A., and M. Hass (1970). *J. Am. Ceram. Soc.*, **53**, 109.
Gissler, W., A. Axmann, and T. Springer (1968). *Neutron Inelastic Scattering*, Vol. I. I.A.E.A. Vienna. p. 245.
Henninger, E. H., R. C. Buschert, and L. Heaton (1967). *Physics Chem. Solids*, **28**. 423.
Kaplow, R., S. L. Strong, and B. L. Averbach (1965). *Phys. Rev.*, **138A**, 1336.
King, S. V. (1967). *Nature*, **215**, 615.
Leadbetter, A. J. (1969). *J. chem. Phys.*, **51**, 779.
Leadbetter, A. J., and D. Litchinsky (1970). *Discuss. Faraday Soc.*, **50**.
Leadbetter, A. J., and A. C. Wright (1970). *J. non-cryst. Solids*, **3**, 239.
Leadbetter, A. J., and A. C. Wright (1971). *To be published.*
Lorch, E. A. (1967). *Ph.D. Thesis*, University of Birmingham.
Lorch, E. A. (1969). *J. Phys. C.*, **2**, 229.
Lorch, E. A. (1970). *J. Phys. C.*, **3**, 1314.
Mozzi, R. L., and B. E. Warren (1969). *J. appl. Cryst.*, **2**, 164.
Placzek, G. (1952). *Phys. Rev.*, **86**, 377.
Egelstaff, P. A. (Ed.) (1965). *Thermal Neutron Scattering.* Academic Press, New York.
Turchin, V. F. (1965). *Israel Program for Scientific Translations*, Jerusalem.

Van Hove, L. (1954). *Phys. Rev.*, **95**, 249.
Waser, J., and V. Schomaker (1953). *Rev. mod. Phys.*, **25**, 671.
Wright, A. C. (1969). *Ph.D. Thesis*, University of Bristol.
Zarzycki, J. (1957). *Verres Réfract.*, **11**, 3.

43

Localization of Phonons in Vitreous Silica and Related Glasses

R. J. BELL* and P. DEAN†

National Physical Laboratory, Teddington, Middlesex

In the past there have been two main approaches to the problem of determining atomic vibrational behaviour in glasses, one using methods developed in the theory of molecular spectroscopy, and the other the techniques of crystal lattice dynamics. The former approach rests upon an implicit assumption that a small unit of the glass structure, containing perhaps only several molecules, can adequately characterize in its vibrational behaviour the whole glass system. Such a method often gives quite a reasonable account of the number and positions of bands in the vibrational spectra. However it can give a meaningful description of the detailed atomic motions only for those vibrational modes (if any) of the full glass system which are intensely localized in regions similar to the units considered. The second approach replaces the glass, not with a small molecular unit, but by an infinite regular crystalline array, and derives the vibrational properties of this array using conventional lattice dynamical procedures. The method gives a fair account of the band positions, but implicit in the approach is the assumption of an extended plane wave-like form for the normal modes of vibration.

Until quite recently, then, the theoretical analyses had provided no really objective information as to the nature and spatial characteristics of normal modes in a glass. The theoretical work reported upon here is aimed at obtaining such information directly, with particular emphasis on the localization properties of the modes. Localization is important not only in the general context of spectral interpretation—it is also central to the problem of heat

* Division of Numerical and Applied Mathematics.

† Division of Quantum Metrology.

conduction in glasses and has extremely important implications in connection with electron behaviour in amorphous solids. Our results show that spatial localization varies considerably throughout the spectrum. At low frequencies the modes are appreciably extended, although rather less so, in general, than the plane wave-like forms for periodic solids. At higher frequencies there is a much stronger tendency towards localization, particularly near the band edges. In addition, certain spectral bands in vitreous SiO_2 and GeO_2, arising from the presence of non-bridging oxygen atoms, contain modes of very intense localization, with the motion restricted to just one or two molecules.

CALCULATION OF THE FREQUENCY SPECTRA AND NORMAL MODES

The procedure for obtaining frequency spectra and normal modes was as follows. Sample atomic arrangements of the vitreous SiO_2 structure were generated by building physical ball and spoke models (cf. Bell and Dean 1966) to comply with short range structural data from X-ray and neutron diffraction experiments. The atomic arrangements were based upon XY_4 tetrahedral units (X = Si, Ge or Be; Y = O or F), linked to form a completely connected three-dimensional random network (Zachariasen 1932, 1935). The particular model upon which the results of this paper are based contained 101 X-atoms or cations and 233 Y-atoms or anions, 62 of the anions being singly coordinated, or 'non-bridging' atoms at the model surface. Atomic positions in the model were measured and, together with a two parameter harmonic central and non-central force field, were used to generate a dynamical matrix from which the normal mode frequencies and atomic amplitudes of vibration can be calculated. For the central force constants in SiO_2, GeO_2 and BeF_2 we have taken Zarzycki and Naudin's (1960) estimates of 400, 330 and 110 N m^{-1} respectively (1 N m^{-1} = 10^3 dynes cm^{-1}). The ratio of the non-central to central force constant was taken to be $\frac{3}{17}$ in all cases; this is the value suggested by Saksena (1963) for SiO_2, no corresponding information being available for GeO_2 and BeF_2. Two types of boundary condition were considered—a fixed-end condition, in which the non-bridging (O or F) atoms at the model surface were regarded as fixed, and the free-end condition in which these non-bridging atoms were free to vibrate.

Frequency distributions were obtained as histograms by repeated application of the negative eigenvalue theorem (cf. Dean and Bacon 1965) to the dynamical matrix (which was typically of order 1000 × 1000, with half band width of around 100). To obtain information about selected normal modes, we first used a bisection-interpolation procedure to evaluate specific eigenvalues accurately, and then computed the corresponding eigenvectors by the method of inverse iteration (Wilkinson 1965).

THE PARTICIPATION RATIO FOR NORMAL MODES

There is no difficulty in distinguishing between the plane wave-like vibrations characteristic of atoms in regular lattices and the intensely localized modes which can arise in connection with point impurities in perfect crystals. However in situations intermediate between these two extremes, it seems necessary to introduce some specific method of parametrizing the degree of spatial localization. We introduce the concept of a 'participation ratio' for a normal mode, as a convenient means of describing the localization properties of the mode.

Let us define, first, the energy moments of a normal mode by

$$M_p(\omega) = \sum_{\text{atoms } i} [\varepsilon_i(\omega)]^p, \tag{1}$$

where $\varepsilon_i(\omega)$, the mean kinetic energy of atom i in the normal mode of frequency ω, can be obtained directly from the appropriate eigenvector of the dynamical matrix. Obviously M_0 is identical with N, the total number of atoms in the system, while M_1 is just the mean kinetic energy of the system in a mode with frequency ω. It is easy to show that the quantity

$$p(\omega) = [M_1]^2/M_0M_2, \tag{2}$$

which lies between $1/N$ and 1, gives a measure of the proportion of atoms which participate effectively in a given normal mode. In the case of a simple monoatomic system, for example, one can show that a mode in which all N atoms vibrate with equal amplitude has $p(\omega) = 1$; if just P atoms are in motion with roughly equal amplitudes we find $p(\omega) \cong P/N$ and, for a mode involving effectively only one atom, $p(\omega) = 1/N$. For sinusoidal plane wave like modes characteristic of regular lattices we expect $p(\omega) \cong 1/2$.

For systems containing several atomic species, $p(\omega)$ is found to be dependent on atomic mass ratios. We can eliminate this mass dependence to some extent by choosing to restrict the summations (1) to just one type of atom. In the present work we have chosen to sum over the cations of the system; working with the anions instead would not appreciably alter our conclusions on localization of the modes.

ANALYSIS OF THE FREQUENCY SPECTRA AND NORMAL MODES

Smoothed versions of the frequency spectra histograms for vitreous SiO_2, GeO_2 and BeF_2, calculated by the method described in the second paragraph, are shown below in the top diagrams, (a), of Figures 1, 2 and 3 respectively. The continuous curves refer to calculations based upon a fixed-end boundary condition, while the broken lines indicate additional spectral features which emerge when the free-end condition is imposed.

Localization properties of the normal modes in the three glasses are also shown in Figures 1, 2 and 3, as plots of participation ratio against frequency. These have been obtained in each case by computing displacement eigenvectors for some 40 modes at approximately equidistant frequencies throughout the spectrum. Diagram (b) in each figure refers to a calculation in which the fixed-end boundary condition was imposed, while (c) refers to the free-end condition.

We now discuss the significance of the individual spectral bands shown in Figures 1, 2 and 3 below, paying particular attention to the localization properties of the normal modes. We consider each of the glasses separately.

SiO_2

In the fixed-end spectrum for vitreous SiO_2, the prominent features at about 1050, 750, 500 (shoulder) and 400 cm^{-1} agree closely with bands observed near 1100, 800, 600 and 500 cm^{-1} in experimental infrared and Raman spectra (Lippincott *et al.* 1958, Miler 1968, Bobovich and Tulub 1958, Harrand 1954, Hass 1970). Detailed analysis of selected normal modes indicates that the 1050 cm^{-1} band in the calculated spectrum is associated with a bond-stretching vibration in which bridging oxygen atoms move parallel to the Si–Si lines joining their immediate silicon neighbours. The modes in the 500 (shoulder) and 750 cm^{-1} regions arise from bond-bending vibrations in which the oxygen atoms move along the bisectors of the Si–O–Si angles. In the 400 cm^{-1} peak, the modes are associated with the bond-rocking motion of bridging oxygens vibrating perpendicular to the Si–O–Si planes. Vibrations associated with the low frequency shoulder below 350 cm^{-1} in the computed spectrum are extremely intricate in character, and probably correspond to overall network deformations.

When the free-end boundary condition is considered, two new features appear in the SiO_2 spectrum—a slight spectral enhancement at approximately 300 cm^{-1} on the low frequency shoulder and, more obviously, an entirely new band at about 850 cm^{-1}. Inspection of atomic displacement eigenvectors indicates that these two features are associated with bond-bending and bond-stretching vibrations respectively of the non-bridging oxygen atoms. It is interesting to note that the band at 850 cm^{-1}, although not normally seen in the spectrum of vitreous silica, occurs in samples of the glass that have been exposed to heavy neutron irradiation. Simon (1957) suggested that the irradiation causes disruption of the silica network, producing non-bridging atoms in the structure. Our present theoretical analysis, in establishing a connection between the 850 cm^{-1} band and the non-bridging oxygens, supports Simon's proposal.

Coming now to the localization properties of the atomic vibrations, for fixed-end boundary conditions we see from Figure 1(b) that the participation ratio rarely exceeds $\frac{1}{2}$, the value for waves in periodic lattices. On the other hand, the modes are not in general intensely localized, except near band edges.

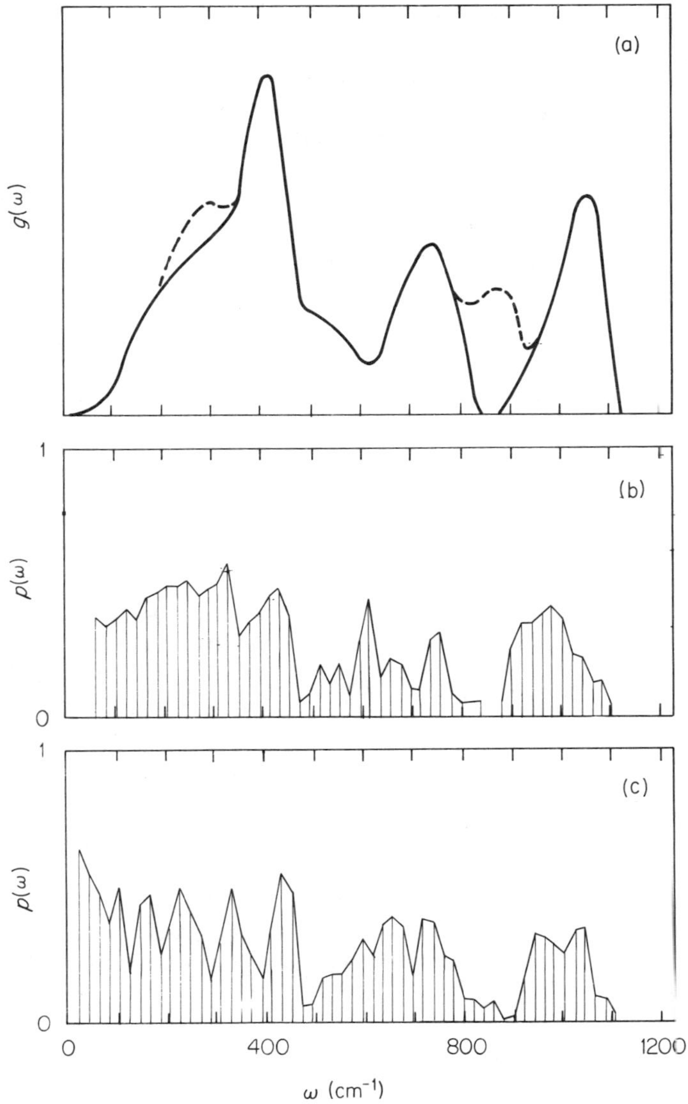

Figure 1. (a) Vibrational frequency spectra and (b) and (c) localization properties of vitreous SiO_2, computed by the method described in the text. In diagram (a) the continuous curve refers to a calculation with the fixed-end boundary condition imposed, while the broken lines indicate additional spectral features which emerge when the free-end condition is considered. Diagram (b) shows the participation ratio $p(\omega)$ for a fixed-end model, while (c) gives $p(\omega)$ for the free-end case. The participation ratio, defined in the text, lies between 0 and 1

where the motion is effectively confined in some cases to just half a dozen molecules.

It is clear, from Figure 1(c), that imposing the free-end condition makes relatively little difference to the localization properties of normal modes over most of the spectrum. The one exception is the non-bridging oxygen band at 850 cm^{-1}; here the localization is particularly intense, effective motion being restricted in some instances to no more than one or two molecules. Such extreme localization is entirely consistent with the position of this spectral band, occurring, as it does, within a gap of the spectrum of the completely connected SiO_2 network.

GeO_2

The spectra obtained for GeO_2, shown in Figure 2(a), closely resemble those for SiO_2. The main features in the fixed-end case are bands occurring at about 200, 370, 500 and 850 cm^{-1}. These are in satisfactory agreement with experimentally observed features in the 200–350, 410, 550–600 and 900 cm^{-1} regions of the infrared and Raman spectra (Cheremisinov 1965, Lippincott *et al.* 1958, Obukhov-Denisov *et al.* 1958, Zarzycki and Naudin 1960, Hass 1970). Inspection of selected eigenvectors throughout the spectrum indicates that modes in these four bands are associated with network deformation, bond-rocking, bond-bending and bond-stretching respectively. Just as with SiO_2, the free-end boundary condition gives rise to two new spectral features —a low frequency enhancement at about 280 cm^{-1}, and a band at 650 cm^{-1}, associated respectively with bond-bending and bond-stretching vibrations of the non-bridging oxygen atoms.

The pattern of localization for vitreous GeO_2, given for the fixed-end boundary condition in Figure 2(b), shows considerable similarities to that for SiO_2, with modes in the low frequency region being most extended in space and with strong localization occurring near band edges. It is noticeable, however, that modes in the higher frequency bands (at 400, 750, 1050 cm^{-1}) tend to be rather less extended in space than their counterparts in SiO_2. This is understandable on the basis that the germanium atoms present a much more effective barrier to energy diffusion than the appreciably lighter silicon atoms in SiO_2.

As in vitreous SiO_2 the use of a free-end boundary condition produces little change in the spatial extension of the modes over the spectrum as a whole. The one point worthy of note is that modes in the extra band at 650 cm^{-1}, associated with bond-stretching vibrations of the non-bridging oxygen atoms, are intensely localized—with the motion effectively restricted, just as in the corresponding SiO_2 case, to just one or two molecular units.

BeF_2

The calculated spectra for vitreous BeF_2, depicted in Figure 3(a), are markedly different from those for SiO_2 and GeO_2. This is due to the reversal of the mass ratio, the X-atom now being the lighter of the two atomic species.

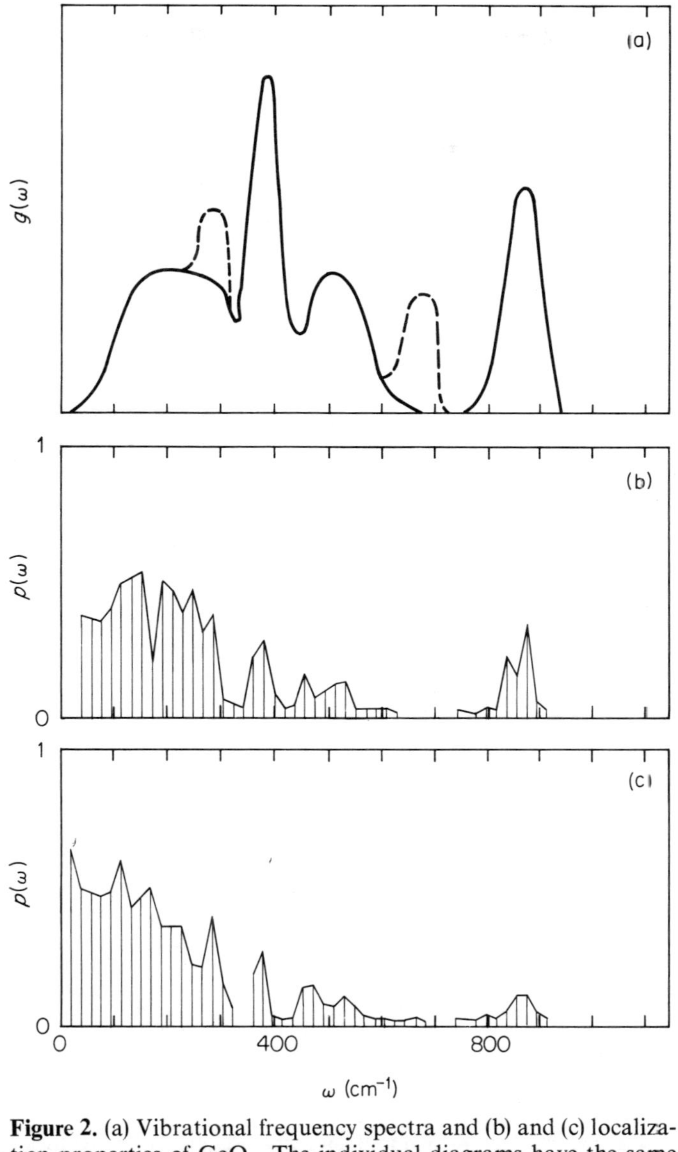

Figure 2. (a) Vibrational frequency spectra and (b) and (c) localization properties of GeO_2. The individual diagrams have the same significance as in Figure 1

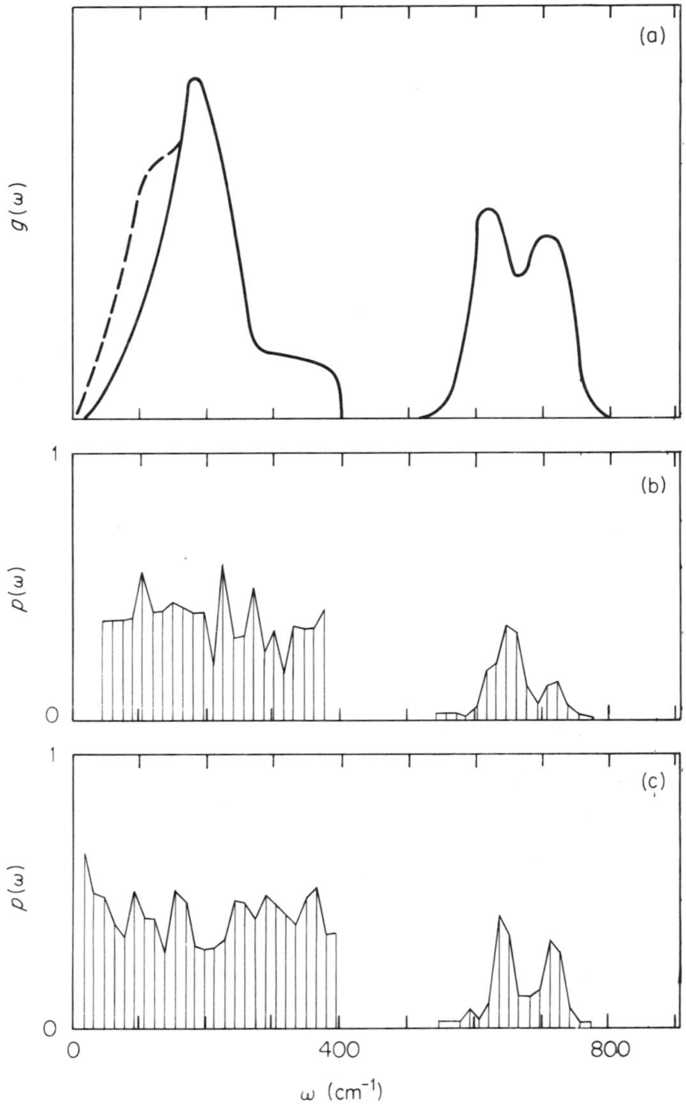

Figure 3. (a) Vibrational frequency spectra and (b) and (c) localization properties of BeF_2. The individual diagrams have the same significance as in Figures 1 and 2

The fixed-end spectrum shows two main features, a sharp low-frequency peak centred at about 200 cm^{-1} (with a shoulder at 350 cm^{-1}) and a double peak in the 700 cm^{-1} range. There is relatively little experimental data on BeF_2, but Zarzycki and Naudin (1960) have observed two bands in the infra-red spectrum of the material at around 430 cm^{-1} and in the 700–1000 cm^{-1} region, and we take these to correspond roughly with our calculated bands. Leadbetter (1970) has also found evidence for bands at about 380 cm^{-1} and above 600 cm^{-1} in his cold neutron scattering experiments. Our study of selected eigenvectors leads us to conclude that normal modes in the low frequency peak are associated with bond-rocking and bond-bending motions of the bridging F atoms, while the shoulder at 350 cm^{-1} arises from bond-stretching vibrations. For modes in the high frequency region, the greater part of the vibrational energy resides with the Be cations. When the free-end boundary condition is imposed on the model, a very prominent new shoulder occurs in the spectrum on the low frequency side of the 200 cm^{-1} peak. Much of the energy of modes in this shoulder arises from bond-bearing and bond-stretching vibrations of the non-bridging fluorine atoms at the edge of the network.

In both fixed and free-end spectra for vitreous BeF_2, the normal modes in the low frequency region are the most extended in space, their participation ratios being typically around 0·4. Modes in the 700–1000 cm^{-1} range tend to be appreciably more localized, especially near band edges, where in some cases effective motion appears to be confined to just two or three molecules. It is worth observing that, in contrast to the situation in SiO_2 and GeO_2, none of the modes arising from the free-end boundary condition in BeF_2 exhibit extreme localization.

CONCLUSION

We have seen that the spatial localization of normal modes varies appreciably throughout the frequency spectra of the glasses considered. In general, modes are fairly extended at low frequencies and show a tendency to greater localization with increasing frequency; this tendency is particularly marked in vitreous GeO_2, with its high mass ratio of atomic species. In most cases the localization tends to be accentuated near the band edges, one notable exception being the low frequency band of vitreous BeF_2.

Bands in the high frequency regions of vitreous SiO_2 and GeO_2, arising from the presence of non-bridging surface atoms, are of particular interest. The modes of these bands are very similar in character to those generated by the substitution of isolated light impurity atoms in a heavy atom host crystalline material.

The fact that there is wide disparity between localization properties of normal modes in the various regions of the spectrum of a glass is a point of considerable importance. One clear consequence is that neither the purely molecular approach, nor that based on an undiscriminating use of crystal lattice dynamics, is likely to be fully successful in yielding accurate information on vibrational modes throughout the spectrum. Only a much more flexible scheme, such as that based on extended atomic models is capable of reproducing the full range of vibrational behaviour.

ACKNOWLEDGEMENT

The work described above has been carried out at the National Physical Laboratory.

REFERENCES

Bell, R. J., and P. Dean (1966). *Nature, Lond.*, **212**, 1354.

Bobovich, Ya. S., and T. P. Tulub (1958). *Usp. fiz. Nauk SSSR*, **66**, 3.

Cheremisinov, V. P. (1965). In *Optical Methods of Investigating Solid Bodies*. Consultants Bureau, New York.

Dean, P., and M. D. Bacon (1965). *Proc. R. Soc.*, A, **283**, 64.

Harrand, M. (1954). *C. r. hebd. Séanc. Acad. Sci.*, Paris, **238**, 784.

Hass, M. (1970). *J. Physics Chem. Solids*, **31**, 415.

Leadbetter, A. J., and A. C. Wright (1970). *J. non-cryst. Solids*, **3**, 239.

Lippincott, E. R., A. V. Valdenberg, C. E. Weir, and E. N. Bunting (1958). *J. Res. natn. Bur. Stand.*, **61**, 61.

Miler, M. (1968). *Czech. J. Phys.*, B, **18**, 354.

Obukhov-Denisov, V. V., N. N. Sobolev, and V. P. Cheremisinov (1958). *Bull. Acad. Sci. USSR phys. Ser.*, **22**, 1073.

Saksena, B. D. (1963). *Trans. Faraday Soc.*, **59**, 276.

Simon, I. (1957). *J. Am. Ceram. Soc.*, **40**, 150.

Zachariasen, W. H. (1932). *J. Am. chem. Soc.*, **54**, 3841.

Zachariasen, W. H. (1935). *J. chem. Phys.*, **3**, 162.

Zarzycki, J., and F. Naudin (1960). *Verres Réfract.*, **14**, 113.

Wilkinson, J. H. (1965). *The Algebraic Eigenvalue Problem.* Clarendon Press, Oxford.

44

Ultra-Low-Expansion Glasses and Their Structure in the SiO_2–TiO_2 System

P. C. SCHULTZ and H. T. SMYTH

Research and Development Laboratories, Corning Glass Works, Corning, New York, U.S.A.

Probably the first report of SiO_2–TiO_2 glasses having an expansion lower than vitreous silica was in a 1939 patent application by M. E. Nordberg (1943). He described a series of binary silicate glasses, produced by a flame hydrolysis process, and noted that their expansion coefficients ($\alpha_{0-300°C}$) generally decreased as the TiO_2 concentration increased in the 5–11 wt % range. Dietzel (1943) attributed this remarkable behaviour to the direct replacement of silicons by titaniums in the glass network, although only a very brief explanation was given.

The subject then apparently lay dormant until late 1966, when Corning introduced a zero-expansion glass on the market. Designated Code 7971 glass, it is a binary titanium silicate having an expansion coefficient of essentially zero in the 5–35 °C range (1969).

Recently, Russian literature (Kozlova, 1968; Ostrizhko, 1970) discuss crucible melted quartz glasses containing TiO_2 additions. The glasses were opaque and, in some cases, contained TiO_2 crystals.

Up to the present time however, neither a good complete description of the thermal expansion characteristics of these binary glasses, nor a well-developed structural interpretation of these anomalous expansion effects has been given. The primary purpose of the present work was to gain some insight into both of these aspects of expansion of SiO_2–TiO_2 glasses.

EXPERIMENTAL PROCEDURE

Sample Preparation

All glasses studied were produced by the flame hydrolysis process using a small two-burner furnace. The procedure used to make the boules is similar to that described in the patents of Nordberg (1943) and Nitsche (1967). A schematic diagram of the furnace is shown in Figure 1. The glass was deposited in the rotating refractory cup by the two burners shown, at temperatures of ~1700 °C. Clear glass boules approximately one inch thick and 6 inches

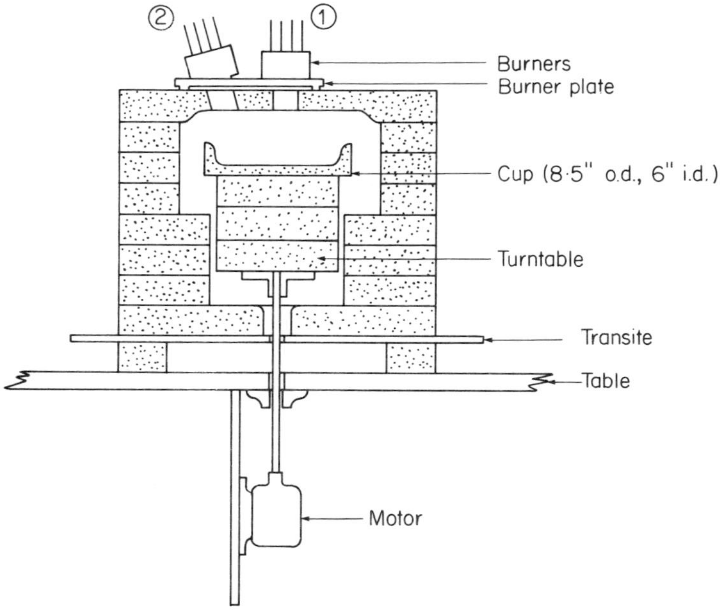

Figure 1. Two-burner hydrolysis furnace (cross-section)

in diameter were produced. After the boules were air-cooled in the hydrolysis furnace, they were removed and thermal expansion bars were cut. These were used as the unannealed samples. Chemical analysis samples were removed from a point adjacent to the expansion bars so that good correlation between expansion and composition could be obtained.

After determining the thermal expansion characteristics of the unannealed bars, they were annealed at 1050 °C and their expansion curves redetermined.

Chemical Analysis

Chemical analyses were performed using X-ray fluorescence techniques on powdered glass samples which had been prepared into small pellets by flux fusion (Stephenson, 1969). The resultant values are good to ±0·1 wt %.

Thermal Expansion

Measurements were made using 3 inch long bar samples. Relative changes in length between 0 °C and 750 °C were measured on the standard fused silica rod dilatometer with a precision of ±10 ppm and accuracy ±25 ppm.

Measurements between −200 °C and +200 °C were made using an extended-range, high-precision dilatometer. Precision of this unit is ±2 ppm and accuracy ±5 ppm. The same dilatometer used in the 5–55 °C range is described by Plummer and Hagy (1968). All expansion data are reported here normalized for zero $\Delta L/L$ at 25 °C.

RESULTS AND DISCUSSION

Glass Quality, Composition and Densities

The compositions produced are listed in Table 1 along with their densities and various expansion coefficients. These properties will be discussed later in this report. Clear, good quality glasses were obtained for each composition listed. The unannealed glasses all had a yellowish-brown tint, while annealed samples were water white. This brown tint was attributed to the presence of small amounts of reduced titanium, Ti^{3+} (i.e., approximately 10 ppm), which were oxidized to Ti^{4+} during the annealing cycle. Magnetic susceptibility measurements and ultraviolet absorption data for both the unannealed and annealed glasses helped to corroborate this theory.

Table 1. Compositions, densities and expansion coefficients of annealed SiO_2–TiO_2 glasses

Glass No.	TiO_2 in wt.%	Density g/cc	α_{25}(5–35 °C) $\times 10^{+7}$/°C	α_{25}(25–700 °C) $\times 10^{+7}$/°C	α_{25}(−200–25 °C) $\times 10^{+7}$/°C
1	3·42	2·201	+3·10	+2·312	−0·89
2	6·00	2·199	+1·25	+0·133	−1·78
3	7·20	2·200	+0·0	−0·741	−2·44
4	7·40	2·199	+0·36	−0·326	−2·58
5	8·36	2·197	n.v.	−1·156	n.v.
6	9·45	2·198	−1·80	−1·926	−3·96
Code 7940[a]	Zero	2·202	+5·30	−4·07	−0·67

n.v.—No value obtained. [a]—Corning Code 7940 fused silica.

Effect of Composition and Expansion

The effect of TiO_2 content on thermal expansion in these binary glasses is best seen in Figure 2 for annealed specimens. It is a plot of the expansion curves over the temperature range of -200 to $+750$ °C for glasses having four different titania concentrations. Corning Code 7940 glass (vitreous silica) is included for comparison. The titania contents of each of the other glasses plotted are listed above their respective curves. It is obvious from this figure that there is a very definite, smooth transition of the shapes of these

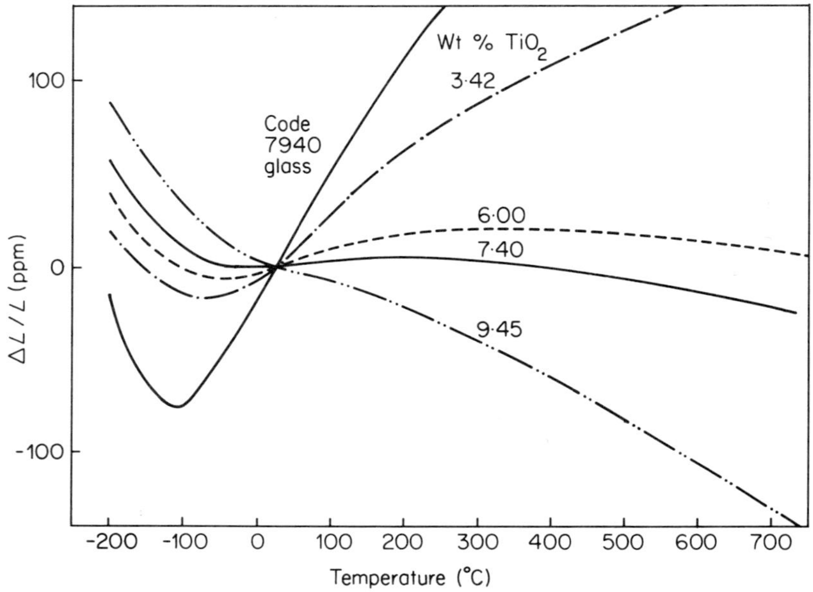

Figure 2. Thermal expansion of annealed vitreous silica and SiO_2–TiO_2 glasses

curves with changes in composition, such that the lower-titania glasses (i.e., 3·42 wt %) approach the expansion characteristics of vitreous silica, while increases in the titania content result in a rapid divergence from vitreous silica behaviour. This divergence occurs in such a way that, as the titania content is increased, the low-temperature trough (-100 °C) in the vitreous silica expansion curve rapidly disappears, and the high-temperature linear positive expansion (0–750 °C) becomes less linear in slope until finally it crosses the zero-expansion point and the slope becomes more and more negative in value. By the time the TiO_2 content has reached 9·45 wt %, the glass exhibits a semilinear negative expansion over the entire temperature range.

Tabulations of expansion coefficients were made for each of the compositions used in this investigation. The coefficients cover three different temperature ranges for each glass, -200 to $+25$, $+5$ to $+35$, and $+25$ to $+750$

°C. Changes in these coefficients with composition give a pretty fair indication of how the overall shape of the expansion curve is changing. In Figure 3 the three coefficients are plotted as a function of titania content. The straight line drawn for each coefficient is the least-squares fit of the respective data points. Extrapolation of the curves to zero TiO_2 comes reasonably close to the actual values for pure vitreous silica (shown here as hexagonal data points). These results indicate that the effect of TiO_2 content on the thermal expansion is of such a nature that the transition is both smooth and linear for all values of TiO_2 between approximately zero and 10 wt %.

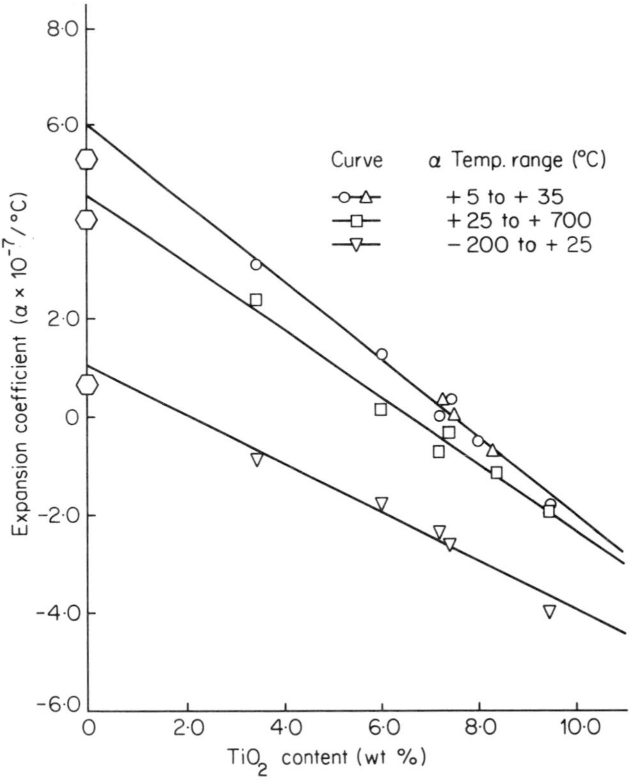

Figure 3. Effect of TiO_2 content on expansion coefficient for annealed glasses

Effect of Annealing on Expansion

The effect of annealing becomes apparent when both unannealed and annealed curves are plotted together for each composition. This can be seen in Figure 4 for the 3·42, 7·40, and 9·45 wt % TiO_2 glasses. In each case, there are two main changes caused by the annealing process. First, the unannealed glasses always exhibit a larger positive relative change in length than the annealed ones over the temperature range −200 to +25 °C, although the difference becomes smaller as the TiO_2 content is increased. Second, the annealed curves tend to cross over the unannealed curves, and veer away in a negative $\Delta L/L$ direction, somewhere between 25 and 750 °C. The crossover point occurs at lower temperatures as the TiO_2 content in the glass is increased. By 9·45 wt % TiO_2 this crossover point virtually disappears, and the annealed curve has a lower positive $\Delta L/L$ in the low temperature range, and a larger negative $\Delta L/L$ in the high temperature range, than the unannealed curve.

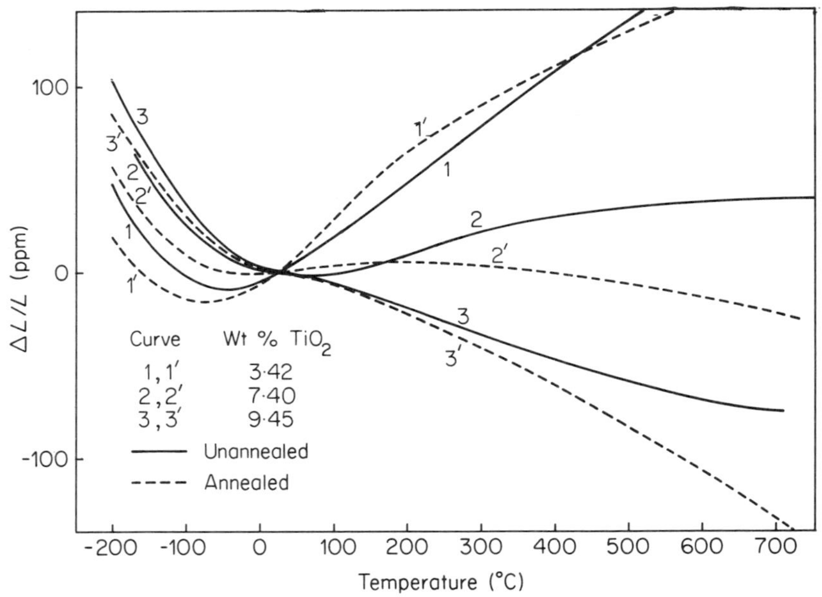

Figure 4. Effect of annealing on termal expansion of SiO_2–TiO_2 glasses

A detailed investigation of the effect of annealing temperatures, cooling rates and cut-off temperatures on the expansion of these glasses was not carried out in the present investigation, for this was intended simply as a precursory look at the effect of annealing over a wide composition range. The results indicate however that thermal history, as well as composition, plays an important role in determining the expansion characteristics of these glasses.

Theoretical Considerations for Low Expansion

In almost all of the crystalline forms of SiO_2 each oxygen is bonded to two silicons and each silicon is bonded to four oxygens. In the crystalline forms of TiO_2 each oxygen is bonded to three titaniums and each titanium to six oxygens. In the glass composition range covered in this paper it can be shown that the probability of finding more than one titanium in the immediate vicinity of any oxygen is very small and that therefore each oxygen is bonded to either two silicons or one silicon and one titanium and that therefore, to keep things balanced, each titanium is bonded to four oxygens. This point will be discussed in detail in a forthcoming publication by the authors. It can then be shown, also in this low titania range, the concentration of oxygens bonded to a silicon and a titanium varies linearly with molar composition.

It can also be shown (Dietzel, 1943) that a mode of vibration with frequency ν makes a contribution to the amount by which the molar volume exceeds that at absolute zero which is proportional to

$$\{1/(e^{h\nu/kT} - 1)\}\frac{d\nu}{dV}$$

This will also be derived and discussed in the same later publication.

In terms of vibrations of individual atoms (as in the Einstein treatment of thermal capacity) it can be shown that most atoms in a solid have a negative value of $d\nu/dV$ and therefore contribute to positive expansion but that an oxygen lying on the straight line between two cations and not bonded to any others has, in its two transverse modes of vibration, a positive value of $d\nu/dV$ and therefore contributes to negative expansion. (Smyth, 1955, 1966, 1969, 1970.) If the oxygen lies off the straight line by small amount then it happens that the vibration parallel to the cation-cation line contributes to negative expansion and the rotation around this line also contributes to negative expansion.

Thermal expansion curves (or molar volume curves) are not very sensitive to the details of the frequency distribution. Almost any thermal expansion curve, including those investigated, can be quite closely approximated using only four frequencies rather than a continuous distribution. Using characteristic temperatures Θ instead of frequencies where

$$\Theta = h\nu/k$$

a set of 100, 300, 600 and 1200 degrees for Θ with a set of four coefficients for the terms in the expression above corresponding to the four frequencies (or characteristic temperatures) allows very good curve fitting in any of the observed curves. The 'theoretical' curves of Figure 5 were pieced together this way.

The shape of these curves would indicate that the frequency of transverse vibration of an oxygen between a silicon and a titanium makes its contributions at a higher temperature than one between two silicons and therefore between expansion and composition would indicate that, using four characteristic temperatures as above, it should be possible to assign four coefficients

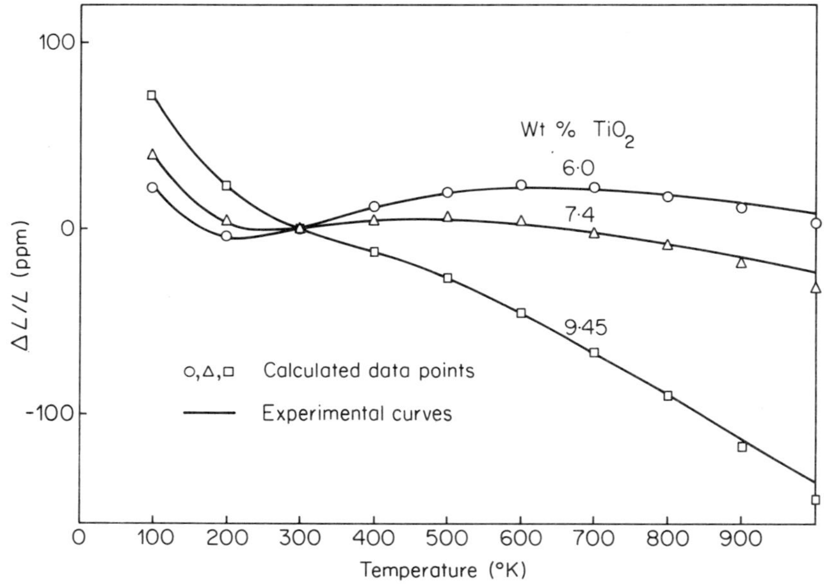

Figure 5. Theoretical expansion of SiO_2–TiO_2 glasses

for each species of atom (e.g. an oxygen bridging between a silicon and a titanium or a titanium surrounded by four oxygens) and predict, knowing the composition, a complete expansion curve for any of the glasses. Whether this procedure would work for a very wide range of compositions remains to be determined. Work is being carried out now on calculating the concentrations of species in different glasses but much remains to be done. If it has a wide application then as well as throwing some light on the atomic processes involved it might provide a very practical means of storing data on the expansion of many families of glasses.

ACKNOWLEDGEMENTS

The authors wish to thank Corning Glass Works for permission to publish this research. We would also like to express our appreciation to W. A.

Plummer for the expansion measurements and D. A. Stephenson for chemical analyses, and numerous other colleagues for their advice and assistance.

REFERENCES

Corning Glass Works (1969). *Low Expansion Materials Bulletin.*

Dietzel, A. (1943). *Naturwissenschaften*, **31**, 22.

Kozlova, L. N., B. V. Tarasov, and K. I. Chepizhnyi (1968). *Steklo*, **3**, 6 (Russ.).

Nitsche, J. E. (1967). *U.S. Patent 3,303,115*. Feb. 7.

Nordberg, M. E. (1943). *U.S. Patent 2,326, 059*. Aug. 3.

Ostrizhko, D. G., and G. A. Pavlova (1970). *Izv. Akad. Nauk. SSSR, Neorg. Mater.*, **6** (1), 74. (Russ.)

Plummer, W. A., and H. E. Hagy (1968). *Appl. Optics*, **7**, 825.

Smith, H. L., and A. J. Cohen (1963). *Physics Chem. Glasses*, **4** (5), 173.

Smyth, H. T., H. S. Skogen, and W. B. Harsell (1953). *J. Am. Ceram. Soc.*, **36** (10), 327.

Smyth, H. T. (1955). *J. Am. Ceram. Soc.*, **38** (4), 502.

(1966). *Theoretical and experimental properties of simple glasses*. National Science Foundation, Grant G24276, Feb. Rutgers University, New Brunswick, N.J., U.S.A.

(1969). Private communications to P. C. Schultz.

(1970). *The Structure of Glass*. Tutorial Symposium on Glass Science, Alfred University, Alfred, N.Y., U.S.A. June 8–19.

Stephenson, D. A. (1969). *An. Chem.*, **41** (7), 966.

45

Pressure Dependence of Elastic Constants and Grüneisen Parameters in Fused SiO_2, GeO_2, BeF_2 and B_2O_3

C. R. KURKJIAN, J. T. KRAUSE

Bell Telephone Laboratories, Incorporated, Murray Hill, New Jersey, U.S.A.

H. J. McSKIMIN, P. ANDREATCH and T. B. BATEMAN

Bell Telephone Laboratories, Incorporated, Holmdel, New Jersey, U.S.A.

A recent note (Krause and Kurkjian, 1968) pointed out and tabulated the various types of anomalous behaviour which have been observed in the vibrational properties of inorganic glasses. From this tabulation and a recent review and discussion by Leadbetter (1968), it would seem that these 'anomalies' can be divided into two categories: (1) those which depend solely on the existence of abnormally low frequency modes, and (2) those which depend on low frequency modes of a specific type; i.e., modes for which $d\omega/dV$ is positive. It was observed that those properties which fall into group 1 (acoustic loss, low frequency Raman scattering and infrared absorption, excess heat capacity) are characteristic of glassy solids in general. On the other hand, those in group 2 (negative expansion coefficient, positive temperature coefficient of elastic constants, negative pressure coefficient of elastic constants), seemed to be present only in those glasses having a silica-type structure. In addition, while those properties in group 1 require the participation of only a small fraction of the total number of modes, those in group 2 seem to be characteristic of a larger number of the modes of the solid.

This paper reports new measurements on the pressure dependence of elastic constants in some simple glasses as well as some low temperature thermal expansion and room temperature acoustic velocity measurements mentioned briefly by Krause and Kurkjian (1968). The glasses studied were SiO_2, GeO_2 and BeF_2 which have fourfold coordinated cations and B_2O_3

in which the boron is threefold coordinated. It is thought that the importance of tetrahedral coordination in the appearance of vibrational anomalies of group 2 above, might be elucidated by this comparison.

THEORY

Anderson (1965), Anderson and Dienes (1960) and more recently Leadbetter (1968), have discussed the long wave vibrational behaviour of chemically simple inorganic glasses. In addition, Dean (1964), Bell, Bird and Dean (1968), Hindley (1967) and Rosenstock (1961), Rosenstock and McGill (1968) and Haas, Rosenstock and McGill (1969) have made model studies and calculations of vibrations in disordered solids. These papers should be consulted for details of the low frequency behaviour of solids in general and disordered solids in particular. Only a brief review will be given here.

Low frequency behaviour can be studied near the origin of the frequency (ω) *vs.* wave number (q) plot, (1) acoustically, at normal temperatures by exciting only low frequency modes, or (2) by studying equilibrium thermodynamic properties at low temperatures, where only low frequency modes are thermally excited.

The usual formalism applied in discussions of heat capacity, thermal expansion and sound velocities is that of Grüneisen γ's. The Grüneisen γ expresses the expected proportionality of heat capacity and thermal expansion:

$$\gamma^{\mathrm{T}} = \frac{\beta V K}{C} \qquad (1)$$

where

C = heat capacity
V = molar volume
β = volume expansion coefficient
K = bulk modulus

For many solids this 'thermal' Grüneisen parameter is essentially a constant as a function of temperature, of the order of 2. In the quasi-harmonic approximation, the frequency of individual modes is taken to be a function of volume alone, thus:

$$\gamma = -\frac{\mathrm{d}\ln\omega}{\mathrm{d}\ln V} \qquad (2)$$

The frequency ω can be replaced by the sound velocity (v) and an acoustic parameter γ^{a} can be obtained. Several somewhat different expressions can be used for acoustic γ's, as discussed by Anderson (1968a) and Scheule and Smith (1964). A general acoustic γ can be expressed as:

$$\gamma^{a} = \tfrac{1}{3} + K\frac{dv}{v dP} \tag{3}$$

At low temperatures, the number of modes that is classically excited is proportional to v^{-3}. Thus

$$\gamma^{a}_{LT} = \left(\frac{\gamma_1}{v_1^3} + \frac{2\gamma_s}{v_s^3}\right) \Big/ \left(\frac{1}{v_1^3} + \frac{2}{v_s^3}\right) \tag{4}$$

At low temperatures (as $T \to 0$), appropriate values of elastic and thermal γ's would be expected to be equal, as would the elastic and thermal values of $(C/T^3)_{T\to 0}$ as discussed by Anderson and Dienes (1960), Flubacher *et al.* (1959) and Antoniou and Morrison (1965), unless there exists a low frequency optical branch, or an acoustic branch with curvature.

Since the quasi-harmonic model assumes frequency changes to be due only to volume changes, the Grüneisen γ's should be a function of temperature only through the effect of thermal expansion. Variations from this behaviour are indications of deviations from the quasi-harmonic approximation.

A direct measure of the deviation from the quasi-harmonic model is the magnitude of the temperature dependence of the acoustic velocity or elastic constants, at constant volume (Lazarus, 1949). If the approximation is valid, $(\partial M/\partial T)_V = 0$, where M is any modulus. By measuring $(\partial M/\partial T)_P$, $(\partial M/\partial P)_T$, β and K, this evaluation can be made according to;

$$\left(\frac{\partial M}{\partial T}\right)_P = -K\beta\left(\frac{\partial M}{\partial P}\right)_T + \left(\frac{\partial M}{\partial T}\right)_V \tag{5}$$

The negative values of thermal expansion at low temperatures given by Krause and Kurkjian (1968) indicate that for glasses in addition to SiO_2 the thermal Grüneisen γ will be negative, and also implies that for these glasses the acoustic γ's and thus $\partial v/v\partial P$ will be negative. The possibility of negative Grüneisen γ's for some modes was recognized by Dayal, (1944) and the occurrence of negative γ's and β's in some tetrahedral structures was noted by Smyth (1955), Barron (1957) and Blackman (1957, 1958). Smyth has made calculations for the specific heat (1953b), thermal expansion (1955) and compressibility (1953a) of SiO_2 using an Einstein-like model and considering only ionic bonding. His model related the anomalies in these parameters to

(1) low frequency vibrations with positive $d\omega/dV$

(2) bond angle, rather than bond length changes

made possible by the rather special transverse oxygen vibrations in the tetrahedral silica structure. Barron (1957) has proposed a mechanical model which simply illustrates a mechanism for producing increasing vibrational frequencies for transverse vibrations as the volume increases.

A somewhat similar model has recently been proposed by Gaskell (1966). He considers the comparable vibrational and expansion behaviour of fused silica and the high temperature forms of quartz and cristobalite. He concludes that the anomalously low and negative expansivities of these open structures 'are all manifestations of a decrease in the value of the Si–O–Si angle ε, accompanying the 'normal' increases in nearest neighbour, silicon-oxygen distances.'

EXPERIMENTAL

The ultrasonic velocity data were obtained at 20 MHz using the pulse–echo and pulse–superposition techniques described by McSkimin *et al.* (1962). Thermal expansion data were obtained at room temperature using an instrument described by Sauer (1968), and at low temperature using a capacitance technique modeled after that of White (1961). The expansivity data are normalized with $\Delta l/l = 0$ at 4·2 °K and the expansion coefficients were obtained by graphical differentiation of $\Delta l/l$ *vs.* T plots. These data are reliable at least up to 20 °K.

RESULTS

A complete tabulation and detailed discussion of the experimental results will be published elsewhere. A summary of the results can be obtained from Figures 1–3 and Table 1. Figure 1 is a plot of the relative longitudinal velocity as a function of temperature. The low temperature portion of the curve represents a relaxation effect, and thus it is clear that except for B_2O_3 these glasses have positive temperature coefficients of acoustic velocity. Figure 2 shows the coefficient of linear thermal expansion for these four glasses at low temperatures. Again B_2O_3 is 'well behaved', while the other three glasses show a range of negative α. Using heat capacity data from the literature (detailed data at low temperatures is lacking only for B_2O_3) values of the thermal Grüneisen parameters have been calculated and plotted in Figure 3. Table 1 summarizes the temperature and pressure coefficients of the shear and bulk moduli near room temperature. As was the case above, the tetrahedral glasses (SiO_2, BeF_2 and GeO_2) are 'anomalous', while B_2O_3 is normal with respect to the pressure derivative of moduli (while

$$\frac{1}{K}\left(\frac{\partial K}{\partial P}\right)_T$$

for GeO_2 is positive, it is quite small.)

Values of

$$\left(\frac{\partial M}{\partial T}\right)_V \Big/ \left(\frac{\partial M}{\partial T}\right)_P$$

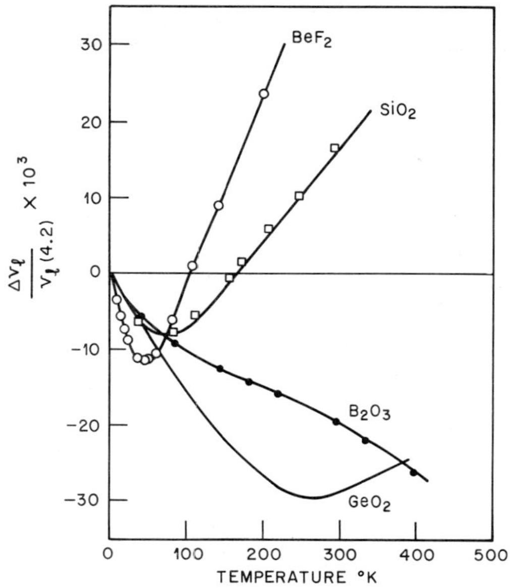

Figure 1. Relative longitudinal velocity (20 MHz) as a function of temperature

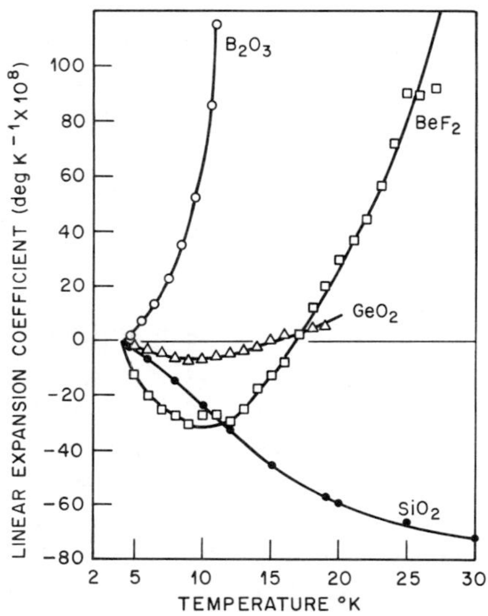

Figure 2. Linear coefficient of thermal expansion as a function of temperature

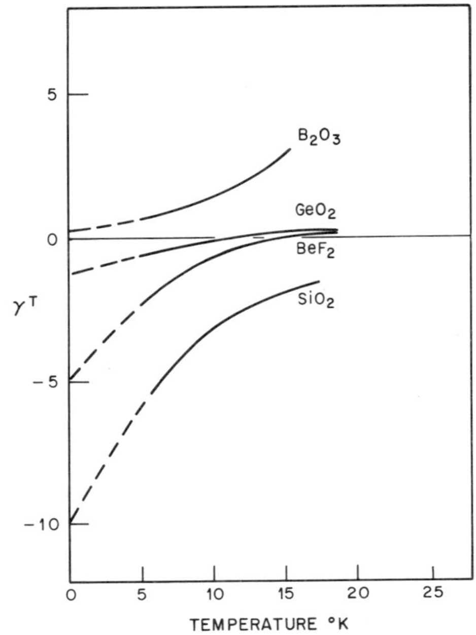

Figure 3. Thermal Grüneisen parameter as a function of temperature

are also listed in Table 1. In the shear mode, for all the glasses except GeO_2, the pure temperature term is a very significant fraction of the total temperature dependence. In terms of the bulk modulus, SiO_2 and BeF_2, are similar, with the pure temperature term accounting for most of the temperature dependence. In the case of GeO_2 and B_2O_3, the peculiarities in bulk behaviour result in values of

$$\left(\frac{\partial K}{\partial T}\right)_V \bigg/ \left(\frac{\partial K}{\partial T}\right)_P$$

of $+1{\cdot}02$ and $-1{\cdot}70$ respectively. The result of this is that for all four glasses studied, except for the shear behaviour of GeO_2, $M = f(T)$ is a better approximation than $M = f(V)$.

The thermal and acoustic Grüneisen γ's are listed in Table 2 along with values of $(C/T^3)^a$ and $(C/T^3)^T$. The low temperature acoustic γ^a_{LT} was calculated from equation (4). The values of γ^T_0 however, were obtained by extrapolation as shown in Figure 3. As a result of the anomalies in $\partial v/\partial T$, $\partial v/\partial P$ and low temperature expansion coefficient for the tetrahedral glasses, all of these show negative limiting Grüneisen parameters, γ^T_0 and γ^a_{LT}. These anomalies in fact derive from modes having positive $d\omega/dV$. In addition

Table 1. Pressure and temperature coefficients of shear and bulk moduli

	$\frac{1}{G}\left(\frac{\partial G}{\partial T}\right)_P$	$\frac{1}{G}\left(\frac{\partial G}{\partial P}\right)_T$	$\frac{1}{G}\left(\frac{\partial G}{\partial T}\right)_V$	$\frac{1}{K}\left(\frac{\partial K}{\partial T}\right)_P$	$\frac{1}{K}\left(\frac{\partial K}{\partial P}\right)_T$	$\frac{1}{K}\left(\frac{\partial K}{\partial T}\right)_V$	$\left(\frac{\partial K}{\partial T}\right)_V \Big/ \left(\frac{\partial K}{\partial T}\right)_P$	$\left(\frac{\partial G}{\partial T}\right)_V \Big/ \left(\frac{\partial G}{\partial T}\right)_P$	ν
SiO_2	$+1.41 \times 10^{-4}$	-1.04×10^{-11}	$+1.33 \times 10^{-4}$	$+3.2 \times 10^{-4}$	-1.69×10^{-11}	$+3.5 \times 10^{-4}$	0·96	0·94	0·167
BeF_2	$+2.78 \times 10^{-4}$	-1.52×10^{-11}	$+2.14 \times 10^{-4}$	$+7.58 \times 10^{-4}$	-1.67×10^{-11}	$+6.88 \times 10^{-4}$	0·91	0·77	0·147
GeO_2	$+0.38 \times 10^{-4}$	-0.61×10^{-11}	$+0.06 \times 10^{-4}$	$+0.97 \times 10^{-4}$	$+0.03 \times 10^{-11}$	$+0.99 \times 10^{-4}$	1·02	0·16	0·209
B_2O_3	-0.97×10^{-4}	$+0.52 \times 10^{-11}$	-0.64×10^{-4}	-0.90×10^{-4}	$+3.88 \times 10^{-11}$	$+1.52 \times 10^{-4}$	−1·70	0·67	0·290

BeF_2 and SiO_2 clearly show inequalities in the thermal and acoustic values in Table 2, while B_2O_3 and GeO_2 show fairly good agreement.

Since the anomalous coefficients of modulus and the negative β at low temperatures have been correlated with low frequency modes in tetrahedral solids, a correlation might be expected between these parameters. The difficulty lies in characterizing the low temperature expansion. Since the 'recovery' of a positive β presumably occurs as the low frequency modes become swamped by the higher frequency modes at high temperatures, some parameter characterizing the expansion in the region below its minimum is required. Figure 4 shows a plot of the coefficients of shear moduli as a function of $(d\alpha/dT)_{T=0}$. It is clear from this figure that all three of these parameters are intimately connected.

Table 2. Low temperature acoustic and thermal Grüneisen parameters and reduced heat capacities

	γ_{LT}^{a}	γ_0^T	$\left(\frac{C}{T^3}\right)_0^a \times 10^5$	$\left(\frac{C}{T^3}\right)_0^T \times 10^5$
SiO_2	−2·15	−10	1·2	2–2·3
BeF_2	−1·40	−5	2·5	5
GeO_2	−0·88	−1·3	~5	~5
B_2O_3	+0·28	+0·2	13	10–15

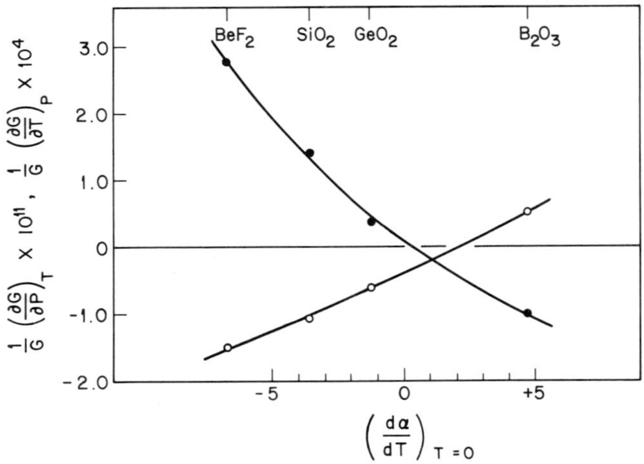

Figure 4. Temperature (●) and pressure (○) coefficients of shear moduli *vs.* $(d\alpha/dT)_{T=0}$

Another interesting correlation was obtained by plotting the pressure coefficients as suggested by Anderson (1968b). He found that for crystalline oxides with the same anion and cation coordination numbers, $(\partial G/\partial P)_T$ was proportional to G/K or *inversely proportional* to v (Poisson's ratio). For the glasses studied here, $(1/M)(\partial M/\partial P)_T$ is found to be *proportional* to v with the derivative zero at $v \sim 0{\cdot}25$. A correlation is also found for Poisson's ratio and the *temperature coefficients*. Only one such plot

$$\left[\frac{1}{G}\left(\frac{\partial G}{\partial P}\right)_T \textit{vs. } v\right]$$

is shown here, Figure 5.

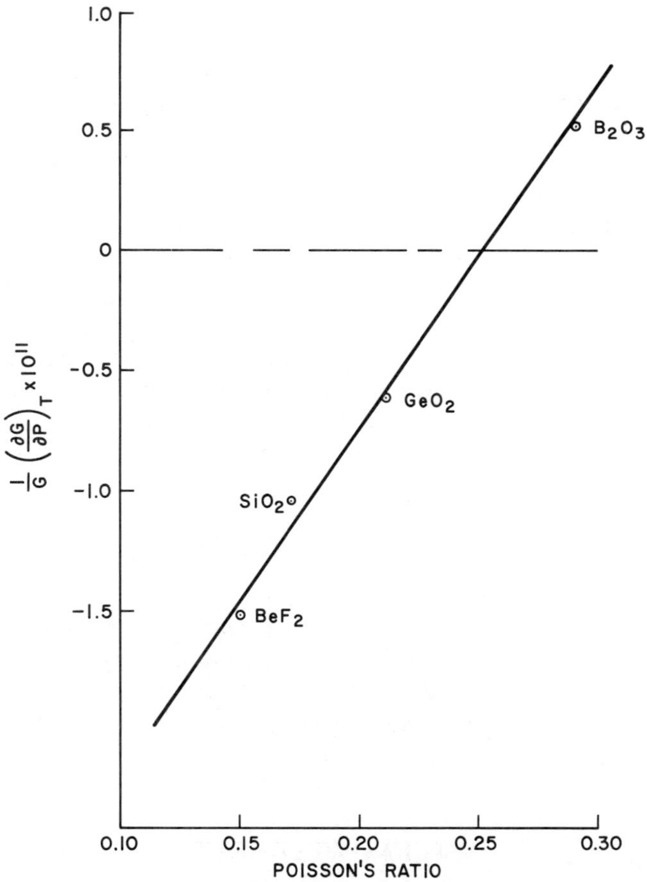

Figure 5. Pressure coefficient of shear modulus *vs.* Poisson's ratio (v)

These relations are reminiscent of the behaviour of the model used by Smyth and coworkers to explain the compressibility (1953a) and Poisson's ratio (1959) in glasses. They found that $\nu = 0{\cdot}25$ was appropriate to an SiO_2 network if only Si–O–Si stretching was allowed, and that values less than 0·25 could be obtained if bond bending was also allowed. In studying the compressibility, they again found that the assumption of bond bending was necessary in order to account for the anomalous pressure dependence.

The anomalous elastic properties can then be accounted for in a qualitative way by Smyth's (1953) or Barron's (1957) models. These both recognize the uniqueness of low frequency transverse vibrations, the frequency of which increases with volume. Changes in pressure and temperature produce changes in the cation–anion–cation bond angle and changes in the transverse anion vibration frequency.

It has been shown here that these elastic and expansion anomalies do not occur in B_2O_3. The 'normal' behaviour of the simple triangular unmodified B_2O_3 can be explained by the lack of 'three dimensionality' of its lattice. This allows distortions to occur which compensate for the effect of the transverse oxygen vibrations.

These anomalous elastic and expansion properties of fourfold networks must clearly be characteristic of the overall 'lattice' vibrations rather than vibrations of localized defects. On the other hand, almost all of the glasses so far studied do show the presence of low frequency modes which can be due to localized defects (tan δ, C_p, Raman and Infrared spectra). Rosenstock (1961) has discussed the possibility of thermal as distinct from elastic anomalies in disordered solids. In his model, nonelastic, low frequency modes result from weakly bonded atoms in 'cavities'. In some cases it is possible that both 'lattice' and 'local' mechanisms are operative; for instance in a heat capacity or mechanical loss measurement.

CONCLUSIONS

(1) Tetrahedrally coordinated glasses show elastic and expansion anomalies while triangular or 'modified' tetrahedral glasses do not. Smyth's (1953) model of bond *angle* changes accounts in a qualitative way for this behaviour.

(2) Most glasses show additional evidence for a more general type of low frequency defect mode. These modes may or may not be associated with those responsible for the elastic anomalies in (1).

ACKNOWLEDGEMENT

The authors would like to thank Mr. E. A. Sigety for help in preparing some of the glasses and assistance in the experimental work, Dr. A. J. Leadbetter for communicating his very fine heat capacity data before publication and Drs. R. N. Thurston and K. Brugger for helpful suggestions.

REFERENCES

Anderson, O. L. (1965). In J. A. Prins (Ed.), *Phys. Noncrystalline Solids.* North-Holland Pub. Co., Amsterdam. p. 179.

Anderson, O. L. (1968a). *J. geophys. Res.*, **73**, 5187.

Anderson, O. L. (1968b). *J. geophys. Res.*, **73**(24), 7707.

Anderson, O. L., and G. J. Dienes (1960). In V. D. Frechett (Ed.), *Non-Crystalline Solids.* John Wiley. N.Y. p. 449.

Antoniou, A. A., and J. A. Morrison (1965). *J. appl. Phys.*, **36**(6), 1873.

Barron, T. H. K. (1957). *Ann. Phys.*, **1**, 77.

Bell, R. J., N. F. Bird, and P. Dean (1968). *J. Phys. C* (*Proc. Phys. Soc.*), *Ser. 2*, **1**, 299.

Blackman, M. (1957). *Proc. phys. Soc. Lond. B.*, **70**, 827.

Blackman, M. (1958). *Phil. Mag.*, *Ser. 8*, **3**, 831.

Dayal, B. (1944). *Proc. Indian Acad. Sci. A.*, **20**, 145.

Dean, P. (1964). *Proc. phys. Soc.*, **84**, 727.

Flubacher, P., A. J. Leadbetter, J. A. Morrison and B. P. Stoicheff (1959). *J. Physics Chem. Solids*, **12**, 53.

Gaskell, P. H. (1966). *Trans. Faraday Soc.*, **62**, 1493.

Hass, M., H. B. Rosenstock, and R. E. McGill (1969). *Solid State Communications*, **7**, 1.

Hindley, N. K. (1967). *Phys. Rev.*, **153**(3), 952.

Krause, J. T., and C. R. Kurkjian (1968). *J. Am. Ceram. Soc.*, **51**(4), 226.

Lazarus, D. (1949). *Phys. Rev.*, **76**, 545.

Leadbetter, A. J. (1968). *Physics. Chem. Glasses*, **9**(1), 1.

McSkimin, H. J., and P. Andreatch (1962). *J. acoust. Soc. Am.*, **34**, 609.

Rosenstock, H. B. (1961). *J. Physics Chem. Solids*, **23**, 659.

Rosenstock, H. B., and R. E. McGill (1968). *Phys. Rev.*, **176**(3), 1004.

Sauer, H. A. (1968). *Rev. scient. Instrum.*, **39**(4), 562.

Schuele, D. E., and C. S. Smith (1964). *J. Physics Chem. Solids*, **25**, 801.

Smyth, H. T. (1955). *J. Am. Ceram. Soc.*, **38**, 140.

Smyth, H. T. (1959). *J. Am. Ceram. Soc.*, **42**, 276.

Smyth, H. T., J. W. Londeree, and G. E. Lorey (1953a). *J. Am. Ceram. Soc.*, **36**, 238.

Smyth, H. T., H. S. Skogen, and W. B. Harsell (1953b). *J. Am. Ceram. Soc.*, **36**, 327.

White, G. K. (1961). *Cryogenics*, **1**, 151.

46

Thermal Properties of Polymers Below 4 °K

G. L. SALINGER

Department of Physics and Astronomy, Rensselaer Polytechnic Institute, Troy, New York, U.S.A.

Even at temperatures below 4 °K, the specific heat of an inorganic glass, as measured calorimetrically, shows significant departures from the Debye theory (Leadbetter, 1968). Although in some cases the predicted T^3 behaviour is found, the measured specific heat is anomalously large compared to that predicted from acoustic data using the Debye theory. In other cases departures from the T^3 behaviour are also observed. In general these 'excess' specific heats are attributed to the existence of low frequency non-acoustic modes in addition to the acoustic modes which give rise to the Debye contributions. Rosenstock (1962) has suggested that these low frequency modes can be explained by the presence of cavities in the disordered structure. The vibrating units which are contained in, or lie adjacent to, these cavities are bound only weakly to the rest of the lattice, thus they vibrate independently and with low frequency. On this model it is expected that excess specific heat should also be found in partly crystalline and completely amorphous organic polymers. Most previous specific heat measurements on polymers (Noer, 1959; Reese, 1965 and 1966) at low temperatures revealed a T^3 dependence. However, for polymethyl methacrylate (PMMA) and polystyrene (PS), Reese (1966b) found an excess specific heat by comparing the calorimetrically measured specific heat data between 1·2 and 4 °K with the specific heat derived from measurements of the velocity of sound extrapolated to 0 °K. Tucker and Reese (1967) also found a deviation from T^3 behaviour in polyethylene (PE). We have extended the specific heat measurements on various polymers to 0·4 °K (Choy, 1970 and unpublished work) and have also made measurements of the velocity of propagation of polarized sound waves between 4 and 150 °K in the same polymers. A considerable excess specific heat is found in the amorphous polymers. The strength and frequency of the localized modes can be readily determined. Evidence for these modes is also found in measurements of the thermal conductivity, internal friction and dielectric constant, (Miller, 1970).

Using the Rosenstock model, the specific heat of the polymers is regarded as the sum of a Debye acoustic term and an excess specific heat due to independent oscillators. The frequency and number of these oscillators can be estimated by invoking the Einstein theory of specific heats. At low temperatures the Debye contribution to the total specific heat is given by

$$\left(\frac{c}{T^3}\right)_{\mathrm{D}} = \frac{2\pi^2 k^4}{5\hbar^3 \rho}\frac{1}{v^3} = \frac{1223 \times 10^{15}}{\rho v^3}\frac{\mathrm{erg}}{\mathrm{gm}\ {}^\circ\mathrm{K}^4} \tag{1}$$

where T, $\hbar$, and k have their usual meanings, ρ is the density of the sample and v is the appropriate average velocity of sound defined by

$$\frac{1}{v^3} = \frac{1}{3}\left(\frac{1}{v_{\mathrm{l}}^3} + \frac{2}{v_{\mathrm{t}}^3}\right). \tag{2}$$

Here the subscript l refers to longitudinal sound waves and t refers to transversely polarized sound waves. The Einstein specific heat is given by

$$c_{\mathrm{E}} = N^*k\left(\frac{\theta}{T}\right)^2 \frac{\mathrm{e}^{\theta/T}}{(\mathrm{e}^{\theta/T} - 1)^2} = N^*kE\left(\frac{\theta}{T}\right) \tag{3}$$

where N^* is the number of modes of frequency ω per unit mass, $\theta(\equiv \hbar\omega/k)$ is the Einstein temperature and $E(\theta/T)$ is the Einstein heat capacity function, (Gopal, 1966). θ and N^* are chosen to give the best fit to the data when the Debye contribution is subtracted out. If there are oscillators of different frequencies, the excess specific heat is the sum of the contributions due to oscillators of each individual frequency.

SOUND VELOCITY

The Debye contribution to the specific heat was determined from equation (1) by measurements of the velocity of propagation of longitudinally and transversely polarized 10 MHz sound waves in the polymers. A single ended pulse echo technique was used. The cylindrical samples of about 2 cm diameter had a coaxial extension about 0·5 cm in diameter and were machined from commercial rod and sheet stock. The end surfaces of these cylinders were lapped parallel to 0·0005 cm and echoes were received by the transducer on the larger surface from the back surfaces of both the larger and smaller cylinder. The time delay between reception of these two echoes was measured using the delay feature of an oscilloscope which was calibrated by a frequency counter. The ratio of the lengths of two cylinders was chosen so that the echoes did not overlap. Since time differences are measured for echoes which traverse the bond the same number of times, time delays in the bond do not affect the measurement (Krause, 1963) and the measurements are reproducible to at least 0·1 %.

The velocities of the polarized sound waves in PMMA and PS between 4 and 150 °K are shown in Figure 1. The thermal contraction of the sample was corrected for by using the NBS Tables (Corruccini, 1961). The 0 °K

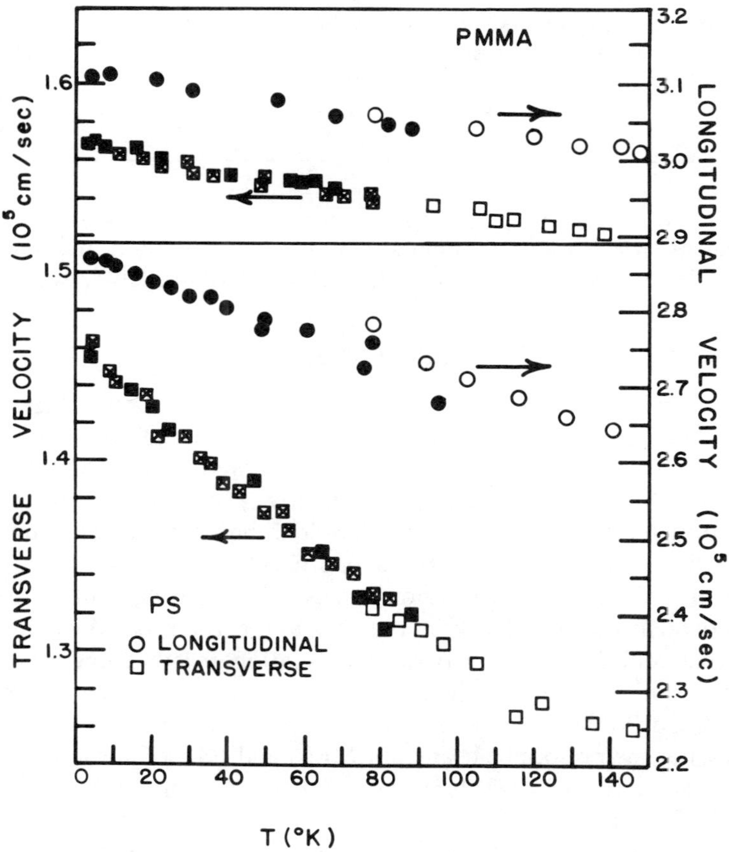

Figure 1. The velocity of longitudinally and transversely polarized 10 MHz sound waves in PMMA and PS below 150 °K

velocities and the Debye contribution to the specific heat as well as the 0 °K values for Young's modulus Y, the shear modulus G, the compressibility, K, and Poisson's ratio σ are given in Table 1. It should be noted that if thermal contraction is neglected and room temperature density, given in Table 1, is used, equation (1) gives the correct value for the Debye specific heat. The temperature dependence of Young's modulus is in excellent

Table 1. Acoustically determined parameters of PMMA and PS

Parameter		Temperature	PMMA	PS
v_l	10^5 cm/sec	0 °K	3·11	2·87
v_t	10^5 cm/sec	0 °K	1·57	1·46
$(c/T^3)_D$	$\frac{\text{erg}}{\text{gm °K}^4}$	0 °K	182	255
ρ	gm/cm^3	300 °K	1·185	1·031
Y	10^{10} dynes/cm^2	0 °K	7·75	4·51
G	10^{10} dynes/cm^2	0 °K	2·92	2·20
K	10^{10} dynes/cm^2	0 °K	7·49	5·58
σ		0 °K	0·327	0·326

agreement with that determined from internal friction measurements (Miller, 1970). The low temperature Grüneisen constants have a value near 3.

SPECIFIC HEAT

The calorimetric specific heat experiments were performed between 0·4 and 4·2 °K, (Choy, 1970b). The heat was applied electrically and the thermometry involved carbon resistors. The samples were connected to the He^3 evaporator by a thin neck of the same material, such that the thermal decay time through the neck was about one hundred times the thermal equilibrium time within the sample. This neck supported the sample and provided the means of cooling the sample without the use of exchange gas. In more recent measurements this neck has been replaced by a carbon rod which facilitates the initial cool down and increases the isolation at low temperatures. The samples were typically rod-shaped and had a mass of about 20 gm. The PMMA and PS samples were machined from commercial cast sheet.

The calorimetric specific heat c of PMMA and PS are shown in Figures 2 and 3 plotted as c/T^3 *vs.* T. It is obvious that c/T^3 is not constant between 0·6 and 4·2 °K. Furthermore the value of c/T^3 above 2 °K is almost twice the acoustically measured Debye value shown as line D in the figures. The calorimetric measurements agree with those of Noer (1958) to within 10%. The thermal diffusivity measurements of Reese (1966) yield values which are 30% higher. Our thermal diffusivity measurements are similarily high (Choy, 1970a).

Equation (1) avoids the difficulties associated with calculation of the Debye temperature for polyatomic solids (Anderson, 1959). However, assuming that a chemical repeat unit is a vibrating unit and that there are four modes of vibration active per repeat unit (the C–C stretching frequency is too high to be activated at these temperatures), the Debye temperatures

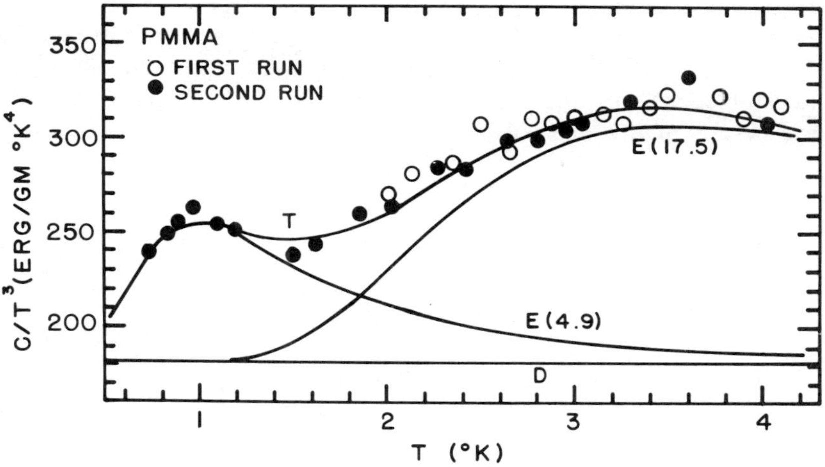

Figure 2. The specific heat of PMMA plotted as c/T^3 *vs. T. D* is the acoustically determined value of c/T^2 $E(4{\cdot}9)$ and $E(17{\cdot}5)$ are the Einstein contributions to c/T^3. T is the sum $D + E(4{\cdot}9) + E(17{\cdot}5)$. (Choy, Hunt and Salinger, 1970b)

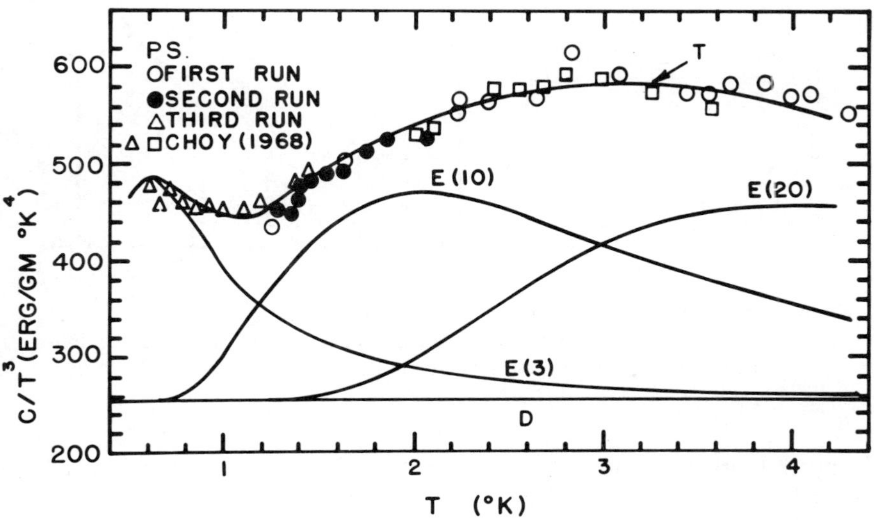

Figure 3. The specific heat of PS plotted as c/T^3 *vs. T.* The curves have the same meaning as for Figure 2. (Choy, Hunt and Salinger, 1970b)

are close to 100 °K. Thus the measurements are made in the region where the T^3 law should be valid.

The deviations of the specific heat from the Debye predictions can be explained by the existence of localized oscillators. As shown schematically in Figure 4, a pendant group can librate at very low frequency if the group is near or in a void. If the void is large enough that the weak forces responsible for interchain bonding no longer bind the group to its neighbours, the group librates independently and the conditions for the Einstein specific heat are met. The resultant frequency distribution of modes is shown schematically in

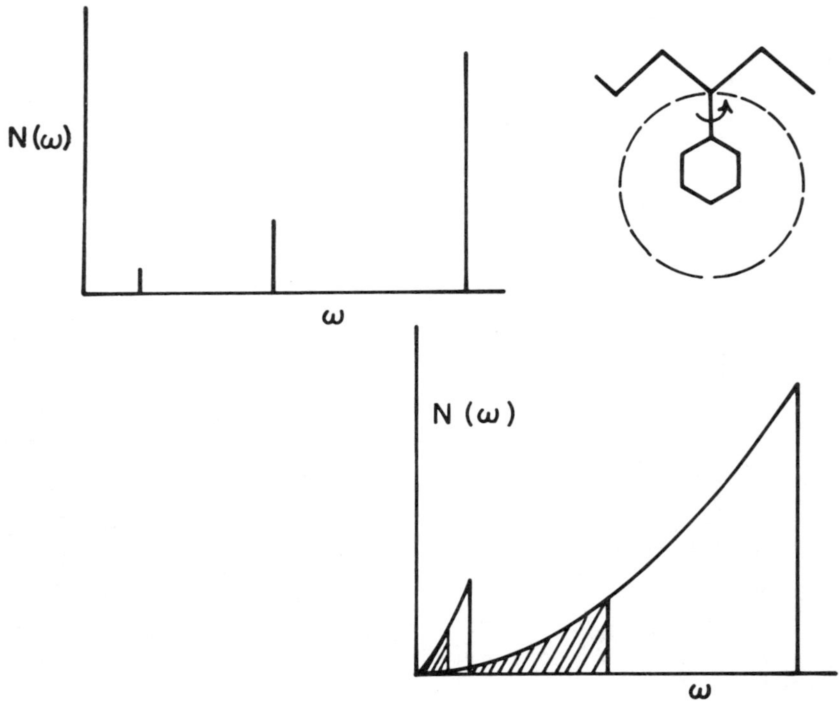

Figure 4. Models for the excess specific heat. The upper graph is an Einstein distribution of modes. The lower graph is a Debye approximation to a distribution of Einstein modes. The figure is a schematic of a librating pendant group

the upper graph of Figure 4. Assuming that the Debye contribution to c/T^3 is 182 and 255 erg $gm^{-1}\,°K^{-4}$ for PMMA and PS respectively, the experimental values for c/T^3 can be fitted by curve T to within a few percent by the addition of two modes for PMMA and three for PS. The contribution to c/T^3 for each mode is shown in Figures 2 and 3 as is the sum of all the contributions, curve T. The values for the Einstein temperatures θ_i and the number of modes per gram N_i^* necessary to fit the data are given in Table 2. Also given is the total number of acoustic modes per gram, $4N_0/M$. A very small number of low frequency modes ($\omega/\theta = 1{\cdot}31 \times 10^{11}$ radians $sec^{-1}\,°K^{-1}$) give rise to a large excess specific heat. The frequency and strength of these modes compares favourably with the modes responsible for relaxation processes in the measurements of internal friction and dielectric constant (Miller, 1970). In an earlier publication (Choy, 1970b) we assigned the modes in polystyrene ignoring the low temperature rise in c/T^3. However the large change in the sound velocity with temperature compared to PMMA (Figure 1), the lower value of the more accurately determined acoustic value of c/T^3, and the modes determined by Miller *et al.* (1970) convince us that indeed there is a lower temperature mode. A similar mode appears in other of our specific heat measurements.

Table 2. Einstein temperatures and strengths of modes contributing to the excess specific heat of PMMA and PS

Parameter	PMMA	PS
θ_1 °K	4·9	3
θ_2 °K	17·5	10
θ_3 °K		20
N_1^* modes/gm	$2{\cdot}9 \times 10^{18}$	$2{\cdot}1 \times 10^{18}$
N_2^* modes/gm	$2{\cdot}3 \times 10^{20}$	$7{\cdot}4 \times 10^{19}$
N_3^* modes/gm		$5{\cdot}5 \times 10^{20}$
$4N_0/M$ modes/gm	$2{\cdot}4 \times 10^{22}$	$2{\cdot}3 \times 10^{22}$

The use of Einstein modes is a convenience. The data can be fitted equally well by a series of Schottky modes. It is also not necessary to have all the oscillators vibrating at the same frequency. In fact it is more probable that there is a distribution of void sizes. Small voids will pin the oscillator. As the void size increases the frequency of the oscillator increases as it becomes increasingly free. When the void reaches a certain size the conditions for the Einstein mode are obtained and there is no further increase in frequency of the oscillator with void size. This is approximated, for mathematical convenience, by a three-dimensional Debye distribution of modes up to some frequency as shown in the bottom graph of Figure 4. The low frequency modes are cancelled out by subtracting the specific heat due to a Debye distribution up to some lower frequency indicated by cross hatching in Figure 4. The data for polystyrene can be fitted equally well by a Debye distribution of modes from 8 to 25 °K containing about $6{\cdot}2 \times 10^{20}$ modes/gm and an additional Debye mode at 5 °K for which no cut off could be determined from the data having a strength of $2{\cdot}7 \times 10^{18}$ modes/gm.

We have also made calorimetric specific heat measurements on polyethylene as a function of crystallinity (Figure 5). The value of the specific heat at 2·5 °K has been plotted as a function of density-determined crystallinity for various samples. The specific heat decreases with increasing crystallinity since the packing voids are associated with the amorphous material. However, the specific heat decreases markedly when the samples are crystallized under pressure, although the density of the sample is virtually unaffected. This is consistent with the increased pressure collapsing the packing voids and stopping the oscillation of pendant group. In polyethylene the pendant groups are due to short branches and impurities.

Pohl and coworkers (Harrison, 1968) have seen localized excess specific heats in doped alkali halides. In well annealed crystalline substances each

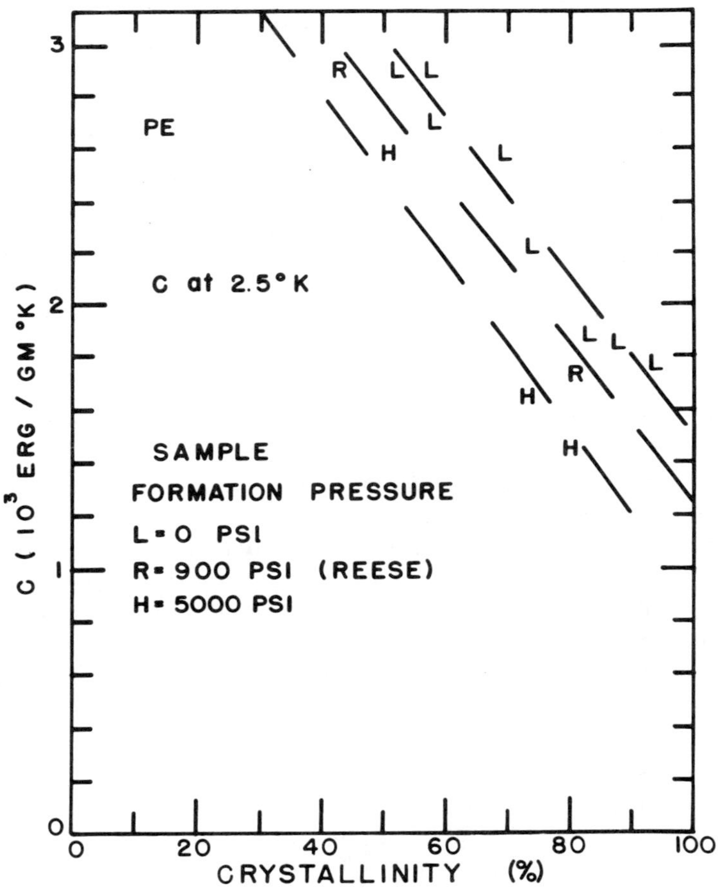

Figure 5. The specific heat of polyethylene at 2·5 °K as a function of density determined crystallinity. The symbols indicate the approximate pressure during crystallization. The points marked R are taken from Tucker and Reese (1967)

impurity sees the same environment and the peaks are much sharper. Thus, although these modes are interesting of themselves, they tell us little about the vitreous state *per se*.

THERMAL CONDUCTIVITY

The thermal conductivity of amorphous PMMA and PS and of polyethylenes of various crystallinities has been measured between 0·4 and 4 °K using a standard steady-state heat flow method (Choy, 1970a). The temperature difference between two points in a rod was measured by carbon resistance thermometry while heat was supplied electrically at one end. The data are shown in Figure 6. The measurements for PMMA and PS agree with those of others to within 10%, (Reese, 1966a; Berman, 1969). Data for SiO_2 are very similar (Fisher, 1968).

Although the temperature range of the measurements is low enough that the wavelength of the dominant thermal disturbance is long compared to the microscopic disorder, the temperature dependence of the thermal conductivity of these samples is probably not characteristic of an amorphous solid. The specific heat measurements indicate the existence of localized modes. These modes do not contribute to the energy flux but instead act as additional scatterers and decrease the heat flow. Berman *et al.* (1969) found that chemically increasing the cross-linking increases the thermal conductivity of PS above 1 °K. Reese (1968) has shown that cross-linking greatly decreases the number of local modes. Thus, it is likely that the explanation of the temperature dependence of the thermal conductivity of PMMA and PS at low temperatures is similar to that for doped alkali halides (Baumann, 1967) and that the Klemens (1951) theory has never been tested. Since the intrinsic temperature dependence of the thermal conductivity of an amorphous solid is not known, we cannot extract information about the scattering power of the local modes. It may be significant that the value of the thermal conductivity of the polymer is two orders of magnitude less than that of the doped alkali halides.

The thermal conductivity of many partly crystalline materials seems to follow a power law behaviour at low temperatures (Anderson, 1963). This behaviour is manifested by the samples of polyethylene shown in Figure 6. The more crystalline material shows a faster temperature dependence. The thermal conductivity of the 84·3% crystalline polyethylene varies as $T^{1·9}$, whereas the variation for the 58·7% crystalline sample varies as $T^{1·4}$ above 1 °K and as $T^{1·1}$ below 1 °K. Similar behaviour has been found in other polyethylene samples (our unpublished work; Giles, 1969) and we have also found this behaviour in carbons.

Similar temperature dependence found in the thermal conductivity of well polished crystal (Thacher, 1967) has been attributed to specular reflection from the surfaces. As polyethylene and carbon have a layered structure it may be that the dominant thermal resistance is specular reflection. The breaks in the curves may occur when the wave length of the thermal disturbance becomes larger than the layer thickness.

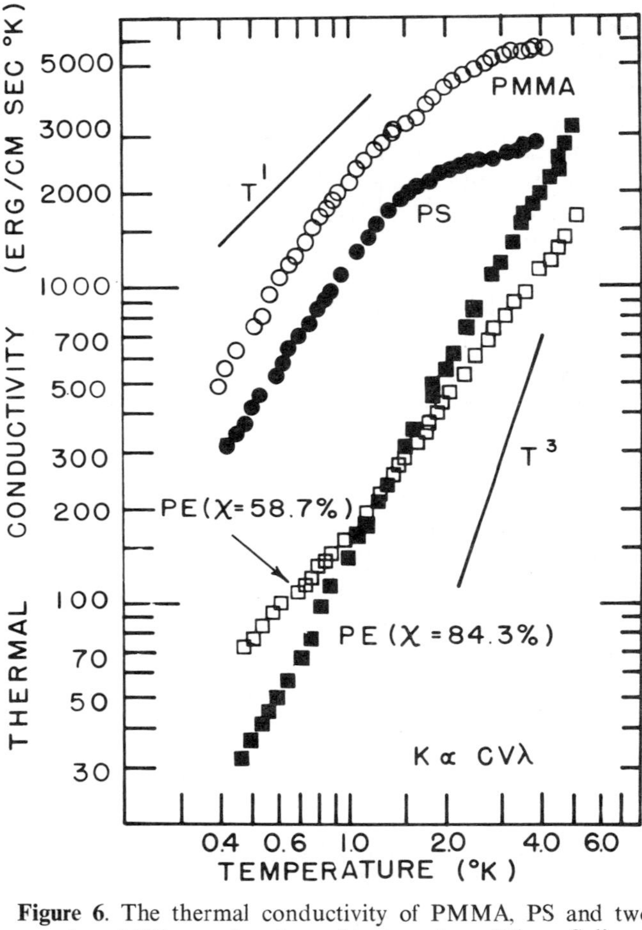

Figure 6. The thermal conductivity of PMMA, PS and two samples of PE as a function of temperature. (Choy, Salinger and Chiang, 1970a)

It appears that impurities are responsible for the apparent thermal properties of polymers and glasses at low temperatures. In order to determine the intrinsic behaviour of amorphous solids, measurements must be made at temperatures low enough to freeze out the local modes.

ACKNOWLEDGEMENTS

I would like to acknowledge the contributions of Y. C. Chiang, C. L. Choy, A. J. Cordell, R. G. Hunt and B. T. Peterson whose measurements are presented here. Dr. C. D. Swartz developed much of the technique for measuring sound velocities while on a subbatical leave from Union College. Discussions with R. K. MacCrone, T. Tomazawa and R. O. Pohl have been especially helpful. We are grateful for the financial support of the National Science Foundation, the Office of Naval Research and the National Aeronautics and Space Administration.

REFERENCES

Anderson, A. C., W. Reese, and J. C. Wheatley (1963). *Rev. scient. Instrum.*, **34**, 1386.
Anderson, O. L. (1959). *J. Physics Chem. Solids*, **12**, 41.
Baumann, F. C., J. P. Harrison, R. O. Pohl, and W. D. Seward (1967). *Phys. Rev.*, **159**, 691.
Berman, B. L., R. P. Madding, and J. R. Dillinger (1969). *Phys. Lett.*, **30A**, 315.
Choy, C. L., G. L. Salinger, and Y. C. Chiang (1970a). *J. appl. Phys.*, **41**, 597.
Choy, C. L., R. G. Hunt, and G. L. Salinger (1970b). *J. chem. Phys.*, **52**, 3629.
Corruccini, R. J., and J. J. Gniewek (1961). *Natl. Bur. Std. (U.S.)*, Monograph 29.
Fisher, R. A., G. E. Brodale, E. W. Hornung, and W. F. Giauque (1969). *Rev. scient. Instrum.*, **40**, 365
Giles, M., and C. Terry (1969). *Phys. Rev. Lett.*, **22**, 882.
Gopal, E. S. R. (1966). *Specific Heats at Low Temperatures.* Plenum Press, New York. p. 207.
Harrison, J. P., P. P. Peressini, and R. O. Pohl (1968). *Phys. Rev.*, **171**, 1037.
Klemens, P. G. (1951). *Proc. R. Soc.*, **A208**, 108.
Krause, J. T. (1963). *J. acoust. Soc. Am.*, **35**, 1.
Leadbetter, A. J. (1968). *Physics Chem. Glasses*, **9**, 1.
Miller, S. C., R. J. MacCrone, and M. Tomazawa (1970). This volume, p. 89.
Noer, R. J., C. W. Dempsey, and J. E. Gordon (1959). *Bull. Am. phys. Soc.*, **4**, 108.
Reese, W., and J. E. Tucker (1965). *J. chem. Phys.*, **43**, 105.
Reese, W. (1966a). *J. appl. Phys.*, **37**, 864.
Reese, W. (1966b). *J. appl. Phys.*, **37**, 3959.
Reese, W., P. J. Higgins, and G. W. Rostine (1968). *J. appl. Phys.*, **39**, 1800.
Rosenstock, H. B. (1962). *J. Physics Chem. Solids*, **23**, 659
Thacher, P. D. (1967). *Phys. Rev.*, **156**, 975.
Tucker, J. E., and W. Reese (1967). *J. chem. Phys.*, **46**, 1388.

FRACTURE

47

Moving Cracks

B. A. BILBY

Department of the Theory of Materials, University of Sheffield, Great Britain

In this paper I shall use some very simple models in an attempt to throw light on the phenomenon of fracture. Mostly, I shall report on the recent work on fast moving cracks carried out in the Department of the Theory of Materials here by my colleagues Drs. Eshelby and Atkinson, and their students, Messrs. Austwick and Fonseca. I am much indebted to them for many valuable discussions, and for making available to me their unpublished results.

Fracture, like fire, is best controlled. Usually we wish to prevent our bridges from collapsing and our ships from falling asunder. But equally it may be vital to make things break correctly, as when cleaving a precious diamond, laying tiles or paving stones, separating a booster from a rocket, or simply, when crushing or grinding. An old art requiring controlled fracture is flint knapping, already highly developed in the Sheffield area thirty thousand years ago at Creswell Crags; as will be shown an analysis of this process poses interesting problems.

CRACK PROPAGATION

What is toughness? If we loosely define it as the ability to resist crack propagation, we are at once led to examine the conditions which determine whether or not a crack will grow. Griffith (1920) was the first to discuss quantitatively the energy changes occurring in an elastic solid undergoing rupture. He recognized that a necessary condition for crack growth was that the total energy of the system consisting of the elastic solid including the crack surfaces, together with the apparatus applying the load, must decrease as the crack extended. In Figure 1 the crack (of length $2c$) extends so that $c \rightarrow c + \delta c$, under a constant stress P at infinity. There is a change $\delta\varepsilon^{\mathrm{POT}} = -\,\delta V$ in the potential of the loading system, where δV is the work done by P. The elastic energy of the body *increases* by $\delta\varepsilon^{\mathrm{EL}}$, and a theorem of linear elasticity shows (e.g. Eshelby, 1956) that

$$\delta\varepsilon^{\mathrm{POT}} = -2\delta\varepsilon^{\mathrm{EL}} \text{ and } \delta\varepsilon^{\mathrm{TOT}} = \delta\varepsilon^{\mathrm{POT}} + \delta\varepsilon^{\mathrm{EL}} = +\tfrac{1}{2}\delta\varepsilon^{\mathrm{POT}}.$$

If γ is the surface energy, the crack extends when

$$\delta\varepsilon^{\mathrm{TOT}} + 4\gamma\,\delta c = 0.$$

Thus, using the value

$$\delta\varepsilon^{\mathrm{TOT}} = -2\pi P^2(1-\nu)\,c\delta c/E$$

appropriate to plane strain, extension occurs when

$$P \geqslant [2E\gamma/\pi(1-\nu^2)\,c]^{\frac{1}{2}} \tag{1}$$

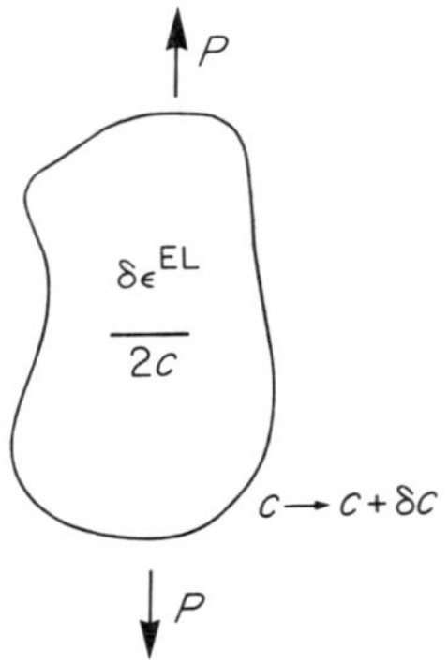

Figure 1. Crack extension

Here ν is Poisson's ratio and E Young's modulus. If γ is interpreted to include plastic work at the crack tip associated with the propagation of the fracture, the theory may be applied to quasi-brittle and other unstable fractures where the fracture energy is many times greater than the true surface energy (Irwin, 1948; Orowan, 1949; Cottrell, 1962).

The necessary condition (1) may not always be sufficient for fracture, for the latter cannot occur until atoms at the crack tip are separated. A brittle crack which is atomically sharp will have a cusp-like tip under stress, with atoms at all stages of separation; all along the crack edge there will be some atoms between which the stress has reached the theoretical maximum value. The stability of such a crack will therefore be governed essentially by the energy condition (1), which is the safest condition, in the sense that crack extension cannot occur if it is not satisfied. The investigation of the resistance of a material to the propagation of the sharpest possible cracks is therefore of great practical importance.

For simplicity we think in the following discussion of cracks with long straight edges parallel to x_3 (normal to the paper in the diagrams), so that their elastic fields are two-dimensional. In fact, if the crack edge is not too sharply curved, the conditions sufficiently near an element of this edge will approximate to those near an appropriate long straight crack lying tangent to this edge. We refer to the edge of the crack as the tip. Near it the stress p and crack opening Δu may be characterized by a constant K, called the stress intensity factor. If s is measured outwards from the crack tip into the material (Figure 2), then

$$p(s) = K(2\pi)^{-\frac{1}{2}}(s)^{-\frac{1}{2}} \tag{2}$$

$$\Delta u(-s) = (K/M)(8/\pi)^{\frac{1}{2}}(-s)^{\frac{1}{2}} \tag{3}$$

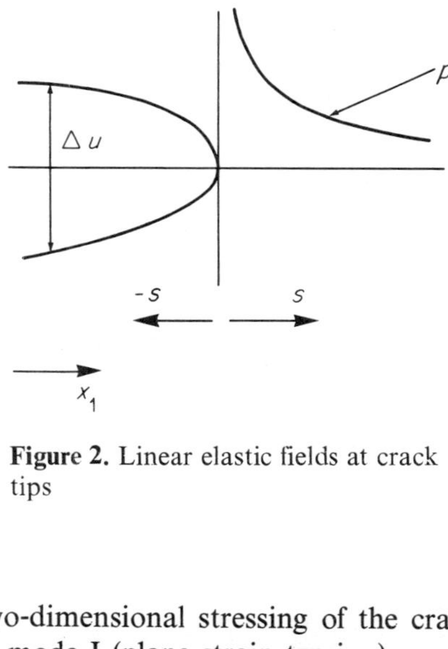

Figure 2. Linear elastic fields at crack tips

Certain types of two-dimensional stressing of the crack are commonly considered. These are mode I (plane strain tension), mode II (plane strain shear) and mode III (anti-plane strain). For these the interpretation of p, Δu and M is given by the following table (e.g. Eshelby, 1969a).

Mode	p	Δu	K	M
I	p_{22}	Δu_2	K_{I}	$E(1 - \nu^2)/2$
II	p_{12}	Δu_1	K_{II}	$E(1 - \nu^2)/2$
III	p_{23}	Δu_3	K_{III}	μ

Here μ is the shear modulus.

The value of K depends on the distribution of applied stress on the crack and on the geometry of the specimen. For a straight crack, $-c < x_1 < c$, $x_2 = 0$, $-\infty < x_3 < +\infty$, in an infinite medium

$$K = (\pi c)^{\frac{1}{2}} \int_{-c}^{+c} \left(\frac{c + x_1'}{c - x_1'}\right)^{\frac{1}{2}} P(x_1')\,\mathrm{d}x_1' \tag{4}$$

where $P(x_1)$ is the distribution of applied stress on $x_2 = 0$ in the absence of the crack (Bilby and Eshelby, 1968). If P is constant, $K = (\pi c)^{\frac{1}{2}}P$.

THE STATIC CRACK EXTENSION FORCE

As with other elastic singularities, the variation of $\varepsilon^{\mathrm{TOT}}$ with the position ξ of the crack tip can be used to define a force tending to move this tip (Irwin 1948). If $\varepsilon^{\mathrm{TOT}}$ refers to unit length of the crack edge (normal to the paper), the quantity

$$G = -\frac{\partial \varepsilon^{\mathrm{TOT}}}{\partial \xi} \tag{5}$$

is the total energy release per unit advance of the crack tip (Figure 3). It is called the *crack extension force*. In fact, since the crack may be represented by

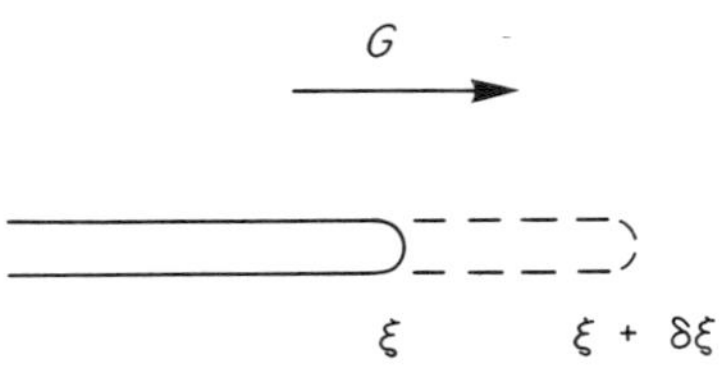

Figure 3. Crack extension force

a continuous distribution of dislocations ($D(x_1)\,\mathrm{d}x_1$ dislocations of Burgers vector b in $\mathrm{d}x_1$, where $bD = -\mathrm{d}(\Delta u)/\mathrm{d}x_1$), G is actually the total force on the dislocation at the tip (Bilby and Eshelby, 1968).

Irwin (1957) has used the following artifice to estimate G (Figure 4). Let unit length of the crack front advance by $\delta\varepsilon$. Then $-\delta\varepsilon^{\mathrm{TOT}}$ is the work required to close the crack front back to ξ again. Thus, using equations (3) and (4),

$$\begin{aligned} \delta\varepsilon^{\mathrm{TOT}} &= -\tfrac{1}{2}\int_0^{\delta\xi} p^{\xi}(s)\,\Delta u^{\xi+\delta\xi}(\delta\xi - s)\,\mathrm{d}s \\ &= -(K^2/2M)\delta\xi \end{aligned}$$

so that G is given in terms of the stress intensity factor K by

$$G = K^2/2M \tag{6}$$

The condition for the crack to extend is now that the crack extension force G be just equal to 2γ, the energy required to create the two new surfaces ahead of the tip (including plastic work if need be):

$$G = 2\gamma \tag{7}$$

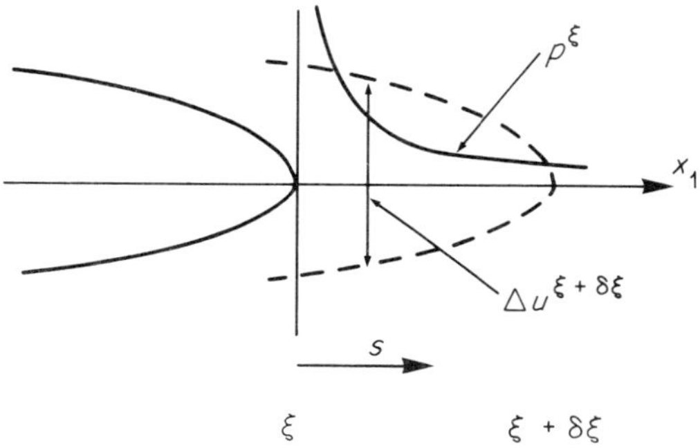

Figure 4. Closing an extended crack

The value $G = K^2/2M = \pi c P^2 (1 - \nu^2)/E$ for plane strain tension (mode I) gives again the formula (1). The fracture toughness of a material is measured by critical value G_c at which a sharp crack begins to extend; this fracture toughness is, of course, merely another name for the surface energy of fracture under the given conditions. Alternatively, K_c values are often quoted, although they have a less direct physical meaning.

If we draw a surface Σ round the crack tip (Figure 5), we see that the energy flowing across Σ from the surroundings is either stored within Σ, or is available at the tip to drive the crack. The general theory of forces on elastic singularities employing the elastic energy–momentum tensor (Eshelby, 1951) enables us to separate out the stored part and so to obtain an expression for G in the form of an integral over Σ. Eshelby shows that if all sources of internal stress and inhomogeneities within any surface Σ (Figure 6) are displaced by $\delta\xi_l$ then

$$-\delta\varepsilon^{\mathrm{TOT}} = F_l \delta\xi_l \tag{8}$$

$$F_l = \int_\Sigma P_{lj}\, \mathrm{d}S_j \tag{9}$$

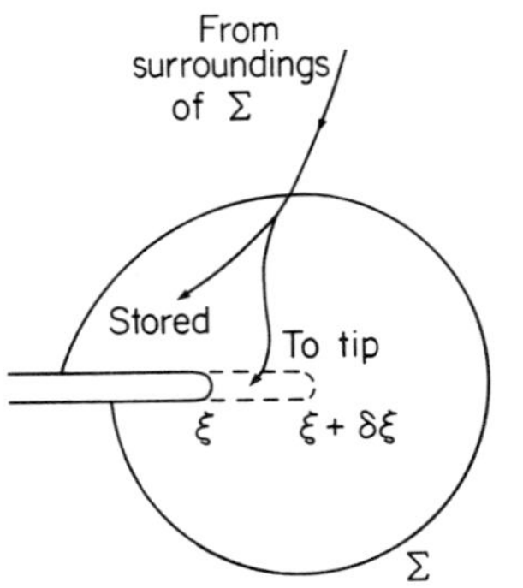

Figure 5. Energy flows

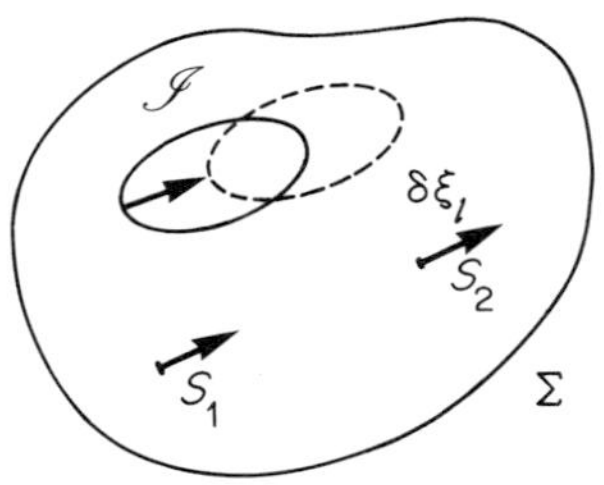

Figure 6. Elastic energy–momentum tensor.
$\mathscr{I}$—inhomogeneity
S_1, S_2—stress singularities

and

$$P_{lj} = W\delta_{lj} - p_{ij}\frac{\partial u_i}{\partial x_l}. \tag{10}$$

Here repeated suffixes are summed over the values 1, 2, 3, W is the elastic energy density, p_{ij} the stress, u_i the displacement and $dS_j = n_j dS$ where n_j is the outward normal to Σ. P_{lj} is the *spatial part of the elastic energy–momentum tensor* and F_l is the resultant force on all singularities within Σ. P_{lj} has the important property that its divergence vanishes

$$\frac{\partial P_{lj}}{\partial x_j} = 0. \tag{11}$$

Thus the integral (9) giving the force F_l is independent of the form of Σ; the surface Σ may be deformed in any manner, provided that the singularities remain inside it. Other integrals for F_l which are independent of Σ are

$$F_l = \frac{1}{2}\int_\Sigma \left(p_{ij}\frac{\partial u_i}{\partial \xi_l} - u_i\frac{\partial p_{ij}}{\partial \xi_l}\right) dS_j \tag{12}$$

$$= -\frac{1}{2}\int_\Sigma \left(p_{ij}\frac{\partial u_i}{\partial x_l} - u_i\frac{\partial p_{ij}}{\partial x_l}\right) dS_j \tag{13}$$

With suitable definitions of the stress and the use of Lagrangian coordinates the expressions (9), (12) and (13) for F_l are valid for non-linear materials undergoing finite deformation (Eshelby, 1956).

To obtain expressions for the crack extension force, we take the crack parallel to the x_1 axis and Σ (which we can now open at A, B) as in Figure 7.

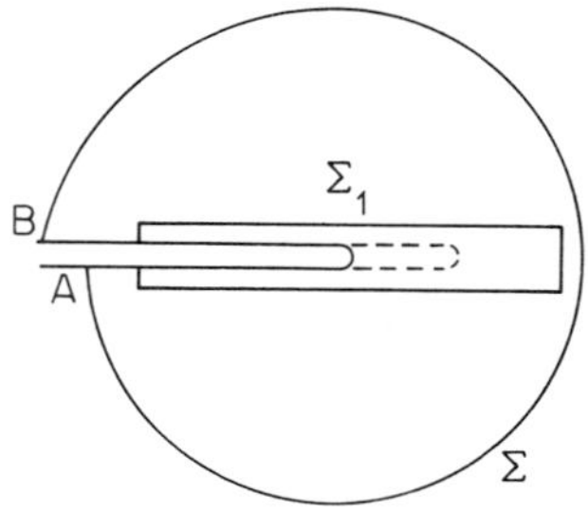

Figure 7. Crack extension force from the elastic energy–momentum tensor

Then (9), (12) or (13) with $l = 1$ give expressions for $G = F_1$ as integrals independent of Σ. Moreover, by taking the form Σ_1 for Σ and carrying out carefully a suitable limiting process we can justify Irwin's procedure (Eshelby, private communication).

The form (12) (with $l = 1$ and applied to a crack) was also given by Sanders (1960) and the form (9), again applied to a crack, by Rice (1968) and by Cherepanov (1968).

THE FORCE ON A MOVING CRACK

When the crack is in motion, rather more care is necessary in discussing the energy balance about its tip. A definitive account has recently been given in papers by Atkinson and Eshelby (1968) and Eshelby (1970). The energy flows are indicated in Figure 8 where we take a surface Σ which moves with the instantaneous velocity v of the crack tip. The total inward rate of energy flow $I(\Sigma)$ across Σ from the surroundings must equal the rate of increase of energy $\dot{\varepsilon}(\Sigma)$ stored within Σ plus the rate vG at which energy flows to the tip:

$$I(\Sigma) = vG + \dot{\varepsilon}(\Sigma) \tag{14}$$

$I(\Sigma)$ is the integral over Σ of the rate of working on the material inside Σ by the surroundings, $\int p_{ij}\dot{u}_i \mathrm{d}S_j$, plus the rate at which potential and kinetic energy is added to the interior of Σ at its leading boundary in region A, minus the similar rate of loss in region B. If the crack tip moves in the 1 direction and T is the kinetic energy density

$$I(\Sigma) = \int_\Sigma (p_{ij}\dot{u}_i + v(W + T)\delta_{1j})\, dS_j \tag{15}$$

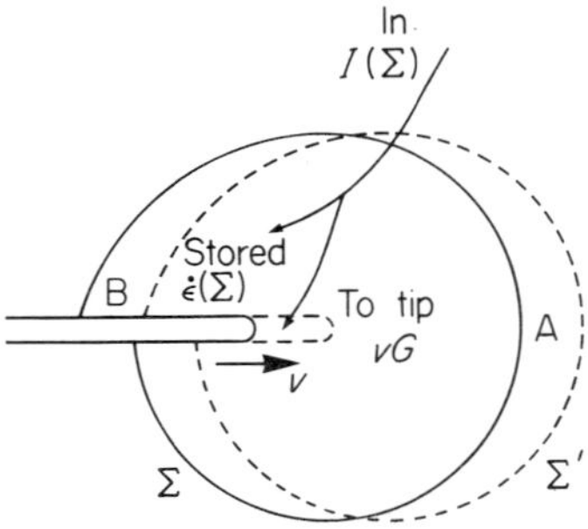

Figure 8. The force on a moving crack

Now suppose the dynamic elastic field is a special one which moves rigidly with the crack tip so that

$$u_i(x_1, x_2, t) = u_i^0(x_1 - vt, x_2) \tag{16}$$

Then $\dot{\varepsilon}(\Sigma) = 0$, $\dot{u}_i = -v\partial u_i/\partial x_1$ and we can write G as

$$G = \int_\Sigma H_{1j}\, dS_j \tag{17}$$

where

$$H_{1j} = (W + T)\,\delta_{1j} - p_{ij}\frac{\partial u_i}{\partial x_1} \tag{18}$$

Since $\partial H_{1j}/\partial x_j = 0$, the integral (17) is again independent of the shape of Σ.

For a general elastic field, the best that can be done is to write the field in the form

$$u_i = u_i^0(x_1 - vt, x_2) + u_i'(x_1, x_2, t)$$

and to arrange if possible that $u_i' \ll u_i^0$ near the tip. Then

$$G = \lim_{\Sigma \to 0} \int_\Sigma H_{1j}\, dS_j \tag{19}$$

Alternatively, if it is possible to show that

$$\lim\, [\dot{\varepsilon}(\Sigma)/I(\Sigma)] = 0,$$

then

$$vG = \lim_{\Sigma \to 0} I(\Sigma) \tag{20}$$

where $I(\Sigma)$ is given by equation (15).

The integral (19) is independent of Σ as $\Sigma \to 0$, but simple special fields can be used to show that it is not possible to obtain for any finite Σ a Σ-independent integral for G in the dynamic case (Atkinson and Eshelby, 1968). These authors have given a discussion equivalent to the above using a fixed surface Σ. They also verify that equation (20), when applied to the solution for a uniformly expanding crack in plane strain given by Broberg (1960, 1964), gives a G which vanishes at the Rayleigh velocity, as found by that author; and further they show that a different conclusion reached by Craggs (1963) arises because he neglected the convected energy terms in (15). The uniformly expanding crack in anti-plane strain has been treated by Austwick (1968); here the expression for G vanishes at the shear wave velocity c.

THE EQUATION OF MOTION OF A CRACK

Many discussions of moving cracks are available, but it is only very recently that a treatment has been given of the sudden starting of a finite crack, and of its subsequent motion with a proper energy balance at the moving tip. The equation of motion is obtained as follows (Eshelby, 1969b, 1969c). The crack is in anti-plane strain and the position of the right hand tip is denoted by $x_1 = \xi(t)$. The tip is at rest ($\xi = 0$) for $t < 0$ and thereafter moves in an arbitrary manner $\xi = \xi(t)$. The elastic field due to this arbitrary motion is calculated, and from it the crack extension force. G is shown to depend only on ξ and $\dot{\xi}$ but not on $\ddot{\xi}$, so that the crack has no inertia. Assuming that the fracture energy γ is known as a function of ξ and $\dot{\xi}$, the equation of motion which determines $\xi(t)$ is then

$$G(\xi, \dot{\xi}) = 2\gamma(\xi, \dot{\xi}) \tag{21}$$

It is convenient to discuss first the 'elementary' semi-inifinite crack, $-\infty < x < \xi, y = 0$ and to write $u_3 = w, x_1 = x, x_2 = y$. (Figure 9). Since w is harmonic, we can write $w = \operatorname{Im} f(z)$, where $z = x + iy$. The static crack with $\xi = 0$ has displacement $w = B \operatorname{Im} z^{\frac{1}{2}}$ or

$$w = BW_0(x, y) = B2^{-\frac{1}{2}} [(x^2 + y^2)^{\frac{1}{2}} - x)]^{\frac{1}{2}} \operatorname{sgn} y = Br^{\frac{1}{2}} \sin \tfrac{1}{2}\theta \tag{22}$$

where $B = (K/\mu)(2/\pi)^{\frac{1}{2}}$, K being the stress intensity factor. A theorem of Bateman states that if $\Omega(x, y)$ is harmonic and homogeneous of degree $\frac{1}{2}$ in x and y, then $\Omega[x - \xi(\tau), y] = w(x, y, t)$ is a solution of the wave equation $c^2\nabla^2 w = \partial^2 w/\partial t^2$. Here τ is the 'retarded time' defined implicitly as $\tau(x, y, t)$

by $c(t - \tau) = + [(x - \xi(\tau))^2 + y^2]^{\frac{1}{2}}$.* Thus $w = BW_0[x - \xi\{\tau(x, y, t)\}, y]$ is the solution for the elementary crack in motion, but it must be made manageable by a development of the implicit τ in terms of t. This may be effected by Lagrange's expansion (Whittaker and Watson, 1927), which gives

$$x - \xi(\tau) = x - \xi(t) + \sum_{n=1}^{\infty} \frac{1}{n!}\left(-\frac{\partial}{c\partial t}\right)^{n-1}\left[\frac{\dot{\xi}(t)}{c}\{R(t)\}^n\right]$$

where $R(t) = [\{x - \xi(t)\}^2 + y^2]^{\frac{1}{2}}$ is the distance from the moving tip.

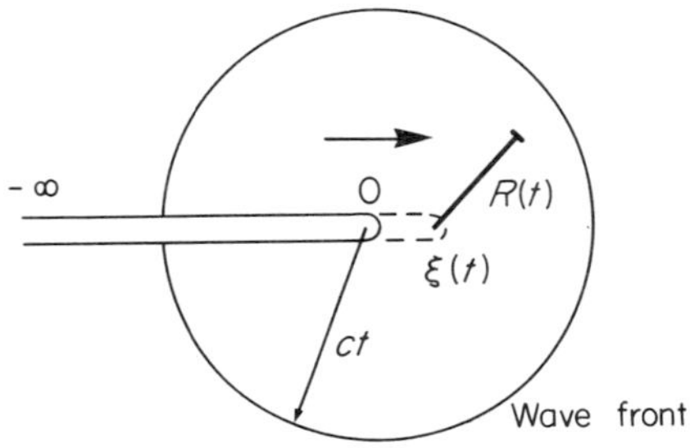

Figure 9. Elementary semi-infinite crack with one end in arbitrary motion ($\xi(0) = 0$)

Finally, then, after some reduction,

$$w = A(\dot{\xi})BW_0\left[\frac{x - \xi(t)}{\beta(\dot{\xi})}, y\right] + O(R^{\frac{3}{2}}) \tag{23}$$

where

$$\beta(\dot{\xi}) = (1 - \dot{\xi}^2/c^2)^{\frac{1}{2}} \tag{24}$$

$$A(\dot{\xi}) = \left(\frac{1 - \dot{\xi}/c}{1 + \dot{\xi}/c}\right)^{\frac{1}{4}} \tag{25}$$

If $\dot{\xi} = v =$ constant, $\xi = vt$ and the first term gives the exact solution.

We now pass to a crack of finite length, $- l < x < \xi(t)$, subjected to an arbitrary applied load (Figure 10). Suppose the static crack, $- l < x < 0$, solved and let $p^0_{yx}(x, 0)$, $x > 0$, be the stress ahead of it. Then the displacement for the crack with one end moving arbitrarily is identical with (23) except that the constant B is now replaced, for $\xi > 0$, by

$$B(\xi) = \frac{2}{\pi\mu}\int_0^{\xi} p^0_{yz}(\xi', 0)(\xi - \xi')^{-\frac{1}{2}}\, d\xi' \tag{26}$$

* If a point moves along Ox so that $x = \xi(t)$, then a signal received at (x, y) at time t was emitted at time τ.

For uniform loading $B(\xi)$ involves complete elliptic integrals.

We now define the dynamic stress intensity factor $K(\xi, \dot{\xi})$ by requiring that

$$p_{yz}(x, 0) = K(\xi, \dot{\xi})/[2\pi(x - \xi)]^{\frac{1}{2}} \tag{27}$$

for $0 < x - \xi \ll l$. Then for the crack $-l < x < \xi(t)$, with arbitrary loading,

$$K(\xi, \dot{\xi}) = (\pi/2)^{\frac{1}{2}} \mu(1 - \dot{\xi}/c)^{\frac{1}{2}} B(\xi). \tag{28}$$

The crack extension force may be calculated using (15) and is

$$G(\xi, \dot{\xi}) = (\pi\mu/4) A^2(\dot{\xi}) B^2(\xi) \tag{29}$$

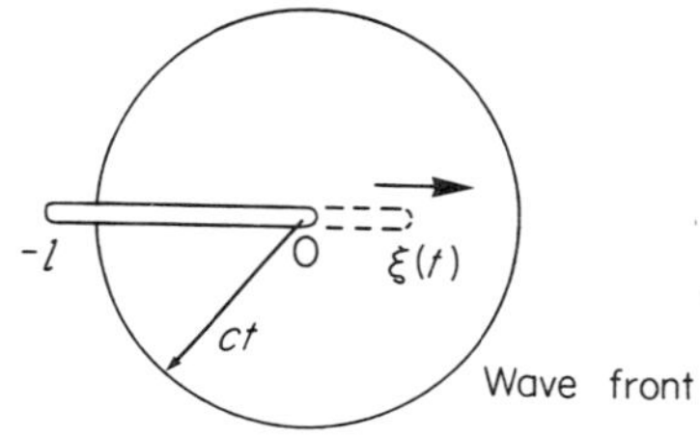

Figure 10. Finite crack, $-l<x<\xi(t)$, under arbitrary load, with one end in arbitrary motion ($\xi(0)=0$)

Because the leading term in w does not involve $\ddot{\xi}$, K and G depend only on ξ and $\dot{\xi}$, as stated above. Moreover, G/K depends on $\dot{\xi}$, and the static relation (6) no longer holds. An equation of motion in which $K(\xi, \dot{\xi})$ = constant, proposed by Kostrov (1966), thus does not give a correct energy balance unless γ varies with velocity so that the equation is equivalent to (21).

The general form of the crack motion predicted by equation (21) can be discerned by considering the crack $-l < x < \xi(t)$ expanding under uniform stress P. For this it may be shown that $B(\xi) = B_{ST}(\xi) f(\xi/l)$ where $B_{ST}(\xi) = P(l + \xi)^{\frac{1}{2}}/\mu$ is the value of B for a static crack with the instantaneous length $(l + \xi)$, and $f(\xi/l)$ is a complicated expression which increases from 1 to 1·27 as ξ/l varies from 0 to ∞. Thus equation (21) becomes $A^2(\dot{\xi}) B^2(\xi) = 8\gamma/\pi\mu$ where $B(\xi)$ increases steadily with ξ. If γ is constant therefore, $A(\dot{\xi})$ must decrease, and hence $\dot{\xi}$ must increase. The crack thus accelerates towards the limiting velocity $\dot{\xi} = c$, when $A(\dot{\xi})$ and G become zero.

The form of G has interesting implications for the theory of crack branching; indeed it provides the first explanation of how branching in a homogeneous medium can occur without violating the conservation of energy. For the overall energy release rate must be 2γ just before branching and, disregarding geometrical factors, approximately 4γ immediately after (Figure 11(a)). The factor $A^2(\dot{\xi})$ in G shows how the crack can achieve this miracle, since $A^2(\dot{\xi})$ increases if the crack velocity drops (Figure 11(b)). The maximum jump in $A^2(\dot{\xi})$ is achieved if $\dot{\xi}$ falls momentarily to zero so that the equation $A^2(0) = 2A^2(\dot{\xi}_b)$ determines a critical velocity $\dot{\xi}_b = 0{\cdot}6\,c$ below which the crack cannot branch. Of course, the geometry and stress field at the tip will doubtless influence the details of the branching process, but any theory of branching must include energy considerations of this kind.

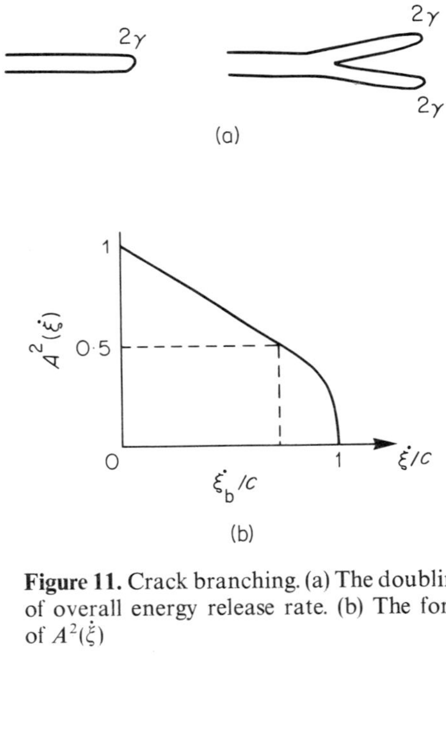

Figure 11. Crack branching. (a) The doubling of overall energy release rate. (b) The form of $A^2(\dot{\xi})$

Before leaving these interesting results on the fields of moving cracks it should be emphasized that the solutions here given are exactly valid only for a limited time—until the original tip field is disturbed by the arrival of the first reflected wave from the far end of the crack. For a discussion of these very complicated effects and of comparisons between results disregarding reflections and results including the effect of an infinite number (which suggest that the factors $A(\dot{\xi})$ and $B(\dot{\xi})$ are not very greatly changed), the reader is referred to the original papers.

FLINT KNAPPING

This is a process not of archaeological interest only; it has implications for crushing, grinding, blasting, impact testing of joints and the phenomenon of shattering. The discussion here is based on unpublished work by Atkinson, Eshelby and Fonseca, shortly to appear in the International Journal of Fracture Mechanics.

A thin shaving is removed from the surface of a flint (Figure 12) either by a steady pressure in *pressure flaking*, or by a sharp blow in *percussion flaking*. In both processes the external force is highly localized and the crack, being very near a free surface, is greatly influenced by reflections of the disturbance it creates.

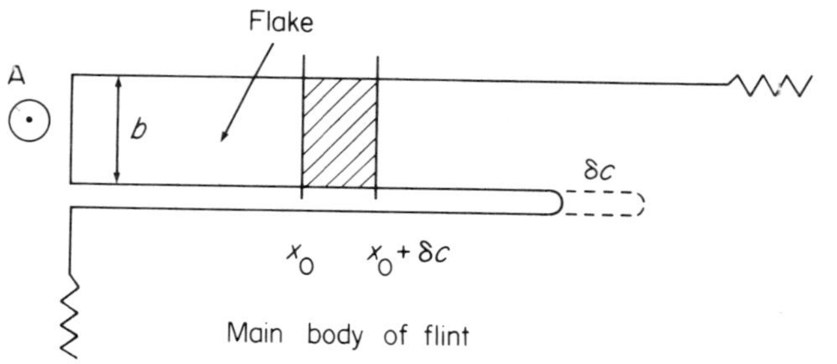

Figure 12. Flaking. (For the moving crack $\delta c = v\delta t$)

An approximate treatment of pressure flaking may be given by elementary arguments. (For simplicity, we limit our discussion to anti-plane strain, so that a force (parallel to $+z$) is applied upwards from the paper at A in Figure 12). Let x_0 be a point about $2b$ to the left of the crack tip. Then when $x < x_0$, there is a uniform stress $p_{xz} = -\sigma$ in the flake and

$$w(x) = -\sigma(x - x_0)/\mu \tag{30}$$

If the crack advances by δc, the point x_0 moves δc to the right and the increase in elastic energy $\delta\varepsilon^{EL}$ is that contained in the shaded volume. If δV is the work done by the external force, which moves through $\delta w = (\partial w/\partial x_0)\,\delta c = \sigma\delta c/\mu$, then

$$\delta V = G\delta c + \delta\varepsilon^{EL}$$

so that, taking unit thickness normal to the paper, $\sigma^2 b\delta c/\mu = G\delta c + (\sigma^2/2\mu)b\delta c$ and we have

$$G = \sigma^2 b/2\mu; \qquad K = (2\mu G)^{\frac{1}{2}} = \sigma b^{\frac{1}{2}} \tag{31}$$

The efficiency η_P is the fraction of the work done used in extending the crack

$$\eta_P = G\delta c/\delta V = \tfrac{1}{2} \tag{32}$$

If the crack moves to the right with velocity v, so will the field (30) in the flake and we can take

$$w\ (x,t) = -\sigma(x - vt)/\mu \tag{33}$$

Repeating the above argument, as the crack advances by $\delta c = v\delta t$, a volume $b\delta c$, with kinetic energy δM (density $T = \frac{1}{2}\rho\dot{w}^2$) and elastic energy $\delta\varepsilon^{EL}$ (density $\sigma^2/2\mu$), is added to the flake, while the external force does work $\sigma^2 b\delta c/\mu$ as before. Since $\rho = \mu/c^2$,

$$G\delta c = \delta V - \delta(\varepsilon^{EL} + M)$$

$$= \frac{\sigma^2 b\delta c}{\mu} - \left(\frac{\sigma^2 b\delta c}{2\mu} + \frac{\sigma^2 v^2 b\delta c}{4\mu c^2}\right),$$

and so,

$$G = \frac{\sigma^2 b}{2\mu}\left(1 - \frac{v^2}{c^2}\right) \tag{34}$$

The efficiency

$$\eta_P = vG/(\sigma^2 vb/\mu) = \frac{1}{2}\left(1 - \frac{v^2}{c^2}\right) \tag{35}$$

decreases with velocity.

It is interesting to confirm the expressions (31) and (34) by integrating the energy–momentum tensor. For the static crack,

$$G = \int_\Sigma \left(W\delta_{1j} - p_{zj}\frac{\partial w}{\partial x}\right) dS_j$$

where we choose Σ to be the contour ABCDE (Figure 13). On ABC the integral must vanish, since it does so if B, C $\to \infty$ and is independent of Σ On CD, $j = 2$ and $p_{zy} = 0$ so that here it vanishes also. On DE, $dS_j = -dS_1 = -dy$ and so

$$G = \int_{DE} \left(W - p_{zx}\frac{\partial w}{\partial x}\right)(-\,dy) = -\frac{\sigma^2 b}{2\mu} + \frac{\sigma^2 b}{\mu} = \frac{\sigma^2 b}{2\mu}.$$

For the dynamic case, (20) gives

$$vG = \lim_{\Sigma\to 0}\int_\Sigma \{v(W + T)\,\delta_{1j} + p_{zj}\dot{w}\}\, dS_j.$$

Since $w = w(x - vt)$, $\dot{w} = -v(\partial w/\partial x)$, and G is given by an integral like (17), independent of Σ Taking again ABCDE for Σ (Figure 13), only DE gives a non-zero contribution and

$$G = \int_{DE} \left(\frac{\sigma^2}{2\mu} + \frac{\sigma^2 v^2}{2\mu c^2} - \frac{\sigma^2}{\mu} \right) (-\,\mathrm{d}y) = \frac{\sigma^2 b}{2\mu} \left(1 - \frac{v^2}{c^2}\right).$$

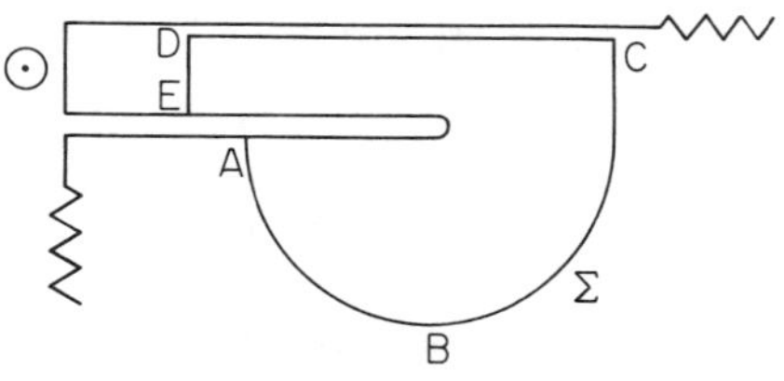

Figure 13. G in flaking from the energy–momentum tensor

The above approximate discussion can be confirmed by exact solutions. For the static crack the analogy between w and the velocity potential ϕ in hydrodynamics enables us to use at once the solution corresponding to the two-dimensional 'Borda's mouthpiece' (Lamb, 1932) so that $w = \mathrm{Re}(\zeta)$ where $z = x + \mathrm{i}y = \zeta + \mathrm{e}^{\zeta}$. Near the crack tip w can be shown to have the form $(\sigma/\mu)(2b/\pi r)^{\frac{1}{2}} \sin \frac{1}{2}\theta$, where r, θ are polar coordinates with the tip as origin; this gives $K = \sigma b^{\frac{1}{2}}$ as in equation (31).

The uniformly moving solution can then be obtained by applying to the static $w(x, y)$ the Lorentz transformation, $x \to (x - ct)/\beta$, $y \to y$, where

$$\beta = \left(1 - \frac{v^2}{c^2}\right)^{\frac{1}{2}},$$

and multiplying the result by β to maintain the applied load at σb. We then find that $K(v) = (1 - v^2/c^2)^{\frac{3}{4}} K(0)$, where $K(0) = \sigma b^{\frac{1}{2}}$, the static value (31). Moreover

$$w = L(v)\,(2\pi)^{-\frac{1}{2}}\,(vt - x)^{\frac{1}{2}} + \mathrm{O}[(vt - x)^{\frac{3}{2}}]$$

and

$$G = \tfrac{1}{8} K(v)\, L(v) \tag{36}$$

where

$$L(v) = (4/\mu)\,(1 - v^2/c^2)^{\frac{1}{4}} K(0),$$

which confirms (34).

The problem of percussion flaking requires rather more elaborate methods.

We suppose that a pulse

$$w_a = -(c/\mu)f\left(t - \frac{x}{c}\right)H\left(t - \frac{x}{c}\right) \tag{37}$$

is applied to the flake and that the crack starts to move with constant velocity v when the pulse first hits it at $t = 0$. Here $H(\lambda) = 1$ for $\lambda > 0$, $H(\lambda) = 0$ for $\lambda < 0$. This problem may be solved using the Wiener–Hopf technique and G obtained in terms of a function $Q(t)$. whose Laplace transform $\bar{Q}(p)$ is an infinite product:

$$G = \frac{2b}{\mu(1 - v^2/c^2)}[Q(t)]^2 \tag{38}$$

An explicit expression for G can only be found by inverting $\bar{Q}(p)$, but the energy E_{ABS} absorbed from the pulse by the crack tip can be obtained without this inversion by using Parseval's theorem. Thus

$$\begin{aligned} E_{ABS} &= \int_0^\infty vG(t)\,dt \\ &= \frac{2bv}{\mu(1 - v^2/c^2)}\int_0^\infty [Q(t)]^2\,dt. \end{aligned}$$

By Parseval's theorem this becomes,

$$E_{ABS} = \frac{2bv}{\mu(1 - v^2/c^2)}\frac{1}{2\pi i}\int_{d-i\infty}^{d+i\infty} \bar{Q}(p)\bar{Q}(-p)\,dp.$$

and finally, after some reduction,

$$E_{ABS} = \frac{2b}{\pi\mu(1 + v/c)}\int_0^\infty |i\lambda\bar{f}(i\lambda)|^2 \exp\left(-\frac{\delta^{\frac{1}{2}}\lambda b}{c}\right)d\lambda \tag{39}$$

where $\delta = (1 - v/c)/(1 + v/c)$. The form (39) is convenient for investigating the effects of different shapes of the pulse $f(t - x/c)$. For a uniform stress pulse $-\sigma$ of length a in space (or a/c in time), we may show that

$$E_{ABS} = \frac{2ab(v/c)}{1 + v/c}\frac{\sigma^2}{\mu}h(a/(b\delta^{\frac{1}{2}}))$$

where the function $h(x)$ tends to x/π for small x and to 1 when x is large. For a long pulse ($a \gg b$), $h(x) = 1$. We can now define an average G, G_{AV}, by setting

$$G_{AV}\bar{x} = E_{ABS}$$

where $\bar{x} = av/(c - v)$ is the distance moved by the crack during the time $a/(c - v)$ that it interacts with the pulse. This gives

$$G_{AV} = \frac{2\sigma^2 b}{\mu} \frac{1 - \dfrac{v}{c}}{1 + \dfrac{v}{c}} \tag{40}$$

Since the injected energy E_I in the pulse is half potential and half kinetic it is

$$ab\left(\frac{\sigma^2}{2\mu} + \frac{\sigma^2}{2\mu}\right) = ab\sigma^2/\mu$$

and the efficiency is

$$\eta_{PER} = \frac{E_{ABS}}{E_I} = \frac{2(v/c)}{1 + (v/c)} h(a/(b\delta^{\frac{1}{2}})) \tag{41}$$

In contrast to η_P (equation (35)), $\eta_{PER} \to 1$ as $v \to c$. This is because in percussion flaking the velocity imparted at A (Figure 12) by the blow depends only on the mechanical impedance of the material, and the injected energy is independent of the crack speed. The efficiency is thus greatest at high crack speeds when the crack absorbs most energy. In pressure flaking on the other hand, the velocity at A is proportional to the crack speed and $\dot{E}_{ABS}$ and vG tend to zero together, η_P remaining finite.

The analysis reviewed here has treated only linear elastic material in antiplane strain. Nevertheless some of the important physical principles governing the behaviour of fast moving cracks have been indicated. We have shown particularly how equations of motion which conserve energy may be set up, and how behaviour may differ under steady forces and under percussion. The extension of the work to other materials and geometries remains an interesting and challenging problem.

REFERENCES

Atkinson, C., and J. D. Eshelby (1968). *Int. J. Frac. Mech.*, **4**, 3.

Austwick, A. (1968). *Dissertation for M.Sc. (Tech.)*, University of Sheffield.

Bilby, B. A., and J. D. Eshelby (1968). In H. Liebowitz (Ed.), *Fracture, An Advanced Treatise*, Volume I. Academic Press. p. 99.

Broberg, K. B. (1960). *Ark. Fys.*, **18**, 159.

Broberg, K. B. (1964). *J. appl. Mech.*, **31**, 546.

Cherepanov, G. P. (1968). *Int. J. Solids Structures*, **4**, 811.

Cottrell, A. H. (1962). In D. J. Littler (Ed.), "Properties of Reactor Materials and the Effects of Radiation Damage". Butterworths, London. p. 5.

Craggs, J. W. (1963). In *Fracture of Solids*. Interscience, New York. p. 51.

Eshelby, J. D. (1951). *Phil. Trans. R. Soc. Ser. A*, **244**, 87.

Eshelby, J. D. (1956). *Solid St. Phys.*, **3**, 79.

Eshelby, J. D. (1969a). In *Fracture Toughness*. ISI Publication 121. p. 13.

Eshelby, J. D. (1969b). In A. S. Argon (Ed.), *Physics of Strength and Plasticity*. MIT Press. p. 263.

Eshelby, J. D. (1969c). *J. Mech. Phys. Solids*, **17**, 177.
Eshelby, J. D. (1970). In M. F. Kanninen *et al.* (Eds.), *Inelastic Behaviour of Solids*. McGraw-Hill. p. 77.
Griffith, A. A. (1920). *Phil. Trans. R. Soc.*, **221**, 163.
Irwin, G. R. (1948). In *Fracturing of Metals*. ASM, Cleveland. p. 147.
Irwin, G. R. (1957). *J. appl. Mech.*, **24**, 361.
Kostrov, B. V. (1966). *J. appl. Math. Mech.* (*PMM*), **30**, 1042.
Lamb, H. (1932). *Hydrodynamics*, C.U.P.
Orowan, E. (1949). *Rep. Prog. Phys.*, **12**, 214.
Rice, J. R. (1968). *J. appl. Mech.*, **35**, 379.
Sanders, J. L. (1960). *J. appl. Mech.*, **27**, 352.
Whittaker, E. T., and G. N. Watson (1927). *Modern Analysis*. C.U.P. p. 132.

48

Mechanisms of Slow Crack Propagation in Glass

J. R. VARNER and V. D. FRECHETTE

S.U.N.Y. College of Ceramics, Alfred University, Alfred, New York, U.S.A.

In previous work on slow fracture, i.e., velocities less than one metre per second, the emphasis has been primarily on developing techniques to measure the fracture surface energies of glasses, polymers, and crystals which depend on the energy balance criterion of Griffith (1920). Examples of such efforts have been reported by Roesler (1956), Benbow and Roesler (1957), Gilman (1960), Benbow (1961), Berry (1963), Linger and Holloway (1968), and Wiederhorn (1966, 1969).

Other research efforts have been directed toward environmental effects upon glass subjected to stress. Such stress corrosion studies have been performed by Wiederhorn (1966, 1967) and by Schönert, Umhauer and Klemm (1969).

Obtaining controlled slow crack propagation in brittle materials is a problem that has been most often solved by using a double cantilever system (Benbow and Roesler, 1957; Benbow, 1961; Gilman, 1960; Berry, 1963; Linger and Holloway, 1968; Wiederhorn, 1966, 1967, 1969). Although this technique has been used with success in various glasses and plastics, a double cantilever system has a serious drawback, because it produces a crack which is inherently directionally unstable (Cotterell, 1966; Berry, 1963). In an isotropic material, therefore, it is necessary to provide some means of controlling the crack direction. Commonly used techniques are (1) guidance by grooves or scratches down the centre (Berry, 1963; Linger and Holloway, 1968; Wiederhorn, 1966, 1967, 1969); (2) the application of compressive loading parallel to the propagation direction (Benbow and Roesler, 1957; Benbow, 1961), and (3) special sample geometries. However, the complication can be avoided by using a tensile device, because Cotterell (1966) has shown that a single edge crack in a plate is inherently directionally stable when subjected to uniaxial tension.

Wiederhorn (1966, 1967) and Schönert, Umhauer and Klemm (1969) in their stress corrosion investigations have shown that there are three distinct regions of crack velocity behaviour when water is present during the experiment. According to Wiederhorn (1966, 1967) and using his nomenclature, in Region I the velocity is exponentially dependent on applied stress, and is controlled by the rate of reaction of water at the crack tip. In Region II, crack propagation is limited by the rate of water vapour transport to the crack tip and is independent of applied stress. Eventually the velocity becomes so great that motion is independent of the water vapour concentration; this is Region III.

Schönert, Umhauer and Klemm (1969) have shown that the crack front changes shape in Region II. Normally, it is concave with respect to the origin of fracture, but it becomes convex when the rate of water vapour transport is the controlling factor for crack propagation.

Most of these past efforts have neglected the details of the mechanism by which a crack propagates through a material at slow velocities. It was the purpose of the present study to determine the mechanisms of slow crack propagation in glass by making use of the techniques of fractography that have been so effectively employed in investigating fast fracture.

EQUIPMENT AND PROCEDURE

It was essential for the success of the experiments to have a device that produced cracks whose velocity could be controlled and whose direction of propagation was stable. A tensile device of new design was built, and proved to be effective in producing directionally stable cracks without the application of external constraints or alterations to the samples. The specimen, which is a rectangular glass sheet, is cemented to two metal bars which are in tight contact with two other vertical metal bars whose axes are parallel to the long axis of the sample. These vertical bars are machined to the same length and can be heated to the same temperature. Normal procedure is to heat the bars to some temperature and then to maintain this temperature throughout the experiment, thus providing a constant strain condition. Control of the propagation rate is possible by adjusting the temperature of the heated bars.

In order to produce controlled propagation, it is also essential to have a single large crack at the start of the experiment. An edge crack was induced by heating the specimen to about 200 °C, followed by water quenching at a prescratched line along one edge. Cracks produced in this manner were usually straight and extended about 1–3 cm into the sample. This technique is identical to that mentioned by Kerkhof and Richter (1969).

Control of the amount of water vapour in the environment surrounding the samples was obtained by enclosing the samples in polyvinyl chloride (PVC) bags. A flow system allowed a variety of environmental conditions to be maintained, the gas being introduced through one opening in the bag and released through another. Normal procedure was to establish sufficient flow to obtain positive pressure in the PVC bag, and then to purge the system for two hours prior to the start of an experiment. Using high-purity nitrogen as the carrier gas, environments with dew points between -85 °C and 23 °C were readily obtained by bubbling the nitrogen through various solutions.

During experiments, crack progress was monitored with a 60× microscope. Velocity was determined by converting the mechanical motion necessary to maintain the crack tip centred at the crosshairs of the eyepiece into an electrical potential which was recorded on a moving chart. The time–distance relationships thus obtained were then used to calculate crack velocities. This system was found useful in the range 10^{-8} to 10^{-3} m/s.

After an experiment was completed aluminium was vapour plated on one of the newly-exposed fracture surfaces. This coating greatly aids microscope observations and is essential if interferometry is to be used to its full potential.

Post-experimental observations were performed using a universal optical microscope (Leitz PANPHOT). A variety of illuminations have been used, including bright- and dark-field incident light and transmitted light (on uncoated samples). Interferometry using Kohaut double-beam and Tolansky multiple-beam systems has also been employed.

Samples were 50 × 150 × 1 mm and 50 × 150 × 2 mm glass plates. The 1 mm samples were a soda-lime-silica window glass, and the 2 mm samples were an aluminosilicate glass (Corning code 0318 unstrengthened CHEMCOR® glass).

RESULTS AND DISCUSSION

The described method of inducing a single edge crack in the sample usually leaves a crack front that is slightly concave with respect to the crack origin. Frequently, partial closing of this crack occurred prior to the start of the experiment. While this had the appearance of crack healing, it of course merely indicated that relaxation of the glass during cooling had caused the distance between crack surfaces to become so small that total internal reflection no longer occurred. As stress was applied, the reclosed portion could be seen to reopen readily until the true end of the crack was reached.

New crack movement did not usually begin along the entire length of the original front at the same time. Extension took place along one edge first, with a gradual progression along the original front until a new front was established across the entire width of the sample. Often, subtle lines on the fracture-exposed surface showed this pattern quite clearly. When this behaviour occurred, the resulting front had a sigmoidal shape, but was still concave with respect to the origin. The velocity could be judged uniform along the front where succeeding indications of the shape of the front had the same slope and form.

Indications of the shape of the fracture front were provided by two types of fracture markings—rib marks and Wallner lines. Rib marks, the more obvious of the two, are sharp details which are easily seen on a fracture surface, often with the unaided eye. When a crack stops and later restarts, the restart is always accompanied with a sharp change in direction of propagation due to a change in the local stress field. The sharp lines which result are the rib marks; because they represent a stationary crack front, they are a certain indicator of the location and shape of the front. Less obvious, and often difficult to see, are Wallner lines, the result of the intersection of the fracture front with a sonic pulse. Such pulses cause a transitory alteration in the direction of principal tension at the crack tip, and, therefore, a momentary change in direction of propagation. Wallner lines are not sharp details, but are wavelike in appearance. In slow fractures they indicate the shape of the front.

Hackle may appear in slow fracture if the axis of principal tension twists, causing the single crack front to split into several parallel, non-coplanar fronts. It indicates the direction of propagation, as hackle is parallel to the direction of movement. When hackle was observed to occur, it was an indication that the direction of principal tension was not parallel to the long axis of the sample, and thus indicated misalignment of the tensile device. Although this situation was guarded against, useful information was sometimes obtained from the presence of hackle.

Rib marks, Wallner lines, and hackle are well-known marks in fractography. However, previously unreported marks were discovered on the fracture-exposed surfaces of the samples. These marks are even more subtle than most Wallner lines, and can only be seen with careful manipulation of illumination. It has been determined that the appearance of these marks is certain evidence that the fracture is in the transition region, that is, in Region II described earlier, where the rate of transport of water vapour to the crack front is the velocity-controlling factor.

As seen in Plate III, these marks appear as relatively broad regions in the central portion of the fracture-exposed surface. Their long axis is parallel to the crack front. Also seen is the parabolic curve which seems to envelope the marks and which is convex with respect to the origin. The entire width of the sample is shown in Plate III, and it is apparent that the parabola intersects both edges.

Velocity calculations show that these marks are formed in the transition region where the rate of transport of water vapour to the crack tip is the controlling factor for crack movement. Although it has been shown that the crack front changes shape in this region (Schönert, Umhauer and Klem, 1969),

Plate III. Fracture-exposed surface of soda-lime-silica glass plate showing elongated marks and enveloping parabola formed in the transition region. Fracture was accelerating from left to right. Dark-field incident illumination. Magnification × 78·5.

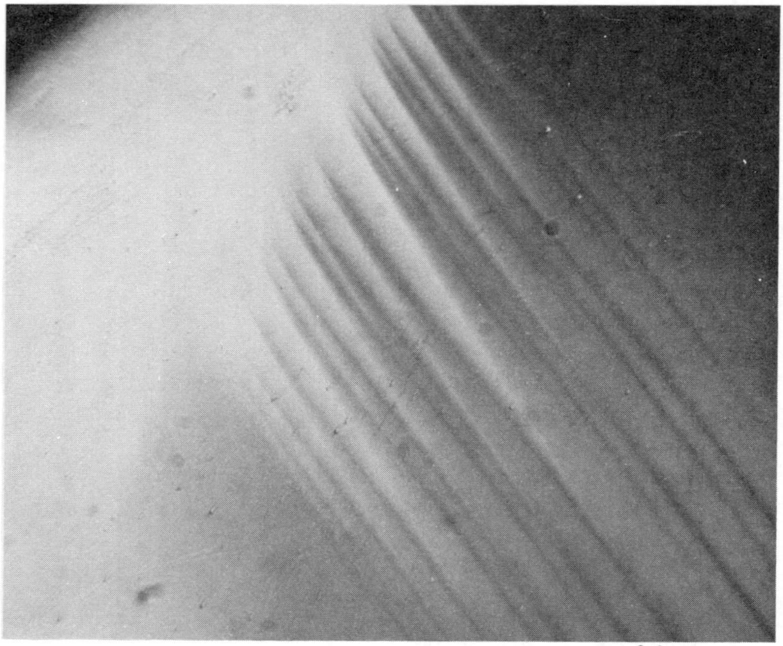

0·1 mm

Plate IV. Fracture-exposed surface of soda-lime-silica glass plate showing shape of marks formed during the transition region. Fracture direction from lower left to upper right; crack was accelerating. Vertical illumination. Magnification × 167.

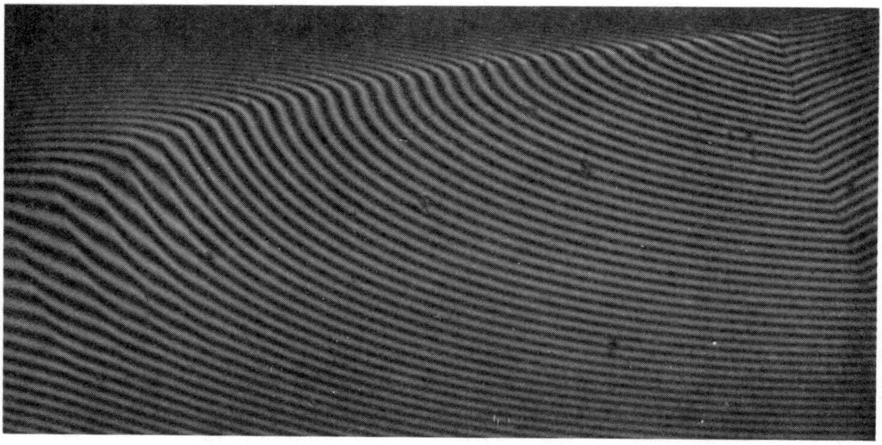

0·1 mm

Plate V. Interferogram of fractured-exposed surface of soda-lime-silica glass plate showing surface contour at the marks and enveloping parabola formed during the transition region. Fracture was accelerating from left to right. Kohaut double-beam interferometer; Tl light source. Magnification × 122.

it has not been previously reported that direct evidence for the change exists on the fracture-exposed surfaces.

The sequence which is postulated to occur when the front changes shape explains the appearance of both the marks and the parabola. Before the parabola is reached, moisture from the atmosphere is transported to the crack front along its entire length in time to participate in the stress-corrosion process. As the fracture velocity increases, the accessibility of the central portion of the advancing front is diminished. At some velocity, the atmospheric water vapour no longer reaches this central portion, and progress there is not assisted by stress corrosion. The central portion stops briefly, but then jumps forward a short distance as advance at the edges, still under the influence of moisture attack, causes the stress in the centre to rise above a critical value. This process of continuous (though not necessarily constant-velocity) fracture advance at the edges, together with stepwise progress in the middle, continues until the velocity at the edges becomes great enough to outstrip any stress corrosion effect. The crack then resumes a normal mode of progress with a front concave toward the origin.

Plate IV shows at higher magnification a region of the marks, in which their distinctive shape is seen. The edges of the marks (parallel to their long axis) indicate rest points in the stepwise progress of the central portion of the crack front. An inteferogram of a similar specimen is shown in Plate V.

By decreasing the stress on the system it was possible to cause deceleration of the crack after it had accelerated through the transition region. On deceleration the parabola and the new marks again appeared on the fracture-exposed surfaces. However, the parabola formed during deceleration was the mirror image of that formed during acceleration. This reversibility corroborated the postulated mechanism of formation; in a decelerating fracture the available moisture first influences the edges of the crack tip, and does not affect the central portion until the velocity has been sufficiently reduced to allow transport of water vapour to the interior region.

The marks have been observed in window glass and in chemically resistant aluminosilicate sheets. They appeared under a wide range of moisture-containing environments including high-purity nitrogen (dew point, -85 °C) where the transition region occurred at about 5×10^{-6} m/s and room atmosphere (dew point, 18 °C) where transition occurred at 2×10^{-4} m/s. The transition-range velocities may be expected to change with glass composition and thickness and with moisture content and temperature.

Work is continuing in which this technique of investigation will be used to study a broader range of environments, glass composition and glass thickness. Future studies should allow a more quantitative explanation to be developed.

These details of fracture behaviour in the transition region have implications in fracture energy determination. The stress field at the crack tip is of prime importance in such considerations. It must be remembered that fracture marks are formed because of localized variations in stress which may occur even if the overall stress condition remains unchanged. Because of this fact fracture may proceed locally at rapid rates even though the overall velocity of propagation (i.e. response to overall stress conditions) is very slow. For example, in order for Wallner lines to form, the fracture front must respond to a pulse travelling at sonic velocity. This response does occur in fractures moving as slowly as 10^{-7} m/s even though sonic pulses move at greater than 3×10^{-3} m/s—a difference of ten orders of magnitude.

Our examinations have shown that continuous crack propagation at velocities less than 10^{-4} m/s is possible to a limited extent. Fracture-exposed surfaces formed at these velocities seldom lack rib marks—an indication of at least momentary stopping. In the transition region, continuous motion along the entire crack front is not possible. It is to be expected that the appearance of these marks in the transition region could markedly affect results of fracture surface energy determinations. Investigators should take their formation and the formation of rib marks into account when discussing their results.

REFERENCES

Benbow, J. J. (1961). *Proc. Phys. Soc.*, **78**, 970.

Benbow, J. J., and F. C. Roesler (1957). *Proc. Phys. Soc.*, **B70**, 201.

Berry, J. P. (1963). *J. appl. Phys.*, **34**, 62.

Cotterell, B. (1966). *Int. J. Frac. Mech.* **2**, 526.

Gilman, J. J. (1960). *J. appl. Phys.*, **31**, 2208.

Griffith, A. A. (1920). *Phil. Trans. R. Soc.* **A221**, 163.

Kerkhof, F. and H. Richter (1969). *Glastech. Ber.*, **42**, 129.

Linger, K. R. and D. G. Holloway (1968). *Phil. Mag.*, **19**, 1269.

Roesler, F. C. (1956). *Proc. Phys. Soc.*, **B59**, 981.

Schönert, K., H. Umhauer, and W. Klemm (1969). In P. L. Pratt (Ed.–in–Chief), *Fracture, Proceedings of the Second International Conference on Fracture, Brighton, April 1969*. Chapman and Hall Ltd., London. p. 474.

Wiederhorn, S. M. (1966). In W. W. Kriegel and H. Palmour III (Eds.), *Materials Science Research*, Vol.3. Plenum Press, New York. p. 503.

Wiederhorn, S. M. (1967). *J. Am. Ceram. Soc.*, **50**, 407.

Wiederhorn, S. M. (1969). *J. Am. Ceram. Soc.*, **52**, 99.

49

Critical Stages in the Fracture of an Organic Glass

R. N. HAWARD

Shell Research Limited, Carrington Plastics Laboratory, Urmston, Manchester, Great Britain

---

Numerous studies on the process of brittle fracture in organic glasses have demonstrated the applicability of established fracture mechanics. For example, Berry (1961) showed that the growth of a crack could be characterized by a surface energy γ required for crack growth and an initial flaw size associated with an assumed natural flaw in the material (Figure 1). Later the work of Kambour (1964, 1965, 1966) demonstrated that the cracks are preceded by the formation of a craze structure at the crack tip and that the energy associated with crack propagation can be largely explained in terms of the plastic and elastic deformation of the craze. It has now also been demonstrated that no special or externally generated flaw is required for craze initiation. On the contrary, this can be observed to take place in apparently homogeneous, isotropic polymer, at right angles to the principal stress whenever it exceeds a certain level (Sternstein *et al.* 1968). In polystyrene this also has been shown to occur at substantially the same stress as that associated by Berry with the extension of a crack from a pre-existing flaw (Murphy *et al.* 1969). Thus the first stage in the fracture of an organic glass lies in the formation of a craze-like structure.

These craze structures have been shown to contain a large number of voids which initially may be very small and rounded, although they may later be distorted by further extension of the craze (Van den Boogart 1966) (Matsuo 1969). Clearly the formation of such a structure must consist essentially of a yield process and evidence to support this concept has also been given (Sternstein *et al.* 1968, Murphy *et al.* 1969, Haward *et al.* 1971). The latter have also suggested that the occurrence of plastic instability, i.e. a decline in engineering stress as the material is extended, is a requirement for the formation of a craze. This requirement provides the essential condition

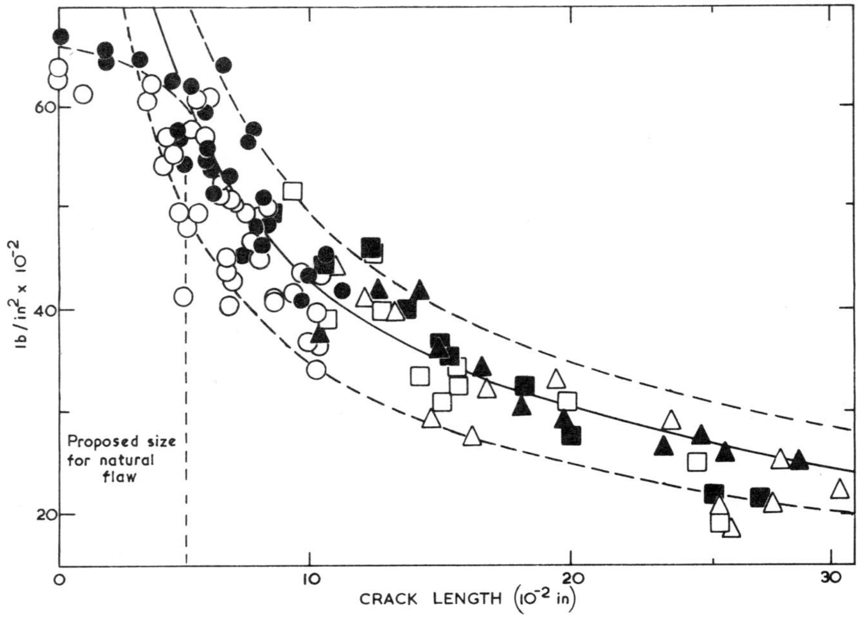

Figure 1. Dependence of tensile strength on crack size for different polystyrene samples (after Berry, 1961) [From *J. Polym. Sci.*, **50**, 107 (1961). Copyright © 1961 John Wiley & Sons Inc. by permission of John Wiley & Sons Inc.]

that less energy is required to cause a large plastic strain in a small volume of material than to give a small strain to a large volume. However, although this condition seems necessary, it does not provide an adequate account of how voids can be formed in a plastic material. As a starting point for the consideration of this problem it is necessary to consider the energy required to form a single void in a polymeric glass.

THE GROWTH OF VOIDS IN A POLYMERIC GLASS

In order for a void to be formed in an isotropic solid with a yield stress Y two criteria must be met.

(1) The stress must be adequate to provide the surface energy to nucleate the void when it is very small.

(2) The stress must also be large enough for the nucleated centre to grow effectively in the elastic–plastic solid.

Following Kambour's (1965) argument that the energy associated with the plastic and elastic deformation necessary for craze formation was larger than the surface energy represented by the voids, we will consider the second of these subjects first.

The hydrostatic tension required to cause steady growth of a void in a plastic solid with yield strength Y, Poisson's ratio ν and Young's modulus E is given by the relation due to Hill (1950), i.e.

$$\sigma_h = \frac{2}{3} Y\left\{\ln \frac{E}{3Y(1-\nu)} + 1\right\}$$

By insertion of suitable constants for polystyrene into this equation, treating the material as a plastic solid with a particular yield stress (in this case two selected values of 1050 and 1570 kilobars were taken), the equation gives σ_h values between (2·4–2·7)Y(Drabble *et al.* 1966). Clearly the growth of single voids at stresses below the yield stress would be most unlikely according to this model. As a next step the adiabatic formation of a void was considered. In this process, which would apply in the case of fast void growth, the energy of plastic deformation is considered to be retained as heat so that the local temperature rises and the yield strength falls as deformation takes place. However, it turns out that this factor will reduce the resistance to cavity growth by a factor of only approximately 15%. Thus the use of an adiabatic model does not allow the growth of voids at stresses below the bulk yield stress to be explained.

A more significant line of approach comes from the consideration of the stress/strain relationship at, and following, the yield point. Here it has been shown by Binder and Muller (1961) and by Haward, Murphy and White (1971) that the true stress in polystyrene falls after yield by between 40% and 25%. A similar 'yield drop' has been observed by Brown and Ward (1968) with polyethylene terephthalate. Thus the existence of this effect will further reduce the work of cavity formation. However, since the yield drops quoted represent the maximum fall in yield stress, the fall in the average value during cavity formation and extension cannot be as great. Thus a reduction of 20–25% in the resistance to cavity growth for this approach seems about the maximum which can be expected. Clearly this factor still cannot reduce the cavity growth stress to a level below the yield stress.

The further factor which now has to be considered is more drastic and also arises more directly from experimental observations. Quite simply, the growth of single voids is not observed. Instead large numbers of holes are developed close together within a limited volume of material. The significance of this factor is illustrated in Figure 2. The energy associated with the growth of a void of radius r is largely due to the energy requirement of the surrounding plastic zone of radius 2·89 r (Drabble *et al.* 1966). However, when two voids are found close together then each will eliminate part of the neighbouring plastic zone and the strains of the interpenetrating plastic zones will also be in part complementary. Thus the energy required to extend each void is reduced. Further, if we consider the energy required to expand unit volume of voids by a small volume δV in an assembly with a void fraction of θ it is clear that the energy of expansion should lie between $\theta\sigma_h\delta V$, where only the energy required in the plastic zone occupied by another void is omitted, and $(1-\theta)\sigma_h\delta V/(\theta \times 2{\cdot}89^3)$ i.e. approximately $(1-\theta)\sigma_h\delta V/24\theta$ where the volume of plastically deformed material is divided by the volume of plastically deformed material associated with the same void volume in isolated voids.

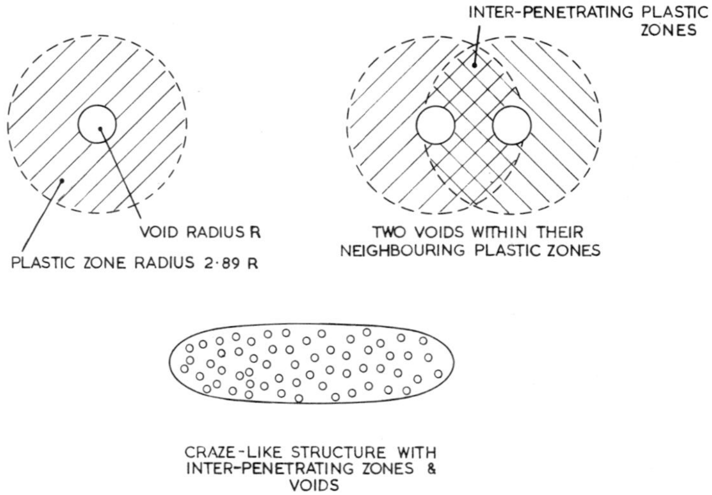

Figure 2. Voids in a polystyrene-like matrix (after Drabble, unpublished)

In each case, but particularly in the latter, the estimated energy required to generate a dispersion of small voids must be substantially reduced and the average stress required to generate the volume increase δV will be reduced by the same factor (stress $\times$ volume $=$ energy). This process may be described as one of the overlapping proliferation of voids.

One difficulty of this approach is that the indefinite increase in the void volume leads to a value of zero for the stress required to form a craze-like structure, thus the formation of structure of intermediate volume is not obvious. To this point there are two answers. In the first place crazes are not particularly stable and easily pass over into cracks, and indeed, where a suitable structure is available to stabilize them (e.g. rubber particles), crazes of very high void volume may be observed (Matsuo 1969). The second point concerns the basic large strain properties in a thermoplastic. In nearly all cases a substantial measure of orientation hardening (similar to work hardening) can be observed at large strains (Vincent 1966, Haward and Thackray

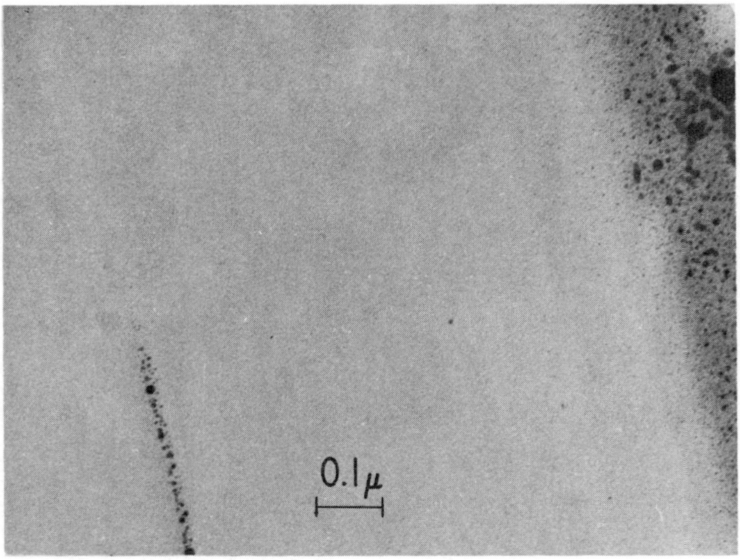

Plate VI. Crazes in polycarbonate (after Kambour, *Polymer*, **5**, 143 (1964)).

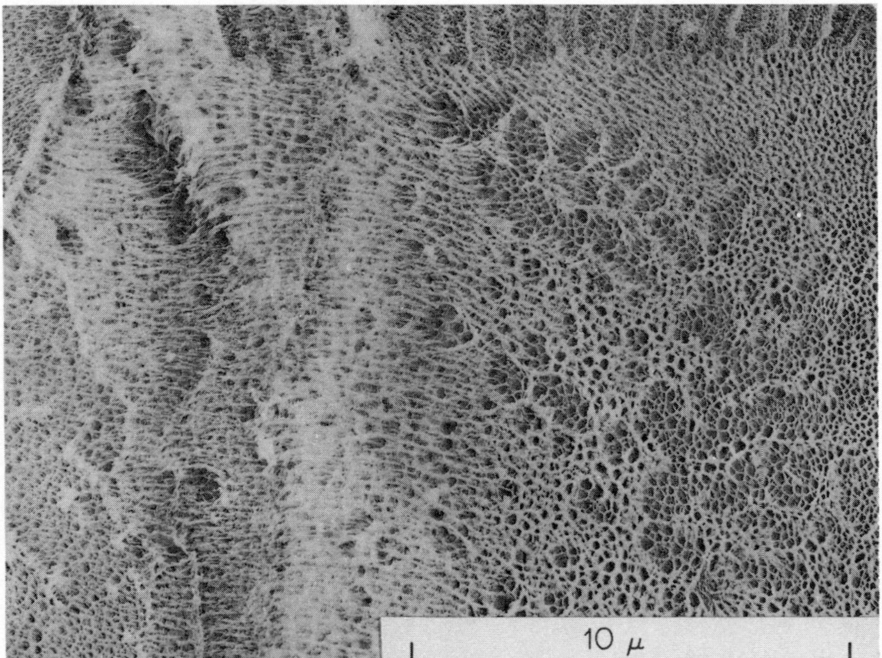

Plate VII. Voids in the crack initiation area of compression moulded polystyrene (after Mann, unpublished).

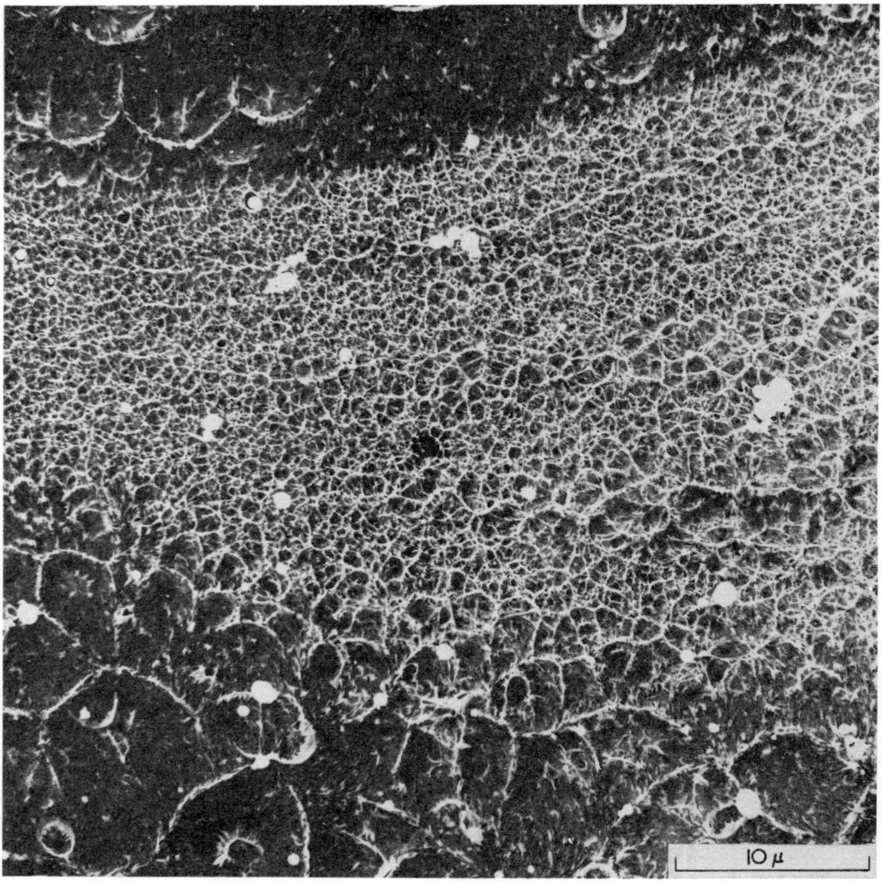

Plate VIII. Void proliferation at the centre of an injection moulded polystyrene test-piece broken in tension. The presence of parabolas indicates that this is the initiation area of the fracture (Photo by Mann, Bird and Rooney).

1968) and this can clearly lead to a measure of stabilization of a craze at a particular void volume characteristic of the polymer concerned. For polystyrene Kambour (1964) has reported a value of 40% for the void volume and this supplies the value of θ which has been used in Table 1. Clearly the concept of overlapping plastic zones does make it possible to envisage the growth of a craze within the experimentally observed stress range. However, the effect of orientation hardening has not been even qualitatively estimated in this very approximate treatment.

Table 1. The contribution of different factors to void growth for a polystyrene-like plastic with yield stress 1050 kbar

Basis for estimate	Pressure or hydrostatic stress for void growth (kbar)
Isothermal growth/isolated void	2900
Adiabatic growth/isolated void	2500
Additional correction due to 'yield drop'	1900 ± 400
Additional correction due to 40% void volume based on void volume only	1140 ± 240
Based on reduction in plastic zone volume	70
Increase in stress due to orientation hardening	Not estimated
Experimental crazing stress	450

Clearly the concept of overlapping proliferation of plastic zones provides a novel approach to the formation of craze-like structures, but one must ask whether the evidence does support any such process. Here we must consider some electron micrographs. In Plate VI, taken from Kambour (1964), we see a small craze extending in polycarbonate. It consists of a series of close-packed nearly spherical voids which follow each other at distances close enough for the plastic zones to overlap. An even more striking illustration is provided by the work of Mann (1966) (Plate VII) and also by Mann, Bird and Rooney (Plate VIII). Each of these pictures shows a clear void structure formed in the initiation area of polystyrene fractures. Taken together these pictures provide rather strong evidence that fracture in organic glasses is initiated by the formation and expansion of voids.

THE NUCLEATION OF VOIDS

If a void, or more particularly a group of voids, is to be formed in a plastic there must be enough energy available for the formation of the new surface.

This requirement has, of course, been known for a long time (Haward 1942) in connection with fracture processes, and is most important when the hole is small. Thus the hydrostatic tension required to expand a hole of radius r is given by $2S/r$ where S is the surface tension of the liquid or plastic solid. This quantity can be shown to be significant only when the holes concerned are very small ($r = 10$–100 Å). The voids observed by Mann (1966) have all gone beyond this size but many of those seen by Kambour (1964) (Plate VI) are indeed of this magnitude. On the other hand, there is always a size below which the void cannot be stable in a particular stress field. To reach this size, energy must first be supplied and the existence of this energy requirement is the factor common to all nucleation phenomena and is believed to be responsible for the 'tensile strength' of liquids. A theoretical treatment of the problem has been given by Irwin (1958) and also by Fisher (1948) who proposed that

$$p = \left[\frac{16\pi}{3kT}\ \frac{S^3}{\ln Nk(T/h)}\right]^{\frac{1}{2}}$$

or, more simply, since the absolute value of the constants are inevitably uncertain

$$p \simeq K\left[\frac{S^3}{T}\right]^{\frac{1}{2}}$$

where p is the hydrostatic tension required to form a void, S the surface tension of the liquid and T the absolute temperature.

This equation may be appropriately applied to measurements of the tensile strength of low viscosity liquids. Unfortunately, these measurements are still extremely uncertain so that the values of K obtained vary widely, as shown in Table 2. Nevertheless, we can take an average value of $K = 3 \times 10^{14}$ and measured values for the surface tension of polystyrene (White and Harford, to be published), (Gray 1965) and use them to estimate the cavitation stresses likely to be required in a glassy polymer like polystyrene. These are given and compared with the yield and crazing stresses (Figure 3). Clearly the estimates for nucleation stress may contain very large errors, though we may note that the measurements of Lindsey (1967), on polyurethane elastomers do fall within our range of estimates. Nevertheless, even allowing for these errors we can reach the following conclusions:

(a) The cavitation stresses are well below the yield stress or the crazing stress at the temperatures at which the plastic is normally used.

(b) At higher temperatures, however, cavitation stresses do not fall to zero as the glass transition temperature is approached, so that under these conditions, crazing should tend to be suppressed and the test-piece should deform as a whole. This is, of course, observed.

Table 2. Tensile strength of liquids

Liquids	Temperature (°C)	Average surface tension (γ) (dynes/cm)	Average tensile cavitation stress (p) (kbar)	K from the equation $p = K[S^3/T]^{\frac{1}{2}}$
Benzene[a]	23	28	30	$1{\cdot}1 \times 10^{14}$
Benzene[b]	0–28	30 (10 °C)	130	$4{\cdot}2 \times 10^{14}$
Aniline[b]	-5 ± 5	39·4	270	$5{\cdot}8 \times 10^{14}$
Carbon tetrachloride	-15 ± 30	27	275	$10{\cdot}3 \times 10^{14}$
Ethanol[c]	18·5	22·7	46·5	$2{\cdot}4 \times 10^{14}$
Ethanol[d]	27	22·7	39	$2{\cdot}4 \times 10^{14}$
Ethyl ether[d]	27	17	72	$5{\cdot}6 \times 10^{14}$

[a] J. J. Donoghue, R. E. Volkath, E. Gerjuoy *J. chem. Phys.*, **19**, 55 (1951).
[b] L. J. Briggs, *J. chem. Phys.*, **19**, 970 (1951).
[c] H. V. Temperley, *Proc. phys. Soc.*, **58**, 420 (1946); **59**, 199 (1947).
[d] J. C. Fisher *J. appl. Phys.*, **19**, 1062 (1948).

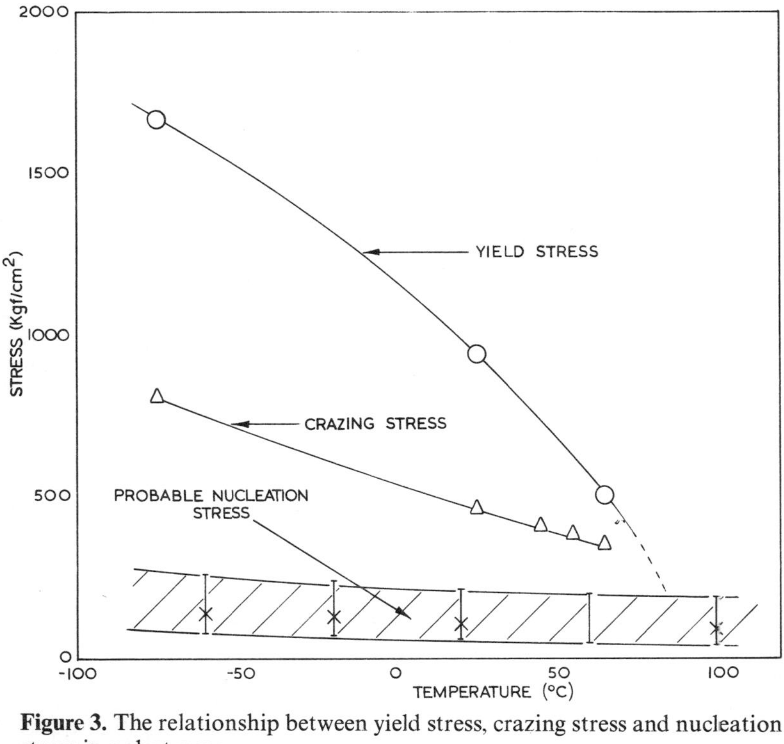

Figure 3. The relationship between yield stress, crazing stress and nucleation stress in polystyrene

(c) As the temperature is reduced both the yield stress and the crazing stress increase much faster than the cavitation stress. Thus nucleation becomes relatively much easier as the temperature falls.

These conclusions may be related to experiments in a number of ways. For example, it is known (Berry 1964) that polymethylmethacrylate fracture surfaces show characteristic hyperbolas formed by the creation of fresh nuclei in front of the main crack. The above theory would suggest that the nucleation process would become relatively easier at lower temperatures, and in Plate IX we show the optical micrographs of the surface of perspex (methylacrylate) Izod impact test pieces broken at three temperatures. It will be seen that the number of independent nuclei increases as the temperature falls.

THE FRACTURE OF BITUMEN

Further light on these processes may be obtained by comparing the fracture surfaces of polystyrene and bitumen. Under certain conditions Marek and Herrin (1968) have shown that bitumen films fracture by the formation of voids. The resulting fracture surfaces show formations which, in spite of the large differences in scale, show a remarkable similarity to those observed in polystyrene (Plate X).

THE FAILURE OF THE CRAZE

The fact that a craze structure is formed does not automatically provide an answer to the question of the strength of a polymer glass. Clearly the highest stresses within the craze will be larger than the average stress on the plastic or on the craze. On the other hand, considerable orientation of the polymer has taken place and at high orientations plastic stability tends to increase (Haward and Thackray 1968, Vincent 1966). On the other hand, as the stress rises the possible adiabatic effects also increase since heat generation is expected to be proportional to the applied stress. Thus, if σ_m is the engineering stress on a part of the craze, then a situation can arise where

$$\left(\frac{d\sigma_m}{d\varepsilon}\right)_{\text{(isothermal, constant strain rate)}} \quad \text{is positive}$$

but

$$\left(\frac{d\sigma_m}{d\varepsilon}\right)_{\text{(adiabatic, constant strain rate)}} \quad \text{becomes negative,}$$

and when this occurs any minor disturbance in the structure could lead to the drawing out of a local element of the plastic of the type shown in Plate X.

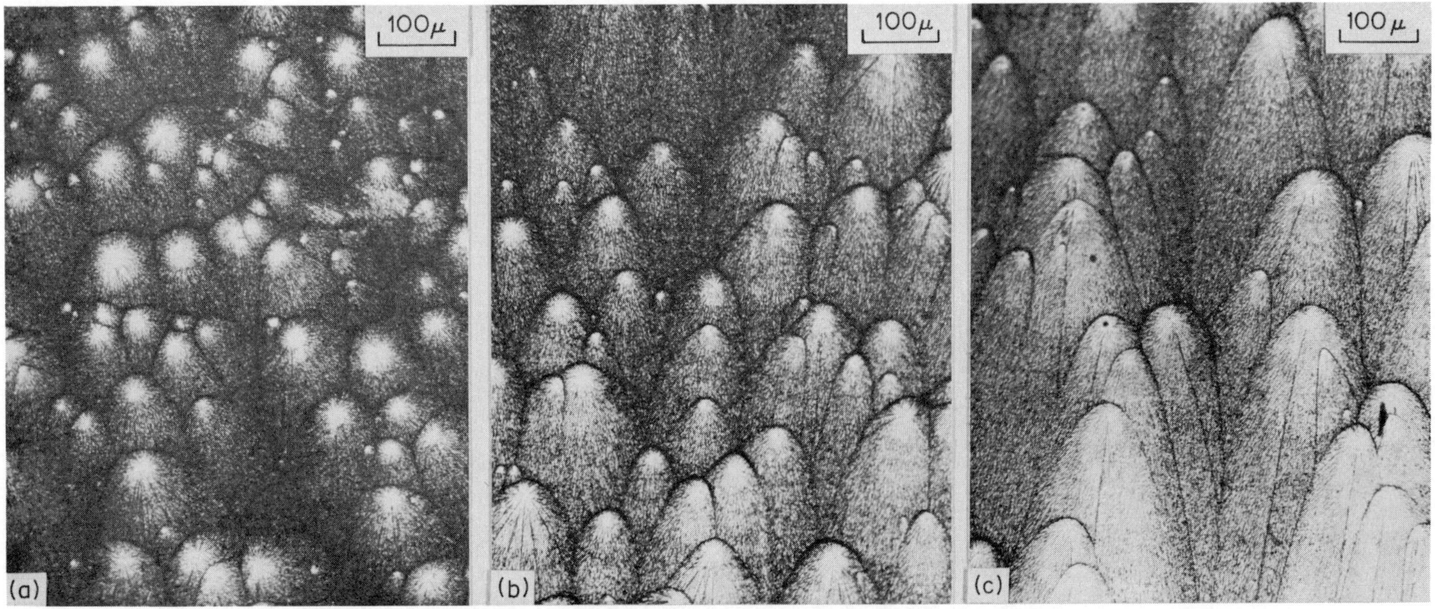

Plate IX. Izod test pieces of polymethylmethacrylate fractured at different temperatures (a) Fractured at −65 °C (100 hyperbolas) (b) Fractured at 23 °C (58 hyperbolas) (c) Fractured at 55 °C (33 hyperbolas).

S

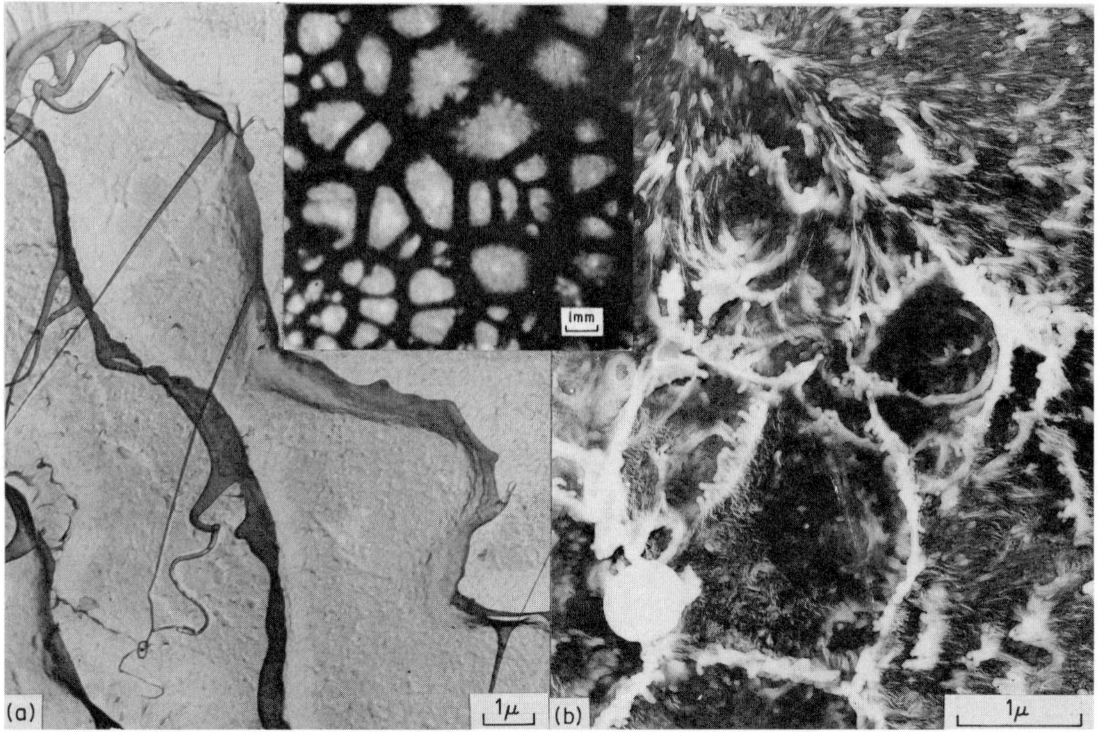

Plate X. (a) Fibrils formed during the fast fracture of compression moulded polystyrene (photo by I. Brough) (b) Void structures formed in the initiation area of an injection moulded test piece (as Plate VIII) (Photo by Mann, Bird and Rooney, to be published) (Inset) Structure of a bitumen film after voiding and fracture (after Marek and Herrin, 1968)

That this process could take place under conditions when the deformation was apparently stable is supported by the work of Vincent (1966), Binder and Muller (1961) and Drabble *et al.* (1966). However, other different concepts appropriate to the failure of an oriented plastic filament may be applied to this stage of the fracture process, so that at the present moment, it is not possible to propose with any certainty the type of mechanism which occurs at the point where the craze fails.

ACKNOWLEDGEMENT

The author wishes to thank Messrs. J. Mann, R. Bird and M. Rooney, Dr. R. Kambour and Professor G. R. Marek for permission to use their fracture photographs either as published or in course of publication.

REFERENCES

Berry, J. P. (1961). *J. Polym. Sci.*, **50**, 107, 313.
Berry, J. P. (1964). 'Brittle Behaviour in Polymeric Solids'. In B. Rosen (Ed.) *Fracture Processes in Polymeric Solids*. Interscience, New York. p. 195.
Binder, G., and P. H. Muller (1961). *Kolloidzeitschrift*, **177**, 129.
Brown, R., and I. M. Ward (1968). *J. Polym. Sci., A2*, **6**, 607.
Drabble, F., R. N. Haward, and W. Johnson (1966). *Br. J. appl. Phys.*, **17**, 241.
Fisher, J. C. (1948). *J. appl. Phys.*, **19**, 1063.
Gray, V. R. (1965). *Chemy. Ind.*, p. 971.
Haward, R. N. (1942). *Trans. Faraday Soc.*, **38**, 394.
Haward, R. N., B. M. Murphy, and E. F. T. White (1971). *J. Polym. Sci., A2*, **9**, 801.
Haward, R. N., and T. G. Thackray (1968). *Proc. Roy. Soc.*, **A302**, 453.
Hill, R. (1950). *The Mathematical Theory of Plasticity*. Clarendon Press, Oxford.
Irwin, G. R. (1958). *Handbuch der Physik* VI. Springer, Berlin, p. 551.
Kambour, R. P. (1964). *J. Polym. Sci., A1*, **2**, 4159.
Kambour, R. P. (1964). *Polymer*, **5**, 143.
Kambour, R. P. (1965). *J. Polym. Sci., A1*, **3**, 1713.
Kambour, R. P. (1966). *J. Polym. Sci., A2*, **4**, 349.
Lindsey, G. H. (1967). *J. appl. Phys.*, **38**, 4843.
Mann, J. (1966). *Chemy. Ind.*, p. 643.
Mann, J., R. Bird, and G. Rooney (private communication). To be published.
Marek, C. R., and M. Herrin (1968). *Proc. Ass. Asph. Pav. Technol.*, **37**, 386.
Matsuo. (1969). *Polym. Eng. and Sci.*, **9**, May, 206.
Moavenzadeh, F. (1967). *Proc. Ass. Asph. Pav. Technol.*, **36**, 51.
Murphy, B. M., R. N. Haward, and E. F. T. White (1969). *Proc. Int. Frac. Conference*, Brighton, England.
Sternstein, S. S., L. Ognchin, and A. Silverman (1968). *Appl. Polym. Symposia.* No. 7, 175.
Van der Boogart. (1966). 'The Physical Basis of Yield and Fracture'. *Inst. Phys. Conf. Proc.*, p. 167, Oxford, England.
Vincent, P. I. (1966). 'The Physical Basis of Yield and Fracture', *Inst. Phys. Conf. Proc.*, p. 155, Oxford, England.
White, E. F. T., and J. R. Harford (private communication).

50

Flaw Distributions on Glass Surfaces and Mechanisms of Surface Damage

B. HAMILTON and H. RAWSON

Department of Glass Technology, University of Sheffield, Great Britain

FLAW DISTRIBUTIONS

In an earlier paper (Hamilton and Rawson, 1970a) we have presented the results of Hertz fracture experiments on a number of glass surfaces and have shown that they can be interpreted with moderate success in terms of the theory of flaw statistics. The analysis of the results involves the determination of parameters characterizing a flaw distribution function for each surface. This function is assumed to be of the form

$$n(\sigma) = \begin{cases} \left(\dfrac{\sigma - \sigma_u}{\sigma_0}\right)^m & \text{if } \sigma \geqslant \sigma_u \\ 0 & \text{if } \sigma < \sigma_u \end{cases} \tag{1}$$

$n(\sigma)$ is the number of flaws per unit surface area which are capable of acting as fracture origins when the applied stress is σ. σ_u, σ_0 and m are the parameters determined from the results.

The samples were plates of glass either $\frac{1}{2}''$ or $1''$ thick, and five surfaces were investigated: polished plate glass as received from the manufacturer, the same material after two etching treatments in dilute hydrofluoric acid and the two surfaces of float glass. The two float glass surfaces differ in strength. They are distinguished here by describing them as the upper and lower surfaces, which refers to their position during the manufacturing process. The flaw distribution functions for the five surfaces are shown in Figure 1.

It is a matter of current controversy whether it is permissible to analyse Hertz fracture experiments on brittle materials in terms of flaw statistics. Frank and Lawn (1967) suggest that it is not, at least when using ball indenters below a certain size, and have presented an alternative theoretical approach.

Our comments on their theory have already been published (Hamilton and Rawson, 1970a, b). We consider that their theory is valid for heavily abraded surfaces containing a high density of deep flaws but not for the stronger surfaces with which we are concerned.

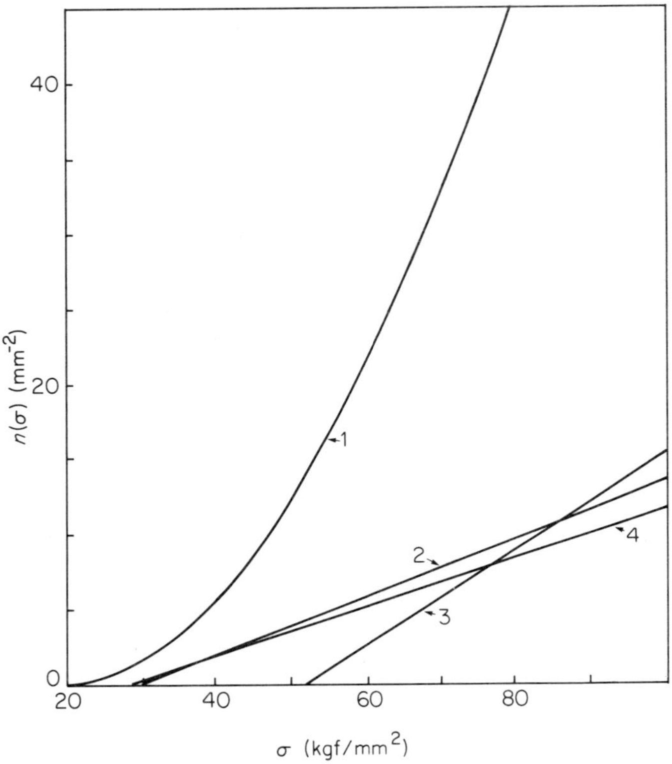

Figure 1. The flaw distribution functions for the surfaces investigated.

1. Polished plate—as received.
2. Polished plate after etching for 4 min in dilute hydrofluoric acid.
3. (a) Polished plate after etching for 7 min in dilute hydrofluoric acid and (b) float glass—upper surface.
4. Float glass—lower surface

[From B. Hamilton and H. Rawson, *J. Mech. Phys. Solids*, **18**, Fig. 11 (1970)]

An alternative method for obtaining information about the number and severity of flaws on a glass surface is the technique developed by Ernsberger (1960, 1962). On immersing a glass containing sodium ions in a fused salt containing lithium ions, base exchange occurs between these ions. A tensile stress is produced in the glass surface, the magnitude of which can be increased by increasing either the time or the temperature of the base exchange process. If this stress is high enough, the flaws propagate as surface cracks which can easily be seen when the glass is subsequently examined under the microscope. The higher the stress, the greater is the number of flaws capable of acting as fracture origins and hence the denser is the resulting crack pattern. Ernsberger has described how fracture origins may be identified by careful examination of the crack pattern. If the surface stress

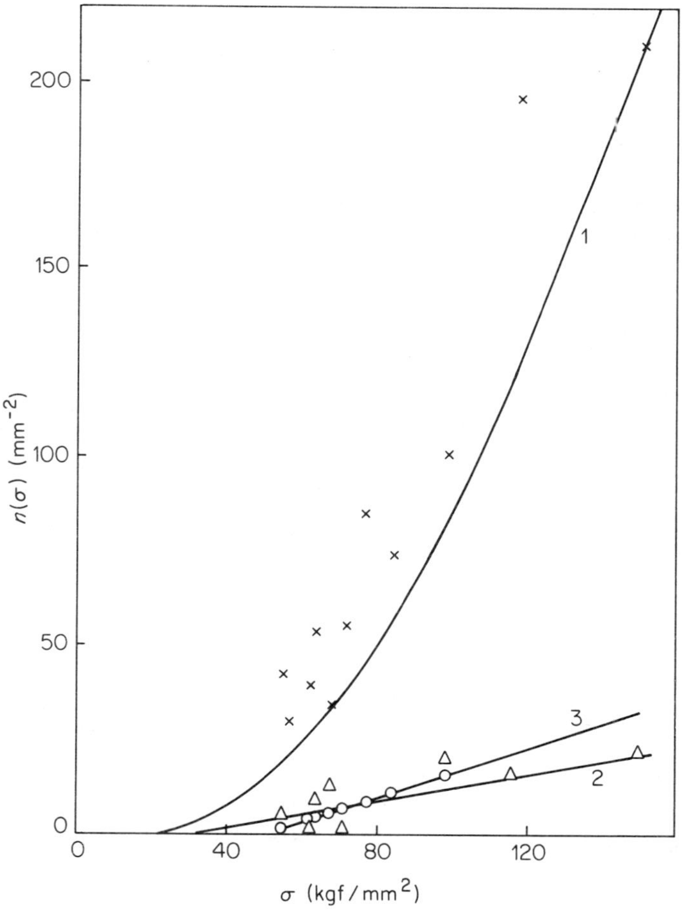

Figure 2. Comparison between flaw densities determined by the Ernsberger technique and values given by the flaw distribution function.

Curve 1 and points marked × —polished plate as received.
Curve 2 and points marked Δ—float glass, lower surface.
Curve 3 and points marked ○—float glass, upper surface

produced by the base exchange process could be measured, one could use the Ernsberger technique to determine the number of flaws per unit area acting as fracture origins at a known value of surface stress. This could then be compared with the $n(\sigma)$ value obtained from the flaw statistics analysis. Compressive surface stresses produced by base exchange have been measured photoelastically (Burggraaf, 1966) and by determination of the curvature of a thin plate specimen after treatment on one face only (Kistler, 1962). Neither method was found to be suitable for measuring the stresses in our samples, possibly because the tensile stresses are relieved by cracking. Thus at present we can use the Ernsberger technique only to compare the flaw densities on two surfaces which have been subjected to the same base-exchange treatment and hence to the same surface stress. The way in which this comparison is made and is related to the flaw distribution functions obtained from the Hertz fracture results is shown in Figure 2. Three surfaces were examined: polished plate glass and the two float glass surfaces. Specimens of these three surfaces were subjected to a series of base exchange treatments for various periods of time at a temperature of 260 °C. In each case one surface, usually the upper surface of float glass is taken as the standard and we are concerned only with a comparison of the flaw densities measured on the other two surfaces with that measured on the standard surface. The three curves in Figure 2 represent the respective $n(\sigma)$ functions determined as above. Having determined the flaw density on the standard surface by the Ernsberger technique, the value is plotted on the $n(\sigma)$ curve for that surface. This in effect establishes the stress value for the particular base exchange treatment given. Points representing the Ernsberger flaw densities for the other two surfaces are then plotted on the diagram at this value of stress. It is clear that both the strength results and the Ernsberger technique give a very similar representation of the way in which the number of active flaws increases as the applied stress is increased.

MECHANISMS OF SURFACE DAMAGE

It is well known that strength-reducing flaws can be produced very easily on a glass surface. Much of the damage is produced by contact with the surfaces of other hard objects. The contact may involve impact or sliding between the two surfaces. Many papers in the literature present results showing the reduction in strength of laboratory specimens or commercial glassware produced by mechanical damage, but relatively little experimental work has been done to determine the conditions under which surface microcracks form when the glass surface is stressed in a controlled way, simulating the type of loading to be expected when a glass article is damaged by impact or by sliding.

We have carried out experiments in which a steel ball is pressed against a glass surface with a constant normal force, *P*, and a steadily increasing force, *T*, is applied to the glass in a direction parallel to the surface. The glass specimen is mounted on a trolley fitted with ball race wheels and so can slide under the loading ball when the force *T* exceeds the frictional force at the ball-glass contact. The force *T* was increased until either the glass fractured under the ball or sliding occurred. The results, which have been discussed in more detail elsewhere (Hamilton and Rawson, 1970c) are shown in Figure 3. The points represent mean values of *T* to produce fracture at the various selected values of *P*. At low values of *P*, the onset of sliding truncates the distribution of *T* values and in this range the mean value of *T* can only be estimated. Curve 1 shows how the value of *T* required to produce fracture would vary if the fracture stress was independent of the method of loading and was equal to the mean fracture stress obtained when $T = 0$. This curve has been calculated using the stress equations derived by Hamilton and Goodman (1960). Thus we find that under combined loading the fracture stress is greater than under simple normal loading. We consider that this result is a reasonable one since the lower the value of *P*, the smaller is the contact circle and the lower is the probability of locating a strength-reducing flaw in the highly stressed region around the contact circle. Our results are qualitatively similar to those reported by Billinghurst *et al.* (1966). Curve 2 shows the variation of *T* to produce fracture calculated by applying the theory of flaw statistics and using values for the parameters of the flaw distribution function determined from normal loading experiments on the same surface. This fits the experimental results reasonably well at high values of *P*. Little can be said about the discrepancy at low values of *P*. The experimental results are less reliable in this range. Also at the very high stresses involved, the flaw distribution function may have a different form from that derived from the normal loading experiments, in which the stresses are somewhat lower. Lawn (1967) has presented a theoretical analysis of fracture under a ball with combined normal and tangential loading. His theory predicts that fracture should occur at values of *T* even lower than values given by curve 1. Probably the most significant point, so far as surface damage is concerned, is that with combined loading, fractures can be produced by very low applied forces. If the normal load is small, the area of contact is small. Application of a tangential force then produces a very large shear stress at the contact. The probability of strength-reducing microcracks being produced at these low forces can, of course, be reduced by treating the glass to reduce the coefficient of friction between its surface and that of other solids with which it comes into contact. The coefficient of friction determines the maximum value of *T* that can be applied.

We have carried out a limited amount of work on damage produced by impact. It can be shown from the Hertz theory of impact stresses that to produce tensile stresses of the order of 500,000 p.s.i., i.e. approaching the

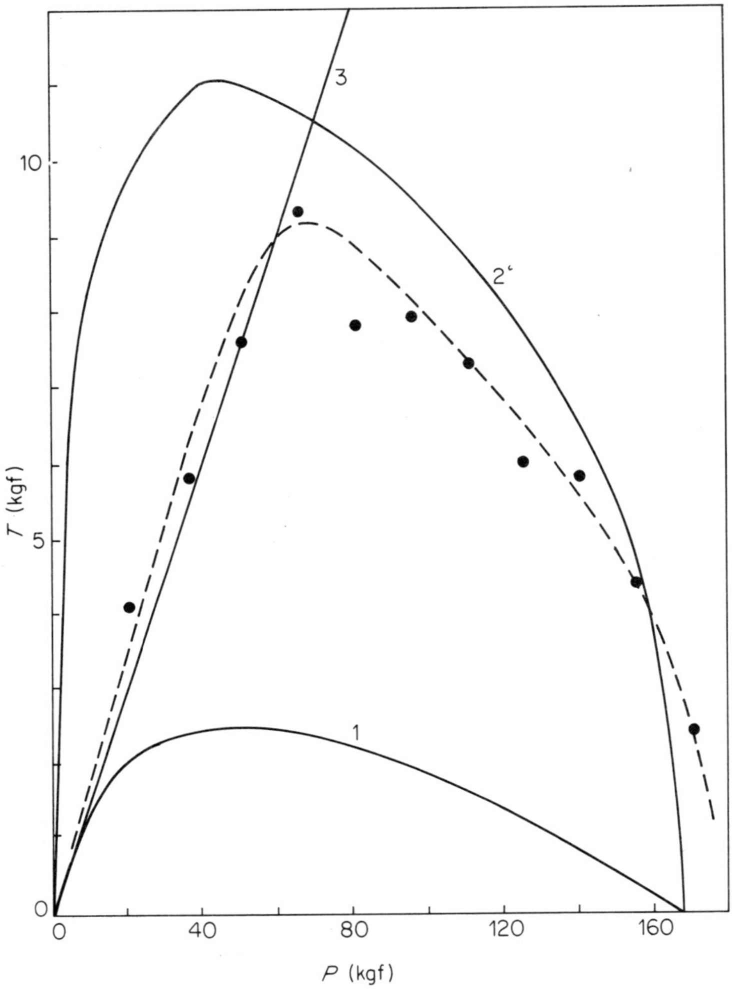

Figure 3. The variation of the tangential force to produce fracture, *T*, with the applied normal force.

1. Calculated assuming a constant fracture stress.
2. Calculated using the theory of flaw statistics.
3. Line representing the onset of sliding, corresponding to a coefficient of friction of approximately 0·18

(by permission of The Institute of Physics and The Physical Society) [From B. Hamilton and H. Rawson, *J. Phys. D*: *Appl. Phys.*, **3**, L40 (1970)]

theoretical strength of glass, equal glass spheres must impact at a relative velocity of 250 ft/sec (Gurney, 1964). This means that one sphere must be dropped onto the other from a height of approximately 1000 ft. Yet Ernsberger (1962) has shown that damage is produced by dropping carborundum grains onto a glass surface from a height of a few centimetres. In our experiments particles of various kinds were dropped onto glass surfaces, which were subsequently examined for damage by the Ernsberger technique. The main purpose of the experiments was to investigate the effects of the angularity and the hardness of the impacting particles. In some experiments spherical particles of alumina 250–500 μm in diameter and of glass 105–210 μm in diameter were used. They were dropped onto relatively flaw-free glass surfaces produced by hydrofluoric acid etching. No damage could be detected with either the alumina or the glass particles dropping from heights of up to 40 cm, the maximum used. Alumina grinding powder having the same particle size as the spheres produced extensive damage when dropped from only 10 cm, but glass granules made by crushing and then sieving to give the same particle size as the glass spheres produced no damage when dropped from heights up to 40 cm.

In a further series of impact experiments, crushed glass fragments of various sizes were used, namely 0·2–0·7 mm, 0·7–1·2 mm, 1·2–1·5 mm, 5 mm and 7 mm. Only fragments of the last two sizes produced damage when dropped from 30 or 40 cm.

It is to be expected that the stresses produced under a sharp edge or point on an angular particle when it strikes a plane surface will be higher than the contact stresses under a spherical particle of the same size falling from the same height. Also impact of an irregular particle is more likely to result in tangential forces being produced across the contact. The hardness of the impacting particle is clearly important. It may be that when an irregular glass particle strikes a glass surface, the sharp edge of the particle is blunted by flow under the very high local stresses produced. This would be far less likely to occur in a harder material like alumina.

The main conclusion of the investigation which is summarized here is that strength measurements made by pressing a hardened steel ball against a glass surface can be satisfactorily interpreted in terms of flaw statistics, provided that the surface does not contain a high concentration of deep flaws. The technique is most suitable for characterizing the surface condition of high strength surfaces which have been only lightly damaged. A limitation of the technique is that it may fail to detect the most severe flaws on the surface which determine the strength when relatively large areas of the glass are subjected to stress. These flaws are widely spaced apart and the probability of locating such a flaw in a ball loading experiment is low.

ACKNOWLEDGEMENTS

B.H. is indebted to Thorn Radio Valves and Tubes Limited, Sunderland and to the Science Research Council for financial support during the period in which this work was done.

REFERENCES

Billinghurst, P. R., C. A. Brookes, and D. Tabor (1966). In *Physical Basis of Yield and Fracture*. The Institute of Physics and The Physical Society, London. pp. 253–258.

Burggraaf, A. J. (1966). *Philips Res. Rep.*, 1966, No. 3.

Ernsberger, F. M. (1960). *Proc. R. Soc. A.*, **257**, 213.

Ernsberger, F. M. (1962). In *Advances in Glass Technology*. Plenum Press, New York. pp. 511–524.

Frank, F. C., and B. R. Lawn (1967). *Proc. R. Soc. A.*, **299**, 291.

Gurney, C. (1964). *Proc. R. Soc. A.*, **282**, 24.

Hamilton, B., and H. Rawson (1970a). *J. Mech. Phys. Solids*, **18**, 127.

Hamilton, B., and H. Rawson (1970b). *J. appl. Phys.*, **41**, 2738.

Hamilton B., and H. Rawson (1970c). *J. Phys. D: appl. Phys.*, **3**, L40.

Hamilton, G. M., and L. E. Goodman (1966). *J. appl. Mech.*, **33**, 371.

Kistler, S. S. (1962). *J. Am. Ceram. Soc.*, **45**, 59.

Lawn, B. R. (1967). *Proc. R. Soc. A*, **299**, 307.

51

The Formation of Micro Shear Bands in Glassy Polymers

P. B. BOWDEN and S. RAHA

Metallurgy Department, Cambridge University, Cambridge, Great Britain

Both PS (polystyrene) and PMMA (polymethyl methacrylate) when deformed plastically can deform inhomogeneously. Instead of the plastic strain being uniform, it is concentrated in fine shear bands each about 1 μm thick and containing a local shear strain of two. The rest of the material is completely undeformed. (Argon, 1968). On a microscopic scale each micron-thick band is made up of a cluster of finer bands with thicknesses less than 0·1 μm. We have investigated this phenomenon and wish to describe here our principal findings. A detailed description of the work is given elsewhere (Bowden and Raha, 1970).

The transition to inhomogeneous deformation occurs at a critical temperature and strain rate. Well-annealed PS tested at a strain rate of $3{\cdot}6 \times 10^{-4}$ $\sec^{-1}$ deforms inhomogeneously below 65 ± 2 °C. For the same material tested at 70°C deformation is inhomogeneous at strain rates greater than 4×10^{-3} $\sec^{-1}$. For PMMA the transition to inhomogeneous deformation occurs around -130°C. The conditions for the transition depend on the thermal history of the material. PS quenched from 100°C in iced water will deform homogeneously at room temperature.

A large number of glassy polymeric materials including PS and PMMA have true-stress *vs.* strain curves that have a maximum at the yield point followed by a strain-softening region where the flow stress falls with increasing strain. If a material exhibits strain softening, then small inhomogeneities in the material are likely to deform preferentially, become softer, and run away to produce regions of high local strain (micro-bands). However, this tendency to run away will be limited if the flow stress of the material increases rapidly with strain rate.

A measure of the rate at which inhomogeneities in strain will develop is the parameter ε^D which is the strain that has to be applied to the sample as a whole for the strain rate of microband nuclei (regions where there is a pre-existing inhomogeneity in strain of $\Delta\varepsilon$) to become double that of the sample as a whole. Microbands are likely to develop if ε^D is small. It can be shown that

$$\varepsilon^D = \varepsilon^* \ln(\varepsilon^*/\Delta\varepsilon)$$

where ε^* depends on the magnitude of the negative slope of the stress *vs.* strain curve in the strain softening region and on the strain rate dependence of the flow stress.

$$\varepsilon^* = -(\partial\sigma/\partial(\ln\dot{\varepsilon}))_{\varepsilon}/(\partial\sigma/\partial\varepsilon)_{\dot{\varepsilon}}$$

It is possible to determine experimental values for ε^* from stress strain curves. For annealed PS ε^* decrease rapidly from 20% at 80 °C at a strain rate of 5×10^{-4} sec^{-1} to 1·6% at the transition, this decrease being mainly due to the rapidly increasing negative slope of the stress strain curve. If ε^* is small, and is as small or smaller than $\Delta\varepsilon$, then ε^D will be very small and microbands will develop rapidly. For PMMA ε^* is 11% at 22 °C decreasing to 4% at − 50 °C; it is not possible to obtain stress strain data at lower temperatures.

Microband nucleation appears to be due to the plastic instability of small pre-existing inhomogeneities in strain ($\Delta\varepsilon \sim 2\%$). Once the bands have been nucleated it is probable that adiabatic heating leads to further local softening and accounts for the extreme fineness of the bands and the very large plastic strains in them.

REFERENCES

Argon, A. S., R. D. Andrews, J. A. Godrick, and W. Whitney (1968). *J. appl. Phys.*, **39**, 1899.

Bowden, P. B., and S. Raha (1970). *Phil. Mag.*, **22**, 463.

52

Adsorption-Sensitive Machining Behaviour of Glass

A. R. C. WESTWOOD, G. H. PARR, Jr. and R. M. LATANISION

Research Institute for Advanced Studies, Martin Marietta Corporation, Baltimore, Maryland, U.S.A.

The fracture behaviour of glass can be markedly influenced by its testing environment, and both crack initiation (Kitaigorodskii and Kopytov, 1963) and crack propagation (Berdennikov, 1934) processes can be affected. Explanations for such effects usually fall into one of two categories (i) those dependent upon adsorption-induced reductions in surface free energy (Rebinder, 1928; Orowan, 1944), and (ii) those which assume the presence of water to be critical, and of which the stress-assisted corrosion mechanism of Charles and Hillig (1962) is most widely accepted. For explanations in category (ii), the role of testing media other than water is usually considered to be that of screening the highly reactive water molecules from the glass surface (Moorthy and Tooley, 1956; Culf, 1957; Mould, 1961).

Ample evidence exists for the deleterious influence of water on the strength of silica-based materials, especially under static fatigue conditions (e.g. see Mould, 1961). However, under conditions where fracture is induced by abrasion or grinding, there is evidence that certain organic environments are considerably more active than water. Engelhardt (1946), for example, has shown that only $3{\cdot}2 \times 10^6$ ergs are required to produce 1 cm^2 of quartz surface when grinding is performed under octanol, whereas $5{\cdot}8 \times 10^6$ ergs are required under water. Thus under these conditions, octanol is more active than water. Likewise, Ramsauer's (1951) work with glass has shown that turpentine and petroleum are superior 'drilling liquids' to water. It would appear that explanations dependent upon the presence and corrosive influence of water are invalid for such examples of environment-sensitive fracture behaviour.

It does not follow, however, that such effects must therefore be caused by adsorption-induced reductions in surface free energy, because a further possibility should also be considered. Recent work has shown that the significant influence of organic environments on the flow, fracture, and drilling behaviour of crystalline ceramics such as MgO is primarily associated with chemisorption-induced variations in the mobility of dislocations in the near surface (1–10 μ) regions of such solids (Westwood, Goldheim and Lye, 1967, 1968; Westwood and Goldheim, 1970). In this case, the environment influences the flow processes which precede and are responsible for crack initiation in these notch-brittle* materials. Whether dislocation-like defects (Gilman, 1969) are involved in the flow or fracture behaviour of glass, or whether its deformation behaviour can be better explained in terms of the weakening influence of mobile sodium ions (Cox, 1969) or by some other mechanism, remains to be determined. Nevertheless, there is convincing evidence that flow can occur in glasses at room temperature (Weyl, 1949; Marsh, 1964; Dick, 1970, Lange, 1970), though its role in the fracture process has yet to be identified. It is possible, therefore, that certain examples of the environment-sensitive fracture behaviour of glass may be a consequence of adsorption-induced variations in the flow and flow-dependent fracture properties of the near-surface regions of this solid.

The present investigation was undertaken to examine this possibility, and the initial results of this work are presented below. Since few, if any, systematic studies of environment effects have been conducted in the past, this programme has concentrated to date on the effects of the homologous series of n-alkanes and n-alcohols on the abrasion hardness, brittle fracture, and drilling behaviour of a conventional soda-lime glass.

EXPERIMENTAL

Materials

The soda-lime glass used was manufactured by Libbey Owens Ford, and was of approximate composition (wt %) SiO_2—72·8; Na_2O—13·0; CaO—8·8; MgO—3·5; Al_2O_3—0·9; and other elements ~ 1·0. The environments utilized were distilled water, toluene, normal alkanes ranging from n-pentane (C_5H_{12}; $N_C = 5$) to n-hexadecane ($C_{16}H_{34}$; $N_C = 16$), and n-alcohols ranging from n-methanol (CH_3OH; $N_C = 3$) to dodecyl alcohol ($C_{12}H_{25}OH$; $N_C = 12$). The water contents of these solvents were determined after use by the Karl Fischer technique, and are given in Figure 1(a).

Pendulum sclerometer studies

To determine the relative influence of various test environments on the hardness, H, of glass, a single fulcrum (tungsten carbide) pendulum sclero-

* A notch-brittle solid is one in which failure is crack-initiation controlled.

meter was constructed similar to that of Kuznetsov and Lavrent'eva (Kuznetsov, 1957). Using this instrument, the magnitude of H can be evaluated from considerations of the damping behaviour of the oscillating pendulum. For a hard and brittle solid, relatively little energy will be absorbed in near-surface flow or fracture processes, and so the rate of pendulum damping will be small. For a softer and more ductile solid, on the other hand, relatively more energy will be absorbed by such processes, and so the rate of damping will be greater.

Before undertaking pendulum sclerometer experiments in the present work, it was necessary to determine the increment of damping introduced by

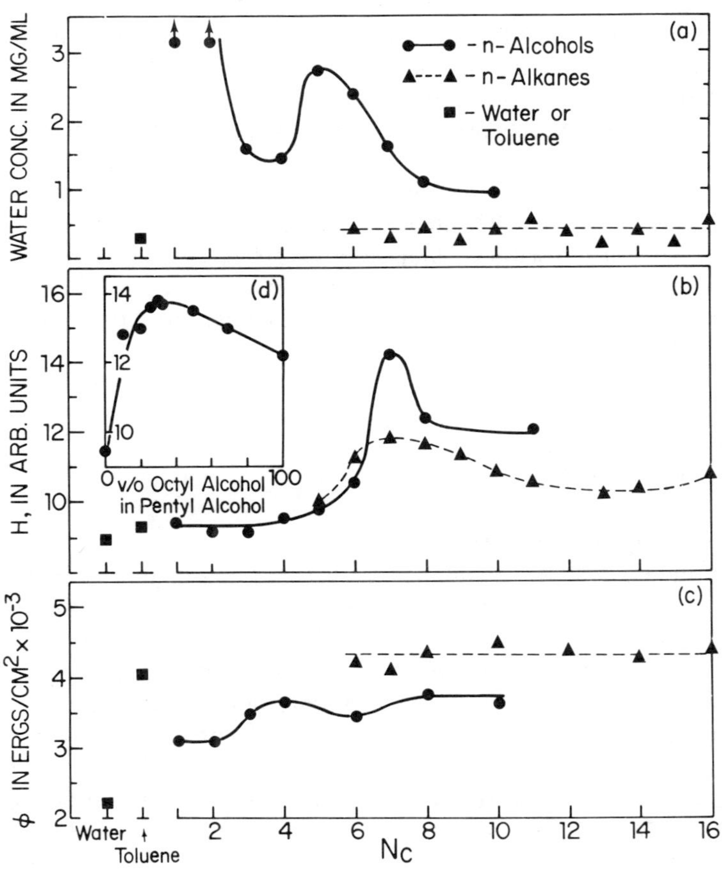

Figure 1. Variation with environmental composition of (a) water content of test environments; (b) and (d) pendulum hardness, H, of a soda-lime glass at 25 °C; and (c) the energy, ϕ, to propagate a crack in the same glass at 25 °C, as determined by the double cantilever technique. N_C is the number of carbon atoms in the n-alcohol or n-alkane molecules

the various test environments themselves. This factor was evaluated by determining the relative damping of the pendulum when oscillating on a polished, hardened steel surface which was either immersed in a representative test environment or exposed to air. It was assumed that any observed differences in behaviour could then be associated with liquid phase damping or lubrication effects, since adsorbed organic species are unlikely to influence the near-surface flow properties of a metallic solid because of free electron screening (Westwood, Preece and Goldheim, 1969). It was found that, under identical testing conditions to those subsequently adopted in the studies with glass, the liquid test environments introduced no significant increase in damping. Indeed, a slight reduction in damping compared to that in air was noted, presumably introduced by lubrication effects. However, since this increment did not vary significantly with either environmental composition or viscosity over the range 1–20 c/s, it was disregarded.

The specific test procedure then adopted was as follows: A glass specimen was fractured in the selected test environment in such a manner as to produce a relatively flat surface. Next, the pendulum (total mass 100 g) was lowered gently on the still-immersed, pristine surface, deflected to its initial amplitude, $A(0)$ and set in oscillation. After 300 sec, the amplitude of oscillation was noted, and its value designated $A(300)$. The relationship derived by Rebinder (see Kuznetsov, 1957, p. 81) was then used to estimate the relative value of H, namely:

$$H = t/2{\cdot}303\,(\log A(0) - \log A(300)) \tag{1}$$

t being the test period.

Fracture energy determinations

A double cantilever technique was used to determine the energy, ϕ, to propagate a pre-existing crack in the glass, the procedure followed being essentially that described in detail by Gilman (1960) and Wiederhorn (1966, 1969). In the present work, however, the test crack was initiated while the specimen was immersed in the appropriate liquid environment, and total immersion was maintained throughout subsequent mounting in an Instron machine and testing to failure. The specimens were $\sim 75 \times 25 \times 1$ mm in size, and ϕ was estimated from the following relationship (Wiederhorn, Shorb and Moses, 1969)

$$\phi = 6P^2L^2\,[1 + 1{\cdot}335\,t/L + 0{\cdot}446\,(t/L)^2]/Ew^2t^3 \tag{2}$$

in which P is the load at which the test crack of length L propagates catastrophically; E is Young's modulus, and w and t are the specimen width and half depth, respectively.

Drilling experiments

The diamond-studded, hemispherical-headed bits used were supplied by Starlite Industries, Inc., and were ~ 3 mm in diameter. The test pieces in this case were ~ 25 × 10 × 10 mm in size, with opposing cut surfaces ground parallel. The experimental procedure was as follows: A precision drill press (1750 r.p.m.) was fitted with a bit, the specimen immersed in the test environment, and the bit lowered gently onto its surface. A fixed load was then

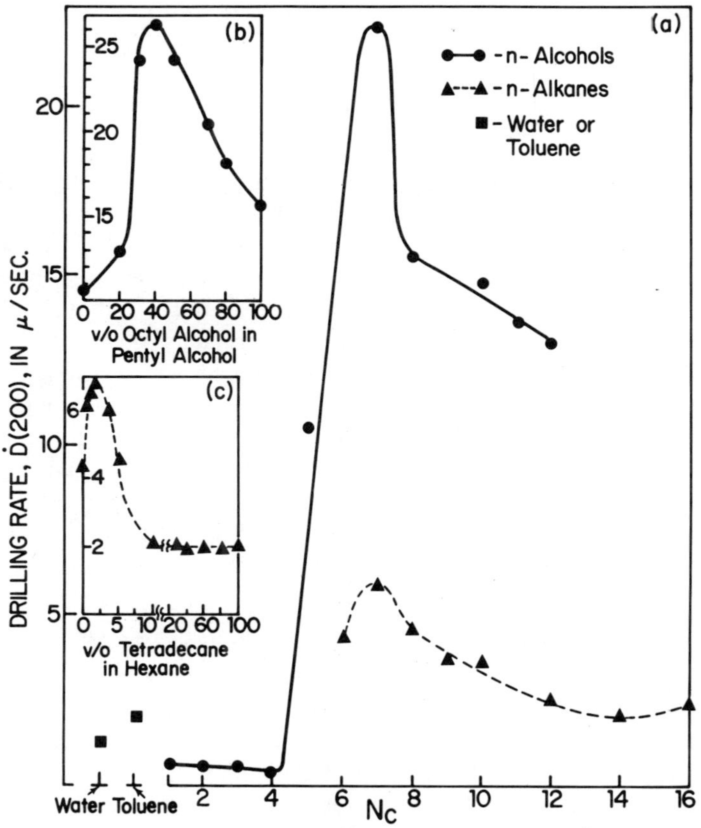

Figure 2. Variation in rate of drilling at time = 200 sec, $\dot{D}(200)$, with environmental composition for a soda-lime glass at 25 °C. The bits used were hemispherical-headed, diamond-studded, and ~ 3 mm in diameter

applied to the drill shaft by means of a lever system, and the drill started. The depth of penetration, *D*, was then monitored as a function of time by means of an LVDT-recorder system.

Once a bit was 'broken-in', it was possible to run many tests reproducibly. It was noted that a relatively constant rate of drilling was established after the first 30 sec of test. Accordingly, the efficiency of drilling was assessed in terms of the average rate of drilling during the period 150–250 sec, this being designated $\dot{D}(200)$. To allow comparison between data produced from a series of different bits, the load for each bit was chosen so that it provided a value of $\dot{D}(200)$ in toluene of $2 \pm 0{\cdot}2\ \mu$/sec. In other words, the load applied to the bit was utilized as the 'normalizing' factor in these experiments. The drilling data so produced, and presented in Figure 2, are averages from between three and six tests for each environment, the scatter being typically $\pm$ 10%.

RESULTS

Pendulum sclerometer experiments

The data from these tests are shown in Figure 1(b). Significant variations occur in the magnitude of H with molecular chain length, particularly for the n-alcohol environments. It is evident that this soda-lime glass is markedly harder in heptyl alcohol ($N_C = 7$) than in any of the other alcohols or alkanes tested. The value of H for a water environment is similar to that for alcohols having N_C between 1 and 4. In alkane environments, however, this glass exhibits values of H lying inbetween those produced by the lower ($N_C < 5$) and higher ($N_C > 7$) alcohols. Note that the variation in H with environment is not related to the water content of the environment, cf. Figures 1(a) and 1(b). Data from tests conducted in solutions of octyl alcohol in pentyl alcohol are given in Figure 1(d). Note particularly the existence of a maximum in H for such solutions, the value of H_{max} being similar to the value of H for an heptyl alcohol environment.

Fracture energy determinations

The variation in energy to propagate a brittle crack in glass with testing environment, as determined by the double cantilever technique, is shown in Figure 1(c). The value of ϕ obtained for this soda-lime glass in water, 2200 ± 300 erg/cm^2, is similar to that of 2000–2500 ergs/cm^2 found in previous work by Linger and Holloway (1968). Likewise, the values of ϕ determined for the alkane series from $N_C = 6$ to $N_C = 16$, 4200 ± 400 ergs/cm^2, are identical to that found by Linger and Holloway (1968) for a 'liquid paraffin' environment, and similar to the values of 4400 ± 300 ergs/cm^2 determined by Wiederhorn and Johnson (1969) for a heptane environment. Note that ϕ does not vary significantly with N_C for these alkane environments. For the n-alcohols, values of ϕ intermediate between that for water and those for the alkanes were obtained. A comparison of the data in Figures 1(a) and and 1(c) suggests that these values reflect the higher concentration of water present in the alcohols. The value of ϕ determined for this glass in relatively water-free toluene was $\sim$ 4000 ergs/cm^2.

Drilling experiments

The variation of $\dot{D}(200)$ with test environment is shown in Figure 2. Note that water and toluene are relatively poor drilling environments, all of the alkanes being superior, and that n-alcohols having $N_C > 5$ are extremely effective drilling environments, $\dot{D}(200)$ for glass in heptyl alcohol being twenty times greater than that for water.

Experiments were also conducted in solutions of one alcohol (or alkane) in another, and the data so obtained are presented in the inserts to Figure 2. These data reveal, as do those in Figure 1(d), that by mixing together alcohols (or alkanes) of $N_C = x$ and $N_C = y$ $(x < y)$, it is possible to reproduce the characteristic variation of $\dot{D}(200)$ for alcohols (or alkanes) having values of N_C between x and y.

DISCUSSION

The results of this study provide a useful basis for evaluating the relevance of certain factors and previously proposed mechanisms to the environment-sensitive drilling (or machining) behaviour of soda-lime glass. For example, a comparison of the data of Figures 1(a) and 2(a) indicates that the efficiency of drilling glass in alcohol or alkane environments is in no way related to the water content of these environments. Thus, any explanation for the observed variations in $\dot{D}(200)$ which is dependent on the presence and dominant influence of water is clearly not relevant. Likewise, the data of Figures 1(c) and 2(a) reveal that, although the energy to propagate cracks in the presence of water is only half of that required in the presence of, say, heptane ($N_C = 7$), heptane is a much more efficient drilling environment (by a factor of six) than is water. Also, while ϕ is independent of N_C for the alkanes. $\dot{D}(200)$ is not. Further, although ϕ for glass in toluene is essentially identical to that for glass in the alkanes, $\dot{D}(200)$ is significantly less. Thus, it may be concluded that the energy (ϕ) to propagate cracks in glass is *not* the controlling factor in the drilling process, and that any explanation dependent upon adsorption-induced reductions in ϕ also cannot be valid. It appears, therefore, that neither of two mechanisms cutomarily relied upon to explain the environment-sensitive mechanical behaviour of glass (see first section) are relevant to the results of this study, and other possibilities must be considered.

With this in mind, it is of interest to first determine whether any of the more obvious environmental variables might be the factor determining the magnitude of $\dot{D}(200)$. Possibilities include dielectric constant (ε) (Berdennikov, 1934; Cox, 1969), dipole moment (p), interfacial energy (γ_{sl}), heat of adsorption (Q_A) or wetting (Q_w) (Weyl, 1949), viscosity (η), or coefficient of friction

(μ_F). Considerations of elementary theory or previous work reveal, however, that for the alcohols or alkanes none of these parameters varies with N_C in such a manner that one might reasonably expect a maximum in $\dot{D}(200)$ at any particular value of N_C such as is evident in Figure 2(a) at $N_C = 7$. For example, as N_C increases, γ_{sl}, ε and μ_F decrease monotonically, Q_A and η increase monotonically, and p does not vary significantly. The further possibility that $\dot{D}(200)$ is related to variations in the effectiveness of these environments as coolants can probably be eliminated in view of the correlation between the data from pendulum sclerometer and drilling experiments, cf. Figures 1(b) and 2(a), and Figures 1(d) and 2(b). Thermal dissipation effects are unlikely to be important in the former case.

It remains, therefore, to consider the possibility that the observed variations in $\dot{D}(200)$ are a consequence of environmentally-induced variations in the near-surface deformation behaviour of this soda-lime glass. This possibility appears reasonable because: (i) previous work has established that adsorbed species *can* influence the anelastic behaviour of glass when tested in bending (Aslanova and Rebinder, 1954), (ii) the observed variations in $\dot{D}(200)$ are reminiscent of the environment-sensitive variations in drilling behaviour of MgO which, as discussed in the Introduction, have a demonstrable relationship with adsorption-induced changes in near-surface flow and fracture behaviour, and (iii) the hardness of a solid is a near-surface or bulk property rather than a true surface property—yet the data of Figures 1(b) and (d) show that this property, as manifested by pendulum sclerometer measurements, can be significantly influenced by the test environment.

The problem then resolves into determining the mechanism by means of which the hardness (i.e. flow or stress-relaxing behaviour) of the near-surface regions of soda-lime glass can be influenced by an environment such that drilling behaviour is significantly affected but ϕ is not. Now it may be assumed that ϕ for soda-lime glasses is relatively insensitive to sodium ion concentration, since the values of ϕ for fused silica and a soda-lime glass containing 14% Na_2O are essentially identical at 77 °K, namely $\sim 4500 \pm 200$ ergs/cm^2 (Wiederhorn, 1969). On the other hand, soda-lime glass is considerably softer than fused silica (typically by about 40%), so that it may be assumed that the hardness of the near-surface regions of glass will be dependent on the concentration and distribution of sodium ions there.

With this in mind, it may be recalled that Volkenstein (1963, 1967) has shown theoretically that species chemisorbed on a semiconductor can, by changing its surface charge, alter the concentration and distribution of any mobile charge-carrying impurity species nearby. It has also been demonstrated experimentally that stress (Weber and Goldstein, 1964) or potential (Weyl, 1960) gradients can cause redestribution of the sodium ions in a glass. Thus, it appears reasonable to conclude that the concentration and distribution of sodium ions in the near-surface regions are the critical variables since these parameters can be influenced by both environmental composition and the complex stresses introduced by the drilling process. Recognizing further that the flow of sodium ions towards an electrode causes an increase in volume of the glass there (Quincke, 1880—see Weyl, 1960), a feasible consequence of any stress-plus-chemisorption-induced flow of sodium ions *towards* the surface of a glass specimen will be some relaxation of any tensile stresses created there (Westwood, Latanision and Lye, 1970) by the drilling process. In such an event the initiation of cracks in the surface layers should become more difficult, and drilling efficiency reduced. Conversely, any environment which tends to induce sodium ions to diffuse *away* from the surface presumably will cause some increase in drilling efficiency. It may be that heptyl alcohol is such an environment.

Though the details of this hypothesis require further study, and the nature of the electronic transitions and resultant changes in surface charge accompanying chemisorption on glass are complex and unknown, the above hypothesis appears to be in general accord with the results of this work. For example, it is known that the capacity of the alcohols and alkanes to donate or accept electrons varies significantly with molecular chain length (Ingold, 1953). Thus the surface charge resulting from their chemisorption will vary in some as yet unpredictable manner with N_C, and so therefore will H. It is also well known that diamond bits drill hard and brittle solids more effectively than they do softer and more ductile materials (an attempt to drill brass or aluminium with a diamond bit will provide convincing proof of this). Thus, the fact that the hardness of this glass is a maximum in heptyl alcohol, Figure 1(b), and that this environment also produces a maximum in drilling efficiency when a diamond bit is used,* Figure 2(a), is consistent with experience, and the similar variation of H and $\dot{D}(200)$ with N_C is intuitively reasonable. The observation that binary solutions of alcohols or alkanes produce drilling efficiencies equivalent to those of pure, intermediate alcohols or alkanes also appears consistent with the present hypothesis, in that such solutions might be expected to produce surface charges intermediate between those of the two participating species, the specific charge depending on the relative surface concentrations of these species.

Finally, on the basis of this hypothesis, one might expect some different variation of $\dot{D}(200)$ with N_C for a glass of significantly different composition. To examine this possibility, a limited number of drilling experiments were performed on fused silica. For this material, the maximum in drilling efficiency in alkane environments occurred at $N_C = 14$ rather than $N_C = 7$.

In conclusion then, it has been shown that the drilling behaviour of a soda-lime glass can be significantly influenced by liquid environments, and

* This point is stressed because different types of bits drill in different ways. For example, a carbide spade bit usually functions more effectively when the solid exhibits a significant degree of plasticity.

that such effects probably are associated with adsorption-dependent variations in hardness which influence the ease with which cracks are initiated in this notch-brittle solid. The possibility that such variations are a consequence of a stress-pulse-chemisorption-induced redistribution of the sodium ions in the near-surface regions of the glass appears to merit consideration.

ACKNOWLEDGEMENTS

It is a pleasure to acknowledge useful discussions with Dr. Carolyn M. Preece, and the expert assistance of R. D. Huntington during the course of this work, which was supported in part by the U.S. Office of Naval Research under Contract N00014–70–C–0330.

REFERENCES

Aslanova, M. S., and P. A. Rebinder (1954). *Dokl. Akad. Nauk. S.S.S.R.*, **96**, 229.

Berdennikov, W. P. (1934). *Zhur. fiz. Khim.*, **5**, 358.

Charles, R. J., and W. B. Hillig (1962). In *Symp. on mech. strength of glass and ways of improving it.* Union Scientifique du Verre, Charleroi. p. 511.

Cox, S. M. (1969). *Physics Chem. Glasses*, **10**, 226.

Culf, C. J. (1957). *J.S.G.T.*, **41**, 157T.

Dick, E. (1970). *Glastech. Ber.*, **43**, 16.

Engelhardt, W. V. (1946). *Naturwissenschaften*, **33**, 195.

Gilman, J. J. (1960). *J. appl. Phys.*, **31**, 2208; (1969). In *Dislocation Dynamics.* McGraw-Hill, New York. p. 3.

Ingold, C. K. (1953). In *Structure and Mechanism in Organic Chemistry.* Cornell Univ. Press, New York.

Kitaigorodskii, I. I., and L. N. Kopytov (1963). *Soviet Phys. Dokl.*, **8**, 308.

Kuznetsov, V. D. (1957). In *Surface Energy of Solids.* Transl. H.M.S.O., London. p. 74.

Lange, F. F. (1970). *Tech. Rept. No. 3 on Contract N000 14-68-C-0323, NR-032-507.* Westinghouse Elec. Co., R. and D. Center, Pittsburgh, Pa.

Linger, K. R., and D. G. Holloway (1968). *Phil Mag.*, **18**, 1269.

Marsh, D. M. (1964). *Proc. R. Soc.*, **279A**, 420.

Moorthy, V. K., and F. V. Tooley (1956). *J. Am. Ceram. Soc.*, **39**, 215.

Mould, R. E. (1961). *J. Am. Ceram. Soc.*, **44**, 481.

Orowan, E. (1944). *Nature*, **154**, 341.

Ramsauer, R. (1951). *Glastech. Ber.*, **24**, 239.

Rebinder, P. A. (1928). In *Proc. Sixth Phys. Conf.* State Press, Moscow. p. 29.

Volkenstein, F. F. (1963). In *Electronic theory of catalysis on semiconductors.* Pergamon Press, Oxford.; (1967). *Soviet Phys. Usp.*, **9**, 743.

Weber, N., and M. Goldstein (1964). *J. chem. Phys.*, **41**, 2898.

Westwood, A. R. C., and D. L. Goldheim (1970). *J. Am. Ceram. Soc.*, **53**, 142.

Westwood, A. R. C., D. L. Goldheim, and R. G. Lye. (1967). *Phil. Mag.*, **16**, 505; (1968). *Phil. Mag.*, **17**, 951.

Westwood, A. R. C., R. M. Latanision, and R. G. Lye (1970). *Phys. Stat. Sol.*, (a)**3**, K 17.

Westwood, A. R. C., C. M. Preece, and D. L. Goldheim (1969). In *Molecular processes at solid surfaces*. McGraw-Hill, New York. p. 591.

Weyl, W. A. (1949). *J. Am. Ceram. Soc.*, **32**, 367; (1960). In *Rheology* **3**. Academic Press, New York. p. 299.

Wiederhorn, S. M. (1966). In *Mater. Sci. Res.*, **3**. Plenum Press, New York. p. 503.; (1969). *J. Am. Ceram. Soc.*, **52**, 99.

Wiederhorn, S. M., and H. Johnson (1969). *Nat. Bur. Stds. Rept. 10145* on Contract NAonr-7-70, NR-032-517.

Wiederhorn, S. M., A. M. Shorb, and R. L. Moses (1969). *J. appl. Phys.*, **39**, 1569.

Index